Carpentry Fundamentals
Level One

Annotated Instructor's Guide
Fourth Edition

Upper Saddle River, New Jersey
Columbus, Ohio

National Center for Construction Education and Research
President: Don Whyte
Director of Product Development: Daniele Stacey
Carpentry Project Manager: Daniele Stacey
Production Manager: Jessica Martin
Product Maintenance Supervisor: Debie Ness
Editor: Brendan Coote
Desktop Publishers: Jessica Martin and Jennifer Jacobs

Writing and development services provided by Topaz Publications, Liverpool, NY.

Pearson Education, Inc.
Product Manager: Lori Cowen
Production Editor: Stephen C. Robb
Design Coordinator: Karrie M. Converse-Jones
Text Designer: Kristina D. Holmes
Cover Designer: Kristina D. Holmes
Copy Editor: Sheryl Rose
Scanning Coordinator: Karen L. Bretz
Scanning Technician: Janet Portisch
Production Manager: Pat Tonneman
Marketing Manager: Derril Trakalo
Senior Marketing Coordinator: Liz Farrell
Marketing Assistant: Les Roberts

This book was set in Palatino and Helvetica by Carlisle Communications, Ltd. It was printed and bound by Courier Kendallville, Inc. The cover was printed by Phoenix Color Corp.

This information is general in nature and intended for training purposes only. Actual performance of activities described in this manual requires compliance with all applicable operating, service, maintenance, and safety procedures under the direction of qualified personnel. References in this manual to patented or proprietary devices do not constitute a recommendation of their use.

Copyright © 2006, 2001, 1998, 1992 by the National Center for Construction Education and Research (NCCER), Gainesville, FL 32614-1104 and published by Pearson Education, Inc., Upper Saddle River, NJ 07458. All rights reserved. Printed in the United States of America. This publication is protected by copyright, and permission should be obtained from NCCER prior to any prohibited reproduction, storage in a retrieval system, or transmission in any form or by any means, electronic, mechanical, photocopying, recording, or likewise. For information regarding permission(s), write to: NCCER Product Development, P.O. Box 141104, Gainesville, FL 32614-1104.

Pearson Prentice Hall™ is a trademark of Pearson Education, Inc.
Pearson® is a registered trademark of Pearson plc
Prentice Hall® is a registered trademark of Pearson Education, Inc.

Pearson Education Ltd.
Pearson Education Singapore Pte. Ltd.
Pearson Education Canada, Ltd.
Pearson Education—Japan

Pearson Education Australia Pty. Limited
Pearson Education North Asia Ltd.
Pearson Educación de Mexico, S.A. de C.V.
Pearson Education Malaysia Pte. Ltd.

10 9 8
ISBN 0-13-228593-2

Preface

TO THE INSTRUCTOR

We are proud to present the fourth edition of *Carpentry Fundamentals Level One*. This Annotated Instructor's Guide offers an array of teaching tips and ideas that can be adapted to suit your instructional objectives and your student's individual needs. The Annotated Instructor's Guide is actually the Trainee Guide enhanced with specific directions to the instructor, space for the instructor's notes, suggestions for session break-outs, a comprehensive materials and equipment list, and teaching tips to correspond to the performance examinations, laboratories, and demonstrations. Additionally, each Annotated Instructor's Guide is packaged with a test booklet that includes module exams, performance tests, and answer keys. Also included in this package are task checklists (job sheets) and worksheets for many of the modules. We hope you will find this book a valuable resource as you prepare your students for rewarding careers as carpenters.

NEW WITH THE FOURTH EDITION

The most significant changes are the addition of two new modules on concrete and reinforcing material and on basic stair layout. Another enhancement is the new, user-friendly design of the fourth edition

Check out the opening pages of each of the ten modules in this Instructor's Guide to see people in the carpentry trade competing at nationally renowned events. SkillsUSA and Associated Builders and Contractors provide methods of national recognition through their yearly competitions. For more information about each organization's events, visit **www.skillsusa.org** and **www.abc.org**.

We invite you to visit the NCCER website at **www.nccer.org** for the latest releases, training information, newsletter, and much more. You can also reference the Contren® product catalog online at **www.crafttraining.com**. Your feedback is welcome. You may email your comments to **curriculum@nccer.org** or send general comments and inquiries to **info@nccer.org**.

CONTREN® LEARNING SERIES

The National Center for Construction Education and Research (NCCER) is a not-for-profit 501(c)(3) education foundation established in 1995 by the world's largest and most progressive construction companies and national construction associations. It was founded to address the severe workforce shortage facing the industry and to develop a standardized training process and curricula. Today, NCCER is supported by hundreds of leading construction and maintenance companies, manufacturers, and national associations. The Contren® Learning Series was developed by NCCER in partnership with Prentice Hall, the world's largest educational publisher.

Some features of NCCER's Contren® Learning Series are as follows:

- An industry-proven record of success
- Curricula developed by the industry for the industry
- National standardization providing portability of learned job skills and educational credits
- Compliance with the Office of Apprenticeship, Training, Employer, and Labor Services (OATELS) requirements for related classroom training *(CFR 29:29)*
- Well-illustrated, up-to-date, and practical information

NCCER also maintains a National Registry that provides transcripts, certificates, and wallet cards to individuals who have successfully completed modules of NCCER's Contren® Learning Series. *Training programs must be delivered by an NCCER Accredited Training Sponsor in order to receive these credentials.*

Special Features of This Book

In an effort to provide a comprehensive user-friendly training resource, we have incorporated many different features for your use. Whether you are a visual or hands-on learner, this book will provide you with the proper tools to get started in the carpentry trade.

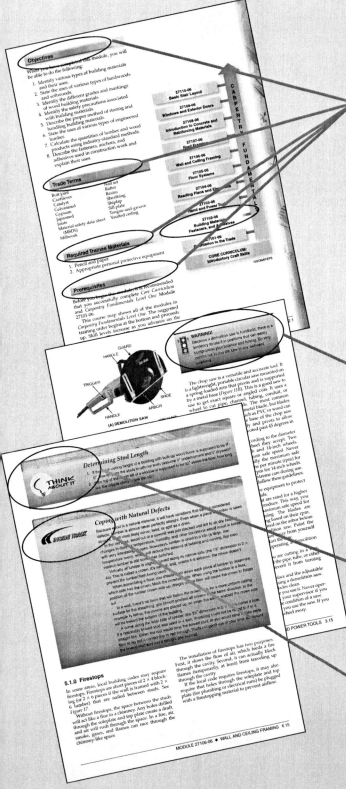

Introduction Page

This page is found at the beginning of each module and lists the Objectives, Trade Terms, Required Trainee Materials, Prerequisites, and Course Map for that module. The Objectives list the skills and knowledge you will need in order to complete the module successfully. The list of Trade Terms identifies important terms you will need to know by the end of the module. Required Trainee Materials list the materials and supplies needed for the module. The Prerequisites for the module are listed and illustrated in the Course Map. The Course Map also gives a visual overview of the entire course and a suggested learning sequence for you to follow.

Notes, Cautions, and Warnings

Safety features are set off from the main text in highlighted boxes and are organized into three categories based on the potential danger of the issue being addressed. Notes simply provide additional information on the topic area. Cautions alert you of a danger that does not present potential injury but may cause damage to equipment. Warnings stress a potentially dangerous situation that may cause injury to you or a co-worker.

Think About It

Think About It features use "What if?" questions to help you apply theory to real-world experiences and put your ideas into action.

Inside Track

Inside Track features provide a head start for those entering the carpentry field by presenting technical tips and professional practices from master carpenters in a variety of disciplines. Inside Tracks often include real-life scenarios similar to those you might encounter on the job site.

Color Illustrations and Photographs

Full-color illustrations and photographs are used throughout each module to provide vivid detail. These figures highlight important concepts from the text and provide clarity for complex instructions. Each figure is denoted in the text in *italic type* for easy reference.

Trade Terms

Each module presents a list of Trade Terms that are discussed within the text, defined in the Glossary at the end of the module, and reinforced with a Trade Terms Quiz. These terms are denoted in the text with **blue bold type** upon their first occurrence. To make searches for key information easier, a comprehensive Glossary of Trade Terms from all modules is provided at the back of this book.

Step-by-Step Instructions

Step-by-step instructions are used throughout to guide you through technical procedures and tasks from start to finish. These steps show you not only how to perform a task but how to do it safely and efficiently.

Review Questions

Review Questions are provided to reinforce the knowledge you have gained. This makes them a useful tool for measuring what you have learned.

Profile in Success

Profiles in Success share the apprenticeship and career experiences of and advice from successful professionals in the carpentry field.

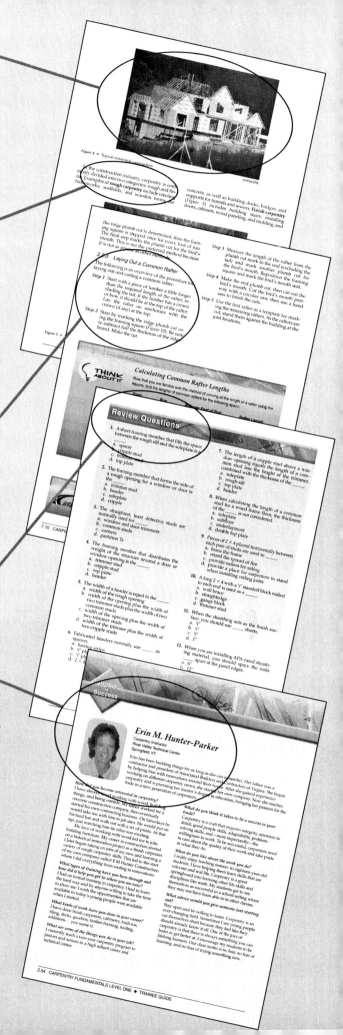

Supplemental Resources

Transparency Masters

Spend more time training and less time at the copier by using the loose, reproducible copies of the overhead transparencies that are referenced in the Annotated Instructor's Guide. The transparency masters package includes most of the Trainee Guide illustrations, enlarged for projection and printed on loose sheets of paper for easy copying onto transparency film using your photocopier. **ISBN 0-13-229157-6**

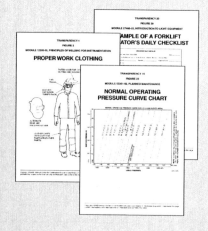

PowerPoint® Presentation Slides

A powerful addition to your classroom discussion, NCCER's Microsoft® PowerPoint® presentation CD features all the drawings and photos from the Trainee Guide. **ISBN 0-13-229136-3**

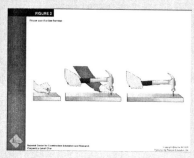

Computerized Testing Software

Ensure test security with NCCER's computerized testing software. This software allows instructors to scramble the module exam questions and answer keys in order to print multiple versions of the same test, customize tests to suit the needs of training units, add questions, or easily create a final exam. **ISBN 0-13-229135-5**

Contren® Connect for Carpentry Fundamentals Level One

Ideal for blended or distance education, Contren Connect® is the perfect online supplement. Access to the e-book gives you and your students the full text online plus a variety of enhanced features including online lectures, audio summaries, additional quizzes, active figures, video clips, and personalization tools. Visit **www.contrenconnect.com** to learn more about these innovative features and to view a demo module from the Trainee Guide.

    Access Code with paperback Trainee Guide **ISBN 0-13-239712-9**
    Access Code with the hardcover Trainee Guide **ISBN 0-13-239711-0**
    Access with paperback Annotated Instructor's Guide **ISBN 0-13-239710-2**

Applied Construction Math: A Novel Approach

Unlike traditional math texts, this book has a unique approach. It presents all the math concepts in the context of a story centering on a group of students tasked with building a single-family home. The journey is even more compelling than the destination. To get there, students must understand and apply math in a variety of real-world situations. To preview the full table of contents, the preface, and the first chapter, please visit **www.crafttraining.com** or contact your Contren® sales specialist.

    Trainee Guide **ISBN 0-13-227298-9**
    Instructor's Edition **ISBN 0-13-227300-4**

From the Ground Up: Class Projects for Forming, Framing, and Finishing, Second Edition

This hands-on, activities-based workbook presents over 30 indoor and outdoor projects in basic building, framing, and electrical. Also featured are fund-raising projects for school or community and updated competition projects to help prepare for statewide and national SkillsUSA events.

    Student Workbook **ISBN 0-13-229164-9**
    Instructor's Package **ISBN 0-13-229165-7**

To order, contact Prentice Hall at **1-800-922-0579.**

Contents

27101-06 Orientation to the Trade 1.i

Reviews the history of the trade, describes the apprentice program, identifies career opportunities for carpentry and construction workers, and lists the responsibilities and characteristics a worker should possess. **(2.5 Hours)**

27102-06 Building Materials, Fasteners, and Adhesives 2.i

Provides an overview of the building materials used in construction work, including lumber, sheet materials, engineered wood products, structural concrete, and structural steel. Also describes the various fasteners and adhesives used in construction work. **(7.5 Hours)**

27103-06 Hand and Power Tools 3.i

Provides detailed descriptions of the hand tools and portable power tools used by carpenters. Emphasis is on safe and proper operation of tools, as well as care and maintenance. **(10 Hours)**

27104-06 Reading Plans and Elevations 4.i

Builds upon the basic information presented in the *Introduction to Blueprints* module studied in the *Core Curriculum*. Trainees will learn the techniques for reading and using blueprints and specifications with an emphasis placed on those drawings and types of information that are relevant to the carpentry trade. Introduces the subject of quantity takeoffs. **(20 Hours)**

27105-06 Floor Systems 5.i

Covers framing basics as well as the procedures for laying out and constructing a wood floor using common lumber as well as engineered building materials. **(25 Hours)**

27106-06 Wall and Ceiling Framing 6.i

Describes the procedures for laying out and framing walls and ceilings, including roughing-in door and window openings, constructing corners and partition Ts, bracing walls and ceilings, and applying sheathing. **(20 Hours)**

27107-06 Roof Framing ... 7.i

Describes the various kinds of roofs and contains instructions for laying out rafters for gable roofs, hip roofs, and valley intersections. Coverage includes both stick-built and truss-built roofs. **(37.5 Hours)**

27108-06 Introduction to Concrete, Reinforcing Materials, and Forms ... 8.i

Describes the ingredients of concrete, discusses the various types of concrete, and describes how to mix concrete. The module also covers basic job-built footing, edge, and wall forms and form ties, and it describes the types and uses of concrete reinforcing materials. **(5 Hours)**

27109-06 Windows and Exterior Doors ... 9.i

Describes the various types of windows, skylights, and exterior doors, and provides instructions for installing them. Also includes instructions for installing weather-stripping and locksets. **(12.5 Hours)**

27110-06 Basic Stair Layout ... 10.i

Introduces the trainee to the various types of stairs and the common building code requirements related to stairs. The module focuses on the techniques for measuring and calculating rise, run, and stairwell openings, laying out stringers, and fabricating basic stairways. **(12.5 Hours)**

Glossary of Trade Terms ... G.1

Figure Credits ... FC.1

Index ... I.1

Contren® Curricula

NCCER's training programs comprise more than 40 construction, maintenance, and pipeline areas and include skills assessments, safety training, and management education.

Boilermaking
Cabinetmaking
Carpentry
Carpentry, Residential
Concrete Finishing
Construction Craft Laborer
Construction Technology
Core Curriculum: Introductory Craft Skills
Currículum Básico
Electrical
Electrical, Residential
Electrical Topics, Advanced
Electronic Systems Technician
Exploring Careers in Construction
Fundamentals of Mechanical and Electrical Mathematics
Heating, Ventilating, and Air Conditioning
Heavy Equipment Operations
Highway/Heavy Construction
Instrumentation
Insulating
Ironworking
Maintenance, Industrial
Masonry
Millwright
Mobile Crane Operations
Painting
Painting, Industrial
Pipefitting
Pipelayer
Plumbing
Reinforcing Ironwork
Rigging
Scaffolding
Sheet Metal
Site Layout
Sprinkler Fitting
Welding

Pipeline

Control Center Operations, Liquid
Corrosion Control
Electrical and Instrumentation
Field Operations, Liquid
Field Operations, Gas
Maintenance
Mechanical

Safety

Field Safety
Orientación de Seguridad
Safety Orientation
Safety Technology

Management

Introductory Skills for the Crew Leader
Project Management
Project Supervision

Acknowledgments

This curriculum was revised as a result of the farsightedness and leadership of the following sponsors:

ABC Heart of America Chapter
ABC National
Associated Training Services–ABC Wisconsin
Brasfield & Gorrie
Construction Education Foundation of Georgia
Greenville Technical College
Guilford Technical Community College
Mid-Maine Technical Center

Onslow-Sheffield, Inc.
Paul Risk Associates, Inc.
River Valley Technical Center
SC Department of Education
The Haskell Company
TKTEK
Turner Construction Company

This curriculum would not exist were it not for the dedication and unselfish energy of those volunteers who served on the Authoring Team. A sincere thanks is extended to the following:

John Ambrosia
Mark Champagne
Shane Harvey
Curtis Haskins
Chuck Hogg
R.P. Hughes
Gary Humphries
Erin M. Hunter

Peter Klapperich
Perry Moore
Mark Onslow
John Payne
Roy Rucks
Duane Sellers
Todd Staub

A final note: This book is the result of a collaborative effort involving the production, editorial, and development staff at Prentice Hall and the National Center for Construction Education and Research. Thanks to all of the dedicated people involved in the many stages of this project.

NCCER PARTNERING ASSOCIATIONS

American Fire Sprinkler Association
American Society for Training & Development
API
Associated Builders & Contractors, Inc.
Associated General Contractors of America
Association for Career and Technical Education
Carolinas AGC, Inc.
Carolinas Electrical Contractors Association
Construction Industry Institute
Construction Users Roundtable
Design-Build Institute of America
Electronic Systems Industry Consortium
Merit Contractors Association of Canada
Metal Building Manufacturers Association
National Association of Minority Contractors
National Association of State Supervisors for Trade and Industrial Education

National Association of Women in Construction
National Insulation Association
National Ready Mixed Concrete Association
National Systems Contractors Association
National Technical Honor Society
National Utility Contractors Association
North American Crane Bureau
North American Technician Excellence
Painting & Decorating Contractors of America
Portland Cement Association
SkillsUSA
Steel Erectors Association of America
Texas Gulf Coast Chapter ABC
U.S. Army Corps of Engineers
University of Florida
Women Construction Owners & Executives, USA

Orientation to the Trade
27101-06

NCCER STANDARDIZED CRAFT TRAINING PROGRAM

The National Center for Construction Education and Research (NCCER) provides a standardized national program of accredited craft training. Key features of the program include instructor certification, competency-based training, and performance testing. The program provides trainees, instructors, and companies with a standard form of recognition through the National Registry. The program is described in full in the *Guidelines for Accreditation*, published by NCCER. For more information on standardized craft training, contact NCCER by writing to P.O. Box 141104, Gainesville, FL 32614-1104; calling 352-334-0911; or e-mailing **info@nccer.org**. More information is available at **www.nccer.org**.

HOW TO USE THIS ANNOTATED INSTRUCTOR'S GUIDE

Each page presents two sections of information. The larger section displays each page exactly as it appears in the Trainee Module. The narrow column ties suggested trainee and instructor actions to each page and provides icons (detailed below) to call your attention to material, safety, audiovisual, or testing requirements. The bottom of each page includes space for your notes.

The **Audiovisual** icon indicates an appropriate time to show a transparency or other audiovisual aid.

The **Classroom** icon prompts you to define a term, stress a point, ask trainees to explain a concept, or give examples.

The **Demonstration** icon directs you to show trainees how to perform tasks.

The **Examination** icon tells you to administer the written module examination.

The **Homework** icon is placed where you may wish to assign reading for the next class, assign a project, or advise trainees to prepare for an examination.

The **Laboratory** icon is used when trainees are to practice performing tasks.

The **Materials** icon is a reminder for you to gather materials needed for classes, laboratories, and testing.

The **Performance Testing** icon tells you to administer a performance test or a portion thereof.

The **Safety** icon is used to emphasize safety issues. It is often keyed to *Caution* and *Warning!* statements in the Trainee Module.

The **Teaching Tip** icon indicates additional guidance is available, such as how to conduct an exercise, get the most educational value from a field trip, or encourage class participation. Teaching Tips may expand on a feature (*Think About It, Did You Know?*) or provide *Quick Quizzes* or similar exercises. You will be referred to the Teaching Tips section at the back of the module if there is additional material.

The **Combination** icon indicates that the laboratory listed corresponds with a performance task. If desired, you can note the proficiency of the trainees during the laboratory, and use it to satisfy performance testing requirements.

PREPARATION

Before teaching this module, you should review the Objectives, Performance Tasks, Materials and Equipment List, and Module Outline. Be sure to allow ample time to prepare your own training or lesson plan and gather all required materials and equipment.

Orientation to the Trade
Annotated Instructor's Guide

Module 27101-06

MODULE OVERVIEW

This module introduces the carpentry trainee to the carpentry trade, including the apprenticeship process and the opportunities within the trade.

PREREQUISITES

Prior to training with this module, it is suggested that the trainee shall have successfully completed *Core Curriculum*.

OBJECTIVES

Upon completion of this module, the trainee will be able to do the following:

1. Describe the history of the carpentry trade.
2. Identify the aptitudes, behaviors, and skills needed to be a successful carpenter.
3. Identify the training opportunities within the carpentry trade.
4. Identify the career and entrepreneurial opportunities within the carpentry trade.
5. Identify the responsibilities of a person working in the construction industry.
6. State the personal characteristics of a professional.
7. Explain the importance of safety in the construction industry.

PERFORMANCE TASKS

This is a knowledge-based module—there is no performance testing.

MATERIALS AND EQUIPMENT LIST

Transparencies
Markers/chalk
Blank acetate sheets
Transparency pens

Pencils and scratch paper
Overhead projector and screen
Whiteboard/chalkboard
Appropriate personal protective equipment
Module Examinations*

*Located in the Test Booklet.

TEACHING TIME FOR THIS MODULE

An outline for use in developing your lesson plan is presented below. Note that each Roman numeral in the outline equates to one session of instruction. Each session has a suggested time period of 2½ hours. This includes 10 minutes at the beginning of each session for administrative tasks and one 10-minute break during the session. Approximately 2½ hours are suggested to cover *Orientation to the Trade*. You will need to adjust the time required for hands-on activity and testing based on your class size and resources.

Topic	Planned Time
Session I. Orientation to the Trade	
A. Introduction	_____
B. History of Carpentry	_____
C. Modern Carpentry	_____
D. Opportunities in the Construction Industry	_____
1. Formal Construction Training	_____
2. Apprenticeship Program	_____
3. Responsibilities of the Employee	_____
4. What You Should Expect from Your Employer	_____

     5. What You Should Expect from a Training Program
     6. What You Should Expect from the Apprenticeship Committee
E. Human Relations
     1. Making Human Relations Work
     2. Human Relations and Productivity
     3. Attitude
     4. Maintaining a Positive Attitude
F. Employer and Employee Safety Obligations
G. Review
H. Module Examination
     1. Trainees must score 70 percent or higher to receive recognition from NCCER.
     2. Record the testing results on Craft Training Report Form 200, and submit the results to the Training Program Sponsor.

Orientation to the Trade
27101-06

Lucas Epperly, a Cabinetmaking contestant at the SkillsUSA 2005 National Championship, uses a router to bevel the sides of a cabinet component. The router was used to create several types of joints called for in the project plans including the dado, rabbet, and shouldered-edge. The router was also used to create beaded edges on the drawer front. Routing the edges of a piece is a crucial task that can upset a project—by wasting time and material—if not done correctly the first time.

Instructor's Notes:

Assign reading of Module 27101-06.

27101-06
Orientation to the Trade

Topics to be presented in this module include:

1.0.0	Introduction	1.1
2.0.0	History of Carpentry	1.1
3.0.0	Modern Carpentry	1.6
4.0.0	Opportunities in the Construction Industry	1.8
5.0.0	Human Relations	1.16
6.0.0	Employer and Employee Safety Obligations	1.17

Overview

The carpentry trade offers numerous career opportunities, from constructing concrete forms to creating fine cabinetry. Carpenters build beautiful structures that can last for centuries and have one of the highest job satisfaction rates of any career in the construction industry. Carpenters have opportunities to work in residential, commercial, and industrial construction. Apprentice carpenters can go on to become master carpenters, estimators, architects, construction managers, and contractors.

Instructor's Notes:

Objectives

When you have completed this module, you will be able to do the following:

1. Describe the history of the carpentry trade.
2. Identify the aptitudes, behaviors, and skills needed to be a successful carpenter.
3. Identify the training opportunities within the carpentry trade.
4. Identify the career and entrepreneurial opportunities within the carpentry trade.
5. Identify the responsibilities of a person working in the construction industry.
6. State the personal characteristics of a professional.
7. Explain the importance of safety in the construction industry.

Trade Terms

Finish carpentry
Rough carpentry
Takeoff

Required Trainee Materials

1. Pencil and paper
2. Appropriate personal protective equipment

Prerequisites

Before you begin this module, it is recommended that you successfully complete *Core Curriculum*.

This course map shows all of the modules in *Carpentry Fundamentals Level One*. The suggested training order begins at the bottom and proceeds up. Skill levels increase as you advance on the course map. The local Training Program Sponsor may adjust the training order.

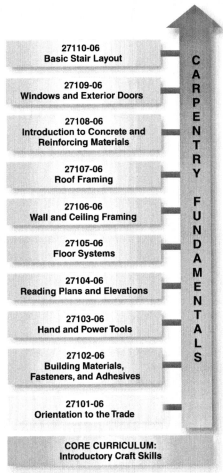

Ensure you have everything required to teach the course. Check the Materials and Equipment List at the front of this module.

Show Transparency 1, Course Objectives.

Explain that terms shown in bold (blue) are defined in the Glossary at the back of this module.

See the general Teaching Tip at the end of this module.

MODULE 27101-06 ◆ ORIENTATION TO THE TRADE 1.1

Discuss the NCCER apprenticeship program. Explain that each trainee's academic record is maintained in the National Registry, and describe how the registry works. Refer the trainees to the *Appendix* in the Trainee Module.

Discuss the history of carpentry.

1.0.0 ◆ INTRODUCTION

Opportunity is driven by knowledge and ability, which are in turn driven by education and training. This program of the National Center for Construction Education and Research (NCCER) was designed and developed by the construction industry for the construction industry. It is the only nationally accredited, competency-based construction training program in the United States. A competency-based program requires that the trainee demonstrate the ability to safely perform specific job-related tasks in order to receive credit. This approach is unlike other apprentice programs that merely require a trainee to put in the required number of hours in the classroom and on the job.

The primary goal of the NCCER is to standardize construction craft training throughout the country so that both employers and employees will benefit from the training, no matter where they are located. As a trainee in an NCCER program, you will become part of a national registry. You will receive a certificate for each level of training you complete. If you apply for a job with any participating contractor in the country, a transcript of your training will be available. If your training is incomplete when you make a job transfer, you can pick up where you left off because every participating contractor is using the same training program. In addition, many technical schools and colleges are using the program.

2.0.0 ◆ HISTORY OF CARPENTRY

Primitive carpentry developed in forest regions during the latter years of the Stone Age, when early humans improved stone tools so they could be used to shape wood for shelters, animal traps,

Tools

Tools are an essential part of carpentry. Although modern tools have advanced beyond the primitive stone tools used by our ancestors, they serve the carpenter's needs in much the same way. Carpentry, like other trades, relies on tools to make difficult tasks easier. If you take the time to learn the proper way to handle and use your tools safely, you'll be able to work much more effectively and produce a high-quality product.

101SA01.EPS

and dugout boats. Between 4000 and 2000 B.C.E, Egyptians developed copper tools, which they used to build vaults, bed frames, and furniture. Later in that period, they developed bronze tools and bow drills. An example of the Egyptians'

1.2 CARPENTRY FUNDAMENTALS LEVEL ONE ◆ TRAINEE GUIDE

Instructor's Notes:
_____
_____
_____

skill in mitering, mortising, dovetailing, and paneling is the intricate furniture found in the tomb of Tutankhamen ("King Tut"). European carpenters did not produce such furniture until the Renaissance (1300 to 1500 C.E.), although they used timber to construct dwellings, bridges, and industrial equipment. In Denmark and ancient Germany, Neolithic people (around 5000 B.C.E.) built rectangular houses from timbers that were nearly 100 feet long. In England, the mortised and fishtailed joints of the stone structures at Stonehenge indicate that advanced carpentry techniques were known in ancient Britain. Before the Roman conquest of Britain (100 C.E.), its carpenters had already developed iron tools such as saws, hatchets, rasps, and knives. They even had turned-wood objects made on primitive pole lathes.

In the Middle Ages, carpenters began a movement toward specialization, such as shipwrights, wheelwrights, turners, and millwrights. However, general-purpose carpenters were still found in most villages and on large private estates. These carpenters could travel with their tools to outlying areas that had no carpenters or to a major building project that required temporary labor. During this period, European carpenters invented the carpenter's brace (a tool for holding and turning a drill bit). The plane, which the Romans had used centuries earlier, reappeared about 1200 C.E. The progress of steelmaking also provided for advancements in the use of steel-edged tools and the advent of crude iron nails. Wooden pegs were used to hold wooden members together before the use of nails. Screws were invented in the 1500s.

The first castles and churches in northern Europe were constructed of timber. When the great stone buildings replaced those made of timber, skilled carpenters built the floors, paneling, doors, and roofs. The erection of large stone buildings also led to the inventions of scaffolding for walls, framework for arch assembly, and pilings to strengthen foundations. Houses and other smaller buildings were still made of timber and thinner wood. Clay was used to fill the gaps between the beams.

The art of carpentry contributed significantly to the grandeur of the great buildings of the Renaissance. Two noted masterpieces of timber construction are the outer dome of St. Paul's Cathedral in London and the 68-foot roof of the Sheldonian Theater in Oxford. After the Renaissance, other examples of architecture requiring skilled carpentry appeared, including the mansard roof with its double slope, providing loftier attics, broad staircases, and sashed windows. These architectural features were incorporated in homes constructed in Colonial America. George W. Snow introduced balloon-frame construction (*Figure 1*) in Chicago in 1840, which proved to be a much cheaper and quicker method because it used machine-made studs and nails. In balloon framing, the studs run from the bottom floor to the uppermost rafters. This method gives the structure exceptional ability to handle strong winds, but requires very long studs that are difficult to manufacture, transport, and store. Because of these problems, balloon framing has almost disappeared. It is used to some extent in Florida to frame the gable ends of buildings in order to provide protection from hurricanes. Today, platform (western) framing (*Figure 2*) has almost completely replaced balloon-frame construction.

This brief history illustrates that carpentry has a long and rich heritage. It also shows that carpentry is an ever-changing trade. You will inevitably discover that learning never ends as you practice the carpentry trade because new and better ways of construction will continue to emerge.

Discuss the types of framing.

Show Transparencies 2 and 3 (Figures 1 and 2).

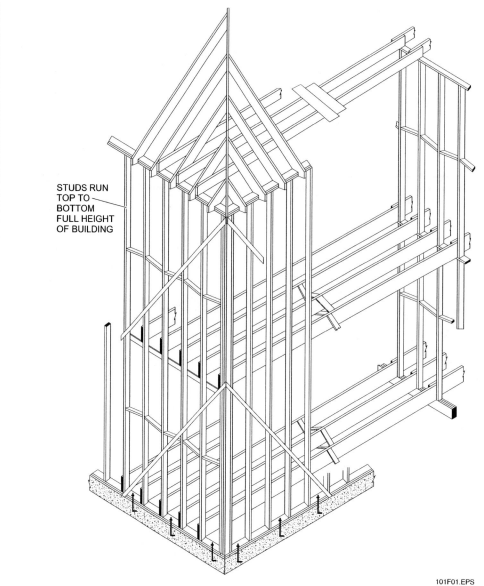

Figure 1 ♦ Balloon framing.

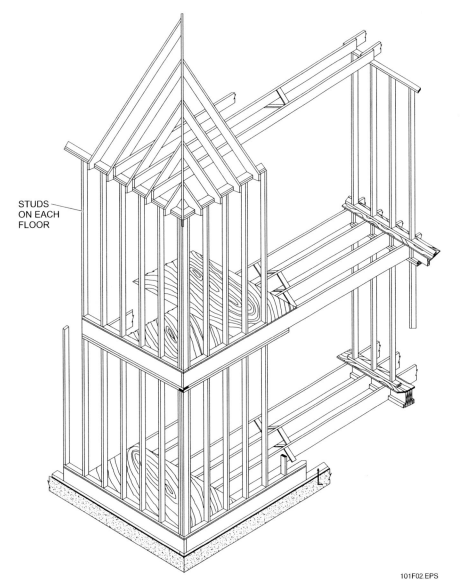

Figure 2 ♦ Western platform framing.

As time permits, take the trainees to local construction sites to view modern construction in progress.

Summarize the duties of carpenters, and point out that carpenters may specialize in various aspects of the trade.

Ancient Construction

Modern carpentry is a continuum from ancient times. Carpenters and other craftsmen were responsible for construction of the world's most celebrated examples of historical architecture. From the lavish interiors of ancient Egypt's great pyramids to the outer dome of London's St. Paul's Cathedral (shown here), the rich heritage of carpentry can be found across the world.

3.0.0 ◆ MODERN CARPENTRY

The scope of carpentry has expanded in modern times with the use of synthetic building materials and ever-improving tools. Today's carpenters must not only know about wood, but also about materials such as particleboard, gypsum wallboard, suspended ceiling tiles, plastics, and laminates. They must also know how to use many modern tools, fasteners, construction techniques, and safety procedures.

The duties of carpenters can vary significantly from one job to another. A carpenter who works for a commercial contractor may work primarily with concrete, steel, and preformed building materials (*Figure 3*). A carpenter who does residential work is more likely to work with wood-frame construction and wood finish materials, but will also encounter an increasing variety of preformed and prefabricated building materials (*Figure 4*).

Figure 3 ◆ Typical commercial construction.

1.6 CARPENTRY FUNDAMENTALS LEVEL ONE ◆ TRAINEE GUIDE

Instructor's Notes:

Figure 4 ◆ Typical residential construction.

In the construction industry, carpentry is commonly divided into two categories: rough and finish. Examples of **rough carpentry** include erecting frameworks, scaffolds, and wooden forms for concrete, as well as building docks, bridges, and supports for tunnels and sewers. **Finish carpentry** (*Figure 5*) includes building stairs; installing doors, cabinets, wood paneling, and molding; and

Figure 5 ◆ Example of finish carpentry.

Explain the professional levels through which a carpenter can progress.

Show Transparency 4 (Figure 6).

putting up acoustical tiles. Skilled carpenters do both rough and finish work.

The duties of carpenters vary even within the broad categories of rough and finish carpentry. The type of construction, size of the company, skill of the carpenter, community size, and other factors affect the carpenter's work. Carpenters who are employed by a large contractor, for example, may specialize in one area, such as laying hardwood floors, while others who are employed by a small firm may build wall frames, put in insulation, and install paneling. They may even perform concrete finishing, welding, and painting. The duties of carpenters also vary because each job is unique.

Carpenters often have great freedom in planning and performing their work. However, carpentry techniques are standard, and most jobs involve the following steps to some extent:

- Use construction drawings to lay out the structure on the site.
- Use drawings and specifications to perform a **takeoff** of materials needed for construction.
- Assemble the materials, tools, and equipment needed for construction.
- Schedule the work.
- Assemble the structure using hand and power tools.
- Check the work using levels, rules, and squares.

Carpenters also use powder-actuated and pneumatic tools and operate power equipment such as personnel lifts, equipment and material lifts, and small earth-moving machines. You will learn more about the various tools and construction methods used by carpenters in other modules of this program.

As in other building trades, the carpenter's work is active and sometimes strenuous. Prolonged standing, climbing, and squatting are often necessary. Many carpenters work outside under adverse weather conditions. Carpenters risk injury from slips or falls, from contact with sharp or rough materials, and from the use of sharp tools and power equipment. Being new to the trade increases the chance of being injured. Therefore, it is essential that you rely on the knowledge of more experienced workers, learn applicable safety procedures, and wear appropriate personal protective equipment.

In order to be successful in the carpentry trade, a person should possess the following:

- Physical strength to lift and move materials
- Hand-eye coordination to use tools
- The ability to perform math calculations in order to estimate materials and lay out the structure
- Attention to detail in order to accurately measure and cut building materials

Above all, a carpenter must be a responsible person with a high degree of concern for the safety of workers and the quality of the work.

4.0.0 ♦ OPPORTUNITIES IN THE CONSTRUCTION INDUSTRY

The construction industry employs more people and contributes more to the nation's economy than any other industry. Our society will always need new homes, roads, airports, hospitals, schools, factories, and office buildings. This means that there will always be a source of well-paying jobs and career opportunities for carpenters and other construction trade professionals. As shown in *Figure 6*, the opportunities are not limited to work on construction projects. A skilled, knowledgeable carpenter can work in a number of areas.

As a construction worker, a carpenter can progress from apprentice through several levels:

- Journeyman carpenter
- Master carpenter
- Foreman/lead carpenter
- Supervisor
- Safety manager
- Project manager/administrator
- Estimator
- Architect
- General contractor
- Construction manager
- Contractor/owner

Journeyman carpenter – After successfully completing an apprenticeship, a trainee becomes a journeyman. The term journeyman originally meant to journey away from the master and work alone. A person can remain a journeyman or advance in the trade. Journeymen may have additional duties such as supervisor or estimator. With larger companies and on larger jobs, journeymen often become specialists.

Master carpenter – A master craftsperson is one who has achieved and continuously demonstrates the highest skill levels in the trade. The master is a mentor and teacher of those to follow. Master carpenters often start their own businesses and become contractors/owners.

Foreman/lead carpenter – This individual is a front-line leader who directs the work of a crew of craft workers and laborers.

Supervisor – Large construction projects require supervisors who oversee the work of crews made up of foremen, apprentices, and journeymen. They are responsible for assigning, directing, and inspecting the work of construction crew members.

1.8 CARPENTRY FUNDAMENTALS LEVEL ONE ♦ TRAINEE GUIDE

Instructor's Notes:

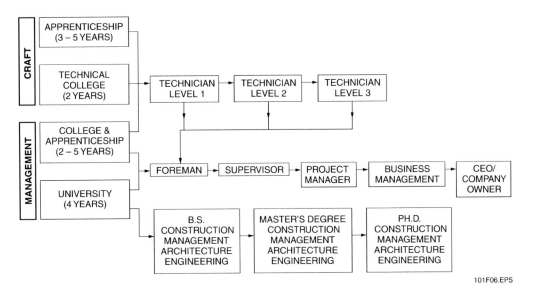

Figure 6 ♦ Opportunities in the construction industry.

As time permits, invite guests who hold positions at different levels and occupations in the carpentry trade to speak to the trainees.

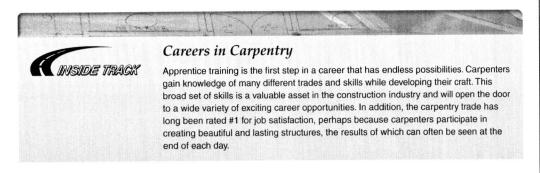

Careers in Carpentry

Apprentice training is the first step in a career that has endless possibilities. Carpenters gain knowledge of many different trades and skills while developing their craft. This broad set of skills is a valuable asset in the construction industry and will open the door to a wide variety of exciting career opportunities. In addition, the carpentry trade has long been rated #1 for job satisfaction, perhaps because carpenters participate in creating beautiful and lasting structures, the results of which can often be seen at the end of each day.

Safety manager – An individual responsible for project safety and health-related issues, including development of the safety plan and procedures, safety training for workers, and regulatory compliance.

Project manager/administrator – Business management and administration deal with controlling the scope and direction of the business and dealing with such concerns as payroll, taxes, and employee benefits. Larger contracting firms may have one or several managers/administrators. This person is responsible for worker output and must determine the best methods to use and the way to apply workers to accomplish the job. A project administrator is responsible for a contractor's support operations, such as accounting, finance, and secretarial work.

Estimator – Estimators work for contractors and building supply companies. They make careful estimates of the materials and labor required for a job. Based on these estimates, the contractor submits bids for jobs. Estimating requires a complete understanding of construction methods as well as the materials and supplies required. Only experienced carpenters who possess good math skills and the patience to prepare detailed, accurate estimates are employed to do this work. This is a highly responsible position since errors in estimates can result in financial losses to the contractor. Depending on the size and type of the business, the job of estimating may be done by the owner, manager, administrator, or an estimating specialist. Today's estimators need solid computer skills because advances in computer software have revolutionized the field of estimating.

Explain the regulations and requirements governing formal construction training.

Emphasize the benefits of apprenticeship programs.

As time permits, introduce the trainees to members of a youth apprenticeship program, and allow them to observe the apprentices in training.

Architect – An architect is a person who is licensed to design buildings and oversee their construction. A person normally needs a specialized degree in architecture to qualify as an architect.

General contractor – A general contractor is an individual or company that manages an entire construction project. The general contractor plans and schedules the project, buys the materials, and usually contracts with carpentry, plumbing, electrical, and other trade contractors to perform the work. The general contractor usually works with architects, engineers, and clients and/or the client's construction manager in planning and implementing a project. General contracting is a natural career path for a master carpenter because, of the many trades involved in a construction project, the carpenter generally plays the largest role and is more likely to have knowledge of the other trades. The general (prime) contractor is also responsible for safety on site.

Construction manager – The role of the construction manager (CM) is different from that of the general contractor. The CM is usually hired by the building owner to represent the owner's interests on the project. The CM is the individual who works with the general contractor and architect to ensure that the building meets the owner's requirements.

Contractor/owner – Construction contractors/owners are those who have established a contracting business. Generally, they hire apprentices, journeymen, and master carpenters to work for them. Depending upon the size of the business, contractors may work with the crew or they may manage the business full-time. Very small contractors may have only one or two people do everything, including managing the business, preparing estimates, obtaining supplies, and doing the work on the job. This group includes specialty subcontractors who perform specialized tasks such as framing, interior trim work, and cabinet installation.

More than any other construction worker, the carpenter is likely to become knowledgeable about many trades. This makes carpentry work interesting and challenging and creates a great variety of career opportunities.

The important thing to learn is that a career is a lifelong learning process. To be an effective carpenter, you need to keep up-to-date with new tools, materials, and methods. If you choose to work your way into management or to someday start your own construction business, you need to learn management and administrative skills on top of keeping your carpentry skills honed. Every successful manager and business owner started the same way you are starting, and they all have one thing in common: a desire and willingness to continue learning. The learning process begins with apprentice training.

As you develop your carpentry skills and gain experience, you will have the opportunity to earn greater pay for your services. There is great financial incentive for learning and growing within the trade. You can't get to the top, however, without learning the basics.

4.1.0 Formal Construction Training

Over the past twenty years, the rate of formal training within the construction industry has been declining. Until the establishment of the NCCER, the only opportunity for formal construction training was through the U.S. Department of Labor, Bureau of Apprenticeship and Training (BAT). The *National Apprenticeship Act* of 1937, commonly referred to as the *Fitzgerald Act*, officially established BAT. The federal government recently created the Office of Apprenticeship, Training, Employer and Labor Services (ATELS) that consolidated both BAT and new employer–labor relations responsibilities.

The federal government established registered apprenticeship training via the *Code of Federal Regulations (CFR) 29:29*, which dictates specific requirements for apprenticeship, and *CFR 29:30*, which dictates specific guidelines for recruitment, outreach, and registration into BAT-approved apprenticeship programs.

Compared to the overall employment in the construction industry, the percentage of enrollment in BAT-style programs has been less than 5 percent for the past decade. BAT programs rely upon mandatory classroom instruction and on-the-job training (OJT). The classroom instruction required is 144 hours per year while the OJT requirement is 2,000 hours per year. A typical BAT program requires 8,000 hours of OJT and 576 hours of related classroom training prior to getting the journeyman certificate dispensed by the BAT.

Craft training via the BAT has not been changed for 30 years, which is believed to be one reason for the lack of use of this program in the construction industry today. Education and training throughout the country is undergoing significant change. As education, political, financial, and student factions argue over the direction and future of education, educators and researchers have been learning and applying new techniques to adjust to how today's students learn and apply their education.

1.10 CARPENTRY FUNDAMENTALS LEVEL ONE ♦ TRAINEE GUIDE

Instructor's Notes:

NCCER is an independent, private educational foundation founded and funded by the construction industry to solve the training problem plaguing the industry today. The basic idea of the NCCER is to supplant governmental control and credentialing of the construction workforce with industry-driven training and education programs. NCCER departs from traditional classroom learning and has adopted a pure competency-based training regimen. Competency-based training means that instead of requiring specific hours of classroom training and set hours of OJT, you simply have to prove that you know what is required and can demonstrate that you can perform the specific skill. NCCER also uses the latest technology, interactive computer-based training, to deliver the classroom portions of the training. All completion information for every trainee is sent to the NCCER and kept within the National Registry. The National Registry can then confirm training and skills for workers as they move from company to company, state to state, or even within their own company (see the *Appendix*).

The dramatic shortage of skills within the construction workforce, combined with the shortage of new workers coming into the industry, is forcing the industry to design and implement new training initiatives to combat the problem. Whether you enroll in a BAT program, an NCCER program, or both, it is critical that you work for an employer who supports a national, standardized training program that includes credentials to confirm your skill development.

4.2.0 Apprenticeship Program

Apprentice training goes back thousands of years; its basic principles have not changed in that time. First, it is a means for individuals entering the craft to learn from those who have mastered the craft. Second, it focuses on learning by doing; real skills versus theory. Although some theory is presented in the classroom, it is always presented in a way that helps the trainee understand the purpose behind the required skill.

4.2.1 Youth Apprenticeship Program

A Youth Apprenticeship Program is also available that allows students to begin their apprentice training while still in high school. A student entering the carpentry program in eleventh grade may complete as much as one year of the NCCER Standardized Craft Training four-year program by high school graduation. In addition, the program, in cooperation with local craft employers, allows students to work in the trade and earn money while still in school. Upon graduation, the student can enter the industry at a higher level and with more pay than someone just starting the apprenticeship program.

This training program is similar to the one used by NCCER learning centers, contractors, and colleges across the country. Students are recognized through official transcripts and can enter the second year of the program wherever it is offered. They may also have the option of applying the credits at a two-year or four-year college that offers degree or certificate programs in the construction trades.

4.2.2 Apprenticeship Standards

All apprenticeship standards prescribe certain work-related or on-the-job training. This on-the-job training is broken down into specific tasks in which the apprentice receives hands-on training during the period of the apprenticeship. In addition, a specified number of hours is required in each task. The total number of hours for the carpentry apprenticeship program is traditionally 8,000, which amounts to about four years of training. In a competency-based program, it may be possible to shorten this time by testing out of specific tasks through a series of performance exams.

In a traditional program, the required OJT may be acquired in increments of 2,000 hours per year. Layoffs or illness may affect the duration.

The apprentice must log all work time and turn it in to the Apprenticeship Committee (discussed later) so that accurate time control can be maintained. Another important aspect of keeping work records up-to-date is that after each 1,000 hours of related work, the apprentice will receive a pay increase as prescribed by the apprenticeship standards.

The classroom-related instruction and work-related training will not always run concurrently due to such reasons as layoffs, type of work needed to be done in the field, etc. Furthermore, apprentices with special job experience or coursework may obtain credit toward their classroom requirements. This reduces the total time required in the classroom while maintaining the total 8,000-hour on-the-job training requirement. These special cases will depend on the type of program and the regulations and standards under which it operates.

Informal on-the-job training provided by employers is usually less thorough than that provided through a formal apprenticeship program. The degree of training and supervision in this

Emphasize the standards to which an apprenticeship must conform.

Explain the requirements for apprenticeship application.

Discuss the responsibilities of the employee.

type of program often depends on the size of the employing firm. A small contractor who specializes in home building may provide training in only one area, such as rough framing. In contrast, a large general contractor may be able to provide training in several areas.

For those entering an apprenticeship program, a high school or technical school education is desirable, as are courses in carpentry, shop, mechanical drawing, and general mathematics. Manual dexterity, good physical condition, a good sense of balance, and a lack of fear of working in high places are important. The ability to solve arithmetic problems quickly and accurately and to work closely with others is essential. You must have a high concern for safety.

The prospective apprentice must submit to the apprenticeship committee certain information. This may include the following:

- Aptitude test (General Aptitude Test Battery or GATB Form Test) results (usually administered by the local Employment Security Commission)
- Proof of educational background (candidate should have school(s) send transcripts to the committee)
- Letters of reference from past employers and friends
- Results of a physical examination
- Proof of age
- If the candidate is a veteran, a copy of Form DD214
- A record of technical training received that relates to the construction industry and/or a record of any pre-apprenticeship training
- High school diploma or General Equivalency Diploma (GED)

The apprentice must:

- Wear proper safety equipment on the job
- Purchase and maintain tools of the trade as needed and required by the contractor
- Submit a monthly on-the-job training report to the committee
- Report to the committee if a change in employment status occurs
- Attend classroom-related instruction and adhere to all classroom regulations such as that for attendance

4.3.0 Responsibilities of the Employee

In order to be successful, the professional must have the skills to use current trade materials, tools, and equipment to produce a finished product of high quality in a minimum period of time.

A carpenter must be adept at adjusting methods to meet each situation. The successful carpenter must continuously train to remain knowledgeable about the technical advancements in trade materials and equipment and to gain the skills to use them. A professional carpenter never takes chances with regard to personal safety or the safety of others.

4.3.1 Professionalism

The word professionalism is a broad term that describes the desired overall behavior and attitude expected in the workplace. Professionalism is too often absent from the construction site and the various trades. Most people would argue that it must start at the top in order to be successful. It is true that management support of professionalism is important to its success in the workplace, but it is more important that individuals recognize their own responsibility for professionalism.

Professionalism includes honesty, productivity, safety, civility, cooperation, teamwork, clear and concise communication, being on time and prepared for work, and regard for one's impact on one's co-workers. It can be demonstrated in a variety of ways every minute you are in the workplace. Most important is that you do not tolerate the unprofessional behavior of co-workers. This is not to say that you shun the unprofessional worker; instead, you work to demonstrate the benefits of professional behavior.

Professionalism is a benefit both to the employer and the employee. It is a personal responsibility. Our industry is what each individual chooses to make of it; choose professionalism and the industry image will follow.

4.3.2 Honesty

Honesty and personal integrity are important traits of the successful professional. Professionals pride themselves on performing a job well and on being punctual and dependable. Each job is completed in a professional way, never by cutting corners or reducing materials. A valued professional maintains work attitudes and ethics that protect property such as tools and materials belonging to employers, customers, and other trades from damage or theft at the shop or job site.

Honesty and success go hand-in-hand. It is not simply a choice between good and bad, but a choice between success and failure. Dishonesty will always catch up with you. Whether you are stealing materials, tools, or equipment from the job site or simply lying about your work, it will

---

1.12 CARPENTRY FUNDAMENTALS LEVEL ONE ◆ TRAINEE GUIDE

Instructor's Notes:

Emphasize the importance of good customer relations.

Ethical Principles for Members of the Construction Trades

Honesty: Be honest and truthful in all dealings. Conduct business according to the highest professional standards. Faithfully fulfill all contracts and commitments. Do not deliberately mislead or deceive others.

Integrity: Demonstrate personal integrity and the courage of your convictions by doing what is right even when there is great pressure to do otherwise. Do not sacrifice your principles for expediency, be hypocritical, or act in an unscrupulous manner.

Loyalty: Be worthy of trust. Demonstrate fidelity and loyalty to companies, employers, fellow craftspeople, and trade institutions and organizations.

Fairness: Be fair and just in all dealings. Do not take undue advantage of another's mistakes or difficulties. Fair people display a commitment to justice, equal treatment of individuals, tolerance for and acceptance of diversity, and open-mindedness.

Respect for others: Be courteous and treat all people with equal respect and dignity regardless of sex, race, or national origin.

Law abiding: Abide by laws, rules, and regulations relating to all personal and business activities.

Commitment to excellence: Pursue excellence in performing your duties, be well-informed and prepared, and constantly endeavor to increase your proficiency by gaining new skills and knowledge.

Leadership: By your own conduct, seek to be a positive role model for others.

not take long for your employer to find out. Of course, you can always go and find another employer, but this option will ultimately run out on you.

If you plan to be successful and enjoy continuous employment, consistency of earnings, and being sought after as opposed to seeking employment, then start out with the basic understanding of honesty in the workplace and you will reap the benefits.

Honesty means more than giving a fair day's work for a fair day's pay; it means carrying out your side of a bargain; it means that your words convey true meanings and actual happenings. Our thoughts as well as our actions should be honest. Employers place a high value on an employee who is strictly honest.

4.3.3 Loyalty

Employees expect employers to look out for their interests, to provide them with steady employment, and to promote them to better jobs as openings occur. Employers feel that they, too, have a right to expect their employees to be loyal to them—to keep their interests in mind; to speak well of them to others; to keep any minor troubles strictly within the plant or office; and to keep absolutely confidential all matters that pertain to the business. Both employers and employees should keep in mind that loyalty is not something to be demanded; rather, it is something to be earned.

4.3.4 Willingness to Learn

Every office and plant has its own way of doing things. Employers expect their workers to be willing to learn these ways. Also, it is necessary to adapt to change and be willing to learn new methods and procedures as quickly as possible. Sometimes the installation of a new machine or the purchase of new tools makes it necessary for even experienced employees to learn new methods and operations. It is often the case that employees resent having to accept improvements because of the retraining that is involved. However, employers will no doubt think they have a right to expect employees to put forth the necessary effort. Methods must be kept up-to-date in order to meet competition and show a profit. It is this profit that enables the owner to continue in business and that provides jobs for the employees.

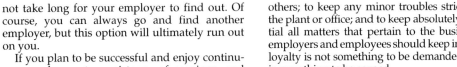

4.3.5 Willingness to Take Responsibility

Most employers expect their employees to see what needs to be done, then go ahead and do it. It is very tiresome to have to ask again and again that a certain job be done. It is obvious that having been asked once, an employee should assume the responsibility from then on. Employees should be alert to see boxes that need to be out of the way, stock that should be stacked, or tools that need to be put away. It is true that, in general, responsibility should be delegated and not assumed; once the responsibility has been delegated, however, the employee should continue to perform the duties without further direction. Every employee has the responsibility for working safely.

4.3.6 Willingness to Cooperate

To cooperate means to work together. In our modern business world, cooperation is the key to getting things done. Learn to work as a member of a team with your employer, supervisor, and fellow workers in a common effort to get the work done efficiently, safely, and on time.

4.3.7 Rules and Regulations

People can work together well only if there is some understanding about what work is to be done, when it will be done, and who will do it. Rules and regulations are a necessity in any work situation and should be so considered by all employees.

4.3.8 Tardiness and Absenteeism

Tardiness means being late for work and absenteeism means being off the job for one reason or another. Consistent tardiness and frequent absences are an indication of poor work habits, unprofessional conduct, and a lack of commitment.

Your work life is governed by the clock. You are required to be at work at a definite time. So is everyone else. Failure to get to work on time results in confusion, lost time, and resentment on the part of those who do come on time. In addition, it may lead to penalties, including dismissal. Although it may be true that a few minutes out of a day are not very important, we must remember that a principle is involved. Our obligation is to be at work at the time indicated. We agree to the terms of work when we accept the job. Perhaps it will help us to see things more clearly if we try to look at the matter from the supervisor's point of view. Supervisors cannot keep track of people if they come in any time they please. It is not fair to others to ignore tardiness. Failure to be on time may hold up the work of fellow workers. Better planning of our morning routine will often keep us from being delayed and so prevent a breathless, late arrival. In fact, arriving a little early indicates your interest and enthusiasm for your work, which is appreciated by employers. The habit of being late is another one of those things that stand in the way of promotion.

It is sometimes necessary to take time off from work. No one should be expected to work when sick or when there is serious trouble at home. However, it is possible to get into the habit of letting unimportant and unnecessary matters keep us from the job. This results in lost production and hardship on those who try to carry on the work with less help. Again, there is a principle involved. The person who hires us has a right to expect us to be on the job unless there is some very good reason for staying away. Certainly, we should not let some trivial reason keep us home. We should not stay up nights until we are too tired to go to work the next day. If we are ill, we should use the time at home to do all we can to recover quickly. This, after all, is no more than most of us would expect of a person we had hired to work for us, and on whom we depended to do a certain job.

If it is necessary to stay home, then at least phone the office early in the morning so that the boss can find another worker for the day. Some

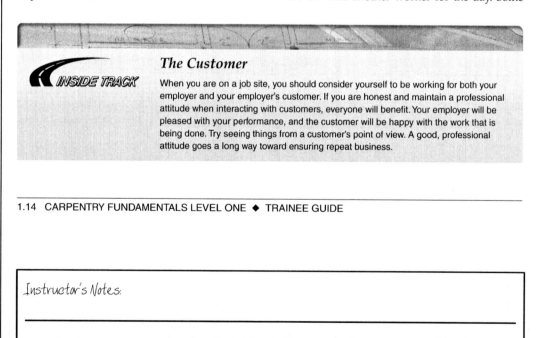

INSIDE TRACK

The Customer

When you are on a job site, you should consider yourself to be working for both your employer and your employer's customer. If you are honest and maintain a professional attitude when interacting with customers, everyone will benefit. Your employer will be pleased with your performance, and the customer will be happy with the work that is being done. Try seeing things from a customer's point of view. A good, professional attitude goes a long way toward ensuring repeat business.

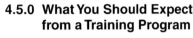

Late for Work

 Showing up on time is a basic requirement for just about every job. Your employer is counting on you to be there at a set time, ready to work. While legitimate emergencies may arise that can cause you to be late for or even miss work, starting a bad habit of consistent tardiness is not something you want to do. What are the possible consequences that you could face as a result of tardiness and absenteeism?

Discuss the "Think About It." Emphasize the negative consequences of tardiness and absenteeism.

As time permits, invite an employer to engage in an interactive discussion of employers' obligations to their employees.

employees remain home without contacting their employer. This is the worst possible way to handle the matter. It leaves those at work uncertain about what to expect. They have no way of knowing whether you have merely been held up and will be in later, or whether immediate steps should be taken to assign your work to someone else. Courtesy alone demands that you let the boss know if you cannot come to work.

The most frequent causes of absenteeism are illness or death in the family, accidents, personal business, and dissatisfaction with the job. Here we see that some of the causes are legitimate and unavoidable, while others can be controlled. One can usually plan to carry on most personal business affairs after working hours. Frequent absences will reflect unfavorably on a worker when promotions are being considered.

Employers sometimes resort to docking pay, demotion, and even dismissal in an effort to control tardiness and absenteeism. No employer likes to impose restrictions of this kind. However, in fairness to those workers who do come on time and who do not stay away from the job, an employer is sometimes forced to discipline those who will not follow the rules.

4.4.0 What You Should Expect from Your Employer

After an applicant has been selected for apprenticeship by the committee, the employer of the apprentice agrees that the apprentice will be employed under conditions that will result in normal advancement. In return, the employer requires the apprentice to make satisfactory progress in on-the-job training and related classroom instruction. The employer agrees that the apprentice will not be employed in a manner that may be considered in violation of the apprenticeship standards. The employer also agrees to pay a prorated share of the cost of operating the apprenticeship program.

4.5.0 What You Should Expect from a Training Program

First and foremost, it is important that the employer you select has a training program. The program should be comprehensive, standardized, and competency-based, not based on the amount of time you spend in a classroom.

When employers take the time and initiative to provide quality training, it is a sign that they are willing to invest in their workforce and improve the abilities of workers. It is important that the training program be national in scope and that transcripts and completion credentials are issued to participants. Construction is unique in that the employers share the workforce. An employee in the trades may work for several contractors throughout their time in the field. Therefore, it is critical that the training program help the worker move from company to company, city to city, or state to state without having to start at the beginning for each move. Ask how many employers in the area use the same program before you enroll. Make sure that you will always have access to transcripts and certificates to ensure your status and level of completion.

Training should be rewarded. The training program should have a well-defined compensation ladder attached to it. Successful completion and mastery of skill sets should be accompanied by increases in hourly wages.

Finally, the curricula should be complete and up-to-date. Any training program has to be committed to maintaining its curricula, developing new delivery mechanisms (CD-ROM, Internet, etc.), and being constantly vigilant for new techniques, materials, tools, and equipment in the workplace.

4.6.0 What You Should Expect from the Apprenticeship Committee

The Apprenticeship Committee is the local administrative body to which the apprentice is

Emphasize the importance of human relations and its impact on productivity.

Discuss the "Think About It." Have trainees share past experiences regarding teamwork and its impact on productivity.

assigned and to which the responsibility is delegated for the appropriate training of the individual. Every apprenticeship program, whether state or federal, is covered by standards that have been approved by those agencies. The responsibility of enforcement is delegated to the committee.

The committee is responsible not only for enforcement of standards, but must see to it that proper training is conducted so that a craftsperson graduating from the program is fully qualified in those areas of training designated by the standards.

Among the responsibilities of the committee are the following:

- Screen and select individuals for apprenticeship and refer them to participating firms for training.
- Place apprentices under written agreement for participation in the program.
- Establish minimum standards for related instruction and on-the-job training and monitor the apprentice to see that these criteria are adhered to during the training period.
- Hear all complaints of violations of apprenticeship agreements, whether by employer or apprentice, and take action within the guidelines of the standards.
- Notify the registration agencies of all enrollments, completions, and terminations of apprentices.

5.0.0 ◆ HUMAN RELATIONS

Many people underestimate the importance of working well with others. There is a tendency to pass off human relations as nothing more than common sense. What exactly is involved in human relations? One response would be to say that part of human relations is being friendly, pleasant, courteous, cooperative, adaptable, and sociable.

5.1.0 Making Human Relations Work

As important as the above-noted characteristics are for personal success, they are not enough. Human relations is much more than just getting people to like you. It is also knowing how to handle difficult situations as they arise.

Human relations is knowing how to work with supervisors who are often demanding and sometimes unfair. It is understanding the personality traits of others as well as yourself. Human relations is building sound working relationships in situations where others are forced on you.

Human relations is knowing how to restore working relationships that have deteriorated for one reason or another. It is learning how to handle frustrations without hurting others. Human relations is building and maintaining relationships with all kinds of people, whether those people are easy to get along with or not.

5.2.0 Human Relations and Productivity

Effective human relations is directly related to productivity. Productivity is the key to business success. Every employee is expected to produce at a certain level. Employers quickly lose interest in an employee who has a great attitude but is able to produce very little. There are work schedules to be met and jobs that must be completed.

All employees, both new and experienced, are measured by the amount of quality work they can safely turn out. The employer expects every employee to do his or her share of the workload.

However, doing one's share in itself is not enough. If you are to be productive, you must do your share (or more than your share) without antagonizing your fellow workers. You must perform your duties in a manner that encourages others to follow your example. It makes little difference how ambitious you are or how capably

---

Teamwork

Many of us like to follow all sorts of different teams: racing teams, baseball teams, football teams, and soccer teams. Just as in sports, a job site is made up of a team. As a part of that team, you have a responsibility to your teammates. What does teamwork really mean on the job?

1.16 CARPENTRY FUNDAMENTALS LEVEL ONE ◆ TRAINEE GUIDE

Instructor's Notes:

you perform. You cannot become the kind of employee you want to be, or the type of worker management wants you to be, without learning how to work with your peers.

Employees must sincerely do everything they can to build strong, professional working relationships with fellow employees, supervisors, and clients.

5.3.0 Attitude

A positive attitude is essential to a successful career. First, being positive means being energetic, highly motivated, attentive, and alert. A positive attitude is essential to safety on the job. Second, a positive employee contributes to the productivity of others. Both negative and positive attitudes are transmitted to others on the job. A persistent negative attitude can spoil the positive attitudes of others. It is very difficult to maintain a high level of productivity while working next to a person with a negative attitude. Third, people favor a person who is positive. Being positive makes a person's job more interesting and exciting. Fourth, the kind of attitude transmitted to management has a great deal to do with an employee's future success in the company. Supervisors can determine a subordinate's attitude by their approach to the job, reactions to directives, and the way they handle problems.

5.4.0 Maintaining a Positive Attitude

A positive attitude is far more than a smile, which is only one example of an inner positive attitude. As a matter of fact, some people transmit a positive attitude even though they seldom smile. They do this by the way they treat others, the way they look at their responsibilities, and the approach they take when faced with problems.

Here are a few suggestions that will help you to maintain a positive attitude:

- Remember that your attitude follows you wherever you go. If you make a greater effort to be a more positive person in your social and personal lives, it will automatically help you on the job. The reverse is also true. One effort will complement the other.
- Negative comments are seldom welcomed by fellow workers on the job. Neither are they welcome on the social scene. The solution: Talk about positive things and be complimentary. Constant complainers do not build healthy and fulfilling relationships.
- Look for the good things in people on the job, especially your supervisor. Nobody is perfect, but almost everyone has a few worthwhile qualities. If you dwell on people's good features, it will be easier to work with them.
- Look for the good things where you work. What are the factors that make it a good place to work? Is it the hours, the physical environment, the people, the actual work being done, or is it the atmosphere? Keep in mind that you cannot be expected to like everything. No work assignment is perfect, but if you concentrate on the good things, the negative factors will seem less important and bothersome.
- Look for the good things in the company. Just as there are no perfect assignments, there are no perfect companies. Nevertheless, almost all organizations have good features. Is the company progressive? What about promotional opportunities? Are there chances for self-improvement? What about the wage and benefit package? Is there a good training program? You cannot expect to have everything you would like, but there should be enough to keep you positive. In fact, if you decide to stick with a company for a long period of time, it is wise to look at the good features and think about them. If you think positively, you will act the same way.
- You may not be able to change the negative attitude of another employee, but you can protect your own attitude from becoming negative.

6.0.0 ♦ EMPLOYER AND EMPLOYEE SAFETY OBLIGATIONS

An obligation is like a promise or a contract. In exchange for the benefits of your employment and your own well-being, you agree to work safely. In other words, you are obligated to work safely. You are also obligated to make sure anyone you happen to supervise or work with is working safely. Your employer is also obligated to maintain a safe workplace for all employees. Safety is everyone's responsibility (*Figure 7*).

Some employers will have safety committees. If you work for such an employer, you are then obligated to that committee to maintain a safe working environment. This means two things:

- Follow the safety committee's rules for proper working procedures and practices.
- Report any unsafe equipment and conditions directly to the committee or your supervisor.

Emphasize the importance of maintaining a positive attitude in the workplace.

As time permits, invite a company's safety officer to speak to the trainees about the importance of job site safety.

Emphasize the importance of following OSHA regulations.

Figure 7 ◆ Safety is everyone's responsibility.

Here is a basic rule to follow every working day:

If you see something that is not safe, REPORT IT! Do not ignore it. It will not correct itself. You have an obligation to report it.

Suppose you see a faulty electrical hookup. You know enough to stay away from it, and you do. But then you forget about it. Why should you worry? It is not going to hurt you. Let somebody else deal with it. The next thing that happens is that a co-worker accidentally touches the live wire.

In the long run, even if you do not think an unsafe condition affects you—it does. Do not mess around; report what is not safe. Do not think your employer will be angry because your productivity suffers while the condition is corrected. On the contrary, your employer will be more likely to criticize you for not reporting a problem.

Your employer knows that the short time lost in making conditions safe again is nothing compared with shutting down the whole job because of a major disaster. If that happens, you are out of work anyway. Do not ignore an unsafe condition. In fact, Occupational Safety and Health Administration (OSHA) regulations require you to report hazardous conditions.

This applies to every part of the construction industry. Whether you work for a large contractor or a small subcontractor, you are obligated to report unsafe conditions. The easiest way to do this is to tell your supervisor. If that person ignores the unsafe condition, report it to the next highest supervisor. If it is the owner who is being unsafe, let that person know your concerns. If nothing is done about it, report it to OSHA. If you are worried about your job being on the line, think about it in terms of your life, or someone else's, being on the line.

The U.S. Congress passed the *Occupational Safety and Health Act* in 1970. This act also created OSHA. It is part of the U.S. Department of Labor. The job of OSHA is to set occupational safety and health standards for all places of employment, enforce these standards, ensure that employers provide and maintain a safe workplace for all employees, and provide research and educational programs to support safe working practices.

OSHA requires each employer to provide a safe and hazard-free working environment. OSHA also requires that employees comply with OSHA rules and regulations that relate to their conduct on the job. To gain compliance, OSHA can perform spot inspections of job sites, impose fines for violations, and even stop work from proceeding until the job site is safe.

According to OSHA standards, you are entitled to on-the-job safety training. As a new employee, you must be:

- Shown how to do your job safely
- Provided with the required personal protective equipment
- Warned about specific hazards
- Supervised for safety while performing the work

OSHA was adopted with the stated purpose "to assure as far as possible every working man and woman in the nation safe and healthful working conditions and to preserve our human resources."

1.18 CARPENTRY FUNDAMENTALS LEVEL ONE ◆ TRAINEE GUIDE

Instructor's Notes:

Drugs and Alcohol

When people use drugs and alcohol, they are putting both themselves and the people around them at serious risk. A construction site can be a dangerous environment, and it is important to be alert at all times. Using drugs and alcohol on the job is an accident waiting to happen. You have an obligation to yourself, your employer, and your fellow employees to work safely. What should you do if you discover someone abusing drugs and/or alcohol at work?

Discuss the "Think About It." Emphasize the impact of drug and alcohol use on employee safety and productivity.

The enforcement of the *Occupational Safety and Health Act* is provided by the federal and state safety inspectors who have the legal authority to make employers pay fines for safety violations. The law allows states to have their own safety regulations and agencies to enforce them, but they must first be approved by the U.S. Secretary of Labor. For states that do not develop such regulations and agencies, federal OSHA standards must be obeyed.

These standards are listed in *OSHA Safety and Health Standards for the Construction Industry (29 CFR, Part 1926)*, sometimes called *OSHA Standards 1926*. Other safety standards that apply to the carpentry trade are published in *OSHA Safety and Health Standards for General Industry (29 CFR, Parts 1900 to 1910)*.

The most important general requirements that OSHA places on employers in the construction industry are:

- The employer must perform frequent and regular job site inspections of equipment.
- The employer must instruct all employees to recognize and avoid unsafe conditions, and to know the regulations that pertain to the job so they may control or eliminate any hazards.
- No one may use any tools, equipment, machines, or materials that do not comply with *OSHA Standards 1926*.
- The employer must ensure that only qualified individuals operate tools, equipment, and machines.

Have trainees complete the Review Questions, and go over the answers prior to administering the Module Examination.

Review Questions

1. The use of screws as fasteners has been common since _____.
   a. the 1960s
   b. the Stone Age
   c. about 1900
   d. the 1500s

2. The position that follows an apprenticeship is _____.
   a. master carpenter
   b. journeyman
   c. general contractor
   d. estimator

3. A competency-based training program is one that requires the student to _____.
   a. receive at least four years of classroom training
   b. receive on-the-job training for at least two years
   c. demonstrate the ability to perform specific job-related tasks
   d. pass a series of written tests

4. The _____ is most likely to handle the day-to-day operations on the job site.
   a. supervisor
   b. project manager
   c. architect
   d. owner

5. The purpose of the Youth Apprenticeship Program is to _____.
   a. make sure all young people know how to use basic carpentry tools
   b. provide job opportunities for people who quit high school
   c. allow students to start in an apprenticeship program while still in high school
   d. make sure that people under 18 have proper supervision on the job

6. The combined total of on-the-job and classroom training needed for a carpentry apprentice to advance to journeyman is _____ hours.
   a. 2,000
   b. 4,000
   c. 6,000
   d. 8,000

7. Which of the following is true with respect to honesty?
   a. It is okay to borrow tools from the job site as long as you return them before anyone notices.
   b. You are doing your company a favor by using lower-grade materials than those listed in the specifications.
   c. It is okay to take materials or tools from your employer if you feel the company owes you for past efforts.
   d. Being late and not making up the time is the same as stealing from your employer.

8. If one of your co-workers complains about your company, you should _____.
   a. contribute your own complaints to the conversation
   b. agree with the person to avoid conflict
   c. suggest that the person look for another job
   d. find some good things to say about the company

9. If you see an unsafe condition on the job, you should _____.
   a. ignore it because it is not your job
   b. tell a co-worker
   c. call OSHA
   d. report it to a supervisor

10. The purpose of OSHA is to _____.
    a. catch people breaking safety regulations
    b. make rules and regulations governing all aspects of construction projects
    c. ensure that the employer provides and maintains a safe workplace
    d. assign a safety inspector to every project

1.20 CARPENTRY FUNDAMENTALS LEVEL ONE ◆ TRAINEE GUIDE

Instructor's Notes:

Summary

There are many job and career opportunities for skilled carpenters. An apprenticeship program that combines competency-based, hands-on training with classroom instruction has proven to be the most effective means for a person to learn and advance in the carpentry craft. Developing job skills is only part of the solution; it is just as important to learn good work habits, convey a positive, cooperative attitude to those around you, and practice good safety habits every day.

Summarize the major concepts presented in the module.

Notes

Have trainees complete the Trade Terms Quiz, and go over the answers.

Administer the Module Examination. Record the results on Craft Training Report Form 200, and submit the results to the Training Program Sponsor.

Trade Terms Quiz

1. The installation of roof trusses is a type of _____.

2. The building materials required for the completion of a project are listed on a _____.

3. The installation of kitchen cabinets is a type of _____.

Trade Terms

Finish carpentry
Rough carpentry
Takeoff

1.22 CARPENTRY FUNDAMENTALS LEVEL ONE ◆ TRAINEE GUIDE

Instructor's Notes:

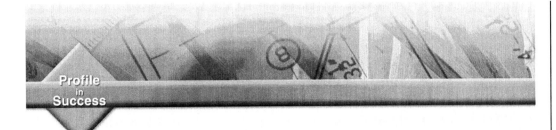

Profile in Success

Clay Kubicek

Clay Kubicek is the education director for Crossland Construction, a midwestern construction company that was recently ranked among the top 400 construction companies in the United States by the *Engineering News-Record*.

Tell us a little bit about what you do for Crossland and its employees.

The Crossland family asked me to join their team to coordinate and build upon their existing education program. As Crossland Construction's education director I work with all in-house training and I assist area public schools in the development of a 5th through 12th grade construction technology program. In these two areas my job is to pull together people and resources to build the education vision of Crossland Construction.

We develop and maintain training programs for all employees. This is done through the Crossland Academy. The Academy currently offers over 30 classes for Crossland employees. These classes range from a one-day policies and procedures class, to a four-year Kansas Department of Labor approved carpentry apprenticeship program. In this four-year program Crossland trainees have the opportunity to earn a journeyman carpentry card. In addition to the journeyman carpentry course, all our safety programs use NCCER curriculum, so our employees also have a transcript at the National Registry. From computer training to project management seminars, the Crossland Academy is actively trying to meet the educational needs of a growing company. In the coming year we will be working to develop a four-year ironworker program in addition to our other programs.

With area public schools my job is to help them develop their construction education programs. Crossland has developed a partnership plan with some area schools in which we provide curriculum, resources, and materials to provide construction career paths for high school students. The Crossland school program is arranged to expose 5th and 6th grade students to commercial construction through field trips; give 7th and 8th graders a chance to explore construction through field trips and one-week modules using Pitsco learning systems, and culminate with 9th through 12th graders using NCCER's Construction Technology curriculum. The students become Core certified in their 9th grade year. In grades 10 and 11 the Construction Technology curriculum has been developed with the Kansas State Department of Education to give students an overview and background in masonry, concrete, carpentry, HVAC, plumbing, and electrical work. The 12th grade option is a dual-credit course in project management.

Interested students are able to earn college credit using this curriculum program. Field trips, guest speakers, subject matter experts are offered to the schools at every grade level as the students move through the program.

The students have the option of taking their NCCER certifications and entering the labor force on their way to becoming a skilled craftsperson, as well as entering a two- or four-year college program in a construction-related field of study. This "Crossland Connection" with area public schools is Crossland's way of showing the students all the opportunities they can find in the construction industry.

The Crossland family is committed to education, and my job is to assist in developing a culture of education. Schools interested in examining this program can find an outline at www.crosslandconstruction.com and use the education link.

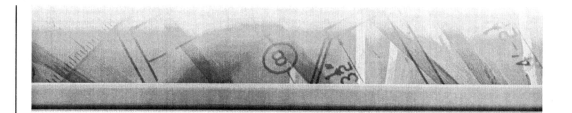

What advice do you have for trainees as they embark on their careers? What should they look for in an employer?
Keeping a positive attitude is one bit of advice I would give to anyone pursuing a goal. I know it is someone else's thunder, but there is power in positive thinking. Have a can-do attitude. Complete your training with superb attendance and participation. Technology is and will continue to change the way training and education take place in the workplace. Stay current with technology. Where the term "lifelong learner" was a catchphrase, it is now a necessity for advancement in any field. Another thing is to be careful not to just focus on today's pay. Look long-term and have vision. When looking for an employer in the construction industry, I believe a trainee would benefit in the long run by looking for a company that wants to invest in the trainee's future, not just their own. Find a company with plans to expand and grow. Look for a company that offers comprehensive training opportunities, a retirement plan, insurance, and a history of looking after their employees. Make sure you are working with people of integrity.

What should trainees do to ensure their progress toward their career goals? What do employers look for in potential employees?
There are multiple things a trainee can do to ensure they reach their career goals. Ask and seek the answers to their questions. If they are curious about how something is done, take the time to find the answer. We are in the information age, and to reach any career goal it is imperative we know how to find needed and relevant information. Keep current on materials and construction trends. Once again technology has created many changes in the construction industry. Stay current. It is also important to plan. They need to have a plan to reach their goals, and follow that plan. Most quality companies reach their goals by having integrity. Trainees should follow that lead. Find a role model and mentor. Find a person you would like to emulate.

Learn from success. Finally, develop your craft skills as well as your human relationship skills.

Crossland looks for people who have a can-do attitude, have a sense of urgency in their performance style, and are willing to learn and grow. Crossland is looking for team players who can cooperate and work within a diverse and growing company. Any person with a strong work ethic could find many opportunities with Crossland Construction.

This book is designed for first-year Carpentry trainees. In a few years, most of them will have moved into their careers and will begin progress toward their goals. What should carpenters do to keep up with industry developments?
One thing is to take advantage of relevant training opportunities available within their company. If their company does not offer the training they desire, seek out another source. Carpenters should inquire of their superintendents about their company's future needs and training options. It is important to stay open-minded and flexible. Once again, we should all keep our eyes on the growing role of technology in the construction industry. As with any profession, staying current with industry trends and practices requires reading and research. Subscribing to a couple of industry-driven magazines and journals helps in keeping up-to-date on industry developments. For tools and techniques there are a number of woodworking journals available. For a larger perspective the *Engineering News-Record* provides quality industry insight. To keep abreast with the business side of the industry *Construction Executive: the Magazine for the Business of Construction* provides good insight. There are numerous sources available on the Internet also. If a trainee is interested in pursuing a professional career in construction, they should be on a constant search for knowledge and information concerning the industry. A carpenter must work with many aspects of a construction job. Exemplary carpenters have a knowledge base of all aspects of the construction process.

1.24 CARPENTRY FUNDAMENTALS LEVEL ONE ♦ TRAINEE GUIDE

Instructor's Notes:

Trade Terms Introduced in This Module

Finish carpentry: The portion of the carpentry trade associated with interior and exterior trim, cabinetry, siding, wall finishes, and decorative work.

Rough carpentry: The portion of the carpentry trade associated with framing and other work that will be covered with finish materials.

Takeoff: A list of building materials obtained by analyzing the project drawings (also known as a material takeoff).

Appendix

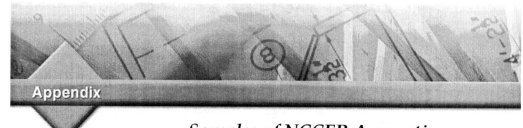

Samples of NCCER Apprentice Training Recognition

101A01.EPS

1.26 CARPENTRY FUNDAMENTALS LEVEL ONE ◆ TRAINEE GUIDE

Instructor's Notes:

April 4, 2006

Student Example
NCCER
3600 NW 43rd St Bldg G
Gainesville, FL 32606

Dear Student,

On behalf of the National Center for Construction Education and Research, I congratulate you for successfully completing the NCCER's standardized craft training program.

As the NCCER's most recent graduate, you are a valuable member of today's skilled construction and maintenance workforce. The skills that you have acquired through the NCCER craft training programs will enable you to perform quality work on construction and maintenance projects, promote the image of these industries and enhance your long-term career opportunities.

We encourage you to continue your education as you advance in your construction career. Please do not hesitate to contact us for information regarding our Management Education and Safety Programs or if we can be of any assistance to you.

Enclosed please find your certificate, transcript and wallet card. If you have any questions regarding your credentials, contact the Registry Department at 352-334-0911. Once again, congratulations on your accomplishments and best wishes for a successful career in the construction and maintenance industries.

Sincerely,

Donald E. Whyte
President, NCCER

NATIONAL CENTER FOR CONSTRUCTION EDUCATION AND RESEARCH
P.O. Box 141104 ■ Gainesville, Florida ■ 32614-1104 ■ PH: 352.334.0911 ■ FX: 352.334.0932 ■ www.nccer.org

101A02.EPS

NATIONAL CENTER FOR CONSTRUCTION EDUCATION AND RESEARCH
P.O. Box 141104 ■ Gainesville, Florida 32614-1104
PH: 352.334.0911 ■ FX: 352.334.0932 ■ www.nccer.org
Affiliated with the University of Florida

4/4/06
Page: 1

Official Transcript

Student Example
NCCER
3600 NW 43rd St Bldg G
Gainesville, FL 32606
R#: 000108319

Current Employer:

Course / Description	Instructor	Training Location	Date Compl.
00101-04 Basic Safety	Don Whyte	NCCER	5/10/05
00102-04 Introduction to Construction Math	Don Whyte	NCCER	5/10/05
00103-04 Introduction to Hand Tools	Don Whyte	NCCER	5/10/05
00104-04 Introduction to Power Tools	Don Whyte	NCCER	5/10/05
00105-04 Introduction to Blueprints	Don Whyte	NCCER	5/10/05
00106-04 Basic Rigging	Don Whyte	NCCER	5/10/05
27101 Orientation to the Trade	Don Whyte	NCCER	10/1/05
27102 Wood Building Materials, Fasteners, & Adhesives	Don Whyte	NCCER	10/1/05
27103 Hand and Power Tools	Don Whyte	NCCER	10/1/05
27104 Floor Systems	Don Whyte	NCCER	10/1/05
27105 Wall and Ceiling Framing	Don Whyte	NCCER	10/1/05
27106 Roof Framing	Don Whyte	NCCER	10/1/05
27107 Windows and Exterior Doors	Don Whyte	NCCER	10/1/05

NO ENTRIES BELOW THIS LINE

Donald E. Whyte
President

101A03.EPS

1.28 CARPENTRY FUNDAMENTALS LEVEL ONE ◆ TRAINEE GUIDE

Instructor's Notes:

MODULE 27101-06 — ANSWERS TO REVIEW QUESTIONS

Answer	Section
1. d	2.0.0
2. b	4.0.0
3. c	1.0.0
4. a	4.0.0
5. c	4.2.1
6. d	4.2.2
7. d	4.3.2
8. d	5.4.0
9. d	6.0.0
10. c	6.0.0

MODULE 27101-06 — ANSWERS TO TRADE TERMS QUIZ

1. Rough carpentry
2. Takeoff
3. Finish carpentry

MODULE 27101-06 — TEACHING TIPS

The following are suggested activities or instructional methods to help you teach the material in this module.

General

When you call on someone to answer a question, the rest of the class relaxes or even tunes out because they expect that the question and answer will take place only between you and the trainee you called on. Instead, use this technique to involve more trainees in answering questions and to keep them on their toes.

1. Ask trainees to define a term or explain a concept.
2. After one trainee has answered, ask a trainee seated nearby if the answer is right. Then ask whether a trainee in the back of the room agrees.
3. Ask trainees to explain why they think an answer is right or wrong.
4. Use the session to clear up incorrect ideas and encourage trainees to learn from their mistakes.

CONTREN® LEARNING SERIES — USER UPDATE

NCCER makes every effort to keep these textbooks up-to-date and free of technical errors. We appreciate your help in this process. If you have an idea for improving this textbook, or if you find an error, a typographical mistake, or an inaccuracy in NCCER's Contren® textbooks, please write us, using this form or a photocopy. Be sure to include the exact module number, page number, a detailed description, and the correction, if applicable. Your input will be brought to the attention of the Technical Review Committee. Thank you for your assistance.

Instructors – If you found that additional materials were necessary in order to teach this module effectively, please let us know so that we may include them in the Equipment/Materials list in the Annotated Instructor's Guide.

Write: Product Development and Revision
National Center for Construction Education and Research
P.O. Box 141104, Gainesville, FL 32614-1104

Fax: 352-334-0932

E-mail: curriculum@nccer.org

Craft _____ Module Name _____

Copyright Date _____ Module Number _____ Page Number(s) _____

Description _____

_____

_____

_____

_____

(Optional) Correction _____

_____

_____

_____

(Optional) Your Name and Address _____

_____

_____

_____

Building Materials, Fasteners, and Adhesives
27102-06

NCCER STANDARDIZED CRAFT TRAINING PROGRAM

The National Center for Construction Education and Research (NCCER) provides a standardized national program of accredited craft training. Key features of the program include instructor certification, competency-based training, and performance testing. The program provides trainees, instructors, and companies with a standard form of recognition through the National Registry. The program is described in full in the *Guidelines for Accreditation*, published by NCCER. For more information on standardized craft training, contact NCCER by writing to P.O. Box 141104, Gainesville, FL 32614-1104; calling 352-334-0911; or e-mailing **info@nccer.org**. More information is available at **www.nccer.org**.

HOW TO USE THIS ANNOTATED INSTRUCTOR'S GUIDE

Each page presents two sections of information. The larger section displays each page exactly as it appears in the Trainee Module. The narrow column ties suggested trainee and instructor actions to each page and provides icons (detailed below) to call your attention to material, safety, audiovisual, or testing requirements. The bottom of each page includes space for your notes.

The **Audiovisual** icon indicates an appropriate time to show a transparency or other audiovisual aid.

The **Classroom** icon prompts you to define a term, stress a point, ask trainees to explain a concept, or give examples.

The **Demonstration** icon directs you to show trainees how to perform tasks.

The **Examination** icon tells you to administer the written module examination.

The **Homework** icon is placed where you may wish to assign reading for the next class, assign a project, or advise trainees to prepare for an examination.

The **Laboratory** icon is used when trainees are to practice performing tasks.

The **Materials** icon is a reminder for you to gather materials needed for classes, laboratories, and testing.

The **Performance Testing** icon tells you to administer a performance test or a portion thereof.

The **Safety** icon is used to emphasize safety issues. It is often keyed to *Caution* and *Warning!* statements in the Trainee Module.

The **Teaching Tip** icon indicates additional guidance is available, such as how to conduct an exercise, get the most educational value from a field trip, or encourage class participation. Teaching Tips may expand on a feature (*Think About It*, *Did You Know?*) or provide *Quick Quizzes* or similar exercises. You will be referred to the Teaching Tips section at the back of the module if there is additional material.

The **Combination** icon indicates that the laboratory listed corresponds with a performance task. If desired, you can note the proficiency of the trainees during the laboratory, and use it to satisfy performance testing requirements.

PREPARATION

Before teaching this module, you should review the Objectives, Performance Tasks, Materials and Equipment List, and Module Outline. Be sure to allow ample time to prepare your own training or lesson plan and gather all required materials and equipment.

Building Materials, Fasteners, and Adhesives
Annotated Instructor's Guide

Module 27102-06

MODULE OVERVIEW

This module introduces the carpentry trainee to wood building materials, fasteners, and adhesives.

PREREQUISITES

Prior to training with this module, it is suggested that the trainee shall have successfully completed *Core Curriculum*; *Carpentry Fundamentals Level One*, Module 27101-06.

OBJECTIVES

Upon completion of this module, the trainee will be able to do the following:

1. Identify various types of building materials and their uses.
2. State the uses of various types of hardwoods and softwoods.
3. Identify the different grades and markings of wood building materials.
4. Identify the safety precautions associated with building materials.
5. Describe the proper method of storing and handling building materials.
6. State the uses of various types of engineered lumber.
7. Calculate the quantities of lumber and wood products using industry-standard methods.
8. Describe the fasteners, anchors, and adhesives used in construction work and explain their uses.

PERFORMANCE TASKS

Under the supervision of the instructor, the trainee should be able to do the following:

1. Calculate the quantities of lumber and wood products using industry-standard methods.
2. Given a selection of building materials, identify a particular material and state its use.

MATERIALS AND EQUIPMENT LIST

Transparencies
Markers/chalk
Blank acetate sheets
Transparency pens
Pencils and scratch paper
Overhead projector and screen
Whiteboard/chalkboard
Appropriate personal protective equipment
Samples of lumber containing:
    Grade stamps
    Natural defects
    Manufacturing defects
Samples of plywood containing grade stamps
Samples of engineered sheet materials
    (OSB, particleboard, etc.)
Samples of engineered lumber
    (LVL, PSL, glulam, etc.)

Samples of various concrete blocks
Samples of metal framing materials
Samples of various kinds of:
    Nails
    Screws
    Bolts
    Anchors
    Construction adhesives
Cross section of a tree trunk (optional)
Drill and bits
Hammer
Screwdriver
Calculator
Module Examinations*
Performance Profile Sheets*

* Located in the Test Booklet.

SAFETY CONSIDERATIONS

Ensure that the trainees are equipped with appropriate personal protective equipment.

ADDITIONAL RESOURCES

This module is intended to present thorough resources for task training. The following reference works are suggested for both instructors and motivated trainees interested in further study. These are optional materials for continued education rather than for task training.

Basic Construction Materials. Upper Saddle River, NJ: Prentice Hall.

Principles and Practices of Light Construction. Upper Saddle River, NJ: Prentice Hall.

Principles and Practices of Commercial Construction. Upper Saddle River, NJ: Prentice Hall.

Building Construction Illustrated. New York, NY: John Wiley & Sons.

Fundamentals of Building Construction: Materials and Methods. New York, NY: John Wiley & Sons.

Buildings in Wood: The History and Traditions of Architecture's Oldest Building Material. New York: Rizzoli/Universe International Publications.

TEACHING TIME FOR THIS MODULE

An outline for use in developing your lesson plan is presented below. Note that each Roman numeral in the outline equates to one session of instruction. Each session has a suggested time period of 2½ hours. This includes 10 minutes at the beginning of each session for administrative tasks and one 10-minute break during the session. Approximately 7½ hours are suggested to cover *Wood Building Materials, Fasteners, and Adhesives.* You will need to adjust the time required for hands-on activity and testing based on your class size and resources. Because laboratories often correspond to Performance Tasks, the proficiency of the trainees may be noted during these exercises for Performance Testing purposes.

Topic   **Planned Time**

Session I. Introduction; Lumber Sources and Uses; Lumber Defects; Lumber Grading; Plywood

    A. Introduction

    B. Lumber Sources and Uses

        1. Lumber Cutting

        2. General Classifications of Lumber

    C. Lumber Defects

        1. Moisture and Warping

        2. Preventing Warping and Splitting

    D. Lumber Grading

        1. Grading Terms

        2. Classification of Manufacturing Defects

        3. Abbreviations

    E. Plywood

        1. Plywood Sheet Sizes

        2. Grading for Softwood Construction Plywood

        3. Plywood Storage

Session II. Building Boards; Engineered Wood Products; Pressure-Treated Lumber; Calculating Lumber Quantities; Concrete Block Construction; Commercial Construction Methods

    A. Building Boards

        1. Hardboard

        2. Particleboard

        3. High-Density Overlay (HDO) and Medium-Density Overlay (MDO) Plywood

        4. Oriented Strand Board (OSB)

        5. Mineral Fiberboards

B. Engineered Wood Products
   1. Laminated Veneer Lumber (LVL)
   2. Parallel Strand Lumber (PSL)
   3. Laminated Strand Lumber (LSL)
   4. Wood I-Beams
   5. Glue-Laminated Lumber (Glulam)
C. Pressure-Treated Lumber
D. Calculating Lumber Quantities
E. Laboratory

Have the trainees calculate the quantities of lumber and wood products required for an instructor-supplied project. This laboratory corresponds to Performance Task 1.

F. Concrete Block Construction
G. Commercial Construction Methods
   1. Floors
   2. Exterior Walls
   3. Interior Walls and Partitions
   4. Metal Framing Materials

Session III. Nails; Staples; Screws; Bolts; Mechanical Anchors; Epoxy Anchoring Systems; Adhesives; Review; Module Examination and Performance Testing

A. Nails
B. Staples
C. Screws
   1. Wood Screws
   2. Sheet Metal Screws
   3. Machine Screws
   4. Lag Screws and Shields
   5. Concrete/Masonry Screws
   6. Deck Screws
   7. Drywall Screws
   8. Drive Screws
   9. Hammer-Driven Pins and Studs
D. Bolts
   1. Stove Bolts
   2. Machine Bolts
   3. Carriage Bolts
E. Mechanical Anchors
   1. Anchor Bolts
   2. One-Step Anchors
   3. Bolt Anchors
   4. Screw Anchors
   5. Self-Drilling Anchors
   6. Guidelines for Drilling Anchor Holes in Hardened Concrete or Masonry
   7. Hollow-Wall Anchors

F. Epoxy Anchoring Systems
G. Adhesives
   1. Glues
   2. Construction Adhesives
   3. Mastics
   4. Shelf Life
H. Laboratory

   Have the trainees identify and state the use of various building materials. This laboratory corresponds to Performance Task 2.
I. Review
J. Module Examination
   1. Trainees must score 70 percent or higher to receive recognition from NCCER.
   2. Record the testing results on Craft Training Report Form 200, and submit the results to the Training Program Sponsor.
K. Performance Testing
   1. Trainees must perform each task to the satisfaction of the instructor to receive recognition from NCCER. If applicable, proficiency noted during laboratory exercises can be used to satisfy the Performance Testing requirements.
   2. Record the testing results on Craft Training Report Form 200, and submit the results to the Training Program Sponsor.

Building Materials, Fasteners, and Adhesives
27102-06

James Cordonier, a Cabinetmaking contestant at the SkillsUSA 2005 National Championship, measures cabinet members for assembly. Each contestant had eight hours to complete a cabinet featuring a door with a panel, a drawer with a mechanical slide, and a case with adjustable shelves and a laminate top. The plans for the cabinet are given to all competitors the morning of the competition, as is the exact amount of wood needed and other supplies. Each competitor at the National Championship is a champion of their state contest held earlier in the year. Mr. Cordonier says that the competition is very tough: "One slight mistake will definitely take you out of the top ten. I won the Wyoming State SkillsUSA Cabinetmaking Contest the last two years and attended the SkillsUSA National Contest both of those years. I placed eighth out of 50 last year. SkillsUSA is an awesome organization and a great experience."

Instructor's Notes:

Assign reading of Module 27102-06.

27102-06
Building Materials, Fasteners, and Adhesives

Topics to be presented in this module include:

1.0.0	Introduction	2.1
2.0.0	Lumber Sources and Uses	2.3
3.0.0	Lumber Defects	2.5
4.0.0	Lumber Grading	2.8
5.0.0	Plywood	2.13
6.0.0	Building Boards	2.16
7.0.0	Engineered Wood Products	2.18
8.0.0	Pressure-Treated Lumber	2.20
9.0.0	Calculating Lumber Quantities	2.20
10.0.0	Concrete Block Construction	2.21
11.0.0	Commercial Construction Methods	2.23
12.0.0	Nails	2.30
13.0.0	Staples	2.34
14.0.0	Screws	2.34
15.0.0	Bolts	2.39
16.0.0	Mechanical Anchors	2.39
17.0.0	Epoxy Anchoring Systems	2.46
18.0.0	Adhesives	2.46

Overview

Carpenters are involved in just about every phase of construction, so it is important that they be familiar with a large variety of building materials and fasteners.

A carpenter working in residential construction will be working primarily with wood products. In commercial construction, however, very little wood is used for structural purposes. Steel and concrete are the common structural materials. A high-rise building, for example, might be framed with structural steel, have poured concrete floors, and have glass and metal curtain wall panels attached to provide the exterior finish.

As you will see, there is an almost endless variety of nails, screws, bolts, and anchors, each of them serving a different purpose. Knowledge of the various fasteners and their applications is important to your success as a carpenter.

There are also many types of adhesives used in construction. Like fasteners, different adhesives are used for different purposes. Selecting the wrong one can result in problems. In addition, adhesives are made of chemicals, some of which are toxic and/or flammable, so safety is an important concern when using construction adhesives.

Instructor's Notes:
_____
_____
_____

Objectives

When you have completed this module, you will be able to do the following:

1. Identify various types of building materials and their uses.
2. State the uses of various types of hardwoods and softwoods.
3. Identify the different grades and markings of wood building materials.
4. Identify the safety precautions associated with building materials.
5. Describe the proper method of storing and handling building materials.
6. State the uses of various types of engineered lumber.
7. Calculate the quantities of lumber and wood products using industry-standard methods.
8. Describe the fasteners, anchors, and adhesives used in construction work and explain their uses.

Trade Terms

Butt joint
Cantilever
Catalyst
Galvanized
Gypsum
Japanned
Joists
Material safety data sheet (MSDS)
Millwork
Nail set
Rafter
Resins
Sheathing
Shiplap
Sill plate
Tongue-and-groove
Vaulted ceiling

Required Trainee Materials

1. Pencil and paper
2. Appropriate personal protective equipment

Prerequisites

Before you begin this module, it is recommended that you successfully complete *Core Curriculum* and *Carpentry Fundamentals Level One* Module 27101-06. This course map shows all of the modules in *Carpentry Fundamentals Level One*. The suggested training order begins at the bottom and proceeds up. Skill levels increase as you advance on the course map. The local Training Program Sponsor may adjust the training order.

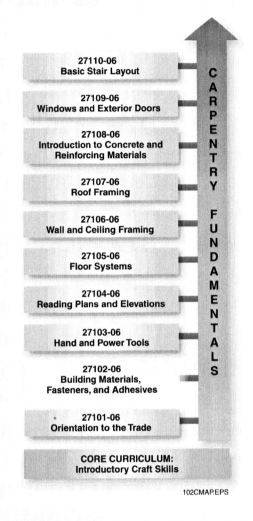

102CMAP.EPS

Ensure you have everything required to teach the course. Check the Materials and Equipment List at the front of this module.

Show Transparency 1, Objectives.

Show Transparency 2, Performance Tasks.

Explain that the terms shown in bold (blue) are defined in the back of this module.

See the general Teaching Tip at the end of this module.

Emphasize that carpenters must be familiar with a number of different building materials, fasteners, and adhesives.

1.0.0 ◆ INTRODUCTION

Carpenters use a wide variety of building materials. Wood framing materials are most common in residential construction, although residential foundations are usually built of concrete and masonry blocks. Carpenters working on commercial building projects are more likely to work with structures made of steel and concrete. This module introduces the different types of building materials used in construction, as well as the fasteners and adhesives used to apply these materials. Many kinds of lumber and wood building products are used in construction. Different wood products suit different purposes, so it is very important for a carpenter to know all the types and grades of wood building materials, along with their uses and limitations. When selecting lumber, the carpenter must know what type best fits the application.

When selecting lumber, it is also important to be able to recognize defects in the material and to know how to properly handle and store wood products to prevent damage. When ordering lumber, the carpenter must know how to specify amounts. Sometimes, material is sold by the linear foot; other times, it is sold by the board foot. Still other material is sold in sheets. The carpenter must know when and how to use the various methods of ordering.

Although wood framing was the norm for many years, the use of metal framing has become increasingly popular in both commercial and residential construction. The modules in this level focus on wood frame construction. However, metal framing is discussed in each of the applicable modules of this level and will be covered in more detail in subsequent levels.

Building a New House
INSIDE TRACK — If wood is used, you can expect to see a lot of it on the job site. It takes approximately 14,000 board feet of lumber to build an average size house.

102SA01.EPS

2.2 CARPENTRY FUNDAMENTALS LEVEL ONE ◆ TRAINEE GUIDE

Instructor's Notes:
_____
_____
_____

2.0.0 ◆ LUMBER SOURCES AND USES

2.1.0 Lumber Cutting

Throughout the ages, people have used wood for a variety of purposes: as fuel, as building materials, in weapons, and in transportation. In this module, it is important to distinguish lumber from wood. Lumber refers to the boards, timbers, etc., produced from sawmills, whereas wood refers to the material itself, which comes from many species of trees.

Wood has several advantages:

- It is easily worked.
- It has durability and beauty.
- It has great ability to absorb shocks from sudden loads.
- It is free from rust and corrosion, comparatively light in weight, and adaptable to a countless variety of purposes.

Wood is useful as a building material because of the manner in which a tree forms its fibers: growing by the addition of new material to the outer layer just under the bark and preserving its old fibers as it adds new ones.

Viewing a cross section of a tree, as shown in *Figure 1*, one can see that the trunk consists of a series of concentric rings covered by a layer of bark. Each annular ring represents one year of tree growth. This growth takes place just under the bark in the cambium layer. The cells that are formed there are long; their tubular fibers are composed mainly of cellulose. They are bound together by a substance known as lignin, which connects them into bundles. In wood, lignin is the material that acts as a binder between the cells, adding strength and stiffness to the cell walls. The bundles of fibers run the length of the tree and carry food from the roots to the branches and leaves.

Running at right angles, the fibers from the outer layer inward are another group of cells known as medullary rays. These rays carry food from the inner bark to the cambium layer and act as storage tanks for food. The rays are more pronounced in some species than others (such as oak) but are present in all trees. As one layer of wood succeeds another, the cells in the inner layer die (cease to function as food storage) and become useful only to give stiffness to the tree. This older wood, known as heartwood, is usually darker in color, drier, and harder than the living layer (sapwood).

Heartwood is more durable than sapwood. If wood is to be exposed to decay-producing conditions without the benefit of a preservative, a minimum percentage of heartwood will be specified.

Sapwood takes preservative treatment more readily than heartwood, and it is equally durable when treated.

Most trees are sawed so that the growth rings form an angle of less than 45 degrees with the surface of the boards produced. Such lumber is called flat-grained in softwoods and plain-sawed in hardwoods (*Figure 2*). This is the least expensive

Discuss the advantages of wood as a building material.

Obtain a cut portion of an actual tree trunk and allow the trainees to examine it. Identify the different portions of the tree's cross section.

Show Transparency 3 (Figure 1).

Explain the impact of different cutting methods on the lumber produced.

Show Transparency 4 (Figure 2).

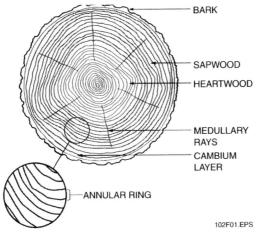

Figure 1 ◆ Cross section of a tree.

MODULE 27102-06 ◆ BUILDING MATERIALS, FASTENERS, AND ADHESIVES 2.3

Review the characteristics and uses of various types of wood. Refer the trainees to *Appendix A* in the Trainee Module.

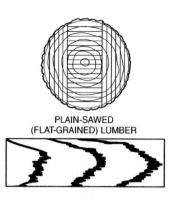
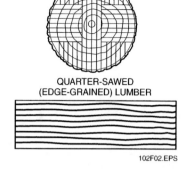

Figure 2 ◆ Common lumber cutting methods.

Ghost Wood

Longleaf pine and bald cypress are praised by carpenters for their outstanding beauty and durability. Unfortunately, clear-cutting that occurred in the late 1800s wiped out the two species from the vast forests in the southern United States. Today, lumber companies have resurrected this wood by retracing the rivers that timber companies used to transport the logs to sawmills. Logs of original-growth pine and cypress have been lying along the bottom of these riverbeds for more than one hundred years. This wood is now being harvested by scuba divers, sawed, dried, and used for various construction purposes.

method, but the lumber produced by this method is more likely to shrink or warp.

There is another method in which the wood is cut with the growth rings at an angle greater than 45 degrees. Lumber cut by this method is known as edge-grained or vertical-grained in softwoods and quarter-sawed in hardwoods. The lumber it produces is usually more durable and less likely to warp or shrink. Quarter-sawed lumber is often used for hardwood floors.

In order to obtain the most lumber from the log, most logs are cut using a combination of the two methods.

2.2.0 General Classifications of Lumber

Lumber falls into one of two general classifications: hardwood or softwood. Hardwoods come from deciduous (leaf-bearing) trees such as oak and maple. Softwoods come from coniferous (cone-bearing) trees such as pine and fir. (The term softwood does not mean it is not strong or durable. On the contrary, some softwoods are harder than certain hardwoods.) Descriptions and uses of various hardwoods and softwoods are provided in *Appendix A*.

Mahogany

Mahogany is used for the finest furniture and millwork. There are many species of mahogany; more than 300 species come from the Philippines alone. South America is a large supplier of mahogany, as are the Fiji Islands in the South Pacific. The most exotic mahogany comes from Africa.

Point out that lumber grading is based largely on the amount and types of defects in lumber.

Discuss the types of natural defects found in lumber.

Hardwoods and Softwoods

As shown here, hardwoods and softwoods vary widely in color and grain.

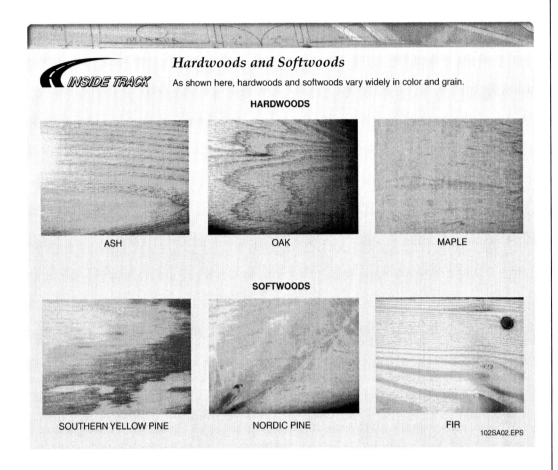

3.0.0 ◆ LUMBER DEFECTS

Before we discuss lumber grading, which is to a large extent based on the amount and types of defects in the lumber, it is necessary to understand the various kinds of defects. Lumber defects can occur naturally (e.g., knots). Defects can also occur during the cutting process. Further defects can occur while the lumber is drying, or because of improper storage or handling. Defects in lumber can affect the lumber's appearance, strength, and usability. In most common uses, a certain amount of defects are permitted; unless it is severely damaged, some use can be found for just about any piece of lumber. That is why lumber is graded and why there are many grades. While

Point out that lumber can warp as it dries.

some lumber defects occur naturally, most defects occur either during the manufacturing and drying process or as a result of improper storage and handling. *Figure 3* shows various types of lumber defects.

3.1.0 Moisture and Warping

Trees contain a large amount of moisture. For example, a newly cut 10' length of 2×10 lumber can contain more than four gallons of water. Because of its high moisture content, freshly cut (green) lumber cannot be used until it is dried. If green lumber is used, it can cause cracked ceilings or floors, squeaky floors, sticking doors, and other problems. Some of the drying is done by heating lumber in kilns at the sawmill; however, the lum-

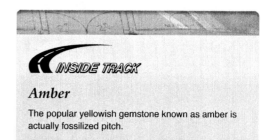

Amber

The popular yellowish gemstone known as amber is actually fossilized pitch.

ber continues to dry for a long time after that. As it dries, the cut lumber will shrink. This can cause warping and splitting because the lumber dries unevenly. A carpenter must be able to tell when

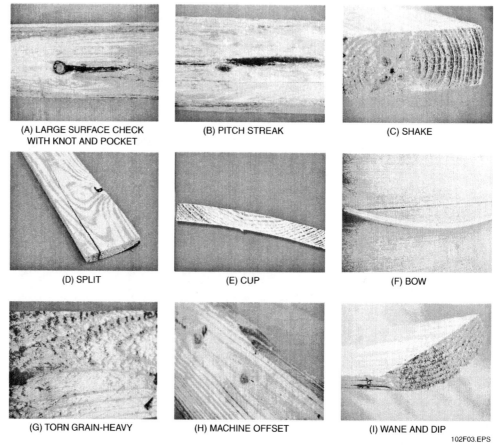

Figure 3 ◆ Common lumber defects.

Wood Defects

When lumber is properly stored at a job site, there is a good chance that most of it will be usable. You may encounter situations in which the lumber may be too damaged to use, at least for its intended purpose. Experienced carpenters can find other uses for damaged and defective lumber. Wood with cosmetic defects, like knots, may be desirable for paneling purposes. Some warped lumber can be used in joists and rafters if the proper precautions are taken to account for the distortion. It is a good idea, however, to consult with an experienced carpenter or supervisor when you encounter these types of situations.

Explain that proper lumber storage and stacking can help to prevent warping and splitting.

Drying Lumber

There are two methods of drying lumber. Kiln drying is the most common method, but air drying is also used. In air drying, the lumber is stacked outdoors and allowed to dry for up to three months, as compared with only a few days for kiln drying. Even though the process is much longer, air-dried lumber is still likely to have a higher moisture content than kiln-dried lumber.

lumber contains too much moisture. Moisture meters can be used on site for that purpose. A simple way to check for moisture is to hit the lumber with a hammer. If moisture comes to the surface in the dent left by the hammer, the lumber is too wet and should not be used.

3.2.0 Preventing Warping and Splitting

Building materials may be delivered to a job site weeks before they are used. Unless the lumber, plywood, and other materials are properly stacked and protected, they can be seriously damaged. Although a contractor allows for a certain amount of waste and spoilage in the estimate, such an allowance does not account for damage due to improper storage. It is up to the work crew to take care of the lumber.

When a complete framing package for a residence is delivered to the job site, the framing package looks great. All the material is neatly packed. Some material is strapped and some neatly tied. However, when the load is dumped, it will shift. What was a neat load suddenly becomes a pile of lumber spread out in different directions. Some lumber is split or broken. Other material may be partially buried in sand, mud, or snow. If the framing package is left the way it is, any of the following may result:

- The lumber is exposed to the elements. Sun, wind, rain, and snow can create serious problems.
- What would have been good lumber may now be crooked, warped, or cupped.
- Some lumber will crack when strain is placed against it; there is a good possibility of splits.
- It will cost the contractor money to replace the materials.
- The time involved in getting additional material delivered to the job site could affect the schedule.
- There is a safety hazard associated with using damaged materials.
- Additional time required in working with defective materials is costly.

A simple way to avoid these problems is to have knowledgeable workers at the site when the materials are delivered. It takes a few hours to stack the lumber and protect it from the elements, saving your employer the cost of replacing the damaged materials. Also, delivery people are more likely to be careful if an observer is present.

There are several factors to consider when unloading and stacking building materials at the job site. These may include the following:

- The order in which the material will be used
- The weight of the material
- The weather conditions

Describe the grading system for lumber, and explain how to interpret lumber grading stamps.

Show Transparency 5 (Figure 4).

You should get instructions from your supervisor before deciding how and where to stack the materials. Here are some general guidelines.

Be sure to consider the order in which the material will be used. For example, do not put the siding on top of the **sheathing**.

All material should be stacked on a level surface or as close to level as possible. If stacking material indoors, the amount of material you stack on the subfloor should not exceed its weight-bearing capacity. If stacking outdoors, try not to block the foundation or slab. The next step is to find some scrap lumber and place it flat on the ground. Two or three 2 × 4s approximately 4' apart from each other will work nicely. Stack the widest lumber on top of this scrap lumber. Be sure the good lumber does not come in contact with the ground. Ensure that there is enough support under the lumber so that it will not sag.

Separate the lumber, keeping the pressure-treated material from the regular material.

Keep your materials banded until it is time to use them. Try to avoid using lumber that has been exposed to heavy rain.

Plywood can be stacked vertically or horizontally. Since some plywood is only ½" thick, it should be laid horizontally on top of scrap pieces as described above. Fiberboard should be stacked in a flat position; do not let the fiberboard touch the ground.

Protect the lumber from the elements by covering the materials with a waterproof material (e.g., roofing paper or polyethylene). Cover the lumber to protect it from the elements, but not to prevent air from circulating around the stack. Preventing air circulation could cause twisting or warping, depending on the moisture content of the air.

Special care is needed for interior finish materials. If you are using knock-down (KD) door frames (either wood or metal) for the doors, these usually come tied together. Always ensure that the door frames (also known as jambs) are stacked flat and off the floor or concrete. Some moldings will come wrapped or tied. Take care to lay them flat and off the floor or concrete. Wall paneling should be treated the same way. Extreme caution should be taken not to damage the edges or mar the face of the panel.

Stacking interior finish materials on top of scrap lumber serves two purposes:

- It prevents moisture in the concrete floor from coming in contact with the bottom piece.
- Although the finish material is under cover, there is always a possibility of a pipe bursting and water flowing on the floor.

4.0.0 ◆ LUMBER GRADING

Lumber is graded by inspectors who examine the lumber after it is planed. Grading is based on the number, size, and types of defects in the lumber, which determine its load-carrying capacity. The more defects, the weaker the wood. The defects considered are those described in the preceding section. Grading methods and standards are established by the American Lumber Standards Committee. Basically, the grade describes the type, size, and number of defects allowed in the worst board of that grade.

Inspection criteria are established by the U.S. Department of Commerce and published as product standards. Inspections are performed by regional agencies and associations, such as the Southern Pine Inspection Bureau, Western Wood Products Association, California Redwood Association, etc. Hardwood grades are regulated by the National Hardwood Lumber Association. Different associations may have slightly different standards for the numbers and types of defects permitted within a particular grade. Inspectors from accredited inspection agencies examine the lumber, determine the grade in accordance with the association's specifications, and apply the applicable grade stamp. Keep in mind that since grading is done right after the board is planed, it cannot account for warping that might occur if the board is improperly stored.

Carpenters must be familiar with the various lumber grades. The architect who designs the building will specify the lumber grades to be used on the job. It is absolutely essential that the carpenter or contractor who buys the lumber use the specified grade. In order to do that, you must be able to read the grade stamp. *Figure 4* shows and explains a typical grade stamp.

Lumber is graded based on its strength, stiffness, and appearance. The highest grades have very few defects; the lowest grades may have knots, splits, and other problems. The primary agencies for lumber grading are the Western Wood Products Association and the Southern Pine Council. Although the end results of the grading processes are the same, their grade designations are somewhat different. It is important that you learn to interpret the grade stamps for the grading association in your area.

2.8 CARPENTRY FUNDAMENTALS LEVEL ONE ◆ TRAINEE GUIDE

Instructor's Notes:

(A) Inspection association trademark.
(B) Mill Identification – firm name, brand, or assigned mill number.
(C) Grade Designation – grade name, number, or abbreviation.
(D) Species Identification – indicates species individually or in combination.
(E) Condition of seasoning at time of surfacing:

S-Dry – 19% maximum moisture content
MC15 – 15% maximum moisture content
S-GRN – over 19% moisture content (unseasoned)

Figure 4 ♦ Typical lumber grade stamp.

Table 1 Board Appearance Grades

	Product	Grades	Uses
High-Quality Appearance Grades	Selects	B and Better Select C Select D Select	Selects are used for walls, ceilings, trim, and other areas where appearance is important but the use of hardwood is not desirable because of cost or other considerations.
	Finish	Superior Prime E	
	Paneling	Clear (any select or finish grade) No. 2 Common (selected for knotty paneling) No. 3 Common (selected for knotty paneling)	
	Siding (bevel, bungalow)	Superior Prime	
General-Purpose Appearance Grades	Boards, Sheathing	No. 1 Common No. 2 Common No. 3 Common No. 4 Common No. 5 Common	Common boards are used for shelving, decking, window trim, and other applications where appearance is less critical, or where on-site selection and cutting is practical when appearance is a consideration.
	Alternate Board	Select merchantable Construction Standard Utility Economy	

Softwood lumber is graded in three major categories: boards, dimension lumber, and timbers. These categories are further divided into specific types of lumber and then classified by appearance (*Table 1*) and strength (*Table 2*). The following is a breakdown of each category:

- *Boards* – Boards, sheathing, and form lumber
- *Dimension lumber* – Light framing, studs, structural light framing, and structural **joists** and planks
- *Timbers* – Beams and **stringers**

Refer to *Appendix B* for a graphic depiction of the category breakdown for softwood yard lumber.

Discuss the classifications used by lumber grading agencies.

As time permits, take the trainees to a lumber yard. Have them identify various grades of lumber.

Table 2 Grades of Dimension Lumber and Timbers

Product	Grades	Uses
Light Framing 2" to 4" thick 2" to 4" wide	Construction Standard Utility Economy	Light framing lumber is used where great strength is *not* required, such as for studs, plates, and sills.
Studs 2" to 4" thick	Stud Economy Stud	This optional grade is intended for vertical use, as in loadbearing walls.
Structural Light Framing 2" to 4" thick 2" to 4" wide	Select Structural No. 1 No. 2 No. 3 Economy	Structural lumber is used for light framing and forming where greater bending strength is required. Typical uses include trusses, concrete pier wall forms, etc.
Structural Joists and Planks 2" to 4" thick 5" to 18" wide	Select Structural No. 1 No. 2 No. 3 Economy	These grades are designed especially for lumber 5 inches and wider and are suitable for use as floor joists, rafters, headers, small beams, trusses, and general flooring applications.
Beams and Stringers 5" and thicker with the width more than 2" greater than the thickness.	Select Structural No. 1 No. 2 No. 3	These grades are used for stringers, beams, posts, and other support members.
Posts and Timbers 5" × 5" with the width not more than 2" greater than the thickness.	Select Structural No. 1 No. 2 No. 3	These grades are used for stringers, beams, posts, and other support members.

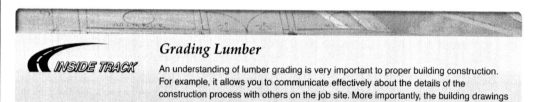

Grading Lumber

An understanding of lumber grading is very important to proper building construction. For example, it allows you to communicate effectively about the details of the construction process with others on the job site. More importantly, the building drawings specify grades of lumber that should be used for various parts of the structure. You will have to interpret those drawings in order to construct the building properly.

4.1.0 Grading Terms

Among the different agencies, criteria for grading may vary somewhat. Generally, all grading agencies use five basic size classifications as follows:

- *Boards (BD)* – Consists of members under 2" thick and 2" and wider.
- *Light framing (L.F.)* – Consists of members 2" to 4" thick and 2" to 4" wide.
- *Joists and planks (J&P)* – Consists of members 2" to 4" thick and 5" to 18" wide.
- *Beams and stringers (B&S)* – Consists of members 5" and thicker with the width more than 2" greater than the thickness.
- *Posts and timbers (P&T)* – Consists of members 5" × 5" with the width not more than 2" greater than the thickness, approximately square.

For each species and size classification, the grading agencies establish several stress-quality grades, for example: Select structural, No. 1, No. 2, No. 3, and assign allowable stresses for each.

Instructor's Notes:

Some other lumber terms used in grading include:

- *Nominal size* – The size by which it is known and sold in the market (e.g., 2 × 4), as opposed to the dressed size.
- *Dressed (actual) size* – The dimensions of lumber after surfacing with a planing machine. The dressed size is usually ½" to ¾" less than the nominal or rough size. A 2 × 4 stud, for example, usually measures about 1½" by 3½". The American Lumber Standard lists standard dressed sizes. *Table 3* compares rough cut (nominal) lumber sizes to dressed sizes for plain-sawn and quarter-sawn lumber. Additional data on softwood and hardwood lumber dimensions are provided in *Appendixes B and D*.
- *Dressed lumber* – Lumber that is surfaced by a planing machine on one side (S1S), two sides (S2S), one edge (S1E), two edges (S2E), or any combination of sides and edges (S1S1E, S1S2E, or S4S). Dressed lumber may also be referred to as planed or surfaced.
- *Dimension lumber* – Lumber is supplied in nominal 2", 3", or 4" thicknesses with standard widths. Light framing, studs, joists, and planks are classified as dimension lumber.
- *Matched lumber* – Lumber that is edge- or end-dressed and shaped to make a close **tongue-and-groove** (T & G) joint at the edges or ends when laid edge to edge or end to end.
- *Patterned lumber* – Lumber that is shaped to a pattern in addition to being dressed, matched, or **shiplapped,** or any combination of these workings.
- *Rough lumber* – Lumber as it comes from the sawmill prior to any dressing operation.
- *Stress-grade lumber* – Lumber grades having assigned working stress and elasticity values in accordance with basic accepted principles of strength grading. Stress is the force exerted on a unit area of lumber, usually expressed in terms of pounds per square inch. Allowable stress is the maximum stress established by the applicable building codes. Allowable stress is always less than the ultimate stress (the amount of stress that can be withstood before the material fails) for obvious structural and safety reasons.
- *Surfaced lumber* – Same as dressed lumber.
- *Framing lumber* – Lumber used for the structural members of a building, such as studs and joists.
- *Finish lumber* – Lumber suitable for **millwork** or for the completion of the interior of a building, chosen because of its appearance or ability to accept a high-quality finish.
- *Select lumber* – In softwoods, a general term for lumber of good appearance and finishing qualities.

Review the common terms associated with lumber grading. Refer the trainees to *Appendixes B* and *D* in the Trainee Module.

Table 3 Nominal and Dressed (Actual) Sizes of Dimension Lumber (in inches)

Nominal	Dressed
2 × 2	1½ × 1½
2 × 4	1½ × 3½
2 × 6	1½ × 5½
2 × 8	1½ × 7¼
2 × 10	1½ × 9¼
2 × 12	1½ × 11¼

Calculating Dressed Softwood Lumber Sizes

As you can see in *Table 3*, nominal lumber sizes don't represent the actual, or dressed, dimensions of the board or lumber. When calculating dressed sizes of softwood boards, there is a general rule that can be applied. Boards with a nominal width of 1" or less usually have dressed dimensions that are ¼" smaller. Boards with a nominal width of 2" to 6" usually have dressed dimensions that are ½" smaller. For boards with widths greater than 6", you can subtract ¾" from the nominal dimensions to get the dressed sizes. Remember, this is a general rule and may not be accurate in every case. The only way to be certain is to perform an actual measurement.

MODULE 27102-06 ◆ BUILDING MATERIALS, FASTENERS, AND ADHESIVES 2.11

Show examples of trimmed lumber. Identify the trimming methods used.

Discuss common lumber manufacturing defects.

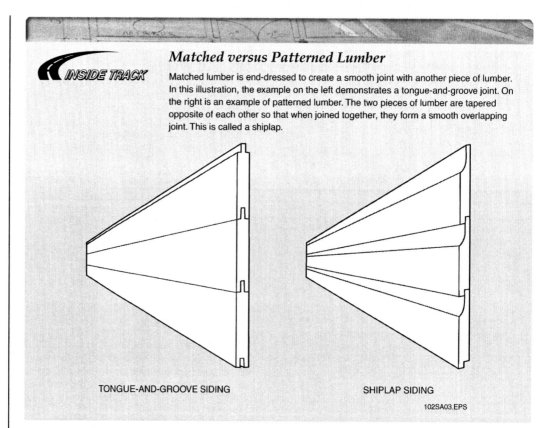

Matched versus Patterned Lumber

Matched lumber is end-dressed to create a smooth joint with another piece of lumber. In this illustration, the example on the left demonstrates a tongue-and-groove joint. On the right is an example of patterned lumber. The two pieces of lumber are tapered opposite of each other so that when joined together, they form a smooth overlapping joint. This is called a shiplap.

TONGUE-AND-GROOVE SIDING

SHIPLAP SIDING

4.1.1 Trim

Trimming is the act of crosscutting a piece of lumber to a given length.

- Double-end-trimmed (DET) lumber is trimmed reasonably square by a saw on both ends.
- Precision-end-trimmed (PET) lumber is trimmed square and smooth on both ends to uniform lengths with a manufacturing tolerance of $\frac{1}{16}$" over or under (+/−) in length in 20 percent of the pieces.
- Square-end-trimmed lumber is trimmed square, permitting only a slight manufacturing tolerance of $\frac{1}{64}$" for each nominal 2" of thickness or width.

4.2.0 Classification of Manufacturing Defects

The various types of defects that can occur while lumber is being cut, dressed, and dried were discussed in an earlier section. The following describes how the defects are considered in grading lumber:

- *Standard A* – Very light torn grain; occasional light chip marks; very slight knife marks.
- *Standard B* – Very light torn grain; very light raised grain; very light loosened grain; slight chip marks; average of one slight chip mark per lineal foot but not more than two in any lineal foot; very slight knife marks; slight mismatch.
- *Standard C* – Medium torn grain; light raised grain; light loosened grain; very light machine bite; very light machine gouge; very light machine offset; light chip marks if well scattered; occasional medium chip marks; very slight knife marks; slight mismatch.
- *Standard D* – Heavy torn grain; medium raised grain; very heavy loosened grain; light machine bite; light machine gouge; light machine offset; medium chip marks; slight knife marks; very light mismatch.

2.12 CARPENTRY FUNDAMENTALS LEVEL ONE ♦ TRAINEE GUIDE

Instructor's Notes:

- *Standard E* – Very heavy torn grain; raised grain; very heavy loosened grain; medium machine bite; machine gouge; medium machine offset; chip marks; knife marks; light wavy dressing; light mismatch.
- *Standard F* – Very heavy torn grain; raised grain; very heavy loosened grain; heavy machine bite; machine gouge; heavy machine offset; chip marks; knife marks; medium wavy dressing; medium mismatch.
- *Standard G* – Loosened grain; raised grain; torn grain; machine bite; machine burn; machine gouge; machine offsets; chip marks; medium wavy dressing; mismatch.

4.3.0 Abbreviations

Most contract documents and construction drawings use abbreviations. *Appendix C* lists the common abbreviations used in the southern pine industry. Many of them also hold for the other wood species.

5.0.0 ◆ PLYWOOD

Plywood is made of layers (plies) of wood veneers. A layer may be $\frac{1}{16}$" to $\frac{3}{8}$" thick. The center layer is known as the core. As layers are added on either side of the core, they are placed alternately at right angles, which increases the strength of the sheet. Layers with the grain at right angles to the core are called crossbands. The outer exposed layer is known as the face veneer.

Veneers are made by the rotary cutting of trees. First, logs are cut to a specific length, then softened with a hot water or steam bath. The bark is then removed and the log locked into its center on a lathe. As the lathe turns the log, a long knife slices off a thin veneer. A steel roller at the rear of the knife assists in keeping the veneer intact and maintains a uniform thickness of approximately $\frac{1}{16}$" to $\frac{5}{16}$".

This continuous veneer is trimmed into smaller sheets that are fed into a hot oven or dryer to reduce the moisture content. The moisture content is reduced to a range of 3 to 8 percent. At this point, the veneers are separated into core and crossband materials. The veneer that will be used for the face now goes through a patching machine to remove and correct defects in the veneer. The patching machine will also match face grains. After all this, the materials go through one final step called splicing prior to becoming a sheet of plywood.

In the final manufacturing process, hot glue is applied by machine to the core, crossbands, faces, and plies. The rough plywood sheet will now go into a hot press. This hot press will apply a great amount of pressure to the pieces to squeeze out the excess glue and will compress the rough plywood to its approximate final thickness. This process takes from two to twenty minutes. Once the process is complete, the sheets are cut and sanded to their final thickness.

5.1.0 Plywood Sheet Sizes

The average or standard size of plywood is 4' × 8'. This is standard in the construction field. A few companies produce plywood from 6' to 8' widths and up to 16' in length. Sheathing-grade plywood is nominally sized by the manufacturer to allow for expansion; that is, 4'0" × 8'0" is really 47¾" × 95¾".

The thickness of plywood will vary from $\frac{3}{16}$" to 1¼". The common sizes are ¼", ½", and ⅝" for finish paneling and ⅜", ½", ⅝", and ¾" for some structural purposes. There are three types of edges on plywood:

- **Butt joints** (two standard pieces joined)
- A shiplap cut or edge
- Interlocking tongue-and-groove

Opposite edges or all four edges may be cut to match.

5.2.0 Grading for Softwood Construction Plywood

After the plywood has been manufactured, the finished product is graded. The grading procedure is similar to the grading of lumber. The trade stamp that appears on each sheet lists the grade of the plywood. *Table 4* shows examples of plywood application data developed by the American Plywood Association (APA). Complete information is available from the APA website (www.apawood.org).

Plywood is rated for interior or exterior use. Exterior-rated plywood is used for sheathing, siding, and other applications where there may be exposure to moisture or wet weather conditions. Exterior plywood panels are made of high-grade veneers bonded together with a waterproof glue that is as strong as the wood itself.

Interior plywood uses lower grades of veneer for the back and inner plies. Although the plies can be bonded with a water-resistant glue, waterproof glue is normally used. The lower-grade veneers reduce the bonding strength, however, which means that interior-rated panels are not suitable for exterior use.

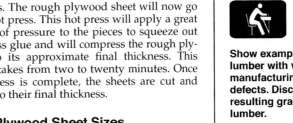

Show examples of lumber with various manufacturing defects. Discuss the resulting grade of the lumber.

Emphasize the importance of being able to interpret common abbreviations. Refer the trainees to *Appendix C* in the Trainee Module.

Show examples of plywood and explain how it is manufactured.

Discuss plywood grading stamps.

Identify the adhesives used in various plywood samples, as indicated by the grading stamp.

Discuss the various types of cores used in plywood.

Show Transparency 6 (Figure 5).

Discuss the types of veneers used in plywood faces.

Table 4 Guide to APA Performance-Rated Plywood Panels

APA Plywood Grade	Description and Use
Rated Sheathing	Unsanded sheathing grade for wall, roof, subflooring, and industrial applications
Structural Rated Sheathing	Unsanded plywood grades to use where shear and cross-panel strength properties are of maximum importance
Rated Sturd-I-Floor	Combination subfloor-underlayment provides smooth surface for application of carpet and pad; available with tongue-and-groove edge
Underlayment	Used as underlayment under carpet and pad; available with tongue-and-groove edge
C-D Plugged	Used for built-ins as well as wall and ceiling tile backing
Appearance Grades	Sanded grade for use where a high-quality surface is required

5.2.1 Plywood Glues

Although most plywood today is rated for interior use, most of the glue used on plywood is exterior glue. The grade stamp will indicate whether the plywood is interior or exterior and if the glue is waterproof or not.

There are plywood sheets designated as structural grade or panels. These panels are manufactured for heavy-duty purposes. This type of plywood is seldom used in residential construction. It may be either interior or exterior plywood.

5.2.2 Plywood Cores

Plywood for industrial use is often called soft plywood. This product has a veneer core. This core is manufactured by the same procedure used to make other plies. It may be of single ply or two plies having the grain running in the same direction and glued together to form a single core or layer (for example, ½" plywood may consist of veneer, but only three plies). Therefore, the number of plies is typically an odd number. The face veneers on sheets of plywood always run in the same direction.

The plywood used in the manufacture of cabinets, doors, furniture, and finished components is known as hardwood plywood. This type of plywood may have any of four different types of core:

- A lumber core
- A particleboard core
- A veneer core (common in construction plywood)
- A fiberboard core

The type of core will not be indicated on the grade stamp, but one can easily tell by looking at the edge of the plywood (*Figure 5*).

5.2.3 Face Veneers

On the top of most plywood grade stamps are two letters separated by a hyphen. Only the letters A, B, C, D, and N are used. These letters indicate the quality of the veneer. The quality or grade of the face of the front panel is indicated by the first letter. The quality or grade of the face of the back panel is indicated by the second letter. The best quality that is free from defects carries an N grade. This grade should be selected to receive a finish that is natural and exposed to view.

LUMBER CORE

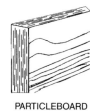
PARTICLEBOARD CORE

VENEER BOARD

FIBERBOARD CORE

102F05.EPS

Figure 5 ♦ Types of plywood.

Instructor's Notes:

Selecting Plywood

Plywood that is expressly manufactured for either interior or exterior use may be used for other purposes in certain situations. Some local codes may require the use of pressure-treated plywood for exterior construction, in bathrooms, or in other high-moisture areas of a house. Pressure-treated plywood can withstand moisture better than interior plywood. Always check the local code(s) before beginning any construction project.

Obtain and allow the trainees to examine a copy of the U.S. Department of Commerce publication, *Plywood Specification and Grade Guide*.

Discuss the types of wood used in the manufacture of plywood.

Emphasize the need for proper handling and storage of plywood.

Assign reading of Sections 6.0.0–11.4.0 for the next class session.

N-grade veneers are typically used in high-quality applications where a natural finish is desired. N-grade veneers are not normally found in construction plywood.

A plywood face that is smooth, has no open defects, but could have some heat repairs is an A grade. This type of veneer accepts paint readily. The next grade is B. This veneer offers a surface that is solid with splits no larger than $\frac{1}{32}$". Some defects may be larger than $\frac{1}{32}$" and may be fixed or repaired with smooth plugs. In most cases, tight knots are allowed. Sanding may create minor flaws that are allowed. This product may be treated like the A grade, but it will not be as smooth. All A and B faces are sanded.

In the C grade, the veneer may have splits to a maximum of $\frac{1}{2}$" and up to $1\frac{1}{2}$" knot holes; however, they must not affect the required strength of the panel. On C grade panels, you may find some sanding defects; these are permitted. The next grade is C Plugged. This veneer is an improved C. The veneer has tighter limits on splits and knot holes. The C Plugged grade has a fully sanded surface. C faces are sanded only if noted.

The poorest grade is the D grade. It has less strength and a poor appearance because of the defects. The D grade is typically used for the cores and backs of interior plywood.

Plywood that is designated for sheathing, subflooring, concrete forms, and sheets used for special structural purposes is called engineered plywood. The veneers are either unsanded or lightly sanded. The plywood sheets carry slightly different grade markings. The face veneers are either C-C or C-D.

5.2.4 Plywood Specification and Grade Guide

The U.S. Department of Commerce *Plywood Specification and Grade Guide* is the standard for plywood grading. Always be sure to check the specifications for grades. Different associations can certify plywood. Some mills certify their own plywood, and this certification may not be accepted by the grading associations or agencies. Be sure to find out who is certifying plywood as a certain grade.

Plywood made from hardwoods such as birch and oak is used in making furniture and cabinets. Hardwood plywood uses a different grading system than the one described previously.

5.3.0 Plywood Storage

Remember, when stacking plywood only two methods may be used. Sheets may be stacked flat on a solid surface and off the floor or ground, or the sheets may be stacked in a full vertical position between two posts. Keep the plywood off the floor or ground. Storing plywood at an angle will cause the material to warp or twist.

Plywood Safety

Plywood is awkward to carry. Remember, carry only one sheet of plywood at a time, and do not hold it over your head. In strong winds, the plywood could act like a wooden sail and injure you and others on the job site.

Discuss the "What's wrong with this picture?" Point out that this plywood will warp because it is being stored improperly.

Explain how building boards are manufactured.

Show examples of hardboard. Describe the three grades of hardboard.

Show examples of particleboard. Describe the two types of particleboard.

What's wrong with this picture?

102SA04.EPS

6.0.0 ◆ BUILDING BOARDS

The ingenuity and technology that helped develop the plywood industry also assisted in the development of other materials in sheet form. The manufacturing processes for these materials is similar to the process for plywood. The main ingredients for these products, known as building boards, are vegetable or mineral fibers. After mixing these ingredients with binder, the mixture becomes very soft.

At this point, the mixture passes through a press, which uses heat and pressure to determine the thickness and density of the finished board.

Sawdust, wood chips, and wood scraps are the major waste materials at a sawmill. These scrap materials are softened with heat and moisture, mixed with a binder and other ingredients, and then run through presses that produce the density and thickness desired by the manufacturer.

The finished wood products that come off the presses are classified as hardboard, particleboard, oriented strand board (OSB), overlay plywood, or fiberboard.

6.1.0 Hardboard

Hardboard is a manufactured building material, sometimes called tempered board or pegboard. Hardboards are extremely dense. The common thicknesses for hardboards are $\frac{3}{16}$", $\frac{1}{4}$", and $\frac{5}{16}$". The standard sheet size for hardboards is 4' × 8'. However, they can be made in widths up to 6' and lengths up to 16' for specialized uses.

These boards are resistant to water, but are susceptible to the edges breaking if they are not properly supported. Holes must be pre-drilled for nailing; direct nailing into the material would cause it to fracture.

Three grades of hardboard are manufactured. The first is known as standard. This hardboard is light brown in color, smooth, and can easily accept paint. Standard hardboard is suitable only for interior use such as cabinets.

The second grade of hardboard is called tempered hardboard (Masonite). This hardboard is the same as standard grade except that in the manufacturing process, the standard board is coated with oils and **resins** and heated or baked to a dark brown color. In going through this process, the tempered hardboard becomes denser, stronger, and more brittle.

Tempered hardboard is suitable for either interior or exterior uses such as siding, wall paneling, and other decorating purposes. One type of tempered hardboard is called perforated hardboard or pegboard. It has holes punched in it 1" apart. Special hooks are made to fit into the holes to support different types of items (shelves, brackets, and hangers for workshops and kitchens, for example).

The third type of hardboard is called service grade. This type of hardboard is not as dense, strong, or heavy as standard grade. The surfaces are not as smooth as standard grade, and it costs less than other hardboards. It can be used for basically everything for which standard or tempered hardboard is used. Service-grade hardboard is manufactured for items such as cabinets, parts of furniture, and perforated hardboard.

6.2.0 Particleboard

The main composition of this type of material is small particles or flakes of wood. Some manufacturing plants may use southern pine materials. Others may use stock selected for its species, fiber characteristics, moisture content, and color. There are two types of particleboard: Type I and Type II.

- Type I is basically a mat-formed particleboard generally made with vera-formaldehyde resin, suitable for interior applications. Refer to the **material safety data sheet (MSDS)** for handling. This type comes in three classes: A, B, and C. The classes are density grades. Class A is a high-density board of 50 lbs per cu ft or over. Class B is a medium-density board of 37 to 50 lbs per cu ft. Class C is a low-density board of 37 lbs per cu ft and lower.
- Type II is a mat-formed particleboard made with durable, moisture- and heat-resistant binders. This makes the particleboard suitable for interior and certain exterior applications when so labeled. This type comes in two classes: Class A is a high-density board of 50 lbs per cu ft or over; Class B is a medium-density board of 37 to 50 lbs per cu ft.

2.16 CARPENTRY FUNDAMENTALS LEVEL ONE ◆ TRAINEE GUIDE

Instructor's Notes:

In addition to the two types and the classes in each type, particleboard is further broken down into still another class of either Class 1 or Class 2. These are strength classifications based on properties of the panels.

Each manufacturer of particleboard has specification sheets that contain the technical information previously discussed. The information sheets are usually available from your local lumber dealer.

Particleboard is pressed under heat into panels. The sheets range in size from ¼" to 1½" in thickness and from 3' to 8' in width. There are also thicknesses of 3" and lengths ranging up to 24' for special purposes. Particleboard has no grain, is smoother than plywood, is more resilient, and is less likely to warp. Particleboard has many uses such as shelving and cabinets. There are cases when ¼" particleboard is used as a plywood core.

Some types of particleboard can be used for underlayment if the local building codes permit. If particleboard is used as underlayment, it is suggested that it be laid with the long dimension across the joists and the edges staggered. Particleboard can be nailed, although some types will crumble or crack when nailed close to the edges.

6.3.0 High-Density Overlay (HDO) and Medium-Density Overlay (MDO) Plywood

High-density overlay (HDO) plywood panels have a hard, resin-impregnated fiber overlay heat-bonded to both surfaces. HDO panels are abrasion- and moisture-resistant and can be used for concrete forms, cabinets, countertops, and similar high-wear applications. HDO also resists damage from chemicals and solvents. HDO is available in five common thicknesses: ⅜", ½", ⅝", ¾", and 1".

Medium-density overlay (MDO) panels are coated on one or both surfaces with a smooth, opaque overlay. MDO accepts paint well and is very suitable for use as structural siding, exterior decorative panels, and soffits. MDO panels are available in eight common thicknesses ranging from 11/32" to 23/32".

Both HDO and MDO panels are manufactured with waterproof adhesive and are thus suitable for exterior use. If MDO panels are to be used outdoors, however, the panels should be edge-sealed with one or two coats of a good-quality exterior housepaint primer. This is easier to do when the panels are stacked.

6.4.0 Oriented Strand Board (OSB)

Oriented strand board (OSB) is a manufactured structural panel used for wall and roof sheathing and single-layer floor construction. *Figure 6* shows two kinds of OSB panels. OSB consists of compressed wood strands arranged in five or more cross-banded layers and bonded with phenolic resin. Some of the qualities of OSB are dimensional stability, stiffness, fastener holding capacity, and no core voids. Some OSB that is used for roof sheathing, for example, is manufactured with a reflective coating on the underside. This gives this particular type of board a higher fire rating. Before cutting OSB, be sure to check the applicable MSDS for safety hazards. The MSDS is the most reliable source for safety information.

6.5.0 Mineral Fiberboards

The building boards just covered are classified as vegetable fiberboards. Mineral fiberboards fall into the same category as vegetable fiberboards. The main difference is that they will not support combustion and will not burn. Glass and **gypsum** rock are the most common minerals used in the manufacture of these fiberboards. Fibers of glass or gypsum powder are mixed with a binder and pressed or sandwiched between two layers of asphalt-impregnated paper, producing a rigid insulation board.

There are some types of chemical foams mixed with glass fibers that will also make a good, rigid insulation. However, this mineral insulation will crush and should not be used when it must support a heavy load.

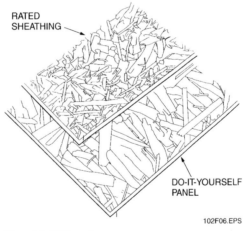

Figure 6 ♦ OSB panels.

Emphasize the safety precautions for working with materials that might contain asbestos.

Identify the different classifications of engineered wood products. Emphasize the benefits of using these products.

Show examples of LVL and PSL and discuss their applications.

> **CAUTION**
>
> Whenever you work with older materials that may be made with asbestos, contact your supervisor for the company's policies on safe handling of the material. State and federal regulations require specific procedures to follow prior to removing, cutting, or disturbing any suspect materials. Also, some materials emit a harmful dust when cut. Check the MSDS before cutting.

7.0.0 ◆ ENGINEERED WOOD PRODUCTS

In the past, the primary source of structural beams, timbers, joists, and other weight-bearing lumber was old-growth trees. These trees, which need many decades to mature, are tall and thick and can produce a large amount of high-quality, tight-grained lumber. Extensive logging of these trees to meet demand resulted in higher prices and conflict with forest conservation interests.

The development of wood laminating techniques by lumber producers has permitted the use of younger-growth trees in the production of structural building materials. These materials are given the general classification of engineered lumber products.

Engineered lumber products fall into five general classifications: laminated veneer lumber (LVL), parallel strand lumber (PSL), laminated strand lumber (LSL), wood I-beams, and glue-laminated lumber (glulam). These materials provide several benefits:

- They can be made from younger, more abundant trees.
- They can increase the yield from a tree by 30 percent to 50 percent.
- They are stronger than the same size of structural lumber.
- Greater strength allows the engineered lumber to span a greater distance.
- A length of engineered wood is lighter than the same length of solid lumber. It is therefore easier to handle.
- They are dimensionally accurate and do not warp, crown, or twist.

Figure 7 shows three of the main types of engineered lumber.

7.1.0 Laminated Veneer Lumber (LVL)

Like plywood, LVL is made from laminated wood veneer. Douglas fir and southern pine are the primary sources. Thin ($\frac{1}{10}$" to $\frac{3}{16}$") sheets are peeled from the tree in widths of 27" or 54". The veneers are laid-up in a staggered pattern with the veneers overlapping to increase strength. Unlike plywood, the grain of each layer runs in the same direction as the other layers. The veneers are bonded with an exterior-grade adhesive, then pressed together and heated under pressure.

LVL is used for floor and roof beams and for the support members called headers that are used over window and door openings. It is also used in scaffolding and concrete forms. No special cutting tools or fasteners are required.

LVL

PSL

LSL

Figure 7 ◆ LVL, PSL, and LSL.

7.2.0 Parallel Strand Lumber (PSL)

PSL is made from long strands of Douglas fir and southern pine. The strands are about $\frac{1}{8}$" or $\frac{1}{10}$"

Instructor's Notes:

thick and are bonded together with adhesive in a special heating process.

PSL is used for beams, posts, and columns. It is manufactured in thicknesses up to 7". Columns can be up to 7" wide; beams range up to 18" in width.

7.3.0 Laminated Strand Lumber (LSL)

LSL can be made from small logs of almost any kind of wood. Aspen, red maple, and poplar that cannot be used for standard lumber are commonly used. In the manufacturing process, the logs are cut into short strands, which are bonded together and pressed into long blocks (billets) up to 5½" thick, 8' wide, and 40' long.

LSL is used for millwork such as doors and windows and any other product that requires high-grade lumber. However, LSL will not support as much of a load as a comparable size of PSL because PSL is made from stronger wood.

7.4.0 Wood I-Beams

Wood I-beams (*Figure 8*) consist of a web with flanges bonded to the top and bottom. This arrangement, which mimics the steel I-beam, provides exceptional strength. The web can be made of OSB or plywood. The flanges are grooved to fit over the web.

Wood I-beams are used as floor joists, **rafters,** and headers. Because of their strength, wood I-beams can be used in greater spans than a comparable length of dimension lumber. Lengths of up to 60' are available.

7.5.0 Glue-Laminated Lumber (Glulam)

Glulam is made from lengths of solid, kiln-dried lumber that have been glued together. It is popular in architectural applications where exposed beams are used (*Figure 9*). Because of its exceptional strength and flexibility, glulam can be used in areas subject to high winds or earthquakes. Glulam is available in three appearance grades: industrial, architectural, and premium. Industrial grade is used in open buildings such as warehouses and garages where appearance is not a priority or where beams are not exposed. Architectural grade is used where beams are exposed and appearance is important. Premium, the highest grade, is used where the highest quality appearance is needed (*Figure 10*).

Glulam beams are available in widths from 2½" to 8¾". Depths range from 5½0" to 28½". They are available in very long lengths. They are used for many purposes, including ridge beams; basement beams; headers of all types; stair treads, supports, and stringers; and **cantilever** and **vaulted ceiling** applications. Because glulam beams are laminated, they can be formed into arches and any number of curved configurations. Glulam beams are especially popular for use in churches.

Provide examples of LSL, wood I-beams, and glulam. Describe their qualities and applications.

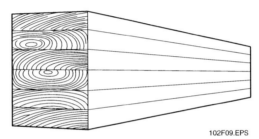

Figure 9 ♦ Glulam beam construction.

Figure 8 ♦ Wood I-beams.

Figure 10 ♦ Glulam beam application.

Show examples of pressure-treated lumber and discuss its applications.

Emphasize the safety precautions necessary when working with pressure-treated lumber.

Have the trainees calculate the quantities of lumber and wood products required for an instructor-supplied project. This laboratory corresponds to Performance Task 1.

Show Transparency 8 (Figure 11).

Use of Engineered Wood Products

Engineered wood products are used in a wide array of applications that were once exclusively served by cut lumber. For example, PSL is used for columns, ridge beams, girders, and headers. LVL is also used for form headers and beams. Wood I-beams are used to frame roofs as well as floors. An especially noteworthy application is the use of LSL studs, top plates, and soleplates in place of lumber to frame walls.

8.0.0 ♦ PRESSURE-TREATED LUMBER

Pressure-treated lumber is softwood lumber protected by chemical preservatives forced deep into the wood through a vacuum-pressure process. Pressure-treated lumber has been used for many years in on-ground and below-ground applications such as landscape timbers, **sill plates**, and foundations. In some parts of the country, it is also extensively used in the building of decks, porches, docks, and other outdoor structures subject to decay from exposure to the elements. It is popular for these uses in areas where structures are exposed to insect or fungus attack. A major advantage of pressure-treated lumber is its relatively low price in comparison with redwood and cedar. When natural woods such as these are used, only the more expensive heartwood will resist decay and insects.

 WARNING!
Because the chemicals used in pressure-treated lumber represent some hazard to people and the environment, special precautions apply:

- When cutting pressure-treated lumber, always wear eye protection and a dust mask.
- Wash any skin that is exposed while cutting or handling the lumber.
- Wash clothing that is exposed to sawdust separately from other clothing.
- Do not burn pressure-treated lumber, because the ash poses a health hazard. Check local regulations for proper disposal.
- Be sure to read and follow the manufacturer's safety instructions.

Pressure-treated lumber is available in three grades. The three grades are designated by their pounds per cubic foot of preservative retention. Above ground (0.25 lb/cu ft) grade is to be used only 18" or more above ground. Ground contact grade (0.40 lb/cu ft) is used when there is contact with water or soil on or below ground. The third grade (0.60 lb/cu ft) is used when structural reliability is required, such as in wooden foundations and power poles.

Fire-Retardant Building Materials

Lumber and sheet materials are sometimes treated with fire-retardant chemicals. The lumber can either be coated with the chemical in a non-pressure process or impregnated with the chemical in a pressure-treating process. Fire-retardant chemicals react to extreme heat, releasing vapors that form a protective coating around the outside of the wood. This coating, known as char, delays ignition and inhibits the release of smoke and toxic fumes.

9.0.0 ♦ CALCULATING LUMBER QUANTITIES

Large quantities of lumber are normally ordered by the board foot. A board foot is equivalent to a piece of lumber that is 1" thick, 12" wide, and 1' long. Each of the boards in *Figure 11* represents one board foot. When calculating board feet, always make sure you use the nominal lumber dimensions. The following formula is used to calculate board feet:

Board feet = number of pieces × thickness in inches × width in inches × length in feet ÷ 12

For example, 20 pieces of 2 × 6 lumber that are each 8' long equals 160 board feet.

$$\frac{20 \times 2 \times 6 \times 8}{12} = \frac{1{,}920}{12} = 160 \text{ board feet}$$

To calculate the amount of sheet materials such as plywood, particleboard, drywall, and sheathing

---

2.20 CARPENTRY FUNDAMENTALS LEVEL ONE ♦ TRAINEE GUIDE

Instructor's Notes:

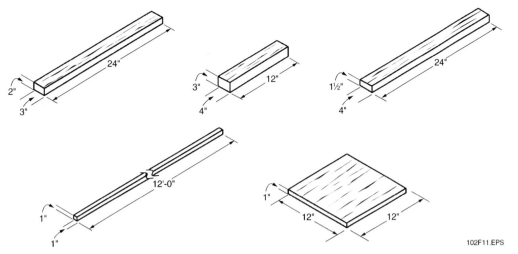

Figure 11 ◆ Examples of a board foot.

Measure Twice, Cut Once

An apprentice carpenter was helping his boss remodel a basement. The boss asked the apprentice to measure and cut the studs for the walls. The apprentice, looking to impress the boss, attacked the job with gusto. Instead of trying out the first stud, he cut all of them at once, based on a quick measurement. When all the studs were cut, they started to place them, only to find out that each stud was 3¼" short. The apprentice had forgotten to add the measuring tape case to the tape reading.

you need, calculate the area to be covered (length × width = area). The area will always be the square of the values used (square feet, square inches, etc.). For example, to figure the amount of plywood decking needed for a 20' by 41' roof:

20 × 41 = 820 square feet

A standard 4' by 8' sheet contains 32 square feet. Therefore, if you divide 820 by 32, it yields 25⅝ sheets, which is then rounded to 26 sheets.

Trim and moldings are priced by the lineal foot and are ordered by the dimension of the piece (for example, 150' of 1" quarter round molding). *Appendix D* contains lumber conversion tables.

10.0.0 ◆ CONCRETE BLOCK CONSTRUCTION

Concrete masonry units (CMUs), commonly referred to as concrete block, are commonly used in building foundations and basement walls. Concrete block was once called cinder block because it contained the cinders left after burning coal. Today, concrete block is constructed using other materials such as sand and gravel.

Blocks are made of water added to Portland cement, filler materials such as sand and gravel, and special additives known as admixtures. Admixtures affect the color and other properties of the cement, such as freeze resistance, weight, and speed of setting.

The block is machine-molded into shape. It is compacted in the molds and cured, typically using live steam. After curing, the blocks are dried and aged. The moisture content is checked. It must be a specified minimum amount before the blocks can be shipped for use.

Concrete block is typically 8" × 8" × 16", with a hollow core. Blocks come in modular sizes and colors determined by their ingredients. A variety of surface and mixing treatments can give block varied and attractive surfaces. Newer finishing

Explain how concrete masonry units are manufactured. Provide examples for the trainees to examine.

Discuss various types of concrete blocks and their applications.

Show Transparency 9 (Figure 12).

Block Construction

Concrete block is often used in commercial construction. In some parts of the country, it is also used in the walls of residential construction in place of wood framing. The block can be painted on the outside or faced with brick, stucco, or other finish material.

101SA05.EPS

techniques can give block the appearance of brick, rough stone, or cut stone. Like clay masonry units, block can be placed in structural pattern bonds. *Figure 12* shows the names of the parts of block.

Block takes up more space than other building units, so fewer are needed. Block bed joints usually need mortar only on the shells and webs, so there is less mortaring as well.

Concrete block comes in three weights: normal, lightweight, and aerated. Normal-weight block can be made of concrete with regular, high, and extra-high strengths. The last two are made with different aggregates and curing times. They are used to limit wall thickness in buildings over ten floors high. Lightweight block is made with lightweight aggregates. The loadbearing and appearance qualities of the first two weights are similar; the major difference is that lightweight block is easier and faster to lay. Aerated block is made with an admixture that generates gas bubbles inside the concrete for a lighter block.

Common hollow units have two or three cores. The hollow cores make it easy to reinforce concrete block walls. Grout alone, or steel reinforcing rods combined with grout, can be used to fill the hollow cores. Reinforcement increases loadbearing strength, rigidity, and wind resistance.

Loadbearing block is used as backing for veneer walls, bearing walls, and all structural uses. Both regular and specially shaped blocks are used for paving, retaining walls, and slope protection. Nonstructural block is used for screening, partition walls, and as a veneer wall for wood, steel, or other backing. Both kinds of blocks come in a variety of shapes and modular sizes.

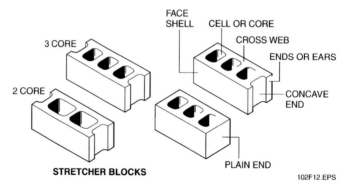

Figure 12 ◆ Concrete blocks.

11.0.0 ◆ COMMERCIAL CONSTRUCTION METHODS

The structural framework of large buildings such as office buildings, hospitals, apartment houses, and hotels is usually made from concrete or structural steel. The exterior finish is often concrete panels that are either prefabricated and raised into place or poured into forms built at the site. Floors are usually made of concrete that is poured at the site using wood, metal, or fiberglass forms. Exterior walls (curtain walls) may also be made of glass in a metal or concrete framework. *Figure 13* shows such a curtain wall section being installed. Before the concrete is poured, provision must be made for electrical, plumbing, and data cabling pathways.

In some buildings, the framework is made of structural steel and the floors of poured concrete. Panels fabricated off site are lifted into place and bolted or welded to the steel (*Figure 14*).

Figure 15 shows the structure of a building in which all the structural framework is made of concrete and steel. Each component of the structure requires a different type of form. In this case,

Figure 14 ◆ Curtain wall under construction.

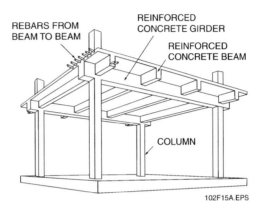

Figure 13 ◆ Installing a corner curtain wall panel.

Figure 15 ◆ Concrete structure.

Describe the types of materials used in commercial construction.

Describe how the floors are constructed in commercial buildings.

Show Transparency 10 (Figure 17).

the floor and beams were made in a single pour using integrated floor and beam forms, which were removed once the concrete hardened.

In some specialized commercial applications, tilt-up concrete construction is used. In tilt-up construction, the wall panels are usually poured on the concrete floor slab, then tilted into place on the footing using a crane (*Figure 16*). The panels are welded together.

The main difference between tilt-up and other types of large commercial construction is that there is no steel or concrete framework in tilt-up construction. The walls and floor slab bear the entire load. Tilt-up is most common in one- or two-story buildings with a slab at grade (no below-grade foundation). It is popular for warehouses, low-rise offices, churches, and a variety of other commercial and multi-family residential applications. Tilt-up panels of 50' in height are not uncommon. They typically range from 5" to 8" thick, but thicker walls can be obtained when using lighter-weight concrete.

11.1.0 Floors

Once the framework is in place, the concrete floors are poured using deck forms. Shoring is placed under the form to support it until the concrete hardens. *Figure 17* shows cellular floors poured over corrugated steel forms, which remain in place, providing channels though which cabling can be run.

A section of metal, plastic, or fiber sleeve is often inserted vertically into the form before the concrete is poured to allow for electrical, communications, and other cabling to pass through the floor.

In some installations, underfloor duct systems are embedded in the concrete floor and are used to provide horizontal distribution of cables. Vertical access ports (handholes) are embedded in the

Figure 16 ◆ Tilt-up panel being lifted into place.

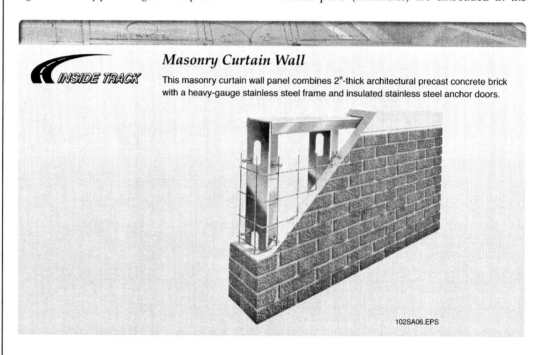

Masonry Curtain Wall

This masonry curtain wall panel combines 2"-thick architectural precast concrete brick with a heavy-gauge stainless steel frame and insulated stainless steel anchor doors.

2.24 CARPENTRY FUNDAMENTALS LEVEL ONE ◆ TRAINEE GUIDE

Tilt-Up Records

The Tilt-Up Concrete Association (www.tilt-up.org) keeps records of tilt-up projects. As of 2005, several amazing statistics had been recorded, including:

- Heaviest tilt-up panel – 310,100 pounds
- Largest tilt-up panel by area – 2,454 square feet
- Tallest tilt-up panel – 101 feet
- Largest number of panels in a single building – 1,310
- Largest tilt-up building (floor area) – 1,883,684 square feet

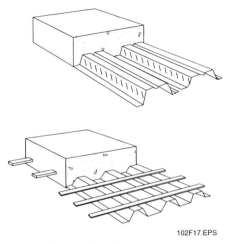

Figure 17 ♦ Corrugated steel forms.

form so that cable can be fished to various locations in the space.

Trench ducts are metal troughs that are embedded in the concrete floor and used as feeder ducts for electrical power and telecommunication lines. *Figure 18* shows a trench duct in a cellular floor.

Access floors consist of modular floor panels supported by pedestals. They may or may not have horizontal bracing in addition to the pedestals. This type of structure is used in computer rooms, intensive care facilities, and other areas where a lot of cabling is required. In some applications, such as a factory, a trench may be formed in the concrete floor to accommodate cabling and other services.

11.2.0 Exterior Walls

When walls are formed of concrete, openings for doors and windows are made by inserting wooden or metal bucks in the form, as shown in *Figure 19*. Openings for services such as piping and cabling are accommodated with fiber, plastic, or metal tubes inserted into the form.

11.3.0 Interior Walls and Partitions

The construction of walls and partitions in commercial applications is driven by the fire and soundproofing requirements specified in local building codes. In some cases, a frame wall with

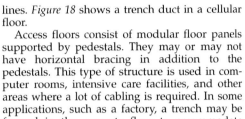

Show Transparency 11 (Figure 18).

Explain how openings are framed in exterior walls in commercial buildings.

Show Transparency 12 (Figure 19).

Discuss the special requirements for partition walls in commercial buildings.

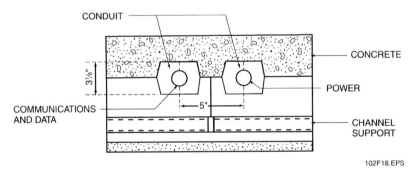

Figure 18 ♦ Trench duct in a cellular floor.

MODULE 27102-06 ♦ BUILDING MATERIALS, FASTENERS, AND ADHESIVES 2.25

Show Transparencies 13 and 14 (Figures 20 and 21).

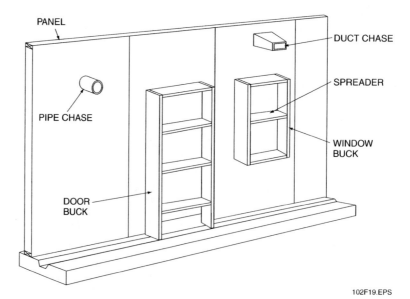

Figure 19 ◆ Framing openings in concrete walls.

½" gypsum drywall on either side is satisfactory. In extreme cases, such as the separation between offices and manufacturing space in a factory, it may be necessary to have a concrete block wall combined with fire-resistant gypsum wallboard, along with rigid and/or fiberglass insulation, as shown in *Figure 20*. This is especially true if there is any explosion or fire hazard.

While they are sometimes used in residential construction, steel studs are the standard for framing walls and partitions for commercial construction. Once the studs are installed, one or more layers of gypsum wallboard and insulation are applied. The type and thickness of the wallboard and insulation depend on the fire rating and soundproofing requirements. Soundproofing needs vary from one use to another and are often based on the amount of privacy required for the intended use. For example, executive and physician offices may require more privacy than general offices.

The requirements for sound reduction and fire resistance can significantly affect the thickness of a wall. For example, a wall with a high sound transmission class (STC) and fire resistance might have a total thickness of nearly 6", while a low-rated wall might have a thickness of only 3" using steel studs and 2¼" using wooden studs. (See *Figure 21*.)

As discussed previously, the fire rating specified by the applicable building code determines the types and amount of material used in a wall or partition. As shown in *Figure 21*, a one-hour rated wall might be made of single sheets of ⅝" gypsum wallboard on wooden or 25-gauge metal studs. A two-hour rated wall requires heavy-gauge metal studs and two layers of fire-resistant gypsum wallboard.

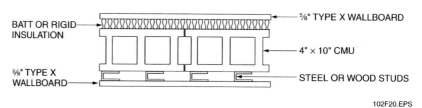

Figure 20 ◆ High fire/noise resistance partition.

2.26 CARPENTRY FUNDAMENTALS LEVEL ONE ◆ TRAINEE GUIDE

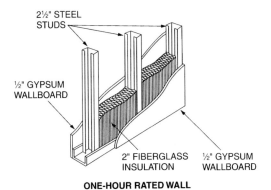

ONE-HOUR RATED WALL

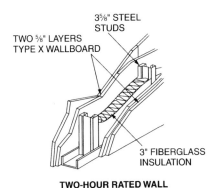

TWO-HOUR RATED WALL

Figure 21 ♦ Partition wall examples.

11.4.0 Metal Framing Materials

Metal framing components include metal studs and, in some cases, metal joists and metal roof trusses (*Figure 22*). The vertical and horizontal framing members serve as structural load-carrying components for a variety of low- and high-rise structures. Metal stud framing is compatible with all types of surfacing materials.

The advantages of metal framing include noncombustibility, uniformity of dimension, lightness of weight, freedom from rot and moisture, and ease of construction. The components of metal frame systems are manufactured to fit together easily. There are a variety of metal framing systems, both loadbearing and nonbearing types. Some nonbearing partitions are designed to be demountable or moveable and still meet the requirements of sound insulation and fire resistance when covered with the proper gypsum system.

Figure 22 ♦ Metal trusses.

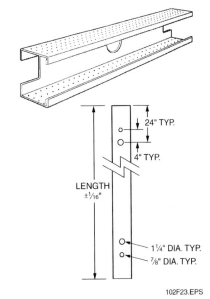

Figure 23 ♦ Standard metal stud stock.

When metal studs are used for drywall framing systems, the channel stock features knurled sides for positive screw settings and comes in two grades. The first grade is the standard drywall stud (*Figure 23*). Standard studs come in widths of 1⅜" to 6". The flanges are 1⅜" × 1¼". Lengths of 6' to 16' are commercially available. The standard drywall stud is 25-gauge steel (the higher the gauge, the lighter the metal). Depending on the product number, lengths range from 8' to 12'; other lengths are available by special order.

Discuss various types of metal framing materials and their applications. Provide examples for the trainees to examine.

Show Transparency 15 (*Figure 23*).

MODULE 27102-06 ♦ BUILDING MATERIALS, FASTENERS, AND ADHESIVES 2.27

Show Transparency 16 (Figure 24).

Describe the construction of a metal stud wall.

Party Walls

When gypsum drywall is used in a party wall, the architect's plans must be followed precisely. If the inspector finds flaws in the construction, a certificate of occupancy will not be issued. This photo illustrates why building codes place so much emphasis on properly constructed party walls.

The second grade is the extra heavy drywall stud (*Figure 24*). These studs have knurled sides for positive screw settings. They also have cutouts and utility knockout holes 12" from each end and at the mid-point of the stud.

The width of the extra heavy studs will also vary from 1⅜" to 6". The flanges are 1⅜" × 1¼". The extra heavy drywall stud is 20 gauge. This type of stud can be ordered in any length that is needed.

Metal studs are also available in greater strengths of 18 gauge, 16 gauge, 12 gauge, and 10 gauge. These strengths are classified as structural steel studs. They are available in widths of 2½" to 8". The flanges are 1⅝" × 2" or 2½". They can be ordered in whatever length is needed.

Although different materials are used, the general approach to framing with metal studs is the same as that used for wooden studs. In fact, metal studs can be used with either wood plates or metal runners.

Like wood framing, metal framing is installed 12", 16", or 24" OC, openings are framed with headers and cripples, and special framing is needed for corners and partition Ts.

Depending on the load, reinforcement may be needed when framing openings. Bracing of walls to keep them square and plumb is also required. The illustrations in this section show examples of common framing techniques. *Table 5* shows the framing spacing for various gypsum drywall applications.

The erection of metal studs typically starts by laying metal tracks in position on the floor and ceiling and securing them (*Figure 25*). If the tracks are being applied to concrete, a low-velocity powder-actuated fastener is generally used. If the tracks are being applied to wood joists, such as in a residence, screws can be driven with a screw gun.

The use of a powder-actuated fastener requires special training and certification. The use of these tools may be prohibited by local codes because of safety concerns.

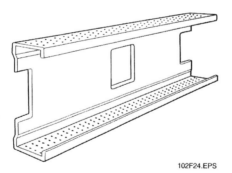

Figure 24 ♦ Heavy-duty stud stock.

2.28 CARPENTRY FUNDAMENTALS LEVEL ONE ♦ TRAINEE GUIDE

Instructor's Notes:

Table 5 Maximum Framing Spacing

	Single-Ply Gypsum Board Thicknesses	Application to Framing	Maximum OC Spacing of Framing
Ceilings	⅜"	Perpendicular	16"
	½"	Perpendicular or Parallel	16"
	½"	Perpendicular	24"
	⅝"	Perpendicular	24"
Sidewalls	⅜"	Perpendicular or Parallel	16"
	½" or ⅝"	Perpendicular or Parallel	24"

Fasteners Only – No Adhesive Between Plies

	Multi-Ply Gypsum Board Thicknesses		Application to Framing		Maximum OC Spacing of Framing
	Base	Face	Base	Face	
Ceilings	⅜"	⅜"	Perpendicular	Perpendicular	16"
	½"	⅜"	Parallel	Perpendicular	16"
		½"	Parallel	Perpendicular	16"
	½"	½"	Perpendicular	Perpendicular	24"
	⅝"	½"	Perpendicular	Perpendicular	24"
	⅝"		Perpendicular	Perpendicular	24"

Sidewalls*

* For two-layer applications with no adhesive between plies, ⅜", ½", or ⅝" gypsum board may be applied perpendicularly (horizontally) or parallel (vertically) on framing spaced a maximum of 24" OC. Maximum spacing should be 16" OC when ⅜" board is used as the face layer.

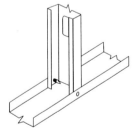

Figure 25 ♦ Metal framing.

Assign reading of Sections 12.0.0–18.4.0 for the next class session.

Explain that only trained and certified personnel may use powder-actuated tools.

Discuss the applications of various types of nails. Pass around examples for the trainees to examine.

Show Transparencies 17 and 18 (Figures 26 and 27).

Identifying Structural Studs

Structural studs are marked with a color code for easy identification. The coding is as follows:

Gauge	Color
20	White
18	Yellow
16	Green
14	Blue
12	Red

Note that there are both light and structural gauge studs made of 20-gauge steel. The difference is in the dimensions.

 WARNING!
The use of a powder-actuated fastener requires special training and certification. The use of these tools may be prohibited by local codes because of safety and seismic concerns.

Once the tracks are in place, the studs and openings are laid out in the same way as a wood frame wall. The studs may be secured to the tracks with screws or they may be welded. In some cases, the entire wall will be laid out on the floor, then raised and secured. When heavy-gauge walls are used, they may be assembled and welded in a shop and brought to the site.

12.0.0 ◆ NAILS

Nailing is the most common method used for attaching two pieces of lumber. Some nails are made specifically to be driven with a hammer; others are made for use with pneumatic or cordless nailers. You will often hear nails referred to as 8-penny, 10-penny, etc. This method of designating the size of a nail dates back many years. In written form, the penny designation appears as a lowercase d (*Figure 26*); an 8-penny nail would read 8d, and so forth. Nails above 16d are referred to as spikes. Spikes range from 20d (4") to 60d (6"). The thickness (gauge) of the nail shank increases as the length increases.

12.1.0 Kinds of Nails

Several kinds of nails are discussed below (*Figure 27*).

- *Common nails* – These are the most frequently used of all nails. They have a flat head, smooth, round shank (shaft), and diamond point. They are available from 2d to 60d in length. They are used when appearance is not of prime importance, such as in rough framing or building concrete forms.
- *Box nails* – These are very similar to the common nail in appearance, except they have a thinner head and shank and less tendency to split the wood. They often have a resin coating to resist rust and create more holding power. They are available from 2d to 40d.
- *Finish nails* – Finish nails have a small barrel-shaped head with a slight indentation in the top to receive a **nail set** tool so the nail can be driven slightly below the surface of the wood. The hole above the head is then filled with a special putty before paint or finish is applied to conceal the nail. Finish nails are used for installing millwork and trim where the final appearance is important. Their small heads reduce their holding power considerably. If holding power is important, finish nails can be obtained with **galvanized** or **japanned** coatings. Finish nails are available in 2d to 20d lengths.

2.30 CARPENTRY FUNDAMENTALS LEVEL ONE ◆ TRAINEE GUIDE

Instructor's Notes:

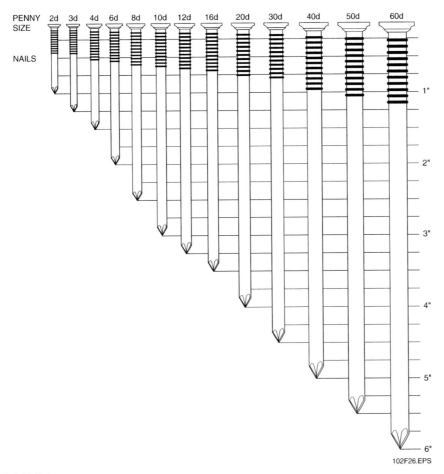

Figure 26 ◆ Nail sizes.

- *Casing nails* – Casing nails look a lot like finish nails; however, they have a conical-shaped head and the shank is larger. Also, they come in lengths from 2d to 40d as opposed to the 20d length of a regular finish nail. The casing nail has more holding power because of the larger shank and the shape of the head.
- *Duplex or scaffold (doublehead) nail* – Duplex nails are designed for building scaffolds, concrete forms, and other temporary work that will later be disassembled. The advantage of this nail is that it can be driven to the first head, which gives it sufficient holding power. They can be easily removed at a later time with a claw hammer or wrecking bar without damaging the wood.

- *T-nails* – This is a specially designed nail used in pneumatic nailing guns. It is coated with resin and comes in strips for insertion into the gun. T-nails come in lengths of 1¼", 1½", 1¾", 2", 2⅜", and 2½". These nails are also available with galvanized coatings. T-nailing reduces labor time as it only takes a pull of the trigger to drive them home.
- *Drywall (ratchet) nails* – Drywall nails have an annular or ring shank. Their principal use is to fasten gypsum or drywall to wood studs. The ring shanks increase holding power and prevent the nail from backing out of the stud. Lengths are from 1" to 2½".

Warn the trainees to wear a face shield when driving cut nails.

A Penny's Worth

Back in the fifteenth century, an Englishman could buy 100 three-inch nails for 10 pennies, so the three-inch nail became known as a 10-penny nail. Other nails were known as 8-penny nails, 12-penny nails, and so on, depending on the cost per hundred. Although a penny isn't worth much these days, the designation is still used to identify the nails. The "d" that is used in the written form (8d, 10d, etc.) is said to represent the denarius, a small Roman coin used in Britain that was equal to a penny.

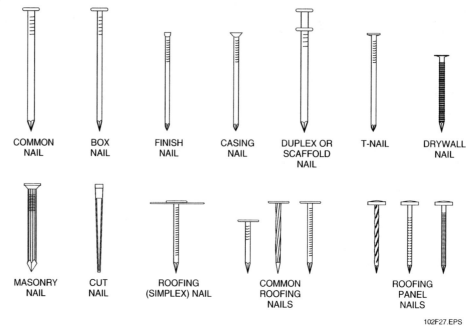

Figure 27 ◆ Kinds of nails.

- *Masonry nails* – Masonry nails are available in lengths of ½" to 4". The shanks on the longer ones have grooves with a sinker head. They are case-hardened and used to fasten metal or wood to masonry or concrete.
- *Cut nails* – Cut nails could also be grouped in with masonry nails, because in today's construction they are very seldom used for other than masonry purposes. They are rectangular in shape and have a blunt point, with little tendency to bend. They are obtainable up to about 5" long.

 WARNING!
Wear a face shield when driving cut nails, because they are subject to breakage, which could do grave damage to an eye.

- *Roofing nails* – Most roofing nails are galvanized to resist rusting. There are several types of roofing nails:
  - Roofing fasteners (often called Simplex fasteners) are commonly used to install roofing felt. They consist of a square metal or plastic plate with a screw driven through the center of the plate.

2.32 CARPENTRY FUNDAMENTALS LEVEL ONE ◆ TRAINEE GUIDE

Selecting the Right Nail

As a rule of thumb, select a nail that is three times longer than the thickness of the material you are fastening. However, the nail length should not exceed the total thickness of the two pieces being fastened together. For example, if you are nailing two 2 × 4 joists where they overlap on a girder, three times the thickness of one joist would be 4½". Since the total thickness of the two joists is only 3", however, it would not make sense to use a 4½" nail.

Common nails are used primarily for rough framing. Box nails, which are thinner and have slightly smaller heads than common nails, are used when working near the edge of the lumber (toenailing studs, for example) because they are less likely to split the wood. Box nails are also used for fastening exterior insulation board and siding.

There are three types of nails used for trim. The finish nail is used primarily for fastening interior trim. The casing nail, which is thicker and has a slightly larger head than the finish nail, is used to fasten heavy trimwork and exterior finish material. The wire brad is thinner than finish and casing nails and is used for very light trimwork such as small moldings. These trim nails all have a small head with a slight indentation, which makes it easy to drive the nail head below the surface of the wood using a nail set (shown here).

102SA08.EPS

Panel Nails

Panel nails are available in a variety of colors to match the paneling. These nails have annular rings to provide more holding capacity. Nails used to fasten paneling should be long enough to penetrate ¾" into the framing.

- Common roofing nails come in lengths of ¾" to 2". All have a thin head but different shanks. They are primarily used to apply asphalt or fiberglass shingles. They are also used to fasten insulation sheathing board to studs. The longer ones are used for re-roofing jobs, while the shorter ones are used on new roofs.
- Roofing panel nails have a neoprene washer. They range in length from 1¾" to 2½", with either a helical (spiral) or plain shank. The helical shank has greater holding power. Roofing panel nails are used on fiberglass, steel, or aluminum roofing panels. The nylon washer serves as a weatherproofing seal.

Listed below are some additional types of nails you will occasionally encounter:

- *Hardboard siding nails* – These nails are designed for installing aluminum, vinyl, and steel siding.
- *Insulated siding nails* – These nails have a zinc coating and come in various colors to match the color of the siding being installed.
- *Panel nails* – These also come in various colors. They come in round- and ring-shank design. The head is small and blends in with the paneling for a good appearance.

Discuss the applications of various types of staples. Pass around examples for the trainees to examine.

Show Transparency 19 (Figure 28).

Discuss the applications of various types of wood screws. Pass around examples for the trainees to examine.

- *Wire brads* – Their appearance is similar to that of regular finish nails. However, they are available with smaller shanks than finish nails, thus reducing splitting tendencies. They are used for fine finish work and come in lengths from ³⁄₁₆" to 3".
- *Escutcheon pins* – These specially designed nails generally come as part of a hardware kit. They have an oval head and vary in length from ³⁄₁₆" to 2". They are used for such things as metal trim on store cabinets or nailing up house numbers. They add a nice appearance to the finish trim and generally match the color of the hardware being installed.

13.0.0 ◆ STAPLES

This module will deal only with the various types of staples used in building construction.

There are hundreds of different types of staples manufactured to do specific jobs. Refer to the different manufacturers' catalogs for this information. The factors that should be considered are:

- The type of point
- The crown (or top) width
- The length
- What type of metal they are made of
- The various coatings that are available

13.1.0 Types of Staples

The type of staple is determined by the type of point it has (*Figure 28*).

- *Chisel* – Recommended for grainy woods; keeps legs parallel to the entire leg length.
- *Crosscut chisel* – Legs penetrate straight and parallel through cross-grain wood. Good for general tacking and nailing purposes.
- *Outside chisel* – Good for clinching inward after penetrating the material being fastened.
- *Inside chisel* – Good for clinching outward after penetrating the material being fastened.
- *Divergent* – After start of penetration, legs diverge to allow use of longer legs in thin material; very good for wallboard insulation.
- *Outside chisel divergent* – Legs diverge, then cross, locking the staple; provides excellent penetration ability.
- *Spear* – Good for penetrating dense materials; will deflect easily when hitting an obstruction.

The most common crown widths vary from ³⁄₈" to 1". The wider crowns are used mostly for millwork, crating, and application of asphalt shingles. The most common wire gauge (thickness) is 14 to 16 and the length is normally ½" to 2".

Staples are glued together in strips for insertion into a stapling tool. Some stapling tools are manually operated. Others are powered by pneumatics or electricity.

14.0.0 ◆ SCREWS

The three types of screws that we will focus on in this module are wood, sheet metal, and machine.

Screws have a neat appearance and can be decorative. They have much more holding power than nails and can easily be removed without damaging the materials involved. However, the cost factor of the screw and the time involved in installation must be considered.

14.1.0 Wood Screws

Wood screws are commonly made in 20 different stock thicknesses and lengths. They range in length from ¼" to 5". The screw gauge (diameter)

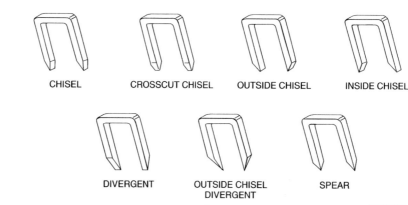

Figure 28 ◆ Staples.

Instructor's Notes:

ranges from 0 to 24. The higher the number, the larger the diameter. (See the wood screw chart in *Figure 29*.)

The two most common types of screw heads are slotted head and Phillips head. Screw heads may have any of several different shapes (*Figure 30*).

When using wood screws, it is generally necessary to drill pilot holes, particularly in hardwoods, to receive the screw. This does four things:

- Ensures pulling the materials tightly together
- Makes the screw easier to drive
- Prevents splitting the wood
- Prevents damage to the screw

For decorative purposes, screws can be obtained in many different finishes (e.g., antique copper or brass, gold- or silver-plated, chromium, lacquered, etc.).

14.2.0 Sheet Metal Screws

Sheet metal screws are thread-cutting or thread-forming screws used to fasten light-gauge sheet metal (*Figure 31*). The threads of sheet metal screws are deeper than those of wood screws. This allows the two pieces of metal being fastened to be drawn tightly together. Because their threads are deeper and hold better, sheet metal screws are also recommended for use with softer building materials like particleboard and hardboard. Sheet metal screws come in about the same range of sizes as wood screws.

Self-drilling sheet metal screws have a cutting edge on their point that eliminates the need for pre-drilling. These screws are driven with a power driver and are commonly used to fasten metal framing.

Show Transparencies 20 and 21 (Figures 29 and 30).

Emphasize the importance of drilling pilot holes.

Discuss the applications of various types of sheet metal screws. Pass around examples for the trainees to examine.

Show Transparency 22 (Figure 31).

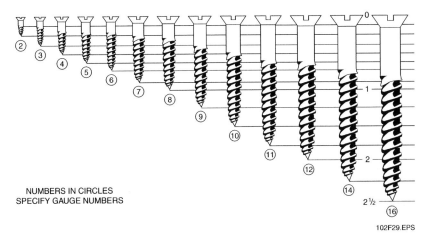

Figure 29 ◆ Wood screw chart.

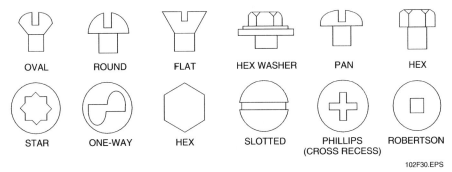

Figure 30 ◆ Common screw heads.

Discuss machine screws and their applications. Pass around examples for the trainees to examine.

Show Transparency 23 (Figure 32).

Point out that a lag screw is basically a heavy-duty wood screw.

Drilling a Pilot Hole

When drilling a pilot hole, use a drill bit that is the same size as the shank of the screw. Hold the drill bit lengthwise over the screw. You should be able to see the wings, or threads, of the screw. If you can't, the drill bit is too large. Coat the screw with soap, paraffin, or beeswax to make it easier to drive.

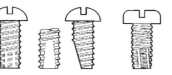

THREAD-CUTTING SCREWS

STANDARD THREAD-FORMING SCREW

SELF-DRILLING SCREW

THREAD-FORMING SCREWS

102F31.EPS

Figure 31 ◆ Sheet metal screws.

14.3.0 Machine Screws

Carpenters use machine screws to fasten butt hinges to metal jambs or door closers to their brackets, and to install lock sets. It is often necessary to drill and tap holes in metal to receive machine screws as fasteners for wood or various kinds of trim. They have four basic head designs, as shown in *Figure 32*.

Machine screws can be obtained with a straight slot or Phillips head. They come in diameters ranging from $\frac{1}{16}$" to $\frac{3}{8}$" and lengths of $\frac{1}{8}$" to 3". The most common machine screws are made of steel or brass, but they can be obtained in other metals as well. They may be obtained with coarse or fine thread.

14.4.0 Lag Screws and Shields

Lag screws (*Figure 33*), or lag bolts, are heavy-duty wood screws with square- or hex-shaped heads that provide greater holding power. Lag screws with diameters ranging between $\frac{1}{4}$" and $\frac{1}{2}$" and lengths ranging from 1" to 6" are common. They are typically used to fasten heavy equipment to wood, but can also be used to fasten equipment to concrete when a lag shield is used.

A lag shield is a metal tube that is split lengthwise but remains joined at one end. It is placed in a predrilled hole in the concrete. When a lag screw is screwed into the lag shield, the shield expands in the hole, firmly securing the lag screw. In hard masonry, short lag shields (typically 1" to 2" long) may be used to minimize drilling time. In soft or weak masonry, long lag shields (typically 1½" to 3" long) should be used to achieve maximum holding strength.

Make sure to use the correct length lag screw to achieve proper expansion. The length of the lag screw used should be equal to the thickness of the component being fastened plus the length of the lag shield. Also, drill the hole in the masonry to a depth approximately ½" longer than the shield being used. If the head of a lag screw rests directly on wood when installed, a flat washer should be placed under the head to prevent the head from

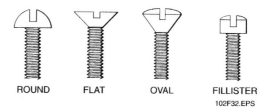

ROUND   FLAT   OVAL   FILLISTER

102F32.EPS

Figure 32 ◆ Machine screws.

2.36 CARPENTRY FUNDAMENTALS LEVEL ONE ◆ TRAINEE GUIDE

Instructor's Notes:

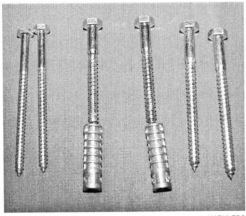

Figure 33 ◆ Lag screws and shields.

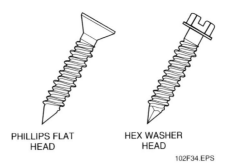

Figure 34 ◆ Concrete screws.

digging into the wood as the lag screw is tightened. Be sure to take the thickness of any washers used into account when selecting the length of the screw.

14.5.0 Concrete/Masonry Screws

Concrete/masonry screws (*Figure 34*), commonly called self-threading anchors, are used to fasten a device or fixture to concrete, block, or brick. No anchor is needed. To provide a matched tolerance anchoring system, the screws are installed using specially designed carbide drill bits and installation tools made for use with the screws. These tools are typically used with a standard rotary drill hammer. The installation tool, along with an appropriate drive socket or bit, is used to drive the screws directly into predrilled holes that have a diameter and depth specified by the screw manufacturer. When being driven into the concrete, the widely spaced threads on the screws cut into the walls of the hole to provide a tight friction fit. Most types of concrete/masonry screws can be removed and reinstalled to allow for shimming and leveling of the fastened device.

14.6.0 Deck Screws

Deck screws (*Figure 35*) are made in a wide variety of shapes and sizes for different indoor and outdoor applications. Some are made to fasten pressure-treated and other types of wood decking to wood framing. Self-drilling types are made to fasten wood decking to different gauges of metal support structures. Similarly, other self-drilling kinds are made to fasten metal decking and sheeting to different gauges and types of metal structural support members. Because of their wide diversity, it is important to follow the manufacturer's recommendations for selection of the proper screw for a particular application. Many manufacturers make a stand-up installation tool for driving their deck screws. This tool eliminates angle driving, underdriven or overdriven screws, and screw wobble. It also reduces operator fatigue.

INSIDE TRACK

Screws
In most applications, either threaded or non-threaded fasteners could be used. However, threaded fasteners are sometimes preferred because they can usually be tightened or removed without damaging the surrounding material.

Driving Wood Screws
To maintain holding power, be careful not to drill your pilot hole too large. It's wise to drill a pilot hole deep enough to equal about two-thirds of the length of the threaded portion of the screw. Additionally, using soap to lubricate screw threads makes the screw easier to drive.

Discuss the applications of drywall screws, drive screws, and hammer-driven pins. Pass around examples for the trainees to examine.

Show Transparencies 26 through 28 (Figures 36 through 38).

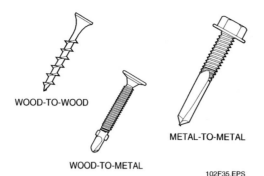

Figure 35 ♦ Typical deck screws.

14.7.0 Drywall Screws

Drywall screws (*Figure 36*) are thin, self-drilling screws with bugle-shaped heads. Depending on the type of screw, it cuts through the wallboard and anchors itself into wood and/or metal studs, holding the wallboard tight to the stud. Coarse thread screws are normally used to fasten wallboard to wood studs. Fine thread and high-and-low thread types are generally used for fastening to metal studs. Some screws are made for use in either wood or metal. A Phillips or Robertson drive head allows the drywall screw to be countersunk without tearing the surface of the wallboard.

14.8.0 Drive Screws

Drive screws do not require the hole to be tapped. They are installed by hammering the screw into a drilled or punched hole of the proper size. Drive screws are mostly used to fasten parts that will not be exposed to much pressure. A typical use of drive screws is to attach permanent name plates on electric motors and other types of equipment. *Figure 37* shows typical drive screws.

14.9.0 Hammer-Driven Pins and Studs

Hammer-driven pins or threaded studs (*Figure 38*) use a special tool to fasten wood or steel to concrete or block without the need to predrill holes. To install these fasteners, insert the pin or threaded stud into the hammer-driven tool point with the washer seated in the recess. Position the pin or stud against the base and tap the drive rod of the tool lightly until the striker pin contacts the pin or stud. Strike the tool's drive rod using heavy blows with about a two-pound engineer's hammer. The force of the hammer blows transmits through the tool directly to the head of the fastener, causing it to be driven into the concrete or block. For best results, the drive pin or stud should be embedded a minimum of ½" in hard concrete to 1¼" in softer concrete block.

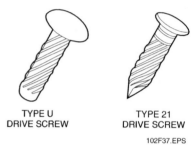

Figure 37 ♦ Drive screws.

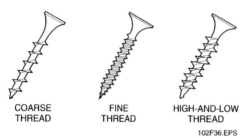

Figure 36 ♦ Drywall screws.

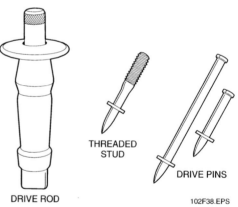

Figure 38 ♦ Hammer-driven pins and installation tool.

15.0.0 ♦ BOLTS

Bolts are often used by the carpenter in attaching one unit or member to another. There are many different designs and types for special jobs, but we will discuss only the most common in this module (*Figure 39*).

15.1.0 Stove Bolts

Stove bolts come in either roundhead or flathead design and lengths from ⅜" to 6". The shanks are threaded all the way to the head on those up to 2" long. If they are longer than 2", they are threaded to a maximum of 2". They can be obtained in several different materials such as steel, brass, and copper. They are used in lighter types of construction assemblies.

15.2.0 Machine Bolts

Machine bolts come with nuts. The nut is normally the same thickness as the diameter of the bolt. Specially designed nuts are obtainable for specific purposes such as self-locking or cap nuts for safety or appearance. The bolts normally have square or hexagon heads with nuts to match. They come in lengths from ¾" to 30". The most common diameters range from ¼" to 2" and are obtainable in either regular or fine thread. For a decorative appearance, they can be ordered with special finishes such as antique or chrome.

15.3.0 Carriage Bolts

The carriage bolt is similar to the machine bolt except for the design of the head. The head is oval and the shank is square. It is designed that way so that the bolt can be driven or drawn into the wood, and the nut tightened without the bolt turning. Also, other members can be fastened over the heads with little or no interference. They come in lengths of ¾" to 20" with either standard or fine threads. Normal diameters range from ¼" to ¾". They can also be obtained in different materials and finishes, such as brass or galvanized.

16.0.0 ♦ MECHANICAL ANCHORS

Mechanical anchors are devices used to give fasteners a firm grip in a variety of materials, where the fasteners by themselves would otherwise have a tendency to pull out. Anchors can be classified in many ways by different manufacturers. In this module, anchors have been divided into five broad categories:

- One-step anchors
- Bolt anchors
- Screw anchors
- Self-drilling anchors
- Hollow-wall anchors

16.1.0 Anchor Bolts

One of the most common anchors in frame construction is the anchor bolt. Anchor bolts are used to anchor the sill plate to the concrete foundation (*Figure 40*). The anchor bolt is set into the wet concrete. The sill plate is then placed over the anchor and a nut and washer are used to secure the sill plate to the foundation. The sill plate acts as the base for the wall framing.

16.2.0 One-Step Anchors

One-step anchors are designed so that they can be installed through the mounting holes in the component to be fastened. This is because the anchor and the hole into which the anchor is installed have the same diameter. They come in various diameters ranging from ¼" to 1¼" with lengths ranging from 1¾" to 12". Wedge, stud, sleeve, one-piece, screw, and nail anchors (*Figure 41*) are common types of one-step anchors.

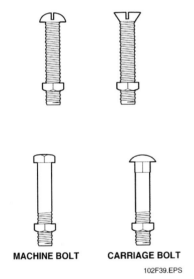

Figure 39 ♦ Bolts.

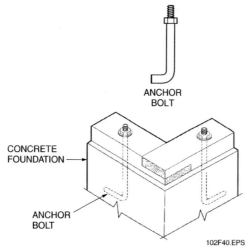

Figure 40 ♦ Anchor bolt.

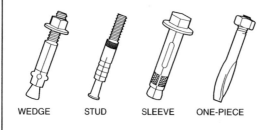

Figure 41 ♦ One-step anchors.

16.2.1 Wedge Anchors

Wedge anchors are heavy-duty anchors supplied with nuts and washers. The drill bit size used to drill the hole is the same diameter as the anchor. The depth of the hole is not critical as long as it meets the minimum recommended by the manufacturer. Blow the hole clean of dust and other material, insert the anchor into the hole, and drive it with a hammer far enough so that at least six threads are below the top surface of the component. Then, tighten the anchor nut to expand the anchor and secure it in the hole.

16.2.2 Stud Bolt Anchors

Stud bolt anchors are heavy-duty threaded anchors. Because this type of anchor is made to bottom in its mounting hole, it is a good choice to use when jacking or leveling of the fastened component is needed. The depth of the hole drilled in the masonry must be as specified by the manufacturer in order to achieve proper expansion. Blow the hole clean of dust and other material, then insert the anchor in the hole with the expander plug end down. Drive the anchor into the hole with a hammer or setting tool to expand the anchor and tighten it in the hole. The anchor is fully set when it can no longer be driven into the hole. Fasten the component using the correct size and thread bolt for use with the anchor stud.

16.2.3 Sleeve Anchors

Sleeve anchors are multi-purpose anchors. The depth of the anchor hole is not critical as long as the minimum length recommended by the manufacturer is drilled. Blow the hole clean of dust and other material, insert the anchor into the hole, and tap it until it is flush with the component. Tighten the anchor nut or screw to expand the anchor and tighten it in the hole.

16.2.4 One-Piece Anchors

One-piece anchors are multi-purpose anchors. They work on the principle that as the anchor is driven into the hole, the spring force of the expansion mechanism is compressed and flexes to fit the size of the hole. Once set, it tries to regain its original shape. The depth of the hole drilled in the masonry must be at least a half-inch deeper than the required embedment. The proper depth is crucial. Overdrilling is as bad as underdrilling. Blow the hole clean of dust and other material, insert the anchor through the component, and drive it with a hammer into the hole until the head is firmly seated against the component. It is important to make sure that the anchor is driven to the proper embedment depth. Note that manufacturers also make specially designed drivers and manual tools that are used instead of a hammer to drive one-piece anchors. These tools allow the anchors to be installed in confined spaces and they help prevent damage to the component from stray hammer blows.

16.2.5 Hammer-Set Anchors

Hammer-set anchors are made for use in concrete and masonry. There are two types: nail and screw. An advantage of the screw-type anchors is that they are removable. Both types have a diameter the same size as the anchoring hole. For both types, the anchor hole must be drilled to the diameter of the anchor and to a depth at least a quarter-inch deeper than that required for embedment. Blow the hole clean of dust and other material, then insert the anchor into the hole through the mounting holes in the component to be fastened. Drive the screw or nail into the anchor body to expand it. It is important to make sure that the head is seated firmly against the component and is at the proper embedment.

16.2.6 Sammy® Anchors

Sammy® anchors are available for installation in concrete, steel, or wood. The Sammy® anchor is designed to support a threaded rod, which is screwed into the head of the anchor after the anchor is installed. A special nut driver is available for installing the screws.

16.3.0 Bolt Anchors

Bolt anchors are designed to be installed flush with the surface of the base material. They are used in conjunction with threaded machine bolts or screws. In some types, they can be used with threaded rod. Drop-in and single- and double-expansion anchors (*Figure 42*) are commonly used types of bolt anchors.

16.3.1 Drop-In Anchors

Drop-in anchors are typically used as heavy-duty anchors. There are two types of drop-in anchors. The first type, made for use in solid concrete and masonry, has an internally threaded expansion anchor with a preassembled internal expander plug. The anchor hole must be drilled to the specific diameter and depth specified by the manufacturer. Blow the hole clean of dust and other material, insert the anchor into the hole, and tap it until it is flush with the surface. Drive the setting tool supplied with the anchor into the anchor to expand it. Position the component to be fastened in place, and fasten it by threading and tightening the correct size machine bolt or screw into the anchor.

The second type, called a hollow set drop-in anchor, is made for use in hollow concrete and masonry base materials. Hollow set drop-in

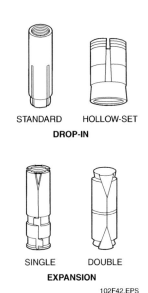

Figure 42 ◆ Bolt anchors.

anchors have a slotted, tapered expansion sleeve and a serrated expansion cone. They come in various lengths compatible with the outer wall thickness of most hollow base materials. They can also be used in solid concrete and masonry. The anchor hole must be drilled to the diameter specified by the manufacturer. When installed in hollow base materials, drill the hole into the cell or void. Blow the hole clean of dust and other material, insert the anchor into the hole, and tap it until it is flush with the surface. Position the component to be fastened in place, then thread the proper size machine bolt or screw into the anchor and tighten it to expand the anchor in the hole.

16.3.2 Single- and Double-Expansion Anchors

Single- and double-expansion anchors are made for use in concrete and other masonry. The double-expansion anchor is used mainly when fastening into concrete or masonry of questionable strength. For both types, the anchor hole must be drilled to the diameter and depth specified by the manufacturer. Blow the hole clean of dust and other material, then insert the anchor into the hole, threaded cone end first. Tap it until it is flush with the surface. Position the component to be fastened in place, then thread the proper size machine bolt or screw into the anchor and tighten it to expand the anchor in the hole.

Discuss the applications and installation of screw and self-drilling anchors. Provide examples for the trainees to examine.

Show Transparency 33 (Figure 44).

Review the guidelines for drilling anchor holes in hardened concrete or masonry.

16.4.0 Screw Anchors

Screw anchors are lighter-duty anchors made to be installed flush with the surface of the base material. They are used in conjunction with sheet metal, wood, or lag screws. Fiber and plastic anchors are common types of screw anchors (*Figure 43*). The lag shield anchor used with lag screws was described earlier in this module.

Fiber and plastic anchors are typically used in concrete and masonry. Plastic anchors are also commonly used in wallboard and similar base materials. The installation of all types is simple. The anchor hole must be drilled to the diameter specified by the manufacturer. The minimum depth of the hole must equal the anchor length. Blow the hole clean of dust and other material, insert the anchor into the hole, and tap it until it is flush with the surface. Position the component to be fastened in place, then drive the proper type and size of screw through the component mounting hole and into the anchor to expand the anchor in the hole.

16.5.0 Self-Drilling Anchors

Some anchors made for use in masonry are self-drilling. *Figure 44* is typical of those in common use. This fastener has a cutting sleeve that is first used as a drill bit and later becomes the expandable fastener itself. A rotary hammer is used to drill the hole in the concrete, using the anchor sleeve as the drill bit. After drilling the hole, pull the anchor out and clean the hole. Then insert the anchor's expander plug into the cutting end of the sleeve. Drive the anchor sleeve and expander plug back into the hole with the rotary hammer until they are flush with the surface of the concrete. As the fastener is hammered down, it hits the bottom, where the tapered expander causes the fastener to expand and lock into the hole. Then snap off the anchor at the shear point with a quick lateral movement of the hammer. The component to be fastened can then be attached to the anchor using the proper size bolt.

16.6.0 Guidelines for Drilling Anchor Holes in Hardened Concrete or Masonry

When selecting masonry anchors, regardless of the type, always take into consideration and follow the manufacturer's recommendations pertaining to hole diameter and depth, minimum embedment in concrete, maximum thickness of the material to be fastened, and the pullout and shear load capacities.

When installing anchors and/or anchor bolts in hardened concrete, make sure the area where the equipment or component is to be fastened is smooth so that it will have solid footing. Uneven footing might cause the equipment to twist, warp, not tighten properly, or vibrate when in operation. Before starting, carefully inspect the rotary hammer or hammer drill and the drill bit(s) to ensure they are in good operating condition. Set the drill or hammer tool depth gauge to the depth of the hole needed. Be sure to use the type of carbide-tipped masonry or percussion drill bits recommended by the drill/hammer or anchor manufacturer because these bits are made to take the higher impact of the masonry materials. The trick to using masonry drill bits is not to force them into the material by pushing down hard on the drill. Use a little pressure and let the

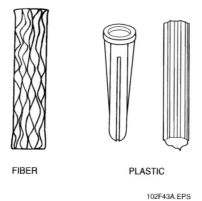

FIBER   PLASTIC

102F43A.EPS

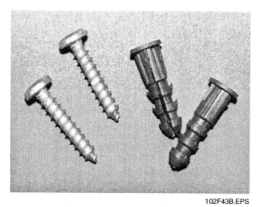

102F43B.EPS

Figure 43 ♦ Screw anchors and screws.

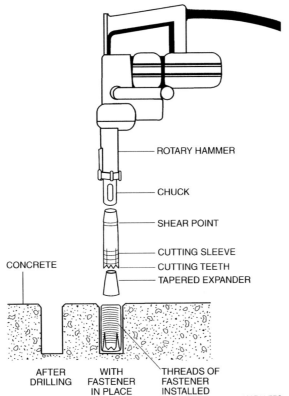

Figure 44 ♦ Self-drilling anchor.

drill do the work. For large holes, start with a smaller bit, then change to a larger bit.

The methods for installing the different types of anchors in hardened concrete or masonry were briefly described in the sections above. Always install the selected anchors according to the manufacturer's directions. Here is an example of a typical procedure used to install many types of expansion anchors in hardened concrete or masonry. Refer to *Figure 45* as you study the procedure.

 WARNING!
Drilling in concrete generates noise, dust, and flying particles. Always wear safety goggles, ear protectors, a dust mask, and gloves. Make sure other workers in the area also wear protective equipment.

Step 1 Drill the anchor bolt hole the same size as the anchor bolt. The hole must be deep enough for six threads of the bolt to be below the surface of the concrete. Clean out the hole using a squeeze bulb.

Step 2 Drive the anchor bolt into the hole using a hammer. Protect the threads of the bolt with a nut that does not allow any threads to be exposed.

Step 3 Put a washer and nut on the bolt, and tighten the nut with a wrench until the anchor is secure in the concrete.

16.7.0 Hollow-Wall Anchors

Hollow-wall anchors are used in hollow materials such as concrete plank, block, structural steel, wallboard, and plaster. Some types can also be used in solid materials. Toggle bolts, sleeve-type

Point out the danger of flying particles when drilling into concrete.

Show Transparency 34 (Figure 45).

Discuss the applications and installation of hollow-wall anchors. Provide examples for the trainees to examine.

Explain how toggle bolts are installed. Pass around examples for the trainees to examine.

Show Transparency 35 (Figure 46).

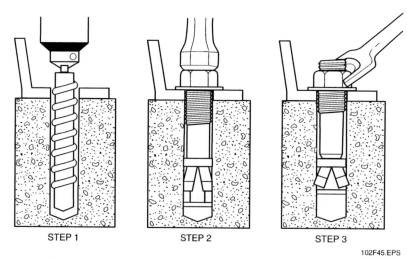

Figure 45 ♦ Installing an anchor bolt in hardened concrete.

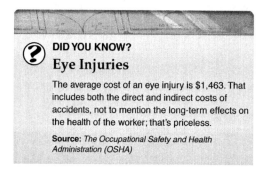

DID YOU KNOW?
Eye Injuries

The average cost of an eye injury is $1,463. That includes both the direct and indirect costs of accidents, not to mention the long-term effects on the health of the worker; that's priceless.

Source: *The Occupational Safety and Health Administration (OSHA)*

wall anchors, wallboard anchors, and metal drive-in anchors are common anchors used when fastening to hollow materials.

When installing anchors in hollow walls or ceilings, regardless of the type, always follow the manufacturer's recommendations pertaining to use, hole diameter, wall thickness, grip range (thickness of the anchoring material), and the pullout and shear load capacities.

16.7.1 Toggle Bolts

Toggle bolts (*Figure 46*) are used to fasten equipment, hangers, supports, and similar items into hollow surfaces such as walls and ceilings. They consist of a slotted bolt or screw and spring-loaded wings. When inserted through the item to be fastened, then through a predrilled hole in the wall or ceiling, the wings spring apart and provide a firm hold on the inside of the hollow wall or ceiling as the bolt is tightened. Note that the hole drilled in the wall or ceiling should be just large enough for the compressed wing-head to pass through. Once the toggle bolt is installed, be careful not to completely unscrew the bolt because the wings will fall off, making the fastener useless. Screw-actuated plastic toggle bolts are also made. These are similar to metal toggle bolts, but they come with a pointed screw and do not require as large a hole. Unlike the metal version, the plastic wings remain in place if the screw is removed.

Toggle bolts are used to fasten a part to hollow block, wallboard, plaster, panel, or tile. The following procedure can be used to install toggle bolts:

WARNING!
Follow all safety precautions to avoid injury.

Step 1 Select the proper size drill bit or punch and toggle bolt for the job.

Step 2 Check the toggle bolt for damaged or dirty threads or a malfunctioning wing mechanism.

Step 3 Drill a hole completely through the surface to which the part is to be fastened.

Step 4 Insert the toggle bolt through the opening in the item to be fastened.

Instructor's Notes:

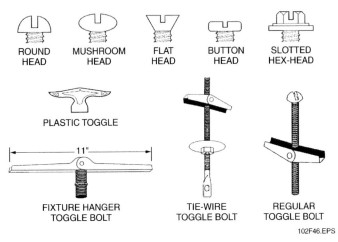

Figure 46 ◆ Toggle bolts.

Step 5 Screw the toggle wing onto the end of the toggle bolt, ensuring that the flat side of the toggle wing is facing the bolt head.

Step 6 Fold the wings completely back and push them through the drilled hole until the wings spring open.

Step 7 Pull back on the item to be fastened to hold the wings firmly against the inside surface to which the item is being attached.

Step 8 Tighten the toggle bolt with a screwdriver until it is snug.

16.7.2 Wall Anchors

Sleeve-type wall anchors (*Figure 47*) are suitable for use in plywood, wallboard, and similar materials. The two types made are standard and drive. The standard type is commonly used in walls and ceilings and is installed by drilling a mounting

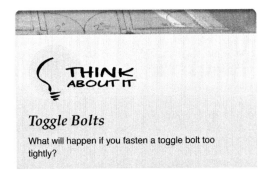

Think About It

Toggle Bolts

What will happen if you fasten a toggle bolt too tightly?

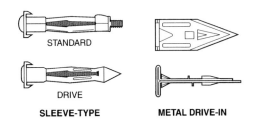

Figure 47 ◆ Sleeve-type, wallboard, and metal drive-in anchors.

hole to the required diameter. Insert the anchor into the hole and tap it until the gripper prongs embed in the base material. Tighten the anchor screw to draw the anchor tight against the inside of the wall or ceiling.

The drive-type anchor is hammered into the material without the need for drilling a mounting hole. After the anchor is installed, remove the anchor screw. Position the component being fastened in place, then reinstall the screw through the mounting hole in the component and into the anchor. Tighten the screw into the anchor to secure the component.

Wallboard anchors are self-drilling medium- and light-duty anchors used for fastening in wallboard. Using a Phillips head manual or cordless

MODULE 27102-06 ◆ BUILDING MATERIALS, FASTENERS, AND ADHESIVES 2.45

Explain how wall anchors are installed. Pass around examples for the trainees to examine.

Show Transparency 36 (Figure 47).

Discuss the "Think About It." Point out that overtightening a toggle bolt can damage both the wall and/or the fastener.

Discuss epoxies and their applications.

Show Transparency 37 (Figure 48).

Emphasize the safety procedures required when working with adhesives.

screwdriver, drive the anchor into the wall until the head of the anchor is flush with the wall or ceiling surface. Position the component being fastened over the anchor, then secure it by driving the proper size sheet metal screw into the anchor.

Metal drive-in anchors are used to fasten light to medium loads to wallboard. They have two pointed legs that stay together when the anchor is hammered into a wall and spread out against the inside of the wall when a No. 6 or 8 sheet metal screw is driven in.

17.0.0 ♦ EPOXY ANCHORING SYSTEMS

Epoxy resin compounds can be used to anchor threaded rods, dowels, and similar fasteners in solid concrete, hollow wall, and brick. For one product, a two-part epoxy is packaged in a two-chamber cartridge that keeps the resin and hardener ingredients separated until use. This cartridge is placed into a special tool similar to a caulking gun. When the gun handle is pumped, the epoxy resin and hardener components are mixed within the gun; then the epoxy is ejected from the gun nozzle.

To use epoxy to install an anchor in solid concrete (*Figure 48*), drill a hole of the proper size in the concrete and clean it using a nylon (not metal) brush. Dispense a small amount of epoxy from the gun to make sure that the resin and hardener have mixed properly. This is indicated by the epoxy being of a uniform color. Then place the gun nozzle into the hole, and inject the epoxy into the hole until half the depth of the hole is filled. Push the selected fastener into the hole with a slow twisting motion to make sure that the epoxy fills all voids and crevices, then set it to the required plumb (or level) position. After the recommended cure time

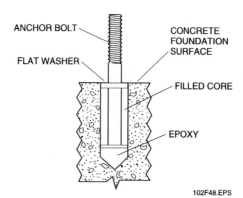

Figure 48 ♦ Fastener anchored in epoxy.

Installation Requirements

In a college dormitory, battery-powered emergency lights were anchored to sheetrock hallway ceilings with sheetrock screws, with no additional support. These fixtures weigh 8–10 pounds each and might easily have fallen out of the ceiling, causing severe injury. When the situation was discovered, the contractor had to remove and replace dozens of fixtures.

The Bottom Line: Incorrect anchoring methods can be both costly and dangerous.

for the epoxy has elapsed, tighten the fastener nut to secure the component or fixture in place.

The procedure for installing a fastener in a hollow wall or brick when using epoxy is basically the same as the one just described. The difference is that the epoxy is first injected into an anchor screen to fill the screen, then the anchor screen is installed into the drilled hole. Use of the anchor screen is necessary to hold the epoxy intact in the hole until the anchor is inserted into the epoxy.

18.0.0 ♦ ADHESIVES

The term adhesives describes a variety of products that are used to attach one surface to another. Construction adhesives are used to attach paneling and drywall to framing, install subflooring, and attach ceiling tiles. Special types of adhesives are used to install ceramic tiles, floor coverings, and carpeting. There are also many types of glues used in a household or workshop environment. Different glues have different characteristics. Some are quick-drying; others are slow. Some are waterproof; some are not. Some must be dried under pressure, etc. Therefore, it is important for the carpenter to know what adhesive to use in a particular situation. At some point in your career, knowing what adhesive to use in a given situation will become second nature. In the meantime, ask a professional such as a building supply specialist. Always read the instructions on the label before using any adhesive. Also keep in mind that new products are being introduced all the time, so it is important to keep up with what is going on in the industry.

Finally, keep in mind that most adhesives are made from chemicals, some of which are toxic

2.46 CARPENTRY FUNDAMENTALS LEVEL ONE ♦ TRAINEE GUIDE

Instructor's Notes:

Adhesive Safety

Here are some general safety precautions to follow when using adhesives:

- Be sure to provide good ventilation.
- Keep all flames and lighted tobacco products out of the area.
- Make sure switches, electric tools, or other sparking devices are not used in the area.
- Close off any area where people that are smoking could wander through.
- Do not breathe vapors for any length of time.
- Immediately remove any adhesive that comes in contact with your skin.
- Wear protective goggles and clothing when required.
- Follow the manufacturer's MSDS instructions and check the label on the container for specific safety precautions.
- When in doubt, ask questions.

Discuss the applications of various types of glues.

and/or flammable. Read and follow the safety instructions on the label and check the MSDS. The MSDS for any toxic or flammable substance must be available at the job site. In general, it is a good idea to work with adhesives in a well-ventilated area away from open flames and to wear a face mask and eye protection.

18.1.0 Glues

Glues are used in laminated construction such as arches, curved members, and building beams, and in the construction of cabinets and millwork. Carpenters may carry a plastic bottle of liquid glue in their toolboxes for making some interior trim joints or cabinet work on the job.

Glues come in two forms: liquid or dry. The dry forms are mixed into a liquid on the job. Some glues come in two or more liquid forms (for example, epoxy glue). When mixed together in directed proportions, they become very strong adhesives. In all cases where mixing is involved, the manufacturer's MSDS instructions must be followed, both for the quality desired and safety.

Some of the most common types of glue the carpenter might use are:

- *Animal or hide glue* – Made from animal hides, hooves, and bones. It has a long setting time, which makes it suitable when extra assembly time is needed. It can be used on furniture and other wood products, but is not waterproof.
- *Polyvinyl or white glue* – Sets quickly, becomes transparent when dry, and is very good for furniture, cabinets, and general interior woodwork. It is a liquid and is obtainable in containers up to a gallon in size, or by special order in larger quantities.
- *Casein glue* – Comes in a powder form and is mixed with water. It can be used in temperatures down to 35°F. It is water resistant. It can be used on oily woods. Under normal circumstances, it will last for years. However, the powder tends to deteriorate with age.
- *Urea formaldehyde or plastic resin glue* – Comes in powder form and is mixed with water. It is slow setting and has good resistance to heat and water. It is not recommended for oily woods. It is often used for laminated timbers and general woodworking where moisture is present.
- *Resorcinol resin or waterproof glue* – An expensive glue that comes in two liquid chemical parts. One part acts as a **catalyst** on the other part when mixed. It is excellent for use in exterior woodwork or laminated timbers that are exposed to extremely cold temperatures. Eye goggles and a respirator are needed when mixing.
- *Contact cement* – A type of rubber cement. It is a liquid and comes ready-mixed in containers from a pint to a 55-gallon drum. It is not a high-strength glue. It is used extensively in applying plastic laminates to cabinets and countertops. It is applied to both surfaces and allowed to dry to the touch before contact is made between the two materials. Upon contact, the bonding is instantaneous and movement or sliding into a different position cannot be done.

WARNING!
Observe all MSDS precautions for this material. Extreme caution must be taken when using contact cement, as it is highly flammable. It should be used in a well-ventilated area, as the fumes can overcome a worker.

Warn the trainees that many glues and cements are flammable, explosive, or otherwise hazardous and must be handled accordingly.

Discuss various types of construction adhesives and their applications.

Show Transparency 38 (Figure 49).

Discuss the applications of various types of mastics.

18.2.0 Construction Adhesives

Construction adhesives are available for many types of heavy-duty construction uses. They are made for specific applications such as installing paneling, structural decking, and drywall. In addition to providing strong bonding between materials, construction adhesives offer such advantages as sound deadening and the bonding of dissimilar materials such as wood, gypsum, glass, metal, and concrete.

Construction adhesives are usually applied from a cartridge with a manual ratchet-type gun or pneumatic gun (*Figure 49*). Cartridges range in capacity from about 10 ounces to 30 ounces.

There are several types of adhesives for fastening sheet materials to framing members. They include the following:

- *Construction adhesive* – Used to apply wood or gypsum to framing. It has a solvent base. The tool needed for the application of construction adhesive is a caulking gun or cartridge.
- *Neoprene adhesive* – Can be used to apply wood to wood and gypsum to wood or metal. Neoprene adhesive will also bond to concrete. Like construction adhesive, neoprene adhesive is also applied with a caulking gun. Neoprene adhesive has a solvent base.
- *Contact cement* – Application is by brush or roller. Contact cement will bond to metal. It can also bond wallboard to framing and wallboard to plywood. It is also used to bond plastic laminates. Contact cement should be nearly dry before joining the two materials together.
- *Drywall adhesive* – Although suitable for lamination, it is used primarily for applying wallboard to framing. A caulking gun is used for application. It is available as a water-based product; however, solvent-based adhesive is more widely used.
- *Instant-bond glue* – Known by trade terms such as Krazy Glue® and Super Glue®. They work well on non-porous materials such as glass, ceramics, metal, and many plastics. Some types of instant-bond glue are designed for porous materials such as wood and paper. When using this glue, do not touch your fingers to each other or to anything else. Acetone or nail polish remover will dissolve the glue, but be sure to keep it away from your eyes.
- *Epoxies* – Made by mixing a resin and a hardening agent at the time of use. Epoxies are good for bonding dissimilar materials. Fast-acting epoxies set almost instantly. They usually come in a special syringe-type applicator that keeps the two components separate. They blend when the plungers are depressed. Slower-setting epoxies are mixed in a container at the job site.

 WARNING!
Many adhesives and cements are flammable and/or explosive. Make sure they are stored in a well-ventilated area and always read all warning labels on the adhesive packaging. Take the proper precautions to avoid an explosion or fire.

18.3.0 Mastics

Mastics are generally used to apply floor coverings, roofing materials, ceramic tiles, or wall paneling. Most mastics have a synthetic rubber or latex base. Some are thick pastes and can be used where moisture is not present. Others are waterproof and can be used in bathrooms, kitchens, and laundries, or in direct contact with concrete. Some require hot application with special equipment to keep them fluid while being applied. Others are

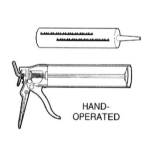
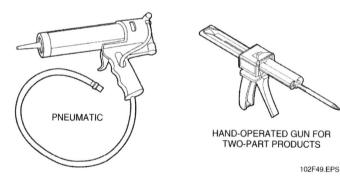

Figure 49 ♦ Adhesive applicators.

2.48 CARPENTRY FUNDAMENTALS LEVEL ONE ♦ TRAINEE GUIDE

Instructor's Notes:

applied with brushes, trowels, or rollers. Some come in tubes and are applied with hand- or air-operated caulking guns.

Keep in mind that good coverage with no voids is necessary for satisfactory bonding. Also, if applying floor tiles or paneling, use caution to prevent mastic from squeezing up between the joints or cracks and thus staining or marring the finish surface. To avoid this, place each piece without sliding after contact is made. When applying the mastic, make sure it is evenly spread. For satisfactory bonding, smooth, clean, dry surfaces are a must. Flaking plaster or old paint must be pretreated with special chemicals or removed. Leveling cements or compounds may be obtained and applied to overcome unevenness in walls or concrete floors. If moisture is likely to be encountered (e.g., concrete on basement walls or floors), a waterproofing material must be applied before the mastic. Only waterproof mastic should be used in these circumstances. Under adverse conditions, check the manufacturer's recommendations for secure installation. Many manufacturers recommend the type of adhesive to be used with their product.

18.4.0 Shelf Life

All adhesives have a limited life span that ranges from 12 to 24 months. If they are not used within that time, they will begin to lose their bonding ability. Most manufacturers claim a shelf life of at least two years. Keep in mind, however, that three to six months or more may have elapsed from the time it was manufactured until it was purchased. When you buy adhesive, write the date on the container and do not use it beyond the shelf life listed on the label.

Explain that all adhesives have a limited shelf life.

Have the trainees identify the correct type of building material for various applications. This laboratory corresponds to Performance Task 2.

Summarize the major concepts presented in the module.

Summary

Lumber and other wood building materials are commonly used in framing and sheathing of single-family residences. Metal framing members, especially studs, are being used more and more as the price grows and the supply of lumber shrinks.

In commercial work, concrete and steel are much more common as building materials. In many instances, entire structures are made of poured concrete. In others, a steel structural frame is combined with concrete floors and curtain wall panels.

It is important to know how to select the right material for any given situation. There are many types and grades of lumber and sheet materials. Each has specific uses. More importantly, each has certain limitations that the user must know in order to avoid harmful or costly errors.

Likewise, there is a correct size and type of fastener for every fastening situation and an adhesive that is just right for occasions when an adhesive is needed. This module provided an introduction to the many types of fasteners and adhesives a carpenter uses on the job. As your training progresses, you will learn more about how and when to use them.

Notes

Instructor's Notes:

Review Questions

1. In the grading stamp shown below, the term S-GRN means that the lumber _____.
   a. is made from short-grain wood
   b. contains more than 19 percent moisture
   c. should be painted with green primer before use
   d. is southern grown

```
12  STAND
ABC  S-GRN   /D\
             /FIR\
         102RQ01.EPS
```

2. The nominal size of a piece of lumber is the _____.
   a. actual size
   b. size before it has fully dried
   c. dressed size
   d. size by which it is known and sold

3. If lumber is graded as double-end-trimmed, its ends are trimmed _____.
   a. square within a tolerance of 1/64" for every 2"
   b. twice at each end
   c. reasonably square on both ends
   d. square and smooth on both ends within a tolerance of 1/16" for the entire length

4. The product known as soft plywood has a _____ core.
   a. lumber
   b. fiber
   c. veneer
   d. particleboard

5. Standard hardboard is suitable only for _____.
   a. cabinet making
   b. siding
   c. exterior sheathing
   d. soundproofing

6. When you want to obtain safety information about an adhesive, the best source of reliable information is _____.
   a. your supervisor
   b. a co-worker
   c. the applicable MSDS
   d. a building supply dealer

7. Which of these engineered wood products is commonly used to make doors and other millwork?
   a. Glulam
   b. LVL
   c. PSL
   d. LSL

8. A glulam beam consists of _____.
   a. laminated veneers bonded together under pressure
   b. solid, kiln-dried lumber glued together
   c. veneers bonded to a core with glue
   d. long strands of veneer bonded together with adhesive

9. Which of these statements about pressure-treated lumber is *not* true?
   a. It is often used to make landscape timbers.
   b. Scrap material should be burned.
   c. Eye protection and a dust mask must be worn when cutting it.
   d. It is less expensive than redwood.

10. How many board feet are there in (30) 2" × 10" × 15' joists?
    a. 480
    b. 640
    c. 750
    d. 1,250

11. How many sheets of plywood would you need in order to provide subflooring for a 25' by 30' room?
    a. 28
    b. 25
    c. 30
    d. 24

Have trainees complete the Review Questions, and go over the answers prior to administering the Module Examination.

Review Questions

12. An 8-penny nail is approximately _____ long.
    a. 1"
    b. 2½"
    c. 4"
    d. 6"

13. The type of staple point that clinches outward after it penetrates the material is the _____.
    a. crosscut chisel
    b. outside chisel
    c. spear
    d. inside chisel

14. The type of bolt that is used to attach the sill plate to the foundation is the _____ bolt.
    a. machine
    b. carriage
    c. anchor
    d. stove

15. The adhesive that requires two components that are mixed together at the time of use is _____.
    a. epoxy
    b. polyvinyl glue
    c. contact cement
    d. mastic

Instructor's Notes:

Trade Terms Quiz

1. The main ingredient in plaster and drywall is _____.
2. After installing finishing nails, use a(n) _____ to recess the nails, then fill the holes with putty.
3. Siding is used to cover the _____.
4. Crown molding is a type of _____.
5. Tapered siding that overlaps from one row to the next is known as _____ siding.
6. _____ siding is an interlocking siding that is attached end-to-end.
7. The _____ describes safe handling methods for a particular substance.
8. The roof sheathing is attached to the _____.
9. The framework of a building is attached to the _____.
10. Use _____ nails to prevent rusting.
11. Nails that have been coated with a glossy finish are said to be _____.
12. A two-part waterproof glue achieves its strength using a(n) _____.
13. A floor is supported by equally spaced _____.
14. When two square-cut pieces of material are placed end-to-end, it is called a(n) _____.
15. A(n) _____ floor is built out to extend beyond the last point of support.
16. A(n) _____ is a popular homebuilding option because the higher ceiling makes the room feel more spacious.
17. Hardboard is coated with protective substances known as _____.

Trade Terms

Butt joint
Cantilever
Catalyst
Galvanized
Gypsum
Japanned
Joists
Material safety data sheet (MSDS)
Millwork
Nail set
Rafter
Resins
Sheathing
Shiplap
Sill plate
Tongue-and-groove
Vaulted ceiling

Have trainees complete the Trade Terms Quiz, and go over the answers.

Administer the Module Examination. Record the results on Craft Training Report Form 200, and submit the form to the Training Program Sponsor.

Administer the Performance Tests, and fill out Performance Profile Sheets for each trainee. If desired, trainee proficiency noted during laboratory sessions may be used to complete the Performance Test. Record the results on Craft Training Report Form 200, and submit the results to the Training Program Sponsor.

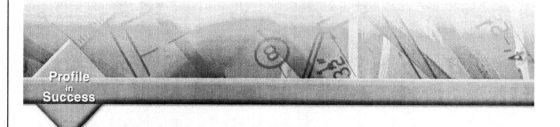

Profile in Success

Erin M. Hunter-Parker
Carpentry Instructor
River Valley Technical Center
Springfield, VT

Erin has been building things for as long as she can remember. Her father was a contractor and president of Associated Builders and Contractors of Virgina. She began by helping him with renovations around their house. After she gained experience working on different carpentry crews, she started her own company. Now she teaches carpentry and is pursuing her master's degree in education, bringing her passion for the trade to a new generation of carpenters.

How did you become interested in carpentry?
I have always loved working with wood, building things, and being outside. My father worked for a concrete construction company, then eventually started his own contracting business. On Saturdays he would take me with him to job sites. He would put on his hard hat and walk out with a set of prints. At that age, just watching him tie rebar was exciting.

My love of working with wood led me to jobs building furniture. My career in construction started on a historical restoration project as a finish carpenter. I later began taking on jobs of my own and learning a variety of rough carpentry skills. This led to the start of my own company called E M Hunter Construction where I did everything from roofing to renovations.

What types of training have you been through and how did it help you get to where you are now?
I had no formal training in carpentry. I had to learn the hard way and by anyone willing to take the time to show me. I wish the opportunities that are available to today's young people were available when I started.

What kinds of work have you done in your career?
I have done finish carpentry, cabinetry, built-ins, tiling, decks, gazebos, timber-framing, roofing, additions ... you name it.

What are some of the things you do in your job?
I currently teach a two-year carpentry program to juniors and seniors in a high school career and technical center.

What do you think it takes to be a success in your trade?
Carpentry is a craft that requires integrity, attention to detail, good people skills, adaptability, problem-solving skills, and—most importantly—the willingness to work. To be successful, carpenters need to care about the quality of their work and take pride in what they do.

What do you like about the work you do?
I really enjoy teaching sixteen- to eighteen-year-old students. I love helping them learn skills that are relevant and real life. Carpentry is a great springboard into learning other life skills and disciplines like math. My students get to see themselves as successful in a school setting where they may not have been able to in other classes.

What advice would you give someone just starting out?
Stay open and be willing to learn. Carpentry is an ever-changing field. Sometimes I see young people cut themselves short because they feel like they should already know it all. One of the joys of carpentry is that there is always something you can learn or get better at. I encourage my students to be lifelong learners. Our class motto is no fear, no fear of learning, and no fear of trying something new.

2.54 CARPENTRY FUNDAMENTALS LEVEL ONE ♦ TRAINEE GUIDE

Instructor's Notes:

Trade Terms Introduced in This Module

Butt joint: The joint formed when one square-cut edge of a piece of material is placed against another material.

Cantilever: A beam, truss, or slab (floor) that extends past the last point of support.

Catalyst: A substance that causes a reaction in another substance.

Galvanized: Protected from rusting by a zinc coating.

Gypsum: A chalky material that is a main ingredient in plaster and drywall.

Japanned: Coated with a glossy finish.

Joist: Generally, equally-spaced framing members that support floors and ceilings.

Material safety data sheets (MSDS): Information that details any toxic, chemical, or potentially harmful substances that are contained in a product.

Millwork: Manufactured wood products such as doors, windows, and moldings.

Nail set: A punch-like metal tool used to recess finishing nails.

Rafter: A sloping structural member of a roof frame to which sheathing is secured.

Resins: Protective natural or synthetic coatings.

Sheathing: Sheet material or boards to which roofing material or siding is secured.

Shiplap: A method of cutting siding in which each board is tapered and grooved so that the upper piece fits tightly over the lower piece.

Sill plate: A horizontal timber that supports the framework of a building on the bottom of a wall or box joist. It is also called a sole plate.

Tongue-and-groove: A joint made by fitting a tongue on the edge of a board into a matching groove on the adjoining board.

Vaulted ceiling: A high, open ceiling that generally follows the roof pitch.

Additional Resources and References

This module is intended to be a thorough resource for task training. The following reference works are suggested for further study. These are optional materials for continued education rather than for task training.

Basic Construction Materials. Upper Saddle River, NJ: Prentice Hall.

Principles and Practices of Light Construction. Upper Saddle River, NJ: Prentice Hall.

Principles and Practices of Commercial Construction. Upper Saddle River, NJ: Prentice Hall.

Building Construction Illustrated. New York: John Wiley & Sons.

Fundamentals of Building Construction: Materials and Methods. New York: John Wiley & Sons.

Buildings in Wood: The History and Traditions of Architecture's Oldest Building Material. New York: Rizzoli/Universe International Publications.

Instructor's Notes:

Appendix A

Common Hardwoods and Softwoods

Tree	Use	Class (H—Hard, S—Soft)
Alder	Used in cabinets and furniture. Alder is a less expensive substitute for cherry. It is also known as red alder and fruitwood.	H
Ash	Coarse-grained, hard, and heavy. The color is reddish brown; the heartwood and sapwood are almost white. Used for furniture, cabinets, trim, tool handles, and plywood.	H
Basswood	One of the softest hardwoods. It has an open, coarse grain. The color is creamy white. Used for veneer core and moldings.	H
Beech	Strong, heavy, hard. Takes a good polish, works easily. The color is white to brown. Also used for tool handles, furniture, and plywood.	H
Birch	Very strong, heavy, and fine-grained. It takes an excellent finish. Can be beautifully hand polished. The sapwood is a soft yellow; the heartwood is brown, sometimes slightly colored with red. Due to the climate, birch has a wide variety of color changes. Used for furniture, cabinets, flooring, and plywood.	H
Western red cedar	The thin sapwood layer is a pale yellow. The heartwood varies from a deep dark brown to a very light pink. The wood is light and soft, but lacks strength. It is straight-grained and has high resistance to changes in moisture. After seasoning, it retains its shape and size exceptionally well. Used for shingles, shakes, siding, planking, building logs, and yard lumber.	S
Tennessee cedar	This product is very light, soft, and close-grained. The sapwood is white in color. The heartwood is reddish brown. This cedar has a very strong odor that acts as an insect repellent. Used for trunk linings and closets.	S
Cherry	Cherry is a close-grained and light wood. It takes a good finish and works easily. The sapwood is an off-white. The heartwood is reddish brown. This product can be used for cabinets, interior trim, and furniture.	H
Chestnut	This beautiful, open-grained wood is worked very easily. The sapwood is light brown in color. The heartwood is a dark tan. Can be used for cabinets or plywood and paneling.	H
Cypress	Cypress is a very durable, light product, and is very easily worked. The heartwood is a brilliant yellow; the sapwood is light brown in color. It is used for interior trim and can also be used for cabinet work and exterior trim.	S
Elm	This strong, majestic tree is very hard and the lumber is very heavy. The color of this tree is light to medium brown, with slight variations of red and gray. Used primarily for tool handles and heavy construction purposes.	H
Douglas fir	Fir is very strong and durable. The sapwood is slightly off-white and the heartwood varies in color from a light red to a deep yellow. Fir is used for all types of construction and in the manufacturing of plywood.	S
Western hemlock	This lumber is straight-grained and fine-textured, and it works easily. It has very good qualities for gluing. It is almost totally free from pitch and is often interchanged with fir. Hemlock is used for general construction.	S
Gum	This soft-textured, close-grained tree is very durable. The color of the heartwood varies from red to brown. The sapwood is yellowish white. It closely resembles walnut in color. This wood works easily, but has a tendency to warp. It is sometimes substituted for walnut in the manufacture of furniture.	H
Larch	This product is strong, fairly hard, and resembles Douglas fir more than any other softwood. This lumber is also suitable for general construction purposes.	H
Mahogany	This lumber is an open-pored, strong, durable product. Mahogany can be worked easily. The color is basically a reddish brown. Poor-grade mahogany has small gray spots or flecks. Much mahogany is from overseas. Can be used for paneling, furniture, plywood, and interior trim.	H
Sugar maple	This lumber is very difficult to work with, but it is very strong and can take an excellent finish. The product is very close-grained and light brown to dark brown in color. Furniture and flooring are some of its uses.	H

Instructor's Notes:

Tree	Use	Class (H—Hard, S—Soft)
Silver maple	This lumber is not very strong or durable. It is soft and light, and it works easily. This lumber is frequently used for turning and interior trim.	H
White oak	White oak is very hard and heavy. It has a close grain with open pores. The product is very difficult to work with, but it takes an excellent finish. The heartwood is tan in color, with a light sapwood. Can be used for flooring, plywood, interior trim, and furniture.	H
Red oak	Red oak has a coarser grain than white oak and is also a little softer. The variation in color is from a light tan to a medium brown. The uses for red oak are furniture, plywood, bearing posts, interior trim, and paneling. It could also be used for cabinets.	H
White pine	This lumber is very easily worked. It is soft, light, and very durable. The heartwood is a very slightly yellowish white. Can be used for construction (code permitting) and is also used in furniture and millwork.	S
Sugar pine	This is a light and very soft lumber. It is uniform in texture and works easily. The sapwood is yellow. The heartwood is light red to medium brown. Sugar pine is mostly quarter-sawed and has specks of reddish brown, which makes it easy to recognize. Can be used for inexpensive furniture and interior millwork.	S
Ponderosa pine	This pine is fairly soft and uniform in texture. Besides being strong, it works smoothly without splintering. The sapwood is a light pale yellow with a darker heartwood. Ponderosa pine is ideal for woodworking in the shop and is common in door and window frames and moldings.	S
Lodgepole pine	This pine lumber seasons very easily and is of moderate durability. This product is soft and straight-grained with a fine, uniform texture. The sapwood is almost white. This product is used largely in making propping timbers for miners.	S
Southern yellow pine	This lumber is also called southern pine or yellow pine. It is a very difficult product to work with. It is strong, hard, and tough. The heartwood is orange-red in color, with the sapwood being lighter in color. This product is used for heavy structural purposes such as floor planks, beams, and timbers.	S
Poplar	This product is uniform in texture, light, and soft. The color varies from white to yellow. This lumber is easy to work with. Can be used for inexpensive furniture, crates, plywood, and moldings.	H
Redwood	The redwood is the largest tree in North America. Redwood is easily worked, light, and coarse-grained. The color is dull red. Although very beautiful, redwood lacks strength. It is used primarily for interior and exterior trim, planking for patio decks, and shingles.	S
White spruce	This lumber is soft, works very easily, and is very lightweight. There is very little contrast between the heartwood and sapwood. The color is a pale yellow. It is not very durable, but among the softwoods it is one of medium strength. In certain situations, it is vulnerable to decay. White spruce has value as pulp wood and is used in light construction.	S
Engelmann spruce	This type of spruce takes a good finish. The lumber is very light and straight-grained. This species is usually larger than white spruce and has a larger percentage of clear lumber with little or no defects. It is used for oars, paddles, sounding boards for musical instruments, and construction work.	S
Sitka spruce	These trees have a straight grain and grow very tall. This species is a very tough, strong wood. The color is a very creamy white with a pinkish or light red tinge. This species resists shattering and splintering. It is used primarily in constructing masts, spars, scaffolding, and general construction (code permitting).	S
Walnut	This product is durable, very hard, strong, and it can easily be finished to a beautiful luster. The sapwood is a light brown to a medium brown. The heartwood is a reddish brown to a dark brown. This product is used in the manufacture of fine cabinets, furniture, flooring, plywood, and interior trim.	H

Appendix B

Softwood Categories

102A01.EPS

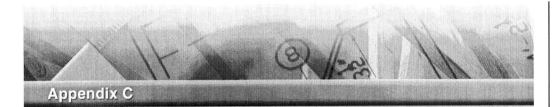

Appendix C

Common Abbreviations Used in the Lumber Industry

AD	Air Dried	E & CB2S	Edge and Center Bead on Two Sides
ADF	After Deducting Freight	E & CV1S	Edge and Center V on One Side
ALS	American Lumber Standard	E & CV2S	Edge and Center V on Two Sides
AST	Anti-Stain Treated		
AVG	Average	FA	Facial Area
B & B	B and Better	Fac.	Factory
B & S	Beams and Stringers	FBM	Feet Board Measure
Bd.	Board	FG	Flat (slash) Grain
BF or Bd. Ft.	Board Feet	Flg.	Flooring
Bdl.	Bundle or Bundled	FOB	Free on Board
B/L	Bill of Lading	FOHC	Free of Heart Centers
Bev.	Bevel or Beveled	FOK	Free of Knots
CB1S	Center Bead on One Side	Frt.	Freight
CB2S	Center Bead on Two Sides	Ft.	Foot or Feet
CC	Cubical Content	G	Girth
Cft. or Cu. Ft.	Cubic Feet or Foot	GM	Grade Marked
CG2E	Center Groove on Two Edges	G/R	Grooved Roofing
CIF	Cost, Insurance, and Freight	HB	Hollow Back
CIFE	Cost, Insurance, Freight, and Exchange	Hrt.	Heart
Clg.	Ceiling	H & M	Hit and Miss
Com.	Common	in.	Inch or Inches
Csg.	Casing	Ind.	Industrial
Ctr.	Center	J & P	Joints and Planks
CV1S	Center V on One Side	KD	Kiln Dried
CV2S	Center V on Two Sides	Lbr.	Lumber
D & M	Dressed and Matched	LCL	Less than Carload
DB Clg.	Double-Beaded Ceiling	Lft.	Lineal Foot or Feet
DB Part	Double-Beaded Partition	Lin.	Lineal
DET	Double-End-Trimmed	LL	Longleaf
Dim.	Dimension	Lng.	Lining
Dkg.	Decking	M	Thousand
D/sdg	Drop Siding	MBM	Thousand (feet) Board Measure
EB1S	Edge Bead on One Side	M.C.	Moisture Content
EB2S	Edge Bead on Two Sides	Merch.	Merchantable
EE	Eased Edges	MG	Medium Grain
EG	Edge Vertical or Rift Grain	MLDG.	Molding
EM	End Matched	Mft.	Thousand Feet
EV1S	Edge V on One Side	No.	Number
EV2S	Edge V on Two Sides	N1E or N2E	Nosed on One or Two Edges
E & CB1S	Edge and Center Bead on One Side	Ord.	Order
		Par.	Paragraph

MODULE 27102-06 ♦ BUILDING MATERIALS, FASTENERS, AND ADHESIVES 2.61

Part.	Partition	STR	Structural
Pat.	Pattern	SYP	Southern Yellow Pine
Pcs.	Pieces	S & E	Side and Edge (surfaced on)
PE	Plain End	S1E	Surfaced on One Edge
P & T	Post and Timbers	S2E	Surfaced on Two Edges
P1S and P2S	See S1S and S2S	S1S	Surfaced on One Side
Reg.	Regular	S2S	Surfaced on Two Sides
Rfg.	Roofing	S1S1E	Surfaced on One Side and One Edge
Rgh.	Rough		
R/L	Random Length	S2S2E	Surfaced on Two Sides and Two Edges
RES	Resawn		
Sdg.	Siding	S2S1E	Surfaced on Two Sides and One Edge
S E	Square Edge		
Sel.	Select	S2S & CM	Surfaced on Two Sides and Center Matched
S E & S	Square Edge and Sound		
SL	Shiplap	S2S - SL	Surfaced on Two Sides and Shiplapped
SM	Surface Measure		
Specs.	Specifications	S2S & SM	Surfaced on Two Sides and Standard Matched
SR	Stress Rated		
Std.	Standard	S4S	Surfaced on Four Sides
Std. Lgths.	Standard Lengths	T & G	Tongued and Grooved
STD M	Standard Matched	Wdr.	Wider
SSND	Sap Stained No Defects (stained)	Wt.	Weight
		YP	Yellow Pine

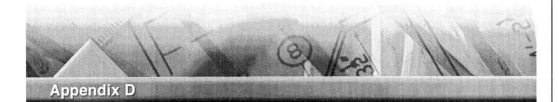

Appendix D

Lumber Conversion Tables

Nominal and Actual Softwood Lumber Sizes	
Nominal (Rough)	Actual (S4S)
1" × 4"	¾" × 3½"
1" × 6"	¾" × 5½"
1" × 8"	¾" × 7¼"
1" × 10"	¾" × 9¼"
1" × 12"	¾" × 11¼"
2" × 4"	1½" × 3½"
2" × 6"	1½" × 5½"
2" × 8"	1½" × 7¼"
2" × 10"	1½" × 9¼"
2" × 12"	1½" × 11¼"
4" × 4"	3½" × 3½"
4" × 6"	3½" × 5½"

Rapid Calculation of Board Measure		
Width	Thickness	Board Feet
3"	1" or less	¼ of the length
4"	1" or less	⅓ of the length
6"	1" or less	½ of the length
9"	1" or less	¾ of the length
12"	1" or less	Same as the length
15"	1" or less	1¼ of the length

Nominal and Actual Hardwood Lumber Sizes		
Nominal (Rough)	Actual (Plain)	Actual (S4S)
½"	⅜"	⁵⁄₁₆"
⅝"	⁷⁄₁₆"	⁷⁄₁₆"
¾"	⁹⁄₁₆"	⁹⁄₁₆"
1"	²⁵⁄₃₂"	²⁵⁄₃₂"
1¼"	1¹⁄₁₆"	1½"
1½"	1⁵⁄₁₆"	1⁹⁄₃₂"
2"	1¾"	1¹¹⁄₁₆"
2½"	2¼"	2¼"
3"	2¾"	2¾"
4"	3¾"	3¾"

Board Feet

Nominal Size (Inches)	Actual Length (in Feet)								
	8	10	12	14	16	18	20	22	24
1 × 2		1⅔	2	2⅓	2⅔	3	3⅓	3⅔	4
1 × 3		2½	3	3½	4	4½	5	5½	6
1 × 4	2¾	3⅓	4	4⅔	5⅓	6	6⅔	7⅓	8
1 × 5		4⅙	5	5⅚	6⅔	7½	8⅓	9⅙	10
1 × 6	4	5	6	7	8	9	10	11	12
1 × 7		5⅝	7	8⅙	9⅓	10½	11⅔	12⅚	14
1 × 8	5⅓	6⅔	8	9⅓	10⅔	12	13⅓	14⅔	16
1 × 10	6⅔	8⅓	10	11⅔	13⅓	15	16⅔	18⅓	20
1 × 12	8	10	12	14	16	18	20	22	24
1¼ × 4		4⅙	5	5⅚	6⅔	7½	8⅓	9⅙	10
1¼ × 6		6¼	7½	8¾	10	11¼	12½	13¾	15
1¼ × 8		8⅓	10	11⅔	13⅓	15	16⅔	18⅓	20
1¼ × 10		10 5/12	12½	14 7/12	16⅔	18¾	20⅚	22 11/12	25
1¼ × 12		12½	15	17½	20	22½	25	27½	30
1½ × 4	4	5	6	7	8	9	10	11	12
1½ × 6	6	7½	9	10½	12	13½	15	16½	18
1½ × 8	8	10	12	14	16	18	20	22	24
1½ × 10	10	12½	15	17½	20	22½	25	27½	30
1½ × 12	12	15	18	21	24	27	30	33	36
2 × 4	5⅓	6⅔	8	9⅓	10⅓	12	13⅓	14⅔	16
2 × 6	8	10	12	14	16	18	20	22	24
2 × 8	10⅔	13⅓	16	18⅔	21⅓	24	26⅔	29⅓	32
2 × 10	13⅓	16⅔	20	23⅓	26⅔	30	33⅓	36⅔	40
2 × 12	16	20	24	28	32	36	40	44	48
3 × 6	12	15	18	21	24	27	30	33	36
3 × 8	16	20	24	28	32	36	40	44	48
3 × 10	20	25	30	35	40	45	50	55	60
3 × 12	24	30	36	42	48	54	60	66	72
4 × 4	10⅔	13⅓	16	18⅔	21⅓	24	26⅔	29⅓	32
4 × 6	16	20	24	28	32	36	40	44	48
4 × 8	21⅓	26⅔	32	37⅓	42⅔	48	53⅓	58⅔	64
4 × 10	26⅔	33⅓	40	46⅔	53⅓	60	66⅔	73⅓	80
4 × 12	32	40	48	56	64	72	80	88	96

Instructor's Notes:

MODULE 27102-06 — ANSWERS TO REVIEW QUESTIONS

Answer	Section
1. b	4.0.0, Figure 4
2. d	4.1.0
3. c	4.1.1
4. c	5.2.2
5. a	6.1.0
6. c	6.2.0
7. d	7.3.0
8. b	7.5.0
9. b	8.0.0
10. c	9.0.0
11. d	9.0.0
12. b	12.0.0; Figure 26
13. d	13.1.0
14. c	16.1.0
15. a	18.2.0

MODULE 27102-06 — ANSWERS TO TRADE TERMS QUIZ

1. Gypsum
2. Nail set
3. Sheathing
4. Millwork
5. Shiplap
6. Tongue-and-groove
7. Material safety data sheet (MSDS)
8. Rafters
9. Sill plate
10. Galvanized
11. Japanned
12. Catalyst
13. Joists
14. Butt joint
15. Cantilever
16. Vaulted ceiling
17. Resins

MODULE 27102-06 — TEACHING TIPS

The following are suggested activities or instructional methods to help you teach the material in this module.

General

When you call on someone to answer a question, the rest of the class relaxes or even tunes out because they expect that the question and answer will take place only between you and the trainee you called on. Instead, use this technique to involve more trainees in answering questions and to keep them on their toes.

1. Ask trainees to define a term or explain a concept.
2. After one trainee has answered, ask a trainee seated nearby if the answer is right. Then ask whether a trainee in the back of the room agrees.
3. Ask trainees to explain why they think an answer is right or wrong.
4. Use the session to clear up incorrect ideas and encourage trainees to learn from their mistakes.

CONTREN® LEARNING SERIES — USER UPDATE

NCCER makes every effort to keep these textbooks up-to-date and free of technical errors. We appreciate your help in this process. If you have an idea for improving this textbook, or if you find an error, a typographical mistake, or an inaccuracy in NCCER's Contren® textbooks, please write us, using this form or a photocopy. Be sure to include the exact module number, page number, a detailed description, and the correction, if applicable. Your input will be brought to the attention of the Technical Review Committee. Thank you for your assistance.

Instructors – If you found that additional materials were necessary in order to teach this module effectively, please let us know so that we may include them in the Equipment/Materials list in the Annotated Instructor's Guide.

Write: Product Development and Revision
National Center for Construction Education and Research
P.O. Box 141104, Gainesville, FL 32614-1104

Fax: 352-334-0932

E-mail: curriculum@nccer.org

Craft _____ Module Name _____

Copyright Date _____ Module Number _____ Page Number(s) _____

Description _____

_____

_____

_____

(Optional) Correction _____

_____

_____

(Optional) Your Name and Address _____

_____

_____

Hand and Power Tools
27103-06

NCCER STANDARDIZED CRAFT TRAINING PROGRAM

The National Center for Construction Education and Research (NCCER) provides a standardized national program of accredited craft training. Key features of the program include instructor certification, competency-based training, and performance testing. The program provides trainees, instructors, and companies with a standard form of recognition through the National Registry. The program is described in full in the *Guidelines for Accreditation*, published by NCCER. For more information on standardized craft training, contact NCCER by writing to P.O. Box 141104, Gainesville, FL 32614-1104; calling 352-334-0911; or e-mailing **info@nccer.org**. More information is available at **www.nccer.org**.

HOW TO USE THIS ANNOTATED INSTRUCTOR'S GUIDE

Each page presents two sections of information. The larger section displays each page exactly as it appears in the Trainee Module. The narrow column ties suggested trainee and instructor actions to each page and provides icons (detailed below) to call your attention to material, safety, audiovisual, or testing requirements. The bottom of each page includes space for your notes.

 The **Audiovisual** icon indicates an appropriate time to show a transparency or other audiovisual aid.

 The **Classroom** icon prompts you to define a term, stress a point, ask trainees to explain a concept, or give examples.

 The **Demonstration** icon directs you to show trainees how to perform tasks.

 The **Examination** icon tells you to administer the written module examination.

 The **Homework** icon is placed where you may wish to assign reading for the next class, assign a project, or advise trainees to prepare for an examination.

 The **Laboratory** icon is used when trainees are to practice performing tasks.

 The **Materials** icon is a reminder for you to gather materials needed for classes, laboratories, and testing.

 The **Performance Testing** icon tells you to administer a performance test or a portion thereof.

 The **Safety** icon is used to emphasize safety issues. It is often keyed to *Caution* and *Warning!* statements in the Trainee Module.

 The **Teaching Tip** icon indicates additional guidance is available, such as how to conduct an exercise, get the most educational value from a field trip, or encourage class participation. Teaching Tips may expand on a feature (*Think About It*, *Did You Know?*) or provide *Quick Quizzes* or similar exercises. You will be referred to the Teaching Tips section at the back of the module if there is additional material.

 The **Combination** icon indicates that the laboratory listed corresponds with a performance task. If desired, you can note the proficiency of the trainees during the laboratory, and use it to satisfy performance testing requirements.

PREPARATION

Before teaching this module, you should review the Objectives, Performance Tasks, Materials and Equipment List, and Module Outline. Be sure to allow ample time to prepare your own training or lesson plan and gather all required materials and equipment.

Hand and Power Tools
Annotated Instructor's Guide

Module 27103-06

MODULE OVERVIEW

This module expands upon the hand and power tool information provided in the *Core Curriculum* and introduces the carpentry trainee to additional tools used in the carpentry trade.

PREREQUISITES

Prior to training with this module, it is suggested that the trainee shall have successfully completed *Core Curriculum; Carpentry Fundamentals Level One*, Modules 27101-06 and 27102-06.

OBJECTIVES

Upon completion of this module, the trainee will be able to do the following:

1. Identify the hand tools commonly used by carpenters and describe their uses.
2. Use hand tools in a safe and appropriate manner.
3. State the general safety rules for operating all power tools, regardless of type.
4. State the general rules for properly maintaining all power tools, regardless of type.
5. Identify the portable power tools commonly used by carpenters and describe their uses.
6. Use portable power tools in a safe and appropriate manner.

PERFORMANCE TASKS

Under the supervision of the instructor, the trainee should be able to do the following:

1. Demonstrate the safe and proper use of the following hand tools:
   - Levels
   - Squares
   - Planes
   - Clamps
   - Saws
2. Demonstrate or describe the safe and proper use of five of the following power tools:
   - Circular saw
   - Portable table saw
   - Compound miter saw
   - Frame and trim saw
   - Drill press
   - Router/laminate trimmer
   - Portable power plane
   - Power metal shears
   - Pneumatic nailer/stapler

MATERIALS AND EQUIPMENT LIST

Transparencies
Markers/chalk
Blank acetate sheets
Transparency pens
Pencils and scratch paper
Overhead projector and screen
Whiteboard/chalkboard
Appropriate personal protective equipment
Soapstone
Yard-long lengths of 1" reinforcing rod
1 × 4 stock about 18" to 24" long
2 × 4s 18" to 24" long
2 × 4s 4' long
6" × 12" pieces of ¾" plywood
Pieces of crown molding 4' long
Angle iron, steel rod, or pipe for cutting
Wood stock of various sizes
Laminate samples
Blocks of scrap wood
Fasteners (nails and staples) designed for the pneumatic fastener being used
Sheet metal stock
Folding rule or steel tape
Levels:
   Line
   Water
   Builder's
   Transit
   Laser
Squares:
   Try
   Sliding T-bevel
   Speed square
   Miter
   Framing
   Adjustable T-square
Planes:
   Block
   Jack
   Smoothing
   Jointer
Clamps:
   Web
   Hand-screw
   Bar
   Spring
   Locking C
   Pipe
Saws:
   Hacksaw and replacement blades
   Backsaw
   Dovetail
   Compass
   Coping
Chalkline
Clamping device
Portable circular saw
Circular saw protractor
Portable table saw
Miter/compound miter saw
Frame and trim saw
Demolition saw
Chop saw
Miter gauge
Ripping fence for portable circular saw
Push stick
Sawhorses or other solid support
Drill press and chuck key
Portable power plane and blades
Power metal shears
Router and router bits
Laminate trimmer and bits
Pneumatic fastener and manufacturer's instruction manual
Electric air compressor with air hose
Copies of Worksheet 1*
Copies of Job Sheets 1 through 7*
Module Examinations**
Performance Profile Sheets**

* Packaged with this Annotated Instructor's Guide.
**Located in the Test Booklet.

SAFETY CONSIDERATIONS

Ensure that the trainees are equipped with appropriate personal protective equipment.

ADDITIONAL RESOURCES

This module is intended to present thorough resources for task training. The following reference works are suggested for both instructors and motivated trainees interested in further study. These are optional materials for continued education rather than for task training.

The Art of Fine Tools. Newton, CT: Taunton Press, Inc.
Field Guide to Tools. Philadelphia, PA: Quirk Publishing.
Measure Twice, Cut Once. Boston, MA: Little, Brown & Company.
Power Tools. Newton, CT: Taunton Press, Inc.
Selecting and Using Hand Tools. Newton, CT: Taunton Press, Inc.
Tools Rare and Ingenious: Celebrating the World's Most Amazing Tools. Newton, CT: Taunton Press, Inc.
Tricks of the Trade: Jigs, Tools and Other Labor-Saving Devices. Newton, CT: Taunton Press, Inc.
Black & Decker. **www.blackanddecker.com**
Bosch Tool Corporation. **www.boschtools.com**
Delta Machinery. **www.deltamachinery.com**
DeWalt Industrial Tool Company. **www.dewalt.com**
Makita Tools USA. **www.makita.com**
Milwaukee Electric Tool Company. **www.milwaukeetool.com**
Porter-Cable Corporation. **www.portercable.com**
Ridge Tool Company. **www.ridgid.com**
The Stanley Works. **www.stanleytools.com**
L.S. Starrett Company. **www.starrett.com**

TEACHING TIME FOR THIS MODULE

An outline for use in developing your lesson plan is presented below. Note that each Roman numeral in the outline equates to one session of instruction. Each session has a suggested time period of 2½ hours. This includes 10 minutes at the beginning of each session for administrative tasks and one 10-minute break during the session. Approximately 10 hours are suggested to cover *Hand and Power Tools*. You will need to adjust the time required for hands-on activity and testing based on your class size and resources. Because laboratories often correspond to Performance Tasks, the proficiency of the trainees may be noted during these exercises for Performance Testing purposes.

Topic   **Planned Time**

Session I. Introduction; Hand Tools
    A. Introduction   _____
    B. Hand Tools   _____
        1. Levels   _____
        2. Laboratory   _____
            Under your supervision, have the trainees practice using various levels. This laboratory corresponds to Performance Task 1.
        3. Squares   _____
        4. Laboratory   _____
            Hand out Job Sheet 27103-1. Under your supervision, have the trainees perform the tasks on the Job Sheet. This laboratory corresponds to Performance Task 1.

5. Planes
6. Laboratory

    Under your supervision, have the trainees practice using various planes. This laboratory corresponds to Performance Task 1.

7. Clamps
8. Laboratory

    Under your supervision, have the trainees practice using various clamps. This laboratory corresponds to Performance Task 1.

9. Saws
10. Laboratory

    Hand out Job Sheet 27103-2. Under your supervision, have the trainees perform the tasks on the Job Sheet. Note the proficiency of each trainee.

Session II. Guidelines for Using All Power Tools; Power Saws

A. Guidelines for Using All Power Tools
   1. Safety Rules Pertaining to All Power Tools
   2. Guidelines Pertaining to the Care of All Power Tools
B. Power Saws
   1. Circular Saws
   2. Laboratory

       Hand out Worksheet 27103-1 and Job Sheets 27103-3 and -4. Under your supervision, have the trainees complete the circular saw safety test prior to performing the related tasks on the Job Sheets. Note the proficiency of each trainee.

   3. Portable Table Saws
   4. Laboratory

       Hand out Worksheet 27103-1 and Job Sheets 27103-3 and -4. Under your supervision, have the trainees complete the table saw safety test before performing the related tasks on the Job Sheets. Note the proficiency of each trainee.

   5. Power Miter Saws/Compound Miter Saws
   6. Laboratory

       Hand out Worksheet 27103-1 and Job Sheet 27103-5. Under your supervision, have the trainees complete the compound miter saw safety test before performing the related tasks on the Job Sheet. Note the proficiency of each trainee.

   7. Frame and Trim Saws
   8. Laboratory

       Hand out Worksheet 27103-1 and Job Sheet 27103-5. Under your supervision, have the trainees complete the frame and trim saw safety test before performing the related tasks on the Job Sheet. Note the proficiency of each trainee.

   9. Abrasive Saws
   10. Power Saw Blades

Session III. Drill Press; Routers/Laminate Trimmers; Portable Power Planes; Power Metal Shears; Pneumatic/Cordless Nailers and Staplers

A. Drill Press
   1. Laboratory

       Hand out Worksheet 27103-1 and Job Sheet 27103-6. Under your supervision, have the trainees complete the drill press safety test prior to performing the related tasks on the Job Sheet. Note the proficiency of each trainee.

B. Routers/Laminate Trimmers

1. Laboratory

   Hand out Worksheet 27103-1. Under your supervision, have the trainees complete the router and laminate trimmer safety test before operating these tools. Note the proficiency of each trainee.

C. Portable Power Planes

1. Laboratory

   Hand out Worksheet 27103-1. Under your supervision, have the trainees complete the power plane safety test before operating these tools. Note the proficiency of each trainee.

D. Power Metal Shears

1. Laboratory

   Hand out Worksheet 27103-1. Under your supervision, have the trainees complete the power metal shears safety test before using the shears. Note the proficiency of each trainee.

E. Pneumatic/Cordless Nailers and Staplers

1. Laboratory

   Hand out Worksheet 27103-1 and Job Sheet 27103-7. Under your supervision, have the trainees complete the pneumatic fasteners safety test before performing the related tasks on the Job Sheet. Note the proficiency of each trainee.

Session IV. Review; Module Examination and Performance Testing

A. Review

B. Module Examination
   1. Trainees must score 70 percent or higher to receive recognition from NCCER.
   2. Record the testing results on Craft Training Report Form 200, and submit the results to the Training Program Sponsor.

C. Performance Testing
   1. Trainees must perform each task to the satisfaction of the instructor to receive recognition from NCCER. If applicable, proficiency noted during laboratory exercises can be used to satisfy the Performance Testing requirements.
   2. Record the testing results on Craft Training Report Form 200, and submit the results to the Training Program Sponsor.

Hand and Power Tools
27103-06

An Associated Builders and Contractors 2005 National Craft Championships contestant displays proper, safe use of a circular saw: The workpiece is elevated; he has assumed a balanced stance; he is wearing eye and hearing protection; and he is gripping the saw with both hands.

Instructor's Notes:

Assign reading of Module 27103-06.

27103-06
Hand and Power Tools

Topics to be presented in this module include:

1.0.0	Introduction	3.2
2.0.0	Hand Tools	3.2
3.0.0	Power Tools	3.9

Overview

A carpenter is only as good as his or her tools. You must have a good set of tools, know how to use them properly, and take good care of them. You must also know the proper tool to use for each task. This is especially important when it comes to selecting saw blades and drill bits. Choosing the wrong tool can damage the material you are working on.

A carpenter uses a variety of power tools including power saws and pneumatic nail guns. These tools are dangerous to the user and others if they are not handled correctly. It is of utmost importance that you be properly trained in using these tools and that you remain conscious of the related safety concerns every time you use them.

Instructor's Notes:

Objectives

When you have completed this module, you will be able to do the following:

1. Identify the hand tools commonly used by carpenters and describe their uses.
2. Use hand tools in a safe and appropriate manner.
3. State the general safety rules for operating all power tools, regardless of type.
4. State the general rules for properly maintaining all power tools, regardless of type.
5. Identify the portable power tools commonly used by carpenters and describe their uses.
6. Use portable power tools in a safe and appropriate manner.

Trade Terms

Bevel cut
Compound cut
Crosscut
Dado
Door jack
Dovetail joint
Kerf
Kickback
Miter box
Miter cut
Mortise
Pocket (plunge) cut
Rabbet cut
Rip cut
Tenon
True

Required Trainee Materials

1. Pencil and paper
2. Appropriate personal protective equipment

Prerequisites

Before you begin this module, it is recommended that you successfully complete *Core Curriculum* and *Carpentry Fundamentals Level One*, Modules 27101-06 and 27102-06.

This course map shows all of the modules in *Carpentry Fundamentals Level One*. The suggested training order begins at the bottom and proceeds up. Skill levels increase as you advance on the course map. The local Training Program Sponsor may adjust the training order.

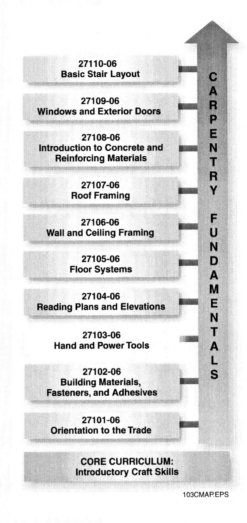

Ensure you have everything required to teach the course. Check the Materials and Equipment List at the front of this module.

Show Transparency 1, Objectives.

Show Transparency 2, Performance Tasks.

Explain that terms shown in bold (blue) are defined in the Glossary at the back of this module.

See the general Teaching Tip at the end of this module.

MODULE 27103-06 ◆ HAND AND POWER TOOLS 3.1

Provide a brief overview of hand tools.

Discuss levels and their applications. Provide examples for the trainees to examine.

1.0.0 ◆ INTRODUCTION

In the *Core Curriculum,* you were introduced to many of the hand and power tools that are commonly used by all trades. This module builds on that information and introduces some new tools widely used by carpenters on a construction site. The focus of this module is on familiarizing you with each of these tools. Safety rules for the use of the power tools are also covered. While under the supervision of your instructor and/or supervisor, you will learn how to properly operate and use each of the tools described in this module.

2.0.0 ◆ HAND TOOLS

In this module, you will be introduced to some additional hand tools and their uses. Keep in mind that this material is introductory; you will receive specific training in the use of these tools as your training progresses.

2.1.0 Levels

In addition to the spirit level covered in the *Core Curriculum,* you will also use the leveling tools and instruments shown in *Figure 1*.

- *Line level* – A line level can be used to level a long span. It consists of a glass tube mounted in a sleeve that has a hook on either end. The hooks are attached at the center of a stretched line, which is moved up or down until the bubble is centered.
- *Water level* – A water level is a simple, accurate tool consisting of a length of clear plastic tubing. It works on the principle that water seeks its own level. The water level is somewhat limited in distance because of the tubing. It can be used effectively for checking the level around obstructions. An example is checking the level from one room to the next when a wall is in the way.
- *Builder's level* – The builder's level is basically a telescope with a spirit level mounted on top. It can be rotated 360 degrees, but cannot be tilted up and down. It is used to check grades and elevations and to set up level points over long distances.
- *Transit level* – A transit level, commonly called a transit, is similar to the builder's level. Unlike the builder's level, the transit level's telescope can be moved up and down 45 degrees, making more operations possible than with a builder's level. When the telescope is locked in a level

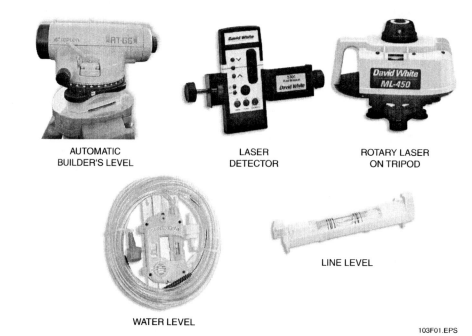

Figure 1 ◆ Levels.

3.2 CARPENTRY FUNDAMENTALS LEVEL ONE ◆ TRAINEE GUIDE

Instructor's Notes:

 *Laser Levels*

A wide variety of handheld laser leveling devices are now on the market. The level shown here combines the traditional spirit level with a laser level. It can be switched on to transmit a fixed laser beam. This particular model extends the level's reference line to 100 feet with an accuracy of ±¼".

103SA01.EPS

Discuss the safety precautions associated with the use of laser levels.

Demonstrate how to use various levels.

Under your supervision, have the trainees practice using various levels. This laboratory corresponds to Performance Task 1.

Discuss various types of squares and their applications. Provide examples for the trainees to examine.

Show Transparency 3 (Figure 2).

position, the transit level can be used to perform all the functions of a builder's level. Movement of the telescope in the up and down positions allows the transit level to be used to plumb columns, building corners, or any other vertical members.

- *Laser level* – A laser level can be used to perform all of the tasks that can be performed with a conventional transit level. The laser level does not depend on the human eye. It emits a high-intensity light beam, which is detected by an electronic laser detector (target) at distances up to 1,000'. Both fixed and rotating laser models are available. Rotating models enable one person, instead of two, to perform any layout operation. When a rotating laser is operated in the sweep mode, the head rotates through 360 degrees, allowing the laser beam to sweep multiple sensors placed at different locations.

 WARNING!
Lasers can cause serious damage to your eyes. NEVER point a laser at anyone or look directly at a laser beam.

2.2.0 Squares

Establishing squareness is an essential part of carpentry; therefore, several types of squares are needed. You have already been introduced to the combination square. *Figure 2* shows a few more. Squares are made from steel, aluminum, and plastic and are available in both standard and metric measurements.

- *Combination square* – A combination square is used to lay out lines at 45-degree and 90-degree angles. The square can be adjusted by sliding the adjustable edge along the length of the blade.
- *Try square* – A try square can be used to lay out a line at a right angle (90 degrees) to an edge or surface and check the squareness of adjoining surfaces and of planed lumber.
- *Miter square* – A miter square has a blade fixed at 45 degrees for laying out and marking lines such as for a **miter cut**. The opposite angle is fixed at 135 degrees.
- *Sliding T-bevel* – A sliding T-bevel is an adjustable gauge for setting, testing, and transferring angles, such as when making a miter or **bevel cut**. The metal blade pivots and can be locked at any angle.

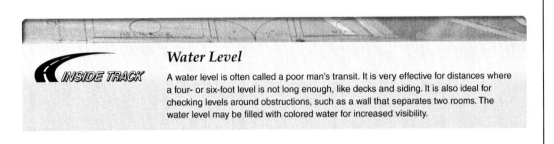 *Water Level*

A water level is often called a poor man's transit. It is very effective for distances where a four- or six-foot level is not long enough, like decks and siding. It is also ideal for checking levels around obstructions, such as a wall that separates two rooms. The water level may be filled with colored water for increased visibility.

Demonstrate how to use a framing square.

Hand out Job Sheet 27103-1. Under your supervision, have the trainees perform the tasks on the Job Sheet. This laboratory corresponds to Performance Task 1.

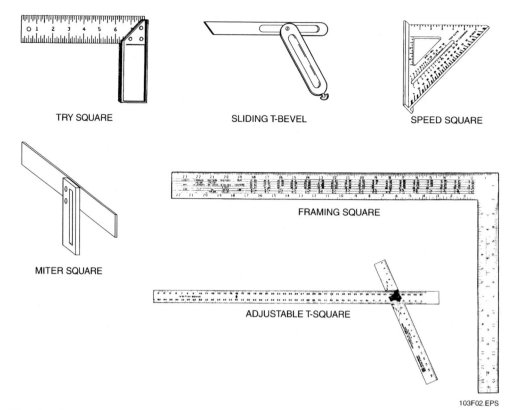

Figure 2 ◆ Squares.

- *Drywall square* – A drywall square, also called a T-square, is used for laying out square and angled lines on large surfaces such as plywood, paneling, wallboard, etc.
- *Framing square* – A framing square, also called a rafter scale or carpenter's square, is used for squaring lines and marking 45-degree lines across wide boards and checking inside corners for squareness. The two surfaces of the square are imprinted with a number of useful tables and formulas, including a rafter table and an Essex board measure table.
- *Speed square* – A speed square, also called a rafter angle square, is used to mark and check 90-degree and 45-degree angles. It is especially useful when laying out angle cuts for roof rafters and stairs. Standard speed squares are 6" triangular tools. The large triangle has a 6" scale on one edge, a full 90-degree scale on another edge, and a T-bar on the third edge. The small triangle has a 2½" scale on one side. There are also 12" speed squares available for stair layout.

3.4 CARPENTRY FUNDAMENTALS LEVEL ONE ◆ TRAINEE GUIDE

Instructor's Notes:

Keeping It Level

Make sure that you are very careful when handling and using a level. Levels are precision instruments and should be treated as such. In order to provide accurate readings, levels need to be calibrated properly. If they aren't, what appears to be a level reading will actually be slightly off. These errors are magnified over long distances, so a small miscalculation can wind up being a very large problem. Avoid this by taking proper care of your instruments.

LASER LEVEL

TRANSIT LEVEL

Square Accuracy

Using tools that don't provide accurate measurements or readings can result in serious errors on a job site. Squares may be damaged accidentally from time to time. Make sure that you periodically check your squares with a known straightedge. This will ensure the accuracy of your measurements, marks, and cuts.

Show various types of planes and discuss their applications. Provide examples for the trainees to examine.

Show Transparency 4 (Figure 3).

Demonstrate how to use various planes.

Under your supervision, have the trainees practice using various planes. This laboratory corresponds to Performance Task 1.

Show various types of clamps and discuss their applications.

2.3.0 Planes

Planes (*Figure 3*) are used to remove excess wood from surfaces. A sharp steel blade protrudes from the flat bottom edge of the plane. On most planes, the depth of the cut can be adjusted by raising or lowering the blade. It is extremely important that the blade be kept sharp.

- *Block plane* – A block plane is used to plane small pieces, edges, and joint surfaces when a small amount of change is needed. Block planes come in 4" to 7" lengths and 1⅜" to 1⅝" widths. Blades are usually adjustable from 12 degrees to 21 degrees. A version of a block plane, called a low-angle block plane, is made for shaping fine trimwork and end grain. Its blade rests at a 12-degree angle for fine cuts on end and cross grains.
- *Jack plane* – A jack plane is a type of bench plane used to plane rough work. Jack planes are 14" long and 2" wide and have an adjustable blade.
- *Smoothing plane* – A smoothing plane is another type of bench plane. Smoothing planes are used to plane small work to a smooth, even surface without removing too much material. Smoothing planes come in lengths of 7" to 10" and widths of 1⅜" to 2".
- *Jointer plane* – A jointer plane is used to plane and **true** long boards such as the edges of doors. They come in 18" to 30" lengths. The cutting edge is ground straight and is set for a fine cut.

2.4.0 Clamps

In addition to the C-clamp, which was introduced in the *Core Curriculum,* you will use several other types of clamps. Some of the most commonly used clamps are shown in *Figure 4*.

- *Hand-screw clamp* – A hand-screw clamp can be used to hold pieces together while adhesive dries. It is useful for beveled or tapered pieces.
- *Locking C-clamp* – Locking C-clamps come in a variety of sizes up to 24". They are based on the same design as clamping pliers; when the handles are squeezed together, the jaws lock into place. The jaw opening is adjustable.
- *Spring clamp* – This simple clamp is spring-loaded so that the jaws will lock down on the work when the handles are released.
- *Quick clamp* – Quick clamps are heavy-duty clamps that are designed to apply extreme pressure. The jaws are tightened by pulling on the trigger and loosened by pulling on the release lever.

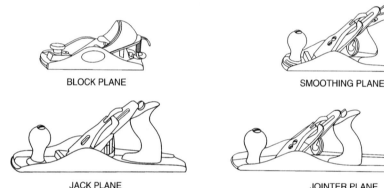

Figure 3 ◆ Planes.

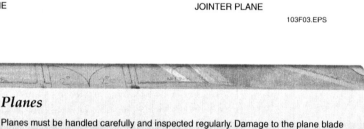

Planes

Planes must be handled carefully and inspected regularly. Damage to the plane blade will result in unsightly grooves in your work. A nail, for example, can cause a sharp burr or take a chunk out of the blade. Always keep your blades sharp and replace them when necessary.

Instructor's Notes:

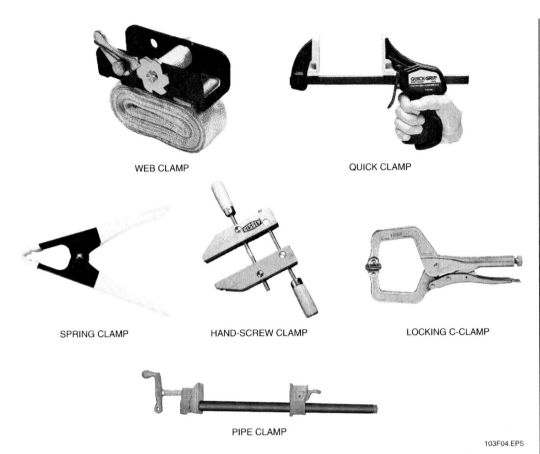

Figure 4 ◆ Clamps.

- *Web (strap) clamp* – Web clamps are designed to secure round, oval, and oddly shaped work. A ratchet assembly is used to tighten the strap.
- *Pipe clamp* – This heavy-duty clamp can also be used as a spreader. Pipe clamps are made to fit ½" or ¾" pipe. The length of the clamp is determined by the length of the pipe. A sliding jaw operates with a spring-locking device, and a middle jaw is controlled by a screw set in a third, fixed jaw. A reversing pipe clamp has a sliding jaw that can also be used in the reverse direction for spreading by applying pressure away from the clamp instead of between the clamp jaws.

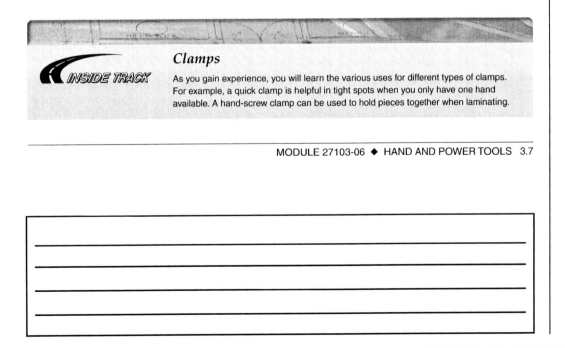

Clamps

As you gain experience, you will learn the various uses for different types of clamps. For example, a quick clamp is helpful in tight spots when you only have one hand available. A hand-screw clamp can be used to hold pieces together when laminating.

MODULE 27103-06 ◆ HAND AND POWER TOOLS 3.7

Discuss various types of saws and their applications. Provide examples for the trainees to examine.

Demonstrate how to use a hacksaw and replace its blade.

Hand out Job Sheet 27103-2. Under your supervision, have the trainees perform the tasks on the Job Sheet. This laboratory corresponds to Performance Task 1.

Assign reading of Sections 3.0.0–3.2.6 for the next class session.

2.5.0 Saws

In the *Core Curriculum,* you were introduced to the common crosscut and rip saws. Carpenters also use many other types of saws, such as those shown in *Figure 5*. Keep your saws sharpened and properly protected when not in use.

- *Hacksaw* – A hacksaw is used to cut metal. It has a replaceable blade.
- *Backsaw (miter saw)* – This saw has a strong back and very fine teeth. It is commonly used with a **miter box** to cut trim.
- *Dovetail saw* – This saw, with its very fine teeth, is used to cut very smooth joints, such as **mortise** and **tenon** and **dovetail joints.** *Figure 6* shows a simple mortise and tenon joint. The blade is extremely thin, but has a band across the top edge to stiffen it.
- *Compass (keyhole) saw* – This saw is used for making rough cuts in drywall, plywood, hardboard, and paneling. It is often used to cut access holes for piping and electrical boxes.
- *Coping saw* – The coping saw has a thin, fine-toothed blade and is ideal for making curved and scroll cuts in moldings and trimwork.

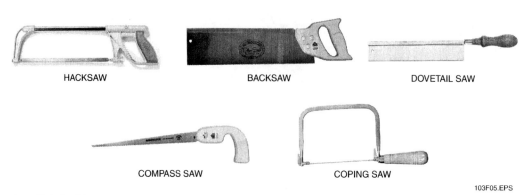

Figure 5 ♦ Special saws.

Inside Track

Miter Box

A backsaw is often used with a miter box. A miter box is either slotted or has holders that guide the saw during a cut. This guided saw movement results in very accurate cuts. Although the miter box was once a necessity for trim work, most carpenters now use a compound miter saw.

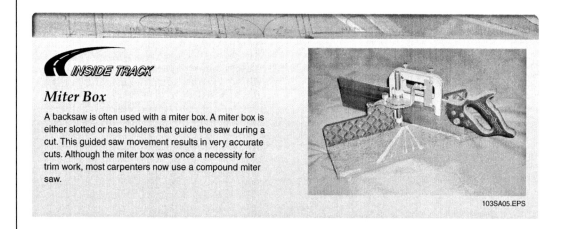

3.8 CARPENTRY FUNDAMENTALS LEVEL ONE ♦ TRAINEE GUIDE

Instructor's Notes:

Figure 6 ♦ Mortise and tenon joint.

3.0.0 ♦ POWER TOOLS

In this section, we will briefly review some of those tools studied earlier in the *Core Curriculum* and introduce you to several new ones and their uses. The intent of this section is to familiarize you with each of the tools and related safety rules. Before you will be allowed to operate a specific power tool, you must be able to show that you know the safety rules for the tool. As your training progresses, you will learn how to operate each of the tools under the supervision of your instructor. Specific operating procedures and safety rules for using a tool are provided in the operator's manual supplied by the manufacturer with each tool. Before operating any power tool for the first time, you should always read this manual to familiarize yourself with the tool. If the manual is missing, you or your supervisor should contact the manufacturer for a replacement.

3.1.0 Guidelines for Using All Power Tools

Before proceeding with our descriptions of power tools, first it is important to overview the general safety rules that apply when using all power tools, regardless of type. It is also important to review the general guidelines that should be followed in order to properly care for power tools.

> **NOTE**
> Power tools may be operated by electricity (AC or DC), air, combustion engines, or explosive powder.

3.1.1 Safety Rules Pertaining to All Power Tools

Rules for the safe use of all power tools include the following:

- Do not attempt to operate any power tool before being checked out by the instructor on that particular tool.
- Always wear eye protection and a hard hat when operating all power tools.
- Wear face and hearing protection when required.
- Wear proper respirator equipment when necessary.
- Wear the appropriate clothing for the job being done. Always wear tight-fitting clothing that cannot become caught in the moving tool. Roll up or button long sleeves, tuck in shirttails, and tie back long hair. Do not wear any jewelry or watches.
- Do not distract others or let anyone distract you while operating a power tool.
- Do not engage in horseplay.
- Do not run or throw objects.
- Consider the safety of others, as well as yourself.
- Do not leave a power tool running while it is unattended.
- Assume a safe and comfortable position before using a power tool.
- Be sure that a power tool is properly grounded and connected to a GFCI circuit before using it.
- Be sure that a power tool is disconnected before performing maintenance or changing accessories.
- Do not use a dull or broken tool or accessory.
- Use a power tool only for its intended use.
- Keep your feet, fingers, and hair away from the blade and/or other moving parts of a power tool.
- Do not use a power tool with guards or safety devices removed or disabled.
- Do not operate a power tool if your hands or feet are wet.
- Keep the work area clean at all times.
- Become familiar with the correct operation and adjustments of a power tool before attempting to use it.
- Keep a firm grip on the power tool at all times.
- Use electric extension cords of sufficient size to service the particular power tool you are using.
- Report unsafe conditions to your instructor or supervisor.

Review the safety guidelines for using power tools.

Review the guidelines for the care of power tools.

Show various types of power saws.

Discuss the uses of a portable circular saw. Provide an example for the trainees to examine.

3.1.2 Guidelines Pertaining to the Care of All Power Tools

The following are some guidelines for the proper care of power tools:

- Keep all tools clean and in good working order.
- Keep all machine surfaces clean and waxed.
- Follow the manufacturer's maintenance procedures.
- Protect cutting edges.
- Keep all tool accessories (such as blades and bits) sharp.
- Always use the appropriate blade for the arbor size.
- Report any unusual noises, sounds, or vibrations to your instructor or supervisor.
- Regularly inspect all tools and accessories.
- Keep all tools in their proper place when not in use.
- Use the proper blade for the job being done.

3.2.0 Power Saws

Power saws, especially the circular saw, are among the carpenter's most commonly used tools. In addition to the reciprocating saw covered in the *Core Curriculum*, there are several types of power saws a carpenter might use on a job site, including the following:

- Circular saws
- Portable table saws
- Power miter saws/compound miter saws
- Frame and trim saws
- Abrasive saws

3.2.1 Circular Saws

Circular saws (*Figure 7*) are versatile, portable saws used to perform the following tasks:

- Ripping (**rip cut**)
- Crosscutting (**crosscut**)
- Mitering
- **Pocket (plunge) cuts**
- Bevel cuts

The size of a circular saw is determined by the diameter of the largest size blade that can be used with the saw, which determines how thick a mate-

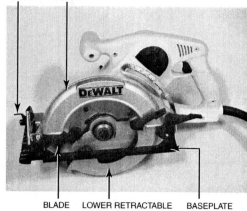

Figure 7 ♦ Circular saw.

rial can be cut. Saws using a 7¼" blade are popular. Circular saws have upper and lower guards that surround the blade. The upper guard is fixed; the lower guard is spring-loaded and retracts as the saw cuts into the workpiece. The saw has a baseplate that rests on the material being cut and can be adjusted to change the depth of the cut or to make bevel cuts ranging from 0 to 45 degrees. The saw is started by pressing the trigger in the handle and stopped by releasing it. It will run only while the trigger is pressed. To make ripping of narrower boards easier, most saws come equipped with an adjustable rip fence that can be attached to the saw baseplate in slots in front of the blade.

 WARNING!

When using a circular saw to cut on an angle, the blade guard may have the tendency to bind. DO NOT reach below the saw to release the guard. Severe injury, such as lost fingers, could result. If you cannot complete the cut, find an alternative saw to finish the job. Never jeopardize your safety for any reason.

Instructor's Notes:

Circular Saws

The circular saw is perhaps the most important tool on a job site. If you are going to be a successful professional carpenter, you must master the use of a circular saw and be able to make all of the different cuts with precision. In order to acquire these skills, you will need to practice. Mastery can only come from extensive practice and experience.

103SA06.EPS

The following are rules for the safe use of a portable circular saw:

- Always wear safety glasses.
- Hold the saw firmly.
- Always start the saw before making contact with the material.
- Keep the electric cord clear of the blade to avoid cutting it.
- Check the condition of the blade and be sure it is secure before starting the saw.
- Be sure the blade guards are in place and working properly.
- Set the blade only deep enough to cut through the stock.
- Check the stock for nails and any other metal before cutting it.
- Allow the saw to reach maximum speed before starting the cut.
- Keep your hands clear of the blade.
- If the blade binds in a cut, stop the saw immediately and hold it in position until the blade stops, then back it out of the cut.
- Stop the saw and lay it on its side after finishing the cut.
- Do not hold stock in your hands while ripping. Secure the material to sawhorses.

 WARNING!
Always unplug any power saw before changing the blade.

Circular saws require frequent blade changes, either to replace a worn-out blade or to change the blade for a different kind of work. The following is a basic procedure for changing a circular saw blade:

Step 1 Using the wrench provided with the saw, remove the arbor bolt and washer by turning the wrench in the direction of the arrow on the blade. Depress the spindle lock, if provided, to keep the blade from turning.

Step 2 Move the blade guard and carefully lift the blade out.

Step 3 Install the new blade, making sure the arrows on the blade point in the same direction as the arrow on the saw.

Step 4 Tighten the arbor bolt and washer, but be careful not to over tighten.

Emphasize the safety precautions associated with the use of circular saws.

Demonstrate how to make straight, angled, and rip cuts using a circular saw.

Hand out Worksheet 27103-1 and Job Sheets 27103-3 and 27103-4. Under your supervision, have the trainees complete the circular saw safety test before performing the related tasks on the Job Sheets. This laboratory corresponds to Performance Task 2.

Discuss portable table saws and their applications. Provide an example for the trainees to examine.

Emphasize the safety precautions associated with the use of portable table saws.

Demonstrate how to make straight, angled, and rip cuts using a portable table saw.

Hand out Worksheet 27103-1 and Job Sheets 27103-3 and 27103-4. Under your supervision, have the trainees complete the table saw safety test before performing the related tasks on the Job Sheets. This laboratory corresponds to Performance Task 2.

3.2.2 Portable Table Saws

Portable table saws (*Figure 8*) are used on job sites to do the following sawing and related woodworking tasks:

- Ripping
- Crosscutting
- Mitering
- Rabbeting (making **rabbet cuts**)
- Dadoing (making **dados**)
- Cutting molding

The table saw consists of a machined flat metal table surface with a slotted table insert in the center through which the circular saw blade extends. The table provides support for the workpiece as it is fed into the rotating blade. The size of a table saw is determined by the diameter of the largest size circular saw blade it can use, with 10" and 12" diameters being common. The depth of the cut is adjusted by moving the blade vertically up or down. The blade can also be tilted to make beveled cuts up to 45 degrees, as indicated on a related degree scale.

For crosscutting, the table is grooved to accept a guide called a miter gauge. The workpiece is held against the vertical face of the miter gauge to keep it at a perfect angle with respect to the blade. Pushing the miter gauge forward moves the workpiece through and past the blade so that each cut is flat and square. Typically, the miter gauge can be turned to any angle up to 60 degrees, with positive stops at 0 and 45 degrees right and left.

For ripping, the saw comes equipped with a rip guide or fence. This clamps to the front and rear edges of the table and has rails that allow it to be moved toward or away from the blade while remaining parallel. It is positioned to the width desired for the work, then locked in place. The workpiece is guided against the fence while making the cut as it is being pushed and/or pulled through the blade.

A blade guard protects the user's hands from the rotating blade and prevents flying chips. At the rear of the blade guard assembly are a **kerf** spreader and an anti-**kickback** device. The kerf spreader acts to keep the kerf in the workpiece open behind the blade. The kerf is the width of the cut made by the saw blade in the workpiece as it is being cut. If the kerf is allowed to close, the blade can become pinched, causing the workpiece to start moving backward while cutting and resulting in possible injury to the operator. If kickback starts to occur, sharp pawls on the anti-kickback device function to grip the workpiece and hold it in place. Rules for the safe use of a table saw include the following:

- Keep the guard over the blade while the saw is being used.
- Do not stand directly in line with the blade.
- Make sure that the blade does not project more than 1/8" above the stock being cut.
- Never reach across the saw blade.
- Use a push stick for ripping all stock less than 4" wide.

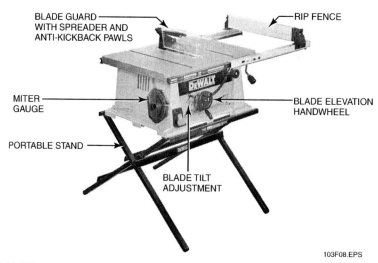

Figure 8 ◆ Portable table saw.

Table Saw Kickback

When making cuts on a table saw, particularly crosscuts and dado cuts, the saw blade may get pinched in the piece you are cutting. This could cause the workpiece to jump back, resulting in an injury. Although most table saws have an anti-kickback device that is designed to hold the lumber in place, it's a good idea to use a featherboard to feed the saw.

Discuss the advantages of power miter saws. Provide an example for the trainees to examine.

Show Transparency 5 (Figure 9).

- Disconnect power when changing blades or performing maintenance.
- Never adjust the fence or other accessories until the saw has stopped.
- Enlist a helper or use a work support when cutting long or wide stock. The weight of the stock may cause the table to lean.
- Before cutting stock to length using a miter gauge and rip fence, clamp on a step block for clearance.
- Do not rip without a fence, or crosscut without a miter gauge.
- When the blade is tilted, be sure that the blade will clear the stock before turning on the machine.
- Be sure that the stock has a straightedge before ripping.
- When using a dado or molding head, take extra care to hold the stock firmly.
- Never remove scraps from the saw table with your hands or while the saw blade is running.
- Use the proper blade for the job being done.

3.2.3 Power Miter Saws/ Compound Miter Saws

Power miter saws/compound miter saws (*Figure 9*) are also commonly called power miter boxes. This saw combines a miter box or table with a circular saw, allowing it to make straight and miter cuts. The saw blade pivots horizontally from the rear of the table and locks in position to cut angles from 0 to 45 degrees right and left. Stops are set for common angles. The difference between the power miter saw and the compound miter saw is that the blade on the compound miter saw can be tilted vertically, allowing the saw to be used to make a **compound cut** (combined bevel and miter cut).

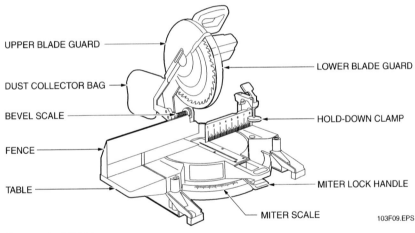

Figure 9 ♦ Compound miter saw.

Explain the features of a compound miter saw and a frame and trim saw. Provide examples for the trainees to examine.

Emphasize the safety precautions associated with the use of these saws.

Show Transparency 6 (Figure 10).

Demonstrate how to make crosscuts, straight miter cuts, bevel cuts, and compound cuts using a compound miter saw and a frame and trim saw.

Hand out Worksheet 27103-1 and Job Sheet 27103-5. Under your supervision, have the trainees complete the compound miter saw and frame and trim saw (sawbuck) safety tests before performing the related tasks on the Job Sheet. This laboratory corresponds to Performance Task 2.

Discuss the applications of demolition and chop saws. Provide examples for the trainees to examine.

The following are rules for the safe use of a power miter saw/compound miter saw:

- Always check the condition of the blade and be sure the blade is secure before starting the saw.
- Keep your fingers clear of the blade.
- Never make adjustments while the saw is running.
- Never leave a saw until the blade stops.
- Be sure the saw is sitting on a firm base and is properly fastened to the base.
- Be sure the saw is locked securely at the correct angle.
- Do not attempt to cut oversized material.
- Turn off the saw immediately after making the cut and use the brake to stop the blade.
- Enlist a helper to support the end of long stock.
- Be sure the blade guards are in place and working properly.
- Allow the saw blade to reach its maximum speed before starting the cut.
- Hold the workpiece firmly against the fence when making the cut.

3.2.4 Frame and Trim Saws

Frame and trim saws (*Figure 10*), commonly called Sawbuck™ saws or sliding compound miter saws, combine the features of a radial arm saw and a power compound miter saw. The table rotates with the track arm, and the saw head can be adjusted to make crosscuts, straight miter cuts, bevel cuts, and compound miter cuts. Because the blade assembly slides along supporting rods, wider pieces of stock can be cut with this saw than with a power miter or compound miter saw. This type of saw is ideal for cutting crown molding.

The following are rules for the safe use of a frame and trim saw:

- Check the condition of the blade and be sure that it is secure before starting the saw.
- Keep your fingers clear of the blade.
- Never reach around or across the blade.
- Never make adjustments with the saw running.
- Allow the saw blade to reach its maximum speed before starting the cut.
- Turn the saw off immediately after making the cut and use the brake to stop the blade.
- Be sure that the blade guards are in place and working properly.
- Never leave the saw until the blade has stopped.
- Be sure the saw is sitting on a level, firm base.
- Be sure the saw is locked securely at the correct angle.
- Hold the material firmly against the fence while making cuts. Clamp the material to the fence when possible.

3.2.5 Abrasive Saws

Abrasive saws use a special wheel that can slice through either metal or masonry. The most common types of abrasive saws are the demolition saw and the chop saw. The main difference between the two is that the demolition saw is not mounted on a base.

Demolition saws (*Figure 11A*) run on either electricity or gasoline. These saws are effective for cutting through most materials found at a construction site. Although all saws are potentially hazardous, the demolition saw is a particularly dangerous tool that requires the full attention and concentration of the operator.

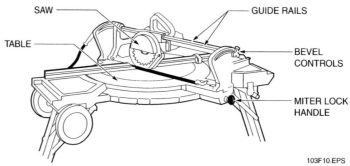

Figure 10 ◆ Frame and trim saw.

3.14 CARPENTRY FUNDAMENTALS LEVEL ONE ◆ TRAINEE GUIDE

Instructor's Notes:

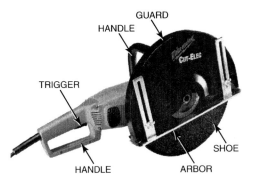

(A) DEMOLITION SAW

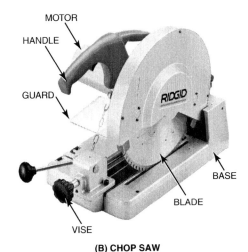

(B) CHOP SAW

Figure 11 ♦ Abrasive saws.

CAUTION
When sawing, be sure to use the appropriate personal protective equipment for your eyes, ears, and hands. If you have long hair, be sure to tie it back or cover it properly.

WARNING!
Gas-powered demolition saws must be handled with caution. Using a gas-powered saw where there is little ventilation can cause carbon monoxide poisoning. Carbon monoxide gas is hard to detect, and it is deadly! Use fans to circulate air, and have a trained person monitor air quality.

WARNING!
Because a demolition saw is handheld, there is a tendency to use it in positions that can easily compromise your balance and footing. Be very careful not to use the saw in any awkward positions.

The chop saw is a versatile and accurate tool. It is a lightweight, portable circular saw mounted on a spring-loaded arm that pivots and is supported by a metal base (*Figure 11B*). This is a good saw to use to get exact square or angled cuts. It uses a wheel to cut pipe, channel, tubing, conduit, or other light-gauge materials. The most common wheel houses an abrasive metal blade, but blades made for other materials such as PVC or wood can also be used. A vise on the base of the chop saw holds the material securely and pivots to allow miter cuts. Some can be pivoted past 45 degrees in either direction.

Chop saws are sized according to the diameter of the largest abrasive wheel they accept. Two common sizes are 12-inch and 14-inch wheels. Each wheel has a maximum safe speed. Never exceed that speed. Typically the maximum safe speeds are 5,000 revolutions per minute (rpm) for 12-inch wheels and 4,350 rpm for 14-inch wheels.

Abrasive saws require extreme care during use. To avoid injuring yourself, follow these guidelines:

- Wear personal protective equipment to protect your eyes, ears, and hands.
- Use abrasive wheels that are rated for a higher rpm than the tool can produce. This way, you will never exceed the maximum safe speed for the wheel you are using. The blades are matched to a safety rating based on their rpm.
- Be sure the wheel is secured on the arbor before you start using a demolition saw. Point the wheel of a demolition saw away from yourself and others before you start the saw.
- Use two hands when operating a demolition saw.
- Secure the materials you are cutting in a vise. The jaws of the vise hold the pipe, tube, or other material firmly and prevent it from turning while you cut.
- Be sure the guard is in place and the adjustable shoe is secured before using a demolition saw.
- Keep the saw and the blades clean.
- Inspect the saw each time you use it. Never operate a damaged saw. Ask your supervisor if you have a question about the condition of a saw.
- Inspect the blade before you use the saw. If you see damage, throw the wheel away.

Emphasize the safety precautions associated with the use of demolition and chop saws.

Identify the applications for various types of saw blades.

Show Transparency 7 (Figure 12).

3.2.6 Power Saw Blades

There are a wide variety of saw blades (*Figure 12*) available for use with circular, table, power miter, and similar saws. Each blade is designed to make an optimum cut in a different type and/or density of material. Generally, blades are standard high-speed steel or tipped with carbide. Carbide-tipped blades stay sharper longer, but they are more brittle and can be damaged if improperly handled. The number of teeth on a blade, the grind of each tooth, and the space between the teeth (gullet depth) determines the smoothness and speed of the cut.

Crosscutting requires the severing of wood fibers, whereas during ripping, the teeth must gouge out lengths of wood fiber. Blades made specifically for crosscutting have small teeth that are alternately beveled across their front edges. The gullets are small because only fine sawdust is produced. Blades made specifically for ripping have large, scraper-like teeth with deep, curved gullets. When both types are combined in a combination blade, the scraper, called a raker, has a much deeper gullet than the crosscut teeth. In chisel-tooth combination blades, the teeth and gullets are a compromise between rip and crosscut designs. They can be used for both types of cutting, but they cut less smoothly than blades designed to do just one job. Blades designed for cutting plywood and other easily splintered or chipped materials have small, scissor-like teeth. The smoothness of the cut is determined by the number of teeth and their sharpness. Use a blade recommended by the blade manufacturer for the type of material being cut. Always make sure that the blade diameter, arbor hole size, and maximum rotation speed are compatible with the saw on which it is to be used. Manufacturers have different names for similar circular saw blades. Some types of circular saw blades widely used for woodworking include the following:

- *Master combination blade (48 to 60 teeth)* – This is an all-purpose blade used for smooth, fast cutting. It can be used for crosscutting, ripping, mitering, and general-purpose work.
- *Chisel-tooth combination blade (24 to 30 teeth)* – This is an all-purpose blade that is free-cutting in any direction. It can be used for crosscutting, ripping, mitering, and general-purpose work.
- *Combination rip blade (36 to 44 teeth)* – This is an all-purpose blade used for crosscutting, ripping, mitering, and general-purpose work.

How to Tell When You Have a Dull Saw Blade

You must keep your saw blades sharp in order to maintain clean, consistent cuts. There are several things that will indicate when a circular saw blade starts to dull. First, you may notice a subtle change in the pitch and sound of the saw motor because the motor has to work harder to compensate for the dullness. Second, crosscuts will become increasingly ragged (feathered). Third, you may notice smoke from your cuts as a result of the increased friction and heat from the saw blade. Sharpen or replace your blades as often as needed.

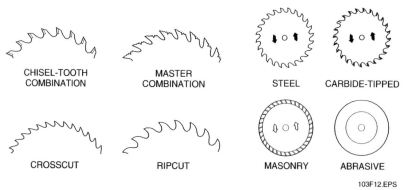

Figure 12 ♦ Circular saw blades.

Instructor's Notes:

- *Carbide-tipped blade* – Most of the combination blades described above can be purchased with carbide-tipped teeth. These blades are popular because of their economy and quality.
- *Hollow-ground plane (miter) blade (48 to 60 teeth)* – Hollow-ground plane blades are used to make smooth finish cuts on all solid woods and are ideal for cabinet work. They can be used for crosscutting, ripping, mitering, and finish work.

3.3.0 Drill Press

Drill presses (*Figure 13*) are versatile tools used for various tasks, including:

- Countersinking
- Drilling
- Sanding
- Mortising
- Cutting holes with a hole saw

Drill presses consist of an electric drill and work table combined into one unit. Both floor and benchtop models are available. The throat capacity of the drill press (the distance between its rear post and the center of the bit) determines the maximum size of the workpiece it can accept. The drill table can be raised and lowered, and on some models it can also be tilted for drilling angled holes. The motor mounts on an adjustable bracket and is controlled by a belt and pulley drive. The belt is shifted between different grooves on the pulley to vary the speed of drilling. A chart is usually mounted on the drill press that gives guidelines for the approximate speed to use for different drilling operations.

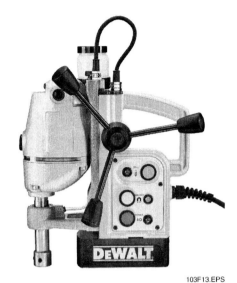

Figure 13 ♦ Drill press.

Cordless Power Tools

Today's cordless tools offer both substantial power capability and convenience. Once limited to only a few types, they are now available to serve a wide variety of applications, including the cordless drill and reciprocating saw shown here.

Assign reading of Sections 3.3.0–3.7.0 for the next class session.

Discuss the various tasks that can be performed by a drill press. Provide an example for the trainees to examine.

Emphasize the safety precautions associated with the use of a drill press.

Demonstrate drilling and boring using a drill press.

Hand out Worksheet 27103-1 and Job Sheet 27103-6. Under your supervision, have the trainees complete the drill press safety test prior to performing the related tasks on the Job Sheet. This laboratory corresponds to Performance Task 2.

Show a router and laminate trimmer.

Drill bits are installed or removed using a chuck key to tighten or loosen the chuck. For drilling, the drill is lowered into the stock by means of a feed handle, which returns to its original position when released. Operations such as sanding and mortising with a drill press require the use of special accessories made for that purpose.

The following are rules for the safe use of a drill press:

- Never leave a drill press until the chuck has stopped.
- Always remove the chuck key before starting the drill.
- Do not make adjustments while the drill is running.
- Secure the lock before drilling.
- Use the correct speed for the job being performed.
- Do not force the drill into the stock.
- Never wear gloves when operating a drill press.
- Be sure the drill or tool is secure in the chuck before starting the drill.
- Even if the power has already been turned off, do not attempt to stop the drill by grabbing the chuck; it may still have enough momentum to cause serious injury. Always wait for all moving parts to come to a complete stop before touching them.

3.4.0 Routers/Laminate Trimmers

A standard power router (*Figure 14*) is a portable tool used to cut joints and patterns in wood. It has a high-speed motor that allows the tool to make clean, smooth cuts. The motor shaft powers a chuck (collet) in which a wide variety of specially designed router bits and cutters can be installed. The motor's vertical position adjusts up or down in order to change the depth of the cut.

Router bits and cutters are made of carbide-tipped steel or high-speed steel. High-speed steel bits are suited for most uses in hard and soft woods. Carbide-tipped bits are good for plastic laminates and plywood. With different bits and available accessories, the router can be used to make many different types of cuts, including the following:

- Shaping edges
- Rabbeting
- Beveling
- Dovetailing
- Dadoing, fluting, and reeding
- Mortising

Spindle Sander

A drill press is valued for its versatility. For example, a drill press can actually be used as a sander. A special adapter is used to connect an abrasive spindle, commonly called a spindle sander, to the drill press.

Figure 14 ♦ Power router.

3.18 CARPENTRY FUNDAMENTALS LEVEL ONE ♦ TRAINEE GUIDE

Instructor's Notes:

A laminate trimmer (*Figure 15*) is a specialized type of router used to trim and shape the edges of plastic laminate materials, such as those used for countertops. It can be used for bevel trimming, flush trimming, and cutting obtuse and acute angles.

The following are rules for the safe use of a router or laminate trimmer:

- Keep your hands clear of router bits.
- Hold the tool firmly.
- Be sure that the material is secured properly before routing.
- Approach the workpiece gently; never gouge the bit into the stock.
- Always use jigs and guides with the router; never route or plow freehand.
- Turn off the tool immediately after making the cut, then wait for the bit to stop before putting the tool down.
- Avoid making cuts too deep. Note that a deep cut can cause kickbacks.

Figure 15 ◆ Laminate trimmer.

Emphasize the safety precautions associated with the use of routers/laminate trimmers.

Demonstrate how to use a router and laminate trimmer.

Hand out Worksheet 27103-1. Under your supervision, have the trainees complete the router and laminate trimmer safety test before practicing with these tools. This laboratory corresponds to Performance Task 2.

Bits

Using a router on certain materials will cause excessive bit wear if the proper technique is not used. Particleboard, for example, will quickly wear out a router bit if you attempt to make a cut in one pass. To avoid this type of wear, make your cut in several passes. You will also run into problems if you haven't selected the correct bit for the type of cut you are making. Double-check to make sure you are using the correct bit for the application. The router shown here is being used to make a dado cut.

Routing with a Straightedge

There are several ways to limit the amount of lateral motion while using a router. One very effective way is to use a straightedge. A clamp holds the straightedge to the stock and provides a solid guide for the carpenter. When making the cut, keep the base of the router firmly in contact with the straightedge. The result will be a cut that is accurate and straight.

Show portable power planes and power metal shears. Discuss their applications.

Emphasize the safety precautions associated with the use of these tools.

Show Transparency 8 (Figure 16).

Demonstrate how to use a portable power plane and power metal shears.

Hand out Worksheet 27103-1. Under your supervision, have the trainees complete the portable power plane and power metal shears safety tests before practicing with these tools. This laboratory corresponds to Performance Task 2.

Discuss pneumatic nailers and staplers. Provide an example for the trainees to examine.

Show Transparency 9 (Figure 18).

3.5.0 Portable Power Planes

Portable power planes (*Figure 16*) are excellent for fitting trim and framing members that have been nailed together. They are commonly used for the following tasks:

- Straightening material
- Edge planing
- Chamfering
- Beveling

There are small block planes used for finish work and jointer planes for heavier work. Jointer planes can be equipped with adjustable fences in order to cut bevels or chamfers.

The following are rules for the safe use of a power plane:

- Hold the plane firmly with both hands.
- Be sure the material is secure before planing.
- Always use a **door jack** when planing door edges.
- Do not try to remove too much material at one time.

3.6.0 Power Metal Shears

Power metal shears (*Figure 17*) are used to make burr-free cuts in sheet metal, metal strips, special lightweight materials, and metal studs. They can cut straight lines, tight right and left curves, and round, square, and irregularly shaped holes. A trigger-operated control allows variable speed operation. Some models have a head that can swivel 360 degrees.

The following are rules for the safe use of power metal shears:

- Keep your fingers away from the sharp edges of the tool and stock being cut.
- Never force the tool into the material.
- Do not attempt to cut material that is too thick.
- Keep the electric cord clear of the blade.

3.7.0 Pneumatic/Cordless Nailers and Staplers

Pneumatic nailers and staplers are fastening tools (*Figure 18*) powered by compressed air, which is fed to the tool through an air hose connected to an air compressor. These tools, known as guns, are widely used for quick, efficient fastening of framing, subflooring, sheathing, etc. Nailers and staplers are made in a variety of sizes to serve different purposes. Under some conditions, staples have some advantages over nails. For example, staples do not split wood as easily as a nail when driven near the end of a board. Staples are also excellent for fastening sheathing, shingles, building paper, and other materials because their

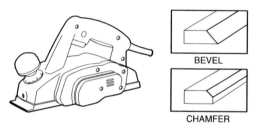

Figure 16 ◆ Power plane.

Figure 17 ◆ Power shears.

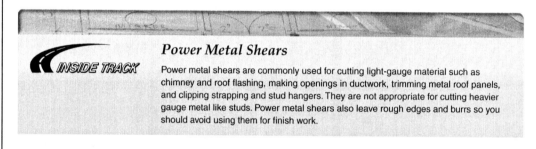

Power Metal Shears

Power metal shears are commonly used for cutting light-gauge material such as chimney and roof flashing, making openings in ductwork, trimming metal roof panels, and clipping strapping and stud hangers. They are not appropriate for cutting heavier gauge metal like studs. Power metal shears also leave rough edges and burrs so you should avoid using them for finish work.

3.20 CARPENTRY FUNDAMENTALS LEVEL ONE ◆ TRAINEE GUIDE

Instructor's Notes:

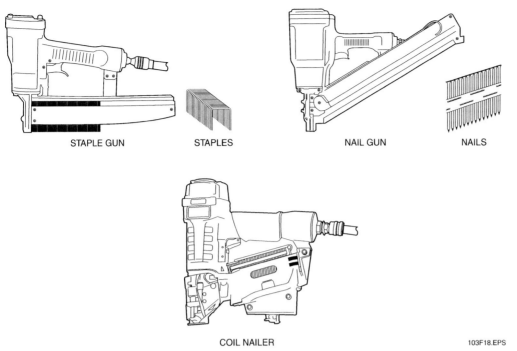

Figure 18 ◆ Pneumatic nailers and stapler.

two-legged design covers more surface area. However, both tools are sometimes used to accomplish the same fastening jobs. Nailers are typically used for the following:

- Applying sheathing
- Applying decking
- Applying roofing
- Installing framing
- Installing finish work
- Constructing cabinets

Staplers are typically used for the following:

- Applying sheathing
- Applying decking
- Applying roofing
- Installing insulation
- Installing ceiling tile
- Installing paneling

For some models of fasteners, the nails or staples come in strips and are loaded into a magazine, which typically holds 100 or more fasteners. Some tools have an angled magazine, which makes it easier for the tool to fit into tight places. Coil-fed models typically use a coil of 300 nails loaded into a circular magazine. Lightweight nailing guns can handle tiny finishing nails. Larger framing nailers can shoot smooth-shank nails of up to 3¼" in length.

The following are rules for the safe use of pneumatic fasteners:

- Be sure all safety devices are in good shape and are functioning properly.
- Use the pressure specified by the manufacturer.
- Never disengage the firing safety mechanism.
- Always assume that the fasteners are loaded.
- Never point a pneumatic fastener at yourself or anyone else.
- If you see someone using these tools improperly, report it to your instructor or supervisor.

 WARNING!
Pneumatic nailers and staplers are extremely dangerous. They are just as dangerous and deadly as a firearm. NEVER point these tools at anyone or in anyone's direction. When nailing or stapling, keep your hands away from the contact area.

Emphasize the safety precautions associated with the use of pneumatic fasteners.

Point out that cordless nailers and staplers are also available.

Show Transparency 10 (Figure 19).

Demonstrate how to use pneumatic nailers and staplers. Emphasize safety precautions.

Hand out Worksheet 27103-1 and Job Sheet 27103-7. Under your supervision, have the trainees complete the pneumatic fasteners safety test before performing the related tasks on the Job Sheet. This laboratory corresponds to Performance Task 2.

- Be sure the fastener is disconnected from the power source before making adjustments or repairs.
- Use caution when attaching the fastener to the air supply because the fastener may discharge.
- Never leave a pneumatic fastener unattended while it is still connected to the power source.
- Use nailers and staplers only for the type of work for which they were intended.
- Use only nails and staples designed for the fastener being used.
- Never use fasteners on soft or thin materials that nails may completely penetrate.

Nailers and staplers are also made in cordless models (*Figure 19*). These use a tiny internal combustion engine powered by a disposable fuel cell and a rechargeable battery. The action of the piston drives the fastener. A cordless stapler can drive about 2,500 staples with one fuel cell. A cordless framing nailer can drive about 1,200 nails on one fuel cell. The battery on a cordless tool must be periodically recharged. It pays to have a spare battery to use while one is being charged. Rules for the safe operation of a cordless nailer or stapler are basically the same as those described previously for pneumatic nailers and staplers.

Figure 19 ♦ Impulse cordless nail gun.

Instructor's Notes:

Summary

Most of the hand and power tools a carpenter will use on the job were covered in the *Core Curriculum*. This module introduced some additional tools and reviewed some of the main tools used by carpenters, such as the circular saw.

Knowing how to choose the proper tool for the job, use it skillfully, and keep it in good condition are essential carpentry skills. The use of power tools allows more work to be done in a shorter period of time. However, these tools can cause serious injury if operated incorrectly and/or unsafely. The carpenter must always be alert to the dangers of operating power tools and must follow applicable safety rules without fail when using power tools. In addition to the safety rules covered in this module, be sure to check the manufacturer's literature for the safety precautions that apply to the specific piece of equipment.

Have trainees complete the Review Questions, and go over the answers prior to administering the Module Examination.

Review Questions

1. A tool that consists of a glass tube in a sleeve with hooks on either end is called a _____ level.
   a. builder's
   b. line
   c. laser
   d. water

2. A large hand plane used for rough work is the _____ plane.
   a. jack
   b. block
   c. smoothing
   d. jointer

3. The handsaw that would be used to make a scroll cut on a trim piece is the _____ saw.
   a. dovetail
   b. back
   c. coping
   d. compass

4. When using any power tool, be sure that it is _____ before performing maintenance or changing accessories.
   a. grounded
   b. turned off
   c. checked by your supervisor
   d. disconnected

5. In general, to properly care for power tools, you should _____.
   a. keep the machined surfaces cleaned and oiled
   b. inspect tools and accessories yearly
   c. report any unusual noises, sounds, or vibrations to your supervisor or instructor
   d. keep tools in your trunk or in the back of a pickup truck when not in use

6. The size of saw blade that is used most often with portable circular saws is _____.
   a. 7¼"
   b. 9"
   c. 10¼"
   d. 12"

7. Which of the following operations is typically done with a table saw?
   a. Dadoing
   b. Circular cuts
   c. Irregular cuts
   d. Mortising

8. When ripping with a table saw, make sure that the blade does not project more than _____ above the stock being cut.
   a. 1/16"
   b. 1/8"
   c. 1/4"
   d. 3/8"

9. Before starting a drill press, always _____.
   a. remove the chuck key
   b. raise the feed handle
   c. insert the chuck key
   d. check the position of the belt on the pulley

10. A laminate trimmer is a specialized type of _____.
    a. router
    b. planer
    c. saw
    d. shaper

3.24 CARPENTRY FUNDAMENTALS LEVEL ONE ◆ TRAINEE GUIDE

Instructor's Notes:

Trade Terms Quiz

1. One of the dangers associated with the use of a table saw is the possibility of _____ as the piece travels through the saw blade.

2. The installation of a strike plate in a door jamb would require a(n) _____ cut.

3. When cutting trim molding around a door frame, you would most likely be performing a(n) _____.

4. If a long board has an imperfect edge, a jointer plane can be used to make it _____.

5. A circular cut made to accommodate water pipes in the back of a vanity cabinet is a(n) _____.

6. Use a(n) _____ to hold a door safely on edge when planing it.

7. A cut that is made across the sloping edge or side of a workpiece at an angle of less than 90 degrees is known as a(n) _____.

8. If you wanted a 2 × 6 board cut down to 2 × 5, you would make a lengthwise cut in the direction of the grain known as a(n) _____.

9. When measuring, you need to account for the amount of material removed by the saw blade, which is known as the _____.

10. A(n) _____ is a triangular joint often used to attach drawer fronts in antique furniture.

11. A cut made across the grain in lumber is known as a(n) _____.

12. A rectangular cut made along the edge or end of a board to receive another board is known as a(n) _____.

13. A(n) _____ can be used to cut lumber at precise angles.

14. A tongue cut on the end of a piece of wood is called a(n) _____.

15. The installation of crown molding often requires that wood be cut with both a bevel and a miter, which is known as a(n) _____.

16. A rabbet cut that is made only partway through a piece of lumber is called a(n) _____.

Trade Terms

Bevel cut
Compound cut
Crosscut
Dado
Door jack
Dovetail joint
Kerf
Kickback
Miter box
Miter cut
Mortise
Pocket (plunge) cut
Rabbet cut
Rip cut
Tenon
True

Have trainees complete the Trade Terms Quiz, and go over the answers.

Administer the Module Examination. Record the results on Craft Training Report Form 200, and submit the form to the Training Program Sponsor.

Administer the Performance Tests, and fill out Performance Profile Sheets for each trainee. If desired, trainee proficiency noted during laboratory sessions may be used to complete the Performance Test. Record the results on Craft Training Report Form 200, and submit the results to the Training Program Sponsor.

Profile in Success

Vincent Giardina
2004 SkillsUSA Championship Bronze Medal Winner

Vince heard about the SkillsUSA Championships from another student who had entered the national competition the year before. He thought he would give it a try and spoke with the local SkillsUSA coordinator. The national competition was held about 100 miles from where he lived. At the competition, he won a bronze medal in cabinetmaking. That gave him the confidence to start his own business. He now owns a business designing and installing kitchen cabinets.

How did you first get interested in carpentry?
I took a class right out of high school on how to build a guitar. The class was taught by John Bolin, one of the world's leading luthiers, who has crafted guitars for Steve Miller, ZZ Top, Lou Reed, Doc Watson, and many others. It took me a year to build an acoustic guitar. I discovered I really enjoyed working with wood and making things. After that, I went to school and got an associate's degree in cabinetry and started working in a cabinet shop.

What do you think it takes to be a success in your trade?
Surround yourself with successful people and learn what they do on a daily basis to be successful. Find a good instructor so that you can learn what you need to get started.

Be on time and do a great job. If you want to be successful, you have to have good work habits and deliver quality products.

What positions have you held and how did they help you get to where you are now?
I worked for many years as a cook and that helped me learn to deal with stress and manage the pace and intensity of the job.

I also worked in a cabinet shop as a sander. I worked on all parts of the cabinets. I also got to see what everyone else was doing. This helped me learn all aspects of cabinetmaking.

What does your current job entail?
I design and build kitchen cabinets from start to finish. This means I organize the job and design the kitchen. I do all the math to prepare a takeoff and order the materials. Then I cut out and build the cabinets and figure out my bottom line, which is how much I am making on the job.

What do you like about the work you do?
I like the satisfaction of overcoming challenges. I enjoy making something beautiful and making it work.

What advice would you give someone just starting out?
Don't be afraid to tackle a job. Ask as many questions as you can until you are comfortable doing that job.

Be persistent and stick with it even when it is difficult. During the SkillsUSA competition, everyone on the floor was nervous. When you are nervous, you do things that you wouldn't normally do. Well, I drilled a hole right through the side of my cabinet, twice. I thought I wouldn't even place. But I said to myself, even when you mess up, keep going and good things will happen. Well, good things did happen. I ended up winning a bronze medal. So that's my advice—even when you mess up, keep going and good things will happen!

Instructor's Notes:

Trade Terms
Introduced in This Module

Bevel cut: A cut made across the sloping edge or side of a workpiece at an angle of less than 90 degrees.

Compound cut: A simultaneous bevel and miter cut.

Crosscut: A cut made across the grain in lumber.

Dado: A rectangular groove or rabbet cut that is made partway through and across the grain in lumber.

Door jack: A holder or stand used to hold a door on edge while planing, routing, or installing hinges.

Dovetail joint: An interlocking joint with a triangular shape like that of a dove's tail.

Kerf: The width of the cut made by a saw blade. It is also the amount of material removed by the blade in a through (complete) cut or the slot produced by the blade in a partial cut.

Kickback: A sharp, uncontrolled grabbing and throwing of the workpiece by a tool as it rejects material being forced into it.

Miter box: A device that is used to cut lumber at precise angles.

Miter cut: A cut made at the end of a piece of lumber at any angle other than 90 degrees.

Mortise: A rectangular cutout or depression made in a piece of wood to receive something such as a hinge, lock, or tenon.

Pocket (plunge) cut: A cut made to remove an interior section of a workpiece or stock (such as in a countertop for a sink) or to make square or rectangular openings in floors and walls.

Rabbet cut: A rectangular cut made along the edge or end of a board to receive another board that has been similarly cut.

Rip cut: A cut made in the direction of the grain in lumber.

Tenon: A tongue that is cut on the end of a piece of wood and shaped to fit into a mortise.

True: Accurately shaped or fitted.

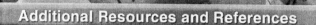

Additional Resources and References

This module is intended to present thorough resources for task training. The following reference works are suggested for further study. These are optional materials for continuing education rather than for task training.

The Art of Fine Tools. Newton, CT: Taunton Press, Inc.

Field Guide to Tools. Philadelphia, PA: Quirk Publishing.

Measure Twice, Cut Once. Boston, MA: Little, Brown & Company.

Power Tools. Newton, CT: Taunton Press, Inc.

Selecting and Using Hand Tools. Newton, CT: Taunton Press, Inc.

Tools Rare and Ingenious: Celebrating the World's Most Amazing Tools. Newton, CT: Taunton Press, Inc.

Tricks of the Trade: Jigs, Tools and Other Labor-Saving Devices. Newton, CT: Taunton Press, Inc.

Black & Decker. www.blackanddecker.com

Bosch Tool Corporation. www.boschtools.com

Delta Machinery. www.deltamachinery.com

DeWalt Industrial Tool Company. www.dewalt.com

Makita Tools USA. www.makita.com

Milwaukee Electric Tool Company. www.milwaukeetool.com

Porter-Cable Corporation. www.portercable.com

Ridge Tool Company. www.ridgid.com

The Stanley Works. www.stanleytools.com

L.S. Starrett Company. www.starrett.com

Instructor's Notes:

MODULE 27103-06 — ANSWERS TO REVIEW QUESTIONS

Answer	Section
1. b	2.1.0
2. a	2.3.0
3. c	2.5.0
4. d	3.1.1
5. c	3.1.2
6. a	3.2.1
7. a	3.2.2
8. b	3.2.2
9. a	3.3.0
10. a	3.4.0

MODULE 27103-06 — ANSWERS TO TRADE TERMS QUIZ

1. Kickback
2. Mortise
3. Miter cut
4. True
5. Pocket (plunge) cut
6. Door jack
7. Bevel cut
8. Rip cut
9. Kerf
10. Dovetail joint
11. Crosscut
12. Rabbet cut
13. Miter box
14. Tenon
15. Compound cut
16. Dado

MODULE 27103-06 — TEACHING TIPS

The following are suggested activities or instructional methods to help you teach the material in this module.

General

When you call on someone to answer a question, the rest of the class relaxes or even tunes out because they expect that the question and answer will take place only between you and the trainee you called on. Instead, use this technique to involve more trainees in answering questions and to keep them on their toes.

1. Ask trainees to define a term or explain a concept.
2. After one trainee has answered, ask a trainee seated nearby if the answer is right. Then ask whether a trainee in the back of the room agrees.
3. Ask trainees to explain why they think an answer is right or wrong.
4. Use the session to clear up incorrect ideas and encourage trainees to learn from their mistakes.

CONTREN® LEARNING SERIES — USER UPDATE

NCCER makes every effort to keep these textbooks up-to-date and free of technical errors. We appreciate your help in this process. If you have an idea for improving this textbook, or if you find an error, a typographical mistake, or an inaccuracy in NCCER's Contren® textbooks, please write us, using this form or a photocopy. Be sure to include the exact module number, page number, a detailed description, and the correction, if applicable. Your input will be brought to the attention of the Technical Review Committee. Thank you for your assistance.

Instructors – If you found that additional materials were necessary in order to teach this module effectively, please let us know so that we may include them in the Equipment/Materials list in the Annotated Instructor's Guide.

Write: Product Development and Revision
National Center for Construction Education and Research
P.O. Box 141104, Gainesville, FL 32614-1104

Fax: 352-334-0932

E-mail: curriculum@nccer.org

Craft _____ Module Name _____

Copyright Date _____ Module Number _____ Page Number(s) _____

Description
_____
_____
_____
_____

(Optional) Correction
_____
_____
_____

(Optional) Your Name and Address
_____
_____
_____

Reading Plans and Elevations
27104-06

NCCER STANDARDIZED CRAFT TRAINING PROGRAM

The National Center for Construction Education and Research (NCCER) provides a standardized national program of accredited craft training. Key features of the program include instructor certification, competency-based training, and performance testing. The program provides trainees, instructors, and companies with a standard form of recognition through the National Registry. The program is described in full in the *Guidelines for Accreditation*, published by NCCER. For more information on standardized craft training, contact NCCER by writing to P.O. Box 141104, Gainesville, FL 32614-1104; calling 352-334-0911; or e-mailing **info@nccer.org**. More information is available at **www.nccer.org**.

HOW TO USE THIS ANNOTATED INSTRUCTOR'S GUIDE

Each page presents two sections of information. The larger section displays each page exactly as it appears in the Trainee Module. The narrow column ties suggested trainee and instructor actions to each page and provides icons (detailed below) to call your attention to material, safety, audiovisual, or testing requirements. The bottom of each page includes space for your notes.

The **Audiovisual** icon indicates an appropriate time to show a transparency or other audiovisual aid.

The **Classroom** icon prompts you to define a term, stress a point, ask trainees to explain a concept, or give examples.

The **Demonstration** icon directs you to show trainees how to perform tasks.

The **Examination** icon tells you to administer the written module examination.

The **Homework** icon is placed where you may wish to assign reading for the next class, assign a project, or advise trainees to prepare for an examination.

The **Laboratory** icon is used when trainees are to practice performing tasks.

The **Materials** icon is a reminder for you to gather materials needed for classes, laboratories, and testing.

The **Performance Testing** icon tells you to administer a performance test or a portion thereof.

The **Safety** icon is used to emphasize safety issues. It is often keyed to *Caution* and *Warning!* statements in the Trainee Module.

The **Teaching Tip** icon indicates additional guidance is available, such as how to conduct an exercise, get the most educational value from a field trip, or encourage class participation. Teaching Tips may expand on a feature (*Think About It*, *Did You Know?*) or provide *Quick Quizzes* or similar exercises. You will be referred to the Teaching Tips section at the back of the module if there is additional material.

The **Combination** icon indicates that the laboratory listed corresponds with a performance task. If desired, you can note the proficiency of the trainees during the laboratory, and use it to satisfy performance testing requirements.

PREPARATION

Before teaching this module, you should review the Objectives, Performance Tasks, Materials and Equipment List, and Module Outline. Be sure to allow ample time to prepare your own training or lesson plan and gather all required materials and equipment.

**Reading Plans and Elevations
Annotated Instructor's Guide**

Module 27104-06

MODULE OVERVIEW

This module reviews and builds on the construction drawing (blueprint) material introduced in the *Core Curriculum*. It also introduces new information and techniques relevant to the carpentry trade for reading construction drawings and specifications.

PREREQUISITES

Prior to training with this module, it is recommended that the trainee shall have successfully completed *Core Curriculum*; and *Carpentry Fundamentals Level One*, Modules 27101-06 through 27103-06.

OBJECTIVES

Upon completion of this module, the trainee will be able to do the following:

1. Describe the types of drawings usually included in a set of plans and list the information found on each type.
2. Identify the different types of lines used on construction drawings.
3. Identify selected architectural symbols commonly used to represent materials on plans.
4. Identify selected electrical, mechanical, and plumbing symbols commonly used on plans.
5. Identify selected abbreviations commonly used on plans.
6. Read and interpret plans, elevations, schedules, sections, and details contained in basic construction drawings.
7. State the purpose of written specifications.
8. Identify and describe the parts of a specification.
9. Demonstrate or describe how to perform a quantity takeoff for materials.

PERFORMANCE OBJECTIVES

Under supervision of the instructor, the trainee should be able to do the following:

1. Interpret selected symbols and abbreviations used on drawings.
2. Read and interpret site/plot plans.
3. Read and interpret foundation, floor, and other plan view drawings.
4. Read and interpret elevation view drawings.
5. Read and interpret section and detail drawings.
6. Read and interpret schedules.
7. Read and interpret written specifications.
8. Perform a quantity takeoff for materials.

MATERIALS AND EQUIPMENT LIST

Transparencies
Markers/chalk
Blank acetate sheets
Transparency pens
Pencils and scratch paper
Overhead projector and screen
Whiteboard/chalkboard
Appropriate personal protective equipment
Set(s) of architect's or general contractor's drawings
Examples of formal and informal construction specifications
Example specification in the Construction Specification Institute (CSI) format
Detailed copy of the Construction Specification Institute (CSI) specification format
Copies of local building codes
Copies of quantity takeoff forms
Architect's and engineer's rule
Calculator
Copies of Worksheets 1 through 4*
Module Examinations**
Performance Profile Sheets**

* Packaged with this Annotated Instructor's Guide.
**Located in the Test Booklet.

SAFETY CONSIDERATIONS

Ensure that the trainees are equipped with appropriate personal protective equipment.

ADDITIONAL RESOURCES

This module is intended to present thorough resources for task training. The following reference works are suggested for both instructors and motivated trainees interested in further study. These are optional materials for continued education rather than for task training.

Architectural Drawing and Light Construction. Upper Saddle River, NJ: Prentice Hall.

Blueprint Reading for the Building Trades. Carlsbad, CA: Craftsman Book Company.

Code Check. Newton, CT: Taunton Press.

Design Drawing. New York, NY: John Wiley & Sons.

Graphic Guide to Frame Construction. Newton, CT: Taunton Press.

International Building Code 2003. Falls Church, VA: International Code Council.

MasterFormat™ 2004 Edition. Alexandria, VA: The Construction Specifications Institute (CSI) and Construction Specifications Canada (CSC).

Measuring, Marking, and Layout. Newton, CT: Taunton Press.

Plan Reading & Material Takeoff. Kingston, MA: R.S. Means Company.

Reading Architectural Plans for Residential and Commercial Construction, Ernest R. Weidhaas. Upper Saddle River, NJ: Prentice Hall, 1998.

The Construction Specifications Institute. An organization that seeks to facilitate communication among all those involved in the building process. **www.csinet.org**

International Code Council. A membership organization dedicated to building safety and fire prevention through development of building codes. **www.iccsafe.org**

TEACHING TIME FOR THIS MODULE

An outline for use in developing your lesson plan is presented below. Note that each Roman numeral in the outline equates to one session of instruction. Each session has a suggested time period of 2½ hours. This includes 10 minutes at the beginning of each session for administrative tasks and one 10-minute break during the session. Approximately 20 hours are suggested to cover *Reading Plans and Elevations.* You will need to adjust the time required for hands-on activity and testing based on your class size and resources. Because laboratories often correspond to Performance Tasks, the proficiency of the trainees may be noted during these exercises for Performance Testing purposes.

Topic **Planned Time**

Session I. Introduction; Drawing Set
   A. Introduction   _____
   B. Drawing Set
      1. Title Sheets, Title Blocks, and Revision Blocks   _____
      2. Plan View Drawings   _____
      3. Elevation Drawings   _____
      4. Section Drawings   _____
      5. Detail Drawings   _____
      6. Schedules   _____
      7. Structural Drawings   _____
      8. Plumbing, Mechanical, and Electrical Plans   _____
      9. Shop Drawings   _____
     10. As-Built Drawings   _____
     11. Soil Reports   _____

Session II. Reading and Interpreting Drawings
  A. Reading and Interpreting Drawings
     1. Lines Used on Drawings
     2. Symbols Used on Drawings
     3. Dimensioning
     4. Abbreviations
     5. Architectural Terms Used in Drawings and Specifications
  B. Laboratory
     Hand out Worksheet 27104-1. Have the trainees complete the Worksheet. This laboratory corresponds to Performance Task 1.

Session III. Guidelines for Reading a Drawing Set
  A. Guidelines for Reading a Drawing Set

Session IV. Laboratory
  A. Laboratory
     Hand out Worksheets 27104-2 and 27104-3. Have the trainees complete the Worksheets. This laboratory corresponds to Performance Tasks 2 through 6.

Session V. Specifications
  A. Specifications
     1. Organization and Types of Specifications
  B. Laboratory
     Hand out Worksheet 27104-4. Have the trainees complete the Worksheet. This laboratory corresponds to Performance Task 7.

Session VI. Building Codes; Quantity Takeoffs
  A. Building Codes
  B. Quantity Takeoffs
  C. Laboratory
     Under your supervision, and using an instructor-supplied drawing set and specifications, have the trainees practice doing a material quantity takeoff for a building, or one room in a building, etc. This laboratory corresponds to Performance Task 8.

Session VII. Project Organization; Working with Other Trades; Project Schedules; Review
  A. Project Organization
  B. Working with Other Trades
  C. Project Schedules
  D. Review

Session VIII. Module Examination and Performance Testing
  A. Module Examination
     1. Trainees must score 70 percent or higher to receive recognition from NCCER.
     2. Record the testing results on Craft Training Report Form 200, and submit the results to the Training Program Sponsor.
  B. Performance Testing
     1. Trainees must perform each task to the satisfaction of the instructor to receive recognition from NCCER. If applicable, proficiency noted during laboratory exercises can be used to satisfy the Performance Testing requirements.
     2. Record the testing results on Craft Training Report Form 200, and submit the results to the Training Program Sponsor.

Reading Plans and Elevation
27104-06

A Carpentry contestant in the SkillsUSA 2005 National Championships studies the plans for the wooden building frame. A rafter square appears on the lumber stack behind him. Careful inspection of blueprints and drawings is essential to good carpentry.

Instructor's Notes:

Assign reading of Module 27104-06.

27104-06
Reading Plans and Elevations

Topics to be presented in this module include:

1.0.0	Introduction	4.2
2.0.0	Drawing Set	4.2
3.0.0	Reading and Interpreting Drawings	4.20
4.0.0	Guidelines for Reading a Drawing Set	4.35
5.0.0	Specifications	4.36
6.0.0	Building Codes	4.43
7.0.0	Quantity Takeoffs	4.43
8.0.0	Project Organization	4.44
9.0.0	Working with Other Trades	4.47
10.0.0	Project Schedules	4.47

Overview

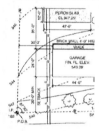

Everything you need to know about the design and construction of a particular structure is documented in its plans and specifications. These documents tell you everything from where to place the structure on the site to what kinds of materials to use.

In order to become a competent carpenter, you must be able to read and interpret plans. Without that skill, you will never be able to work independently, and you will never be able to realize the growth opportunities that exist in the construction trades.

As a carpenter, you will constantly refer to floor plans and elevations. The floor plans will tell you the locations of walls, partitions, windows, doors, stairways, and other building elements. The elevation drawings provide construction details.

The carpenter's work precedes that of other trades such as electricians, plumbers, and HVAC installers, all of whom need spaces set aside for their equipment, wiring, piping, and ductwork. The drawings show how and where to construct these spaces, and the carpenter must make sure it is done right. Otherwise, costly rework and schedule delays can result.

Instructor's Notes:

Objectives

When you have completed this module, you will be able to do the following:

1. Describe the types of drawings usually included in a set of plans and list the information found on each type.
2. Identify the different types of lines used on construction drawings.
3. Identify selected architectural symbols commonly used to represent materials on plans.
4. Identify selected electrical, mechanical, and plumbing symbols commonly used on plans.
5. Identify selected abbreviations commonly used on plans.
6. Read and interpret plans, elevations, schedules, sections, and details contained in basic construction drawings.
7. State the purpose of written specifications.
8. Identify and describe the parts of a specification.
9. Demonstrate or describe how to perform a quantity takeoff for materials.

Trade Terms

Bench mark
Contour lines
Easement
Elevation view
Front setback
Monuments
Nominal size
Plan view
Property lines
Quantity takeoff
Riser diagram
Topographical survey

Required Trainee Materials

1. Pencil and paper
2. Appropriate personal protective equipment

Prerequisites

Before you begin this module, it is recommended that you successfully complete *Core Curriculum* and *Carpentry Fundamentals Level One*, Modules 27101-06 through 27103-06.

This course map shows all of the modules in *Carpentry Fundamentals Level One*. The suggested training order begins at the bottom and proceeds up. Skill levels increase as you advance on the course map. The local Training Program Sponsor may adjust the training order.

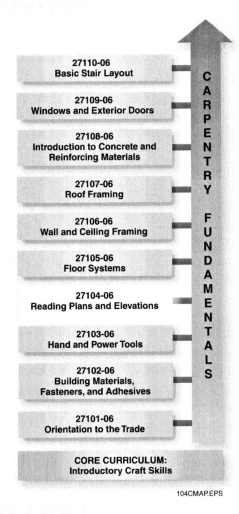

104CMAP.EPS

Ensure you have everything required to teach the course. Check the Materials and Equipment List at the front of this module.

Show Transparency 1, Objectives.

Show Transparency 2, Performance Tasks.

Explain that terms shown in bold (blue) are defined in the Glossary at the back of this module.

See the general Teaching Tip at the end of this module.

Discuss the types of drawings in a typical set of plans, and provide examples for the trainees to examine.

Show Transparency 3 (Figure 1).

Discuss the types of information contained in title and revision blocks.

1.0.0 ◆ INTRODUCTION

This module introduces techniques for reading construction drawings and specifications. Construction drawings tell carpenters and other skilled tradespeople how to build a specific building or structure. A specification is a contractual document used along with the construction drawings. It contains detailed written instructions that supplement the set of drawings. As a carpenter, it is extremely important that you interpret construction drawings and specifications correctly. Failure to do so may result in costly rework and unhappy customers. Depending on the severity of the mistake, it can also expose you and your employer to legal liability.

2.0.0 ◆ DRAWING SET

The set of detailed drawings or plans drawn to scale by an architect and/or engineer show all of the information and dimensions necessary to build or remodel a structure. Copies of the architect's original drawings are made for contractors and others to use. These copies are commonly called blueprints. The term blueprint is derived from a method of reproduction that was used in the past. A true blueprint shows the details of the structure as white lines on a blue background. Today, most drawing reproduction methods produce a black, blue, or brownish-colored line on a white background. These copies or prints are typically made using a diazo or similar copying machine. However, the term blueprint is still widely used when referring to any copies made from original drawings.

Construction drawings or architect's plans consist of several different kinds of drawings assembled into a set (*Figure 1*). In order for complete information about a structure to be conveyed to the reader, various drawings in the set illustrate the structure using a variety of different views. A set of drawings also includes sheets that contain relevant written information, such as window or door schedules.

The types of written information and views normally contained in a drawing set include:

- Title sheets, title blocks, and revision blocks
- Architectural drawings consisting of:
  – Plan views
  – Elevations
  – Sections
  – Details
  – Schedules
- Structural drawings
- Plumbing plans
- Mechanical plans
- Electrical plans

2.1.0 Title Sheets, Title Blocks, and Revision Blocks

A title sheet is normally placed at the beginning of a set of drawings or at the beginning of a major section of drawings. It provides an index to the other drawings; a list of abbreviations used on the drawings and their meanings; a list of symbols used on the drawings and their meanings; and various other project data such as the project location, the size of the land parcel, and the building size. It is important that you use the title sheet(s)

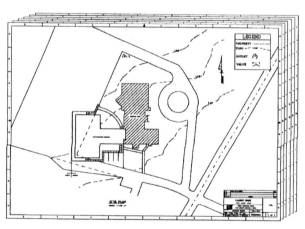

Figure 1 ◆ Typical format of a working drawing set.

4.2 CARPENTRY FUNDAMENTALS LEVEL ONE ◆ TRAINEE GUIDE

Instructor's Notes:

Architectural Plans

Architectural plans have been used over the centuries to pictorially describe buildings and structures before they are actually built. In the past, draftspersons would draw these plans by hand. Today, most drawings for buildings and other structures are generated by computer using a process called computer-aided design (CAD). Working with a variety of architectural software, the drafter creates the electronic drawings for the building or structure on the computer. Then, with a command to the computer, the electronic drawings are sent to a printer or plotter to be output on paper.

104SA01.EPS

Presentation Drawings

Presentation drawings are three-dimensional drawings that show a building from a desirable vantage point in order to display its most interesting features. Presentation drawings do not provide detailed information for construction purposes. They are used mainly by an architect or a contractor to sell the design of a house or building to a prospective customer. As a carpenter, you should be able to visualize and draw a rough three-dimensional sketch of a building after interpreting the architectural plans for the building.

104SA02.EPS

Show Transparency 4 (Figure 2).

Discuss the types of information contained in plan view drawings.

that come with the drawing set so that you understand the specific symbols and abbreviations used throughout the drawings. These symbols and abbreviations may vary from plan to plan.

A title block or box (*Figure 2*) is normally placed on each sheet in a set of drawings. It is usually located in the bottom right-hand corner of the sheet, but this location can vary.

The title block serves several purposes in terms of communicating information. Generally, it contains the name of the firm that prepared the drawings, the owner's name, and the address and name of the project. It also gives locator information, such as the title of the sheet, the drawing or sheet number, the date the sheet was prepared, the scale, and the initials or names of the people who prepared and checked the drawing.

A revision block is normally shown on each sheet in a set of drawings. Typically, it is located in the upper right-hand corner or bottom right-hand corner of the drawing near or within the title block. It is used to record any changes (revisions) to the drawing. An entry in the revision block usually contains the revision number or letter, a brief description of the change, the date, and the initials of the person making the revision(s). When using drawings, it is essential to note the revision designation on each drawing and use only the latest issue; otherwise, costly mistakes will result. If in doubt about the revision status of a drawing, check with your supervisor to make sure that you are using the most recent version of the drawing.

2.2.0 Plan View Drawings

Plan view drawings are drawings that show the structure looking down from above. The object is projected from a horizontal plane. Typically, plan

Discuss the types of information contained in site plans, and provide examples for the trainees to examine.

Show Transparencies 5 and 6 (Figures 3 and 4).

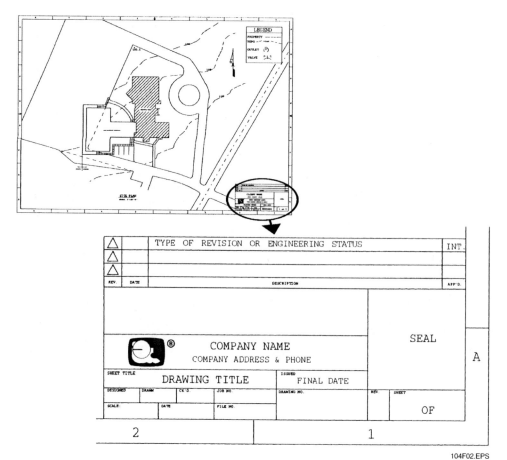

Figure 2 ♦ Title and revision blocks.

view drawings are made to show the overall construction site (plot or site plan), the structure's foundation (foundation plan), and the structure's floor plans.

2.2.1 Site Plans

Man-made and topographic (natural) features and other relevant project information are shown on a site plan, including the information needed to correctly locate the structure on the site. Man-made features include roads, sidewalks, utilities, and buildings. Topographical features include trees, streams, springs, and existing contours. Project information includes the building outline, general utility information, proposed sidewalks, parking areas, roads, landscape information, proposed contours, and any other information that will convey what is to be constructed or changed on the site. A prominently displayed north direction arrow is included for orientation purposes on site plans. Sometimes a site plan contains a large-scale map of the overall area that indicates where the project is located on the site. *Figures 3* and *4* show examples of basic site plans.

Typically, site plans show the following types of detailed information:

- Coordinates of control points or property corners
- Direction and length of **property lines** or control lines
- Description, or reference to a description, for all control and property **monuments**
- Location, dimensions, and elevation of the structure on site

4.4 CARPENTRY FUNDAMENTALS LEVEL ONE ♦ TRAINEE GUIDE

Instructor's Notes:

Define elevation and bench mark.

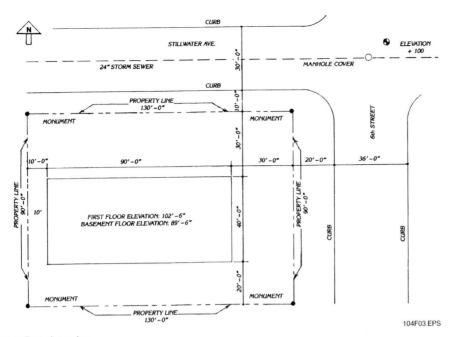

Figure 3 ◆ Typical site plan.

- Finish and existing grade contours
- Location of utilities
- Location of existing elements such as trees and other structures
- Location and dimensions of roadways, driveways, and sidewalks
- Names of all roads shown on the plan
- Locations and dimensions of any **easements**

Like other drawings, site plans are drawn to scale. The scale used depends on the size of the project. A project covering a large area typically will have a small scale, such as 1" = 100', while a project on a small site might have a large scale, such as 1" = 10'.

Normally, the dimensions shown on site plans are stated in feet and tenths of a foot (engineer's scale). However, some site plans state the dimensions in feet, inches, and fractions of an inch (architect's scale). Dimensions to the property lines are shown to establish code requirements. Frequently, building codes require that nothing be built on certain portions of the land. For example, local building codes may have a **front setback** requirement that dictates the minimum distance that must be maintained between the street and the front of a structure. Normally, side yards have a minimum width specified from the property line to allow for access to rear yards and to reduce the possibility of fire spreading to adjacent buildings.

A property owner cannot build on an area where an easement has been identified. Examples of typical easements are the right of a neighbor to build a road; a public utility to install water, gas, or electric lines on the property; or an area set aside for drainage of groundwater.

Site plans show finish grades (also called elevations) for the site based on data provided by a surveyor or engineer. It is necessary to know these elevations for grading the lot and for construction of the structure. Finish grades are typically shown for all four corners of the lot as well as other points within the lot. Finished grades or elevations are also shown for the corners of the structure and relevant points within the building.

All the finish grade references shown are keyed to a reference point, called a **bench mark** or job datum. This is a reference point established by the surveyor on or close to the property, usually at one corner of the lot. At the site, this point may be marked by a plugged pipe driven into the ground, a brass marker, or a wood stake. The location of the bench mark is shown on the plot plan with a grade figure next to it. This grade figure may indicate the actual elevation relative to sea level, or it

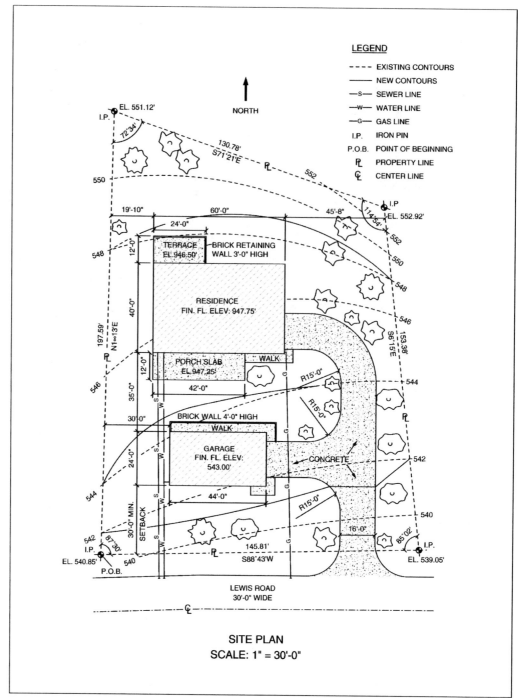

Figure 4 ◆ Typical site plan showing topographical features.

may be an arbitrary elevation of 100.00', 500.00', etc. All other grade points shown on the site plan, therefore, will be relative to the bench mark.

A site plan usually shows the finish floor elevation of the building. This is the level of the first floor of the building relative to the job-site bench mark. For example, if the bench mark is labeled 100.00' and the finish floor elevation indicated on the plan is marked 105.00', the finished floor elevation is 5' above the bench mark. During construction, many important measurements are taken from the finish floor elevation point.

2.2.2 Foundation Plans

As applicable, foundation plans (*Figure 5*) give information about the location and dimensions of footings, grade beams, foundation walls, stem walls, piers, equipment footings, and windows and doors. The specific information shown on the plan is determined by the type of construction involved: full-basement foundation, crawl space, or a concrete slab-on-grade level (*Figure 6*).

The types of information normally shown on foundation plans for full-basement and crawl space foundations include:

- Location of the inside and outside of the foundation walls
- Location of the footings for foundation walls, columns, posts, chimneys, fireplaces, etc.
- Wall openings for windows, doors, crawl space access, and vents
- Walls for entrance platforms (stoops)
- Floor joist direction, size, and spacing
- Location of stairways
- Notations for the strength of concrete used for various parts of the foundation and floor
- Notations for the composition, thickness, and underlaying material of the basement floor or crawl space surface
- Location of furnaces and other equipment

The types of information normally shown on foundation plans for slab-on-grade foundations include:

- Size and shape of the slab
- Exterior and interior footing locations
- Location of fireplaces, floor drains, HVAC ductwork, etc.
- Notations for slab thickness
- Notations for wire mesh reinforcing, fill, and vapor barrier materials

Discuss the types of information contained in foundation plans, and provide examples for the trainees to examine.

Show Transparencies 7 and 8 (Figures 5 and 6).

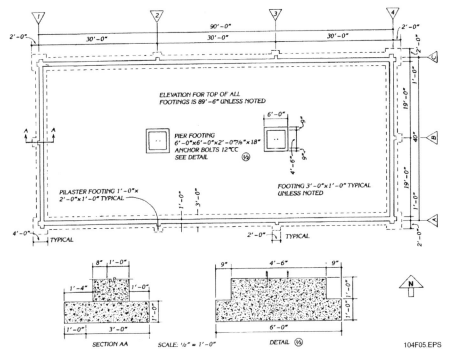

Figure 5 ◆ Foundation plan (NTS).

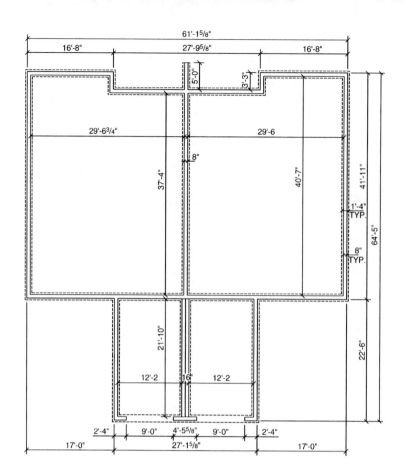

Figure 6 ♦ Slab-on-grade foundation plan (NTS).

4.8 CARPENTRY FUNDAMENTALS LEVEL ONE ♦ TRAINEE GUIDE

2.2.3 Floor Plans

The floor plan is the main drawing of the entire set. For a floor plan view, an imaginary line is cut horizontally across the structure at varying heights so all the important features such as windows, doors, and plumbing fixtures can be shown. For multistory buildings, separate floor plans are normally drawn for each floor. However, if several floors have the same layout, one drawing may be used to show all the floors that are similar. *Figure 7* shows an example of a basic floor plan. The types of information normally shown on floor plans include:

- Outside walls, including the location and dimensions of all exterior openings
- Types of construction materials
- Location of interior walls and partitions
- Location and swing of doors
- Stairways
- Location of windows
- Location of cabinets, electrical and mechanical equipment, and fixtures
- Location of cutting plane line

Each door or window shown on a floor plan for a commercial building is typically accompanied by a number, letter, or both. This number/letter is an identifier that refers to a door or window schedule that describes the corresponding size, type of material, model number, etc., for the specific door or window. Door and window schedules are discussed in more detail later in this section. Residential floor plans often show door sizes directly on the plan drawing.

2.2.4 Roof Plans

When supplied, roof plans provide information about the roof slope, roof drain placement, and other pertinent information regarding ornamental sheet metal work, gutters and downspouts, etc. Where applicable, the roof plan may also show information on the location of air conditioning units, exhaust fans, and other ventilation equipment.

Some drawing sets may also contain a ceiling plan that shows the location of supply diffusers, exhaust grilles, access panels, and the location of

Discuss the types of information contained in floor plans, and provide examples for the trainees to examine.

Show Transparency 9 (Figure 7).

Discuss the types of information contained in roof and ceiling plans, and provide examples for the trainees to examine.

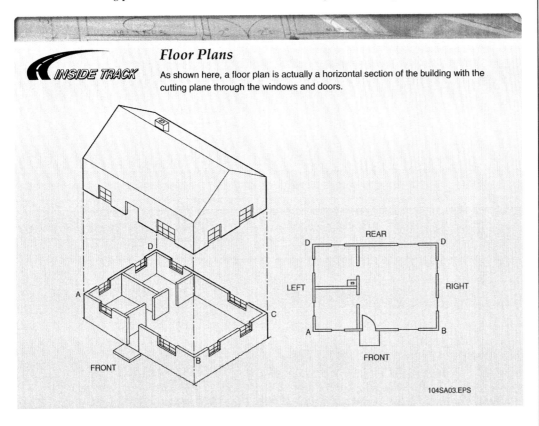

Floor Plans

INSIDE TRACK — As shown here, a floor plan is actually a horizontal section of the building with the cutting plane through the windows and doors.

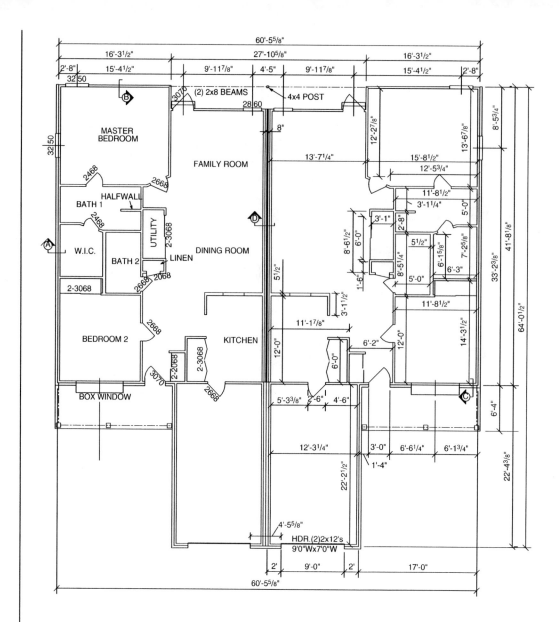

Figure 7 ♦ Basic floor plan (NTS).

structural components and other mechanicals. Some have reflected ceiling plans. These show the details of the ceiling as though it were reflected by a mirror on the floor. Ceiling plans show features of the ceiling while keeping those features in proper relation to the floor plan. For example, if a vertical pipe runs from floor to ceiling in a room and is drawn in the upper left corner of the floor plan, it is also shown on the upper left corner of the reflected ceiling plan of that same room.

2.3.0 Elevation Drawings

Elevation drawings are views that look straight ahead at a structure. The object is projected from a vertical plane. Typically, elevation views are used to show the exterior features of a structure so that the general size and shape of the structure can be determined. Elevation drawings clarify much of the information on the floor plan. For example, a floor plan shows where the doors and windows are located in the outside walls; an elevation view of the same wall shows actual representations of these doors and windows. *Figure 8* shows an example of a basic elevation drawing. The types of information normally shown on elevation drawings include:

- Grade lines
- Floor height
- Window and door types
- Roof lines and slope, roofing material, vents, gravel stops, and projection of eaves
- Exterior finish materials and trim
- Exterior dimensions

Unless one or more views are identical, four elevation views are generally used to show each exposure. With very complex buildings, more than four views may be required. The various views are usually labeled in one of two ways: they may be labeled as front, right side, left side, and rear; or they may be designated by compass direction. For example, if the front of the building faces north, then this becomes the north elevation. The other elevations are then labeled accordingly (east, south, and west).

Drawing sets may also have interior elevation views for the walls in each partitioned area, especially for walls that have special features.

2.4.0 Section Drawings

A section view or drawing (*Figure 9*) shows how a particular feature looks inside or how something is put together internally. The feature is drawn as if a cut has been made through the middle of it at a certain point. The location of the cut and the direction to be viewed are shown on the related plan view.

Section drawings that show a view made by cutting through the length of a structure are referred to as longitudinal sections, while those showing the view of a cut through the width or narrow portion of the structure are referred to as traverse sections.

To show greater detail, section views are normally drawn to a larger scale than that used in plan views.

The types of information normally shown by a section view include:

- Details of construction and information about stairs, walls, chimneys, or other parts of construction that may not show clearly on a plan view
- Floor levels in relation to grade
- Wall thickness at various locations
- Anchors and reinforcing steel

2.5.0 Detail Drawings

Details are enlargements of special features of a building or of equipment installed in a building. They are drawn to a larger scale in order to make the details clearer. The detail drawings are often placed on the same sheet where the feature appears in the plan. *Figure 10* shows a series of typical detail views.

Typically, details may be drawn for the following objects or situations:

- Footings and foundations, including anchor bolts, reinforcing, and control joints
- Beams, floor joists, bridging, and other support members
- Sills, floor framing, exterior walls, and vapor barriers
- Floor heights, thickness, expansion, and reinforcing
- Interior walls
- Windows, exterior and interior doors, and door frames
- Roofs, cornices, soffits, and ceilings
- Gravel stops, fascia, and flashing
- Fireplaces and chimneys
- Staircases and stair assemblies
- Millwork, trim, ornamental iron, and specialty items

Discuss the types of information contained in elevation drawings, and provide examples for the trainees to examine.

Show Transparencies 10 through 14 (Figures 8 through 10).

Explain the types of information contained in section drawings, and provide examples for the trainees to examine.

Discuss the types of information contained in detail drawings, and provide examples for the trainees to examine.

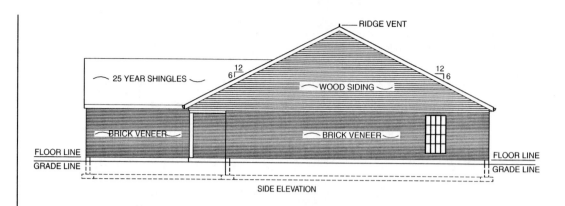
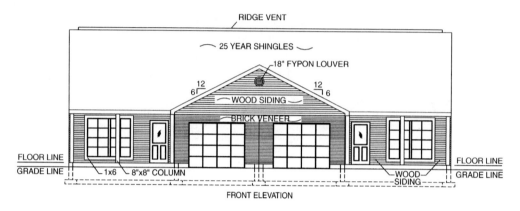
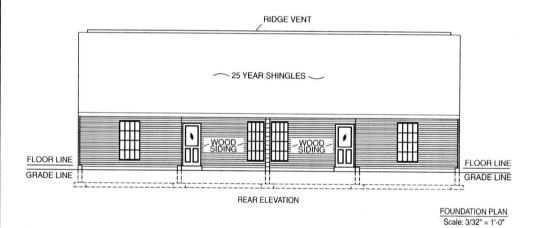

Figure 8 ◆ Elevation drawing (NTS).

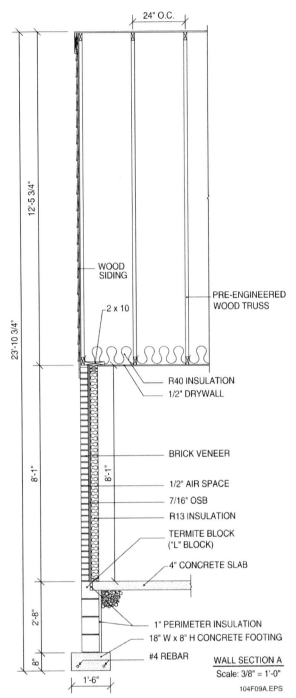

Figure 9 ♦ Section drawing (NTS, 1 of 3).

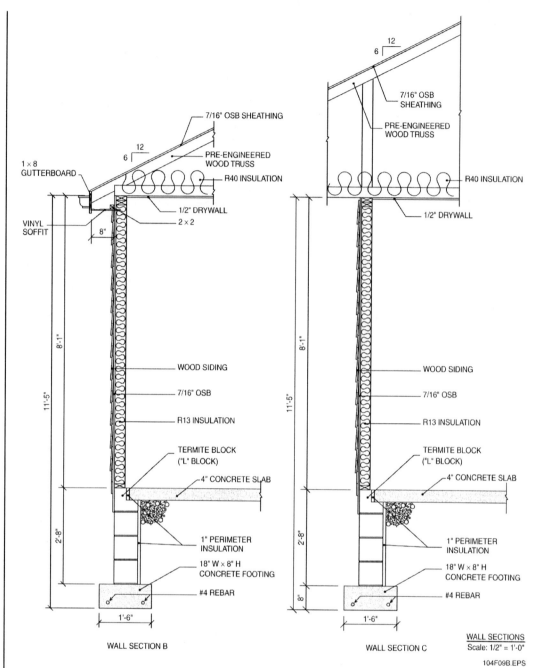

Figure 9 ♦ Section drawing (NTS, 2 of 3).

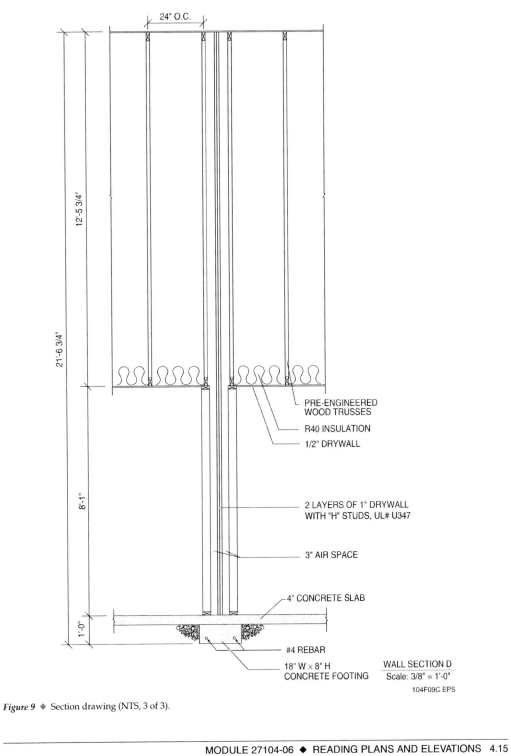

Figure 9 ♦ Section drawing (NTS, 3 of 3).

Discuss the types of information contained in schedules, and provide examples for the trainees to examine. Discuss:

- Door and window schedules
- Finish schedules

Show Transparencies 15 and 16 (Figures 11 and 12).

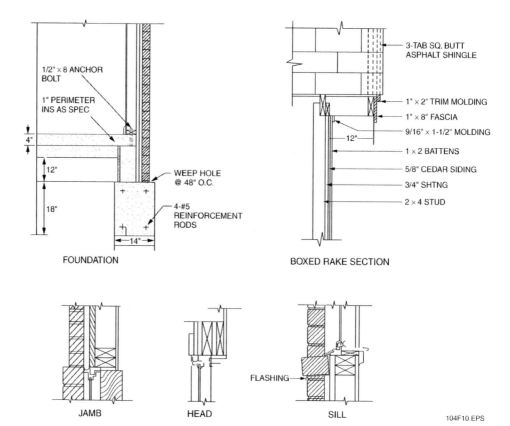

Figure 10 ◆ Detail views.

2.6.0 Schedules

Schedules in a drawing set are tables that describe and specify the various types and sizes of materials used in the construction of a building. Commonly, there are door schedules, window schedules, and finish schedules (see *Figure 11*). For commercial jobs, additional schedules are provided for mechanical equipment and controls, plumbing fixtures, lighting fixtures, and any other equipment that needs to be listed separately.

Door, window, header, and finish schedules are of particular importance to the carpenter. Door and window types are identified on the various plan and elevation drawings by numbers, letters, or both. The door and window schedules list these identifier numbers or letters and describe the corresponding size, type of material, and model number for each different type of door or window used in the structure.

In the finish schedule for a structure (*Figure 12*), each room is identified by name or number. The material and finish for each part of the room (walls, floor, ceiling, base, and trim) are designated, along with any clarifying remarks.

4.16 CARPENTRY FUNDAMENTALS LEVEL ONE ◆ TRAINEE GUIDE

Instructor's Notes:

DOOR SCHEDULE

DOOR	WIDTH	HEIGHT	THICK-NESS	MAT'L	TYPE	STORM DOOR	QTY.	THRES-HOLD	REMARKS	MANUFACTURER
2068	2'-0"	6'-8"	1 3/8"	Wood-Ash	Hollow-core	NO	5	None	Oil Stain	LBJ Door Co.
2468	2'-4"	6'-8"	1 3/8"	Wood-Ash	Hollow-core	NO	1	None	Oil Stain	LBJ Door Co.
2668	2'-6"	6'-8"	1"	Wood-Ash	Cafe	NO	1 pr.	None	Oil Stain	LBJ Door Co.
2668	2'-6"	6'-8"	1 3/8"	Wood-Ash	Sliding Pocket	NO	1	None	Oil Stain	LBJ Door Co.
2668	2'-6"	6'-8"	1 3/8"	Wood-Ash	Hollow-core	1 Screen	6	Alum.	Screen door in garage	LBJ Door Co.
2868	2'-8"	6'-8"	1 3/4"	Metal Clad	Fireproof	YES	1	Alum.	Paint	LBJ Door Co.
3068	3'-0"	6'-8"	1 3/4"	Wood-Ash	Solid-core	NO	1	None	Oil Stain	LBJ Door Co.
3668	3'-6"	6'-8"	1 3/4"	Wood-Ash	Solid-core	YES	1	Alum.	Marine Varnish	LBJ Door Co.
6066	6'-0"	6'-6"	1/2"	Glass/Metal	Sliding	YES	1 pr.	Alum.	Sliding Screen	LBJ Door Co.
6068	6'-0"	6'-8"	1 1/4"	Wood-Ash	Bi-Fold	NO	2 sets	None	Oil Stain	LBJ Door Co.
1956	1'-9"	5'-6"	1/2"	Glass/Metal	Sliding Shower door	NO	2 sets	None	Frosted Glass	LBJ Door Co.

WINDOW SCHEDULE

SYMBOL	WIDTH	HEIGHT	MAT'L	TYPE	SCREEN & DOOR	QUANTITY	REMARKS	MANUFACTURER	CATALOG NUMBER
A	3'-8"	3'-0"	ALUM.	DOUBLE HUNG	YES	2	4 LIGHTS, 4 HIGH	LBJ Window Co.	141 PW
B	3'-8"	5'-0"	ALUM.	DOUBLE HUNG	YES	1	4 LIGHTS, 4 HIGH	LBJ Window Co.	145 PW
C	3'-0"	5'-0"	ALUM.	STATIONARY	STORM ONLY	2	SINGLE LIGHTS	H & J Glass Co.	59 PY
D	2'-0"	3'-0"	ALUM.	DOUBLE HUNG	YES	1	4 LIGHTS, 4 HIGH	LBJ Window Co.	142 PW
E	2'-0"	6'-0"	ALUM.	STATIONARY	STORM ONLY	2	20 LIGHTS	H & J Glass Co.	37 TS
F	3'-6"	5'-0"	ALUM.	DOUBLE HUNG	YES	1	16 LIGHTS, 4 HIGH	LBJ Window Co.	143 PW

HEADER SCHEDULE

HEADER SIZE	EXTERIOR		INTERIOR	
	26' + UNDER	26' TO 32'	26' + UNDER	26' TO 32'
(2) 2 x 4	3'-6"	3'-0"	USE (2) 2 x 6	
(2) 2 x 6	6'-6"	6'-0"	4'-0"	3'-0"
(2) 2 x 8	8'-6"	8'-0"	5'-6"	5'-0"
(2) 2 x 10	11'-0"	10'-0"	7'-0"	6'-6"
(2) 2 x 12	13'-6"	12'-0"	8'-6"	8'-0"

104F11.EPS

Figure 11 ♦ Examples of door, window, and header schedules.

Discuss the types of information contained in structural drawings, and provide examples for the trainees to examine.

Discuss the types of information contained in plumbing, mechanical, and electrical plans. Emphasize that these plans often contain information important to the carpentry trade.

Show Transparency 17 (Figure 13).

ROOM FINISH SCHEDULE																
ROOMS	FLOOR			CEILING			WALL				BASE		TRIM			REMARKS
	CARPET	CERAMIC TILE	RUBBER TILE	CONCRETE	ACOUSTIC TILE	DRYWALL	PAINT	CERAMIC TILE	DRYWALL	PAINT	WALLPAPER	CERAMIC TILE	WOOD	RUBBER	CERAMIC TILE	STAIN	WOOD	STAIN	PAINT
ENTRY		✓			✓				✓	✓	✓		✓			✓	✓	✓		See owner for all painting
HALL	✓				✓				✓	✓			✓			✓	✓	✓		
BEDROOM 1	✓				✓				✓	✓	✓		✓			✓	✓	✓		See owner for grade of carpet
BEDROOM 2	✓				✓				✓	✓			✓			✓	✓	✓		See owner for grade of carpet
BEDROOM 3	✓				✓				✓	✓			✓			✓	✓	✓		See owner for grade of carpet
BATH 1	✓	✓			✓			✓	✓	✓	✓	✓				✓	✓	✓		Wallpaper 3 walls around vanity
BATH 2		✓			✓			✓	✓	✓	✓				✓		✓	✓		Water-seal tile Wallpaper w/wall
UTIL + CLOSETS	✓		✓			✓	✓		✓				✓			✓	✓	✓	✓	Use off-white flat latex
KITCHEN			✓		✓				✓	✓				✓			✓	✓		
DINING	✓				✓				✓	✓	✓		✓			✓	✓	✓		
LIVING	✓				✓				✓	✓			✓			✓	✓	✓		See owner for grade of carpet
GARAGE				✓		✓	✓					✓		✓			✓	✓		

Figure 12 ♦ Example of a finish schedule.

2.7.0 Structural Drawings

Structural drawings are created by a structural engineer and accompany the architect's plans. They are usually drawn for large structures such as office buildings or factories. They show requirements for structural elements of the building, including columns, floor and roof systems, stairs, canopies, and bearing walls. They contain details such as:

- Heights of finished floors and walls
- Height and bearing of bar joist or steel joist
- Locations of bearing steel materials
- Height of steel beams, concrete plank, concrete Ts, and poured-in-place concrete
- Bearing plate locations
- Location, size, and spacing of anchor bolts
- Stairways

2.8.0 Plumbing, Mechanical, and Electrical Plans

For many construction jobs, the details for the plumbing, mechanical, and electrical work are included on the basic floor plans (*Figure 13*). However, for complex commercial work, they are generally shown on separate plan drawings.

Plumbing plans show the layout of water supply and sewage disposal systems and fixtures. Plumbing system plans generally have a separate plumbing **riser diagram** that shows the layout and identification of the piping and fixtures. A plumbing legend is usually located on the plan.

Mechanical plans show details for installation of the HVAC and other equipment. Such plans usually include a refrigerant piping schematic that shows the types and sizes of the piping and

4.18 CARPENTRY FUNDAMENTALS LEVEL ONE ♦ TRAINEE GUIDE

Instructor's Notes:

identification of the fittings. The plan may also include a legend defining the HVAC symbols used on the drawing.

Electrical plans show the details for installation of the electrical system and equipment. The locations of the meter, distribution panel, fixtures, switches, and special items are indicated, along with specifications for load capacities and wire sizes. The plan may also include a legend defining the symbols used on the drawing.

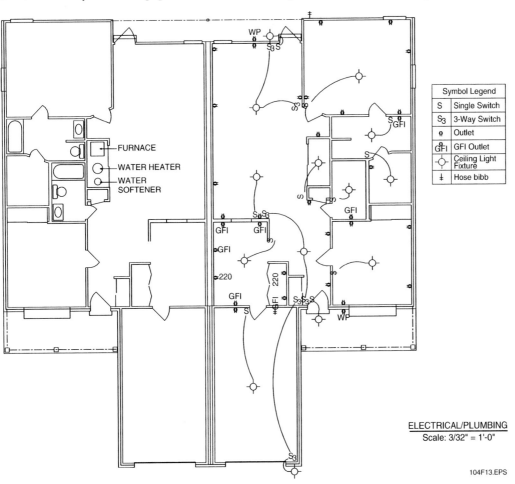

Figure 13 ♦ Example of an electrical/plumbing plan (NTS).

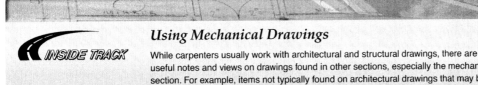

Using Mechanical Drawings

While carpenters usually work with architectural and structural drawings, there are useful notes and views on drawings found in other sections, especially the mechanical section. For example, items not typically found on architectural drawings that may be found on mechanical drawings include exposed HVAC ductwork, heating convectors, and fire sprinkler piping. Any one or all of these items may require the carpenter to build special framing, so make a habit of reviewing all drawings.

Discuss the types of information contained in shop drawings, and provide examples for the trainees to examine. Cover:
· Detail drawings
· Submittals

Discuss the use of as-built drawings and change orders.

Explain that soil conditions play a part in determining the type of foundation used.

Assign reading of Sections 3.00–3.5.0 for the next class session.

Show examples of and discuss the purpose of each type of line commonly used on drawings.

Show Transparency 18 (Figure 14).

Using a set of construction drawings, identify a change that must be made, such as the addition of a window. Have the trainees mark the drawings, including schedules, to reflect the change.

2.9.0 Shop Drawings

Shop drawings are specialized drawings that show how to fabricate and install components of a construction project. One type of shop drawing that a drafter creates after the engineer designs the structure is a detail drawing. It shows the locations of all holes and openings, and provides notes specifying how the part is to be made. Assembly instructions may be included. This type of drawing is used principally for structural steel members.

Another type of shop drawing (or submittal) pertains to the purchase of special items of equipment for installation in a building. This kind of shop drawing is usually prepared by the equipment manufacturer. This drawing shows overall sizes, details of construction, methods of securing the equipment to the structure, and all pertinent data that the architect and contractor need to know for the final placement and installation of the equipment.

Shop drawings produced by a contractor or fabricator are usually submitted to the owner or architect for approval and revisions or corrections. The design drawing is often put on the same sheet as the shop drawing.

2.10.0 As-Built Drawings

As-built drawings are drawings that are formally incorporated into the drawing set to record changes. These drawings are marked up by the various trades to show any differences between what was originally shown on a plan by the architect or engineer and what was actually built. Such changes result from the need to relocate equipment to avoid obstructions; to alter the location of a door, window, wall, etc., for some reason; or because the architect has changed a certain detail in the building design in response to customer preferences. On many jobs, any such changes to the design can only be made after a formal document called a change order has been generated and approved by the architect or other designated person. Depending on the complexity of the change, as-built drawings are typically outlined with a unique design or marked in red ink to make sure they stand out. Changes must be dated and initialed by the responsible party.

2.11.0 Soil Reports

Soil conditions are one of the factors that determine the type of foundation best suited for a structure. Building a structure on soil where the soil conditions can cause a large amount of uneven settlement to occur can result in cracks in the foundation and structural damage to the rest of the building. Therefore, in designing the foundation for a structure, an architect must consider the soil conditions of the building site. Typically, the architect consults a soil engineer who makes test bores of the soil on the building site and analyzes the samples. The results of the soil analysis are summarized in a soil report issued by the engineer. This report is often included as part of the drawing set.

3.0.0 ♦ READING AND INTERPRETING DRAWINGS

In order to read and interpret the information on drawings, you need to learn the special language used in construction drawings. This section of the module describes the different types of lines, dimensioning, symbols, and abbreviations used on drawings.

3.1.0 Lines Used on Drawings

Many different types of lines are used to draw and describe a structure. Lines are drawn wide, narrow, dark, light, broken, and unbroken, with each type of line conveying a specific meaning. *Figure 14* shows the most common lines used on

---

4.20 CARPENTRY FUNDAMENTALS LEVEL ONE ♦ TRAINEE GUIDE

Instructor's Notes:

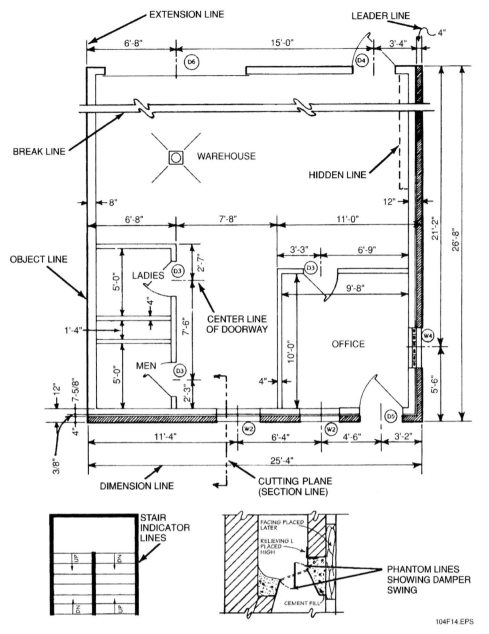

Figure 14 ♦ Drawing lines.

Show Transparency 19 (Figure 15).

construction drawings. The description of each type of line is as follows:

- *Object lines* – Heavier-weight lines used to show the main outline of the structure, including exterior walls, interior partitions, porches, patios, sidewalks, parking lots, and driveways.
- *Dimension and extension lines* – Used to provide the dimensions of an object. An extension line is drawn out from an object at both ends of the part to be measured. Extension lines are not supposed to touch the object lines. This is so they cannot be confused with the object lines. A dimension line is drawn at right angles between the extension lines and a number placed above, below, or to the side of it to indicate the length of the dimension line. Sometimes a gap is made in the dimension line and the number is written in the gap.
- *Center line* – Used to designate the center of an area or object and to provide a reference point for dimensioning. Center lines are typically used to indicate the centers of objects such as columns, posts, footings, and door openings.
- *Cutting plane (section line)* – Used to indicate an area that has been cut away and shown in a section view so that the interior features can be seen. The arrows at the ends of the cutting plane indicate the direction in which the section is viewed. Letters identify the cross-sectional view of that specific part of the structure. More elaborate methods of labeling section reference lines are used in larger, more complicated sets of plans (*Figure 15*). The sectional drawing may be on the same page as the reference line or on another page.
- *Break line* – Used to show that an object or area has not been drawn in its entirety.
- *Leader line* – Used to connect a note or dimension to a related part of the drawing. Leader lines are usually curved or at an angle from the feature being distinguished to avoid confusion with dimension and other lines.
- *Hidden line* – Used to indicate an outline that is invisible to an observer because it is covered by another surface or object that is closer to the observer.
- *Phantom line* – Used to indicate alternative positions of moving parts, such as a damper's swing or adjacent positions of related parts. It may also be used to represent repeated details.
- *Stair indicator line* – A short line with an arrowhead that shows the ascent or descent of stairs on a floor plan.
- *Contour lines* – **Contour lines** are drawn on a plot/site plan (covered later) and **topographical survey** to show the changes in elevation and the contour of the land. The lines may be dashed or solid. Dashed lines are used to show the natural grade, while solid lines show the finish grade to be achieved during construction. Each line across the plot of land represents a different elevation relative to some point such as sea level or a local feature. Each line is drawn at a uniform change in elevation, called a contour interval, such as every 10'. The closer together the contour lines, the steeper the terrain.

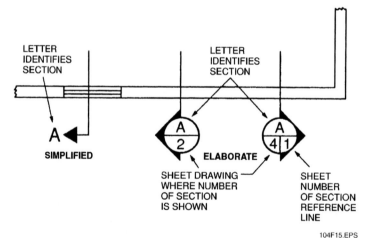

Figure 15 ♦ Methods of labeling section reference lines.

4.22 CARPENTRY FUNDAMENTALS LEVEL ONE ♦ TRAINEE GUIDE

Instructor's Notes:

3.2.0 Symbols Used on Drawings

Symbols are used in architectural plans and drawings to pictorially show different kinds of materials, fixtures, and structural members. The meaning of symbols and the types used are not standardized and can vary from location to location. A set of drawings generally includes a sheet that identifies the specific symbols used and their meanings.

> **NOTE**
> When using any drawing set, you should always refer to this sheet of symbols in order to avoid making mistakes when reading the drawings.

Some symbols are not intended to show part of an object, but are provided to guide you through the drawings themselves. Examples of this type include door and window designators that refer to door and window schedules where the different types are described. Other symbols are used to show the orientation of the object, showing the direction (north, south, front, back, and so on).

For electrical, plumbing, and HVAC, there are symbols that show the types of equipment to be installed, such as switches, lavatories, and warm air supply ducts. Each trade has its own symbols and abbreviations.

Symbols used on drawings generally fall into the following categories:

- Material (general plan) symbols
- Window and door symbols
- Electrical symbols
- Plumbing symbols
- HVAC symbols
- Plot plan and survey symbols
- Structural member symbols
- Welding symbols

3.2.1 Material Symbols

Typically, materials are shown on drawings in two ways (*Figure 16*). One way they are shown is as a **plan view,** where the material is presented as you would see it when looking down on it. The second way materials are shown is as an **elevation view** or section view, where the symbol shows the material roughly as it would look when you are facing it.

3.2.2 Window and Door Symbols

Floor plans show a great deal of information in a small drawing area; therefore, you have to rely on symbols for much of the information. Window and door symbols are usually shown with only a center line measurement to locate them from some reference point. However, more information about windows and doors can be found in separate window and door schedules contained in the drawing set. Window and door symbols are shown on drawings as either elevation or plan views (*Figure 17*).

3.2.3 Electrical Symbols

Electrical symbols are used to show the electrician the location of outlets and switches. They are also used to show auxiliary hardware such as buzzers, telephones, and computer hookups. *Figure 18* shows some common electrical symbols.

Switching arrangements are also indicated on the floor plans by using symbols. *Figure 19* shows some symbolic notations commonly used with electrical symbols on a floor plan.

Figure 19(A) shows the electrician that two outlets are controlled by two three-way switches.

Point out that symbols and their meanings are not standardized and can vary from architect to architect and region to region.

Explain that material symbols are depicted differently depending on whether they appear in a plan view or an elevation view.

Review window and door symbols. Explain that detailed information can be found on the associated schedules.

Review electrical symbols.

Show Transparencies 20 through 23 (Figures 16 through 19).

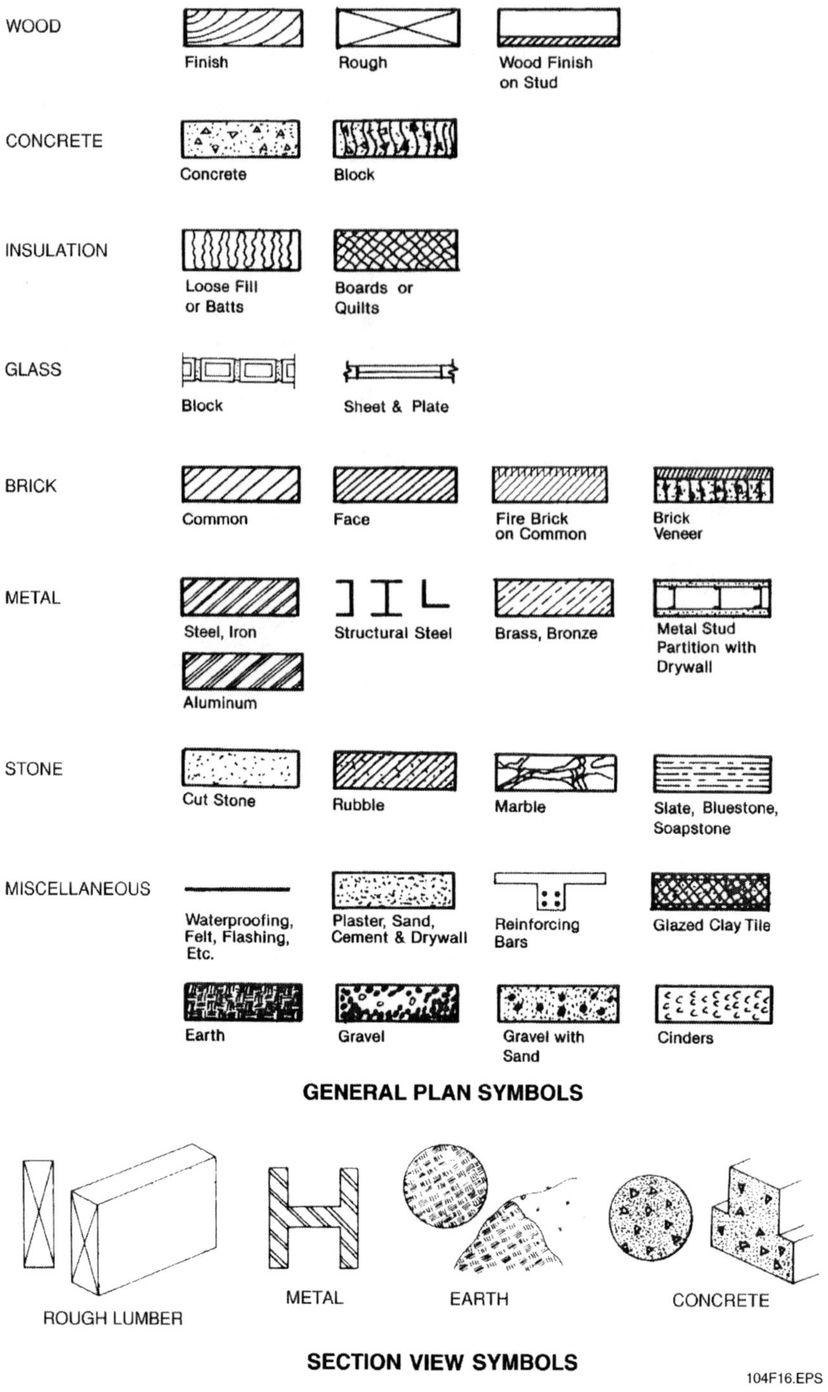

Figure 16 ◆ Typical material symbols.

Instructor's Notes:

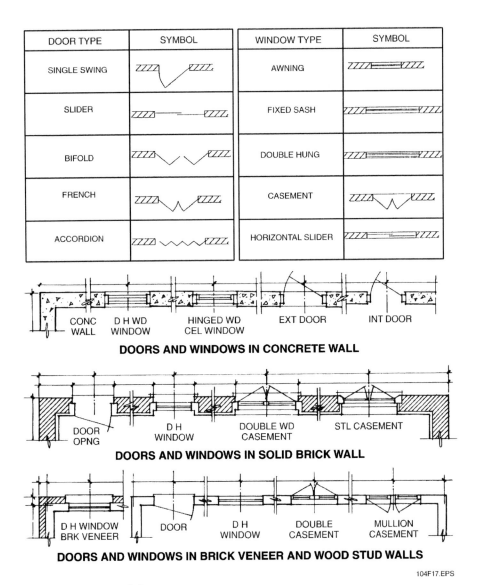

Figure 17 ♦ Window and door symbols.

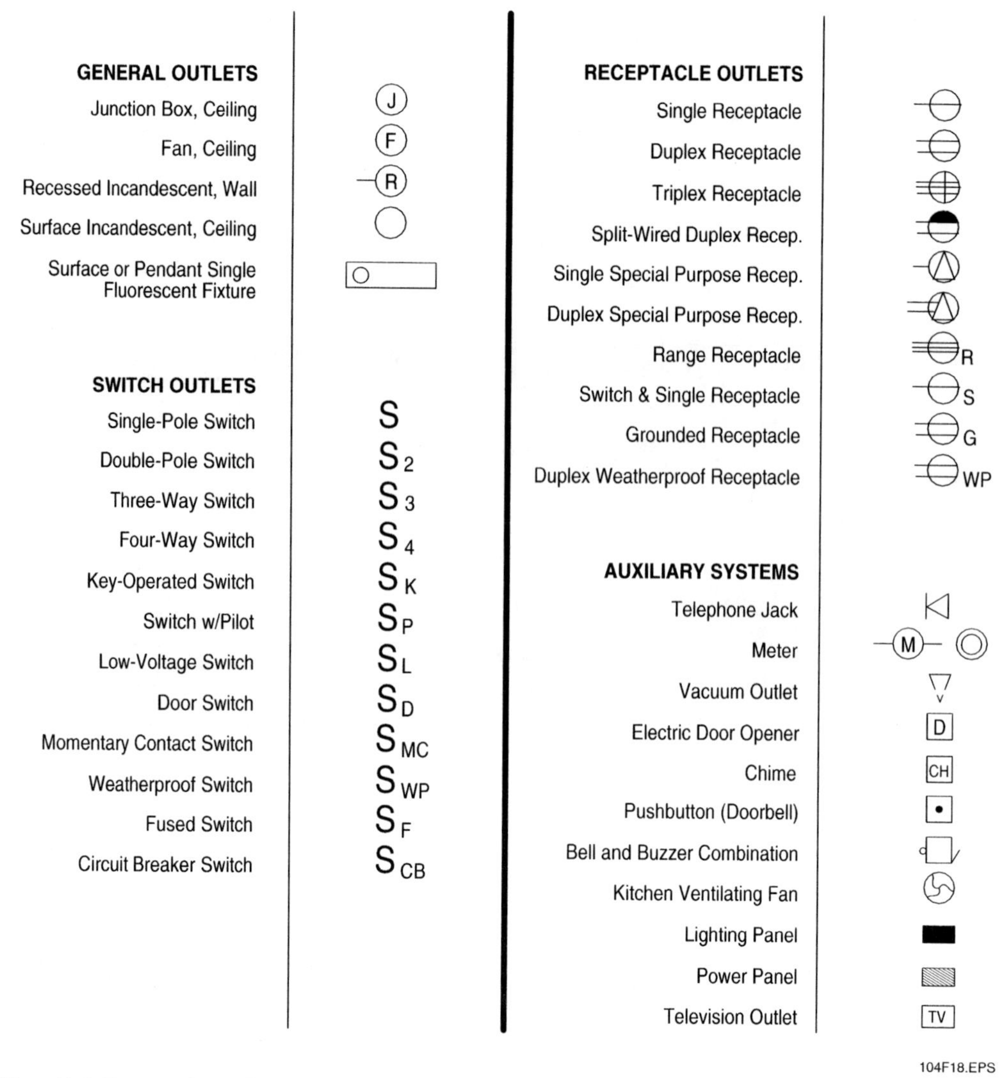

GENERAL OUTLETS

Junction Box, Ceiling	(J)
Fan, Ceiling	(F)
Recessed Incandescent, Wall	–(R)
Surface Incandescent, Ceiling	○
Surface or Pendant Single Fluorescent Fixture	[○]

SWITCH OUTLETS

Single-Pole Switch	S
Double-Pole Switch	S_2
Three-Way Switch	S_3
Four-Way Switch	S_4
Key-Operated Switch	S_K
Switch w/Pilot	S_P
Low-Voltage Switch	S_L
Door Switch	S_D
Momentary Contact Switch	S_{MC}
Weatherproof Switch	S_{WP}
Fused Switch	S_F
Circuit Breaker Switch	S_{CB}

RECEPTACLE OUTLETS

Single Receptacle	
Duplex Receptacle	
Triplex Receptacle	
Split-Wired Duplex Recep.	
Single Special Purpose Recep.	
Duplex Special Purpose Recep.	
Range Receptacle	R
Switch & Single Receptacle	S
Grounded Receptacle	G
Duplex Weatherproof Receptacle	WP

AUXILIARY SYSTEMS

Telephone Jack	
Meter	–(M)– ○
Vacuum Outlet	▽
Electric Door Opener	[D]
Chime	[CH]
Pushbutton (Doorbell)	[•]
Bell and Buzzer Combination	
Kitchen Ventilating Fan	
Lighting Panel	■
Power Panel	▨
Television Outlet	[TV]

Figure 18 ♦ Common electrical symbols.

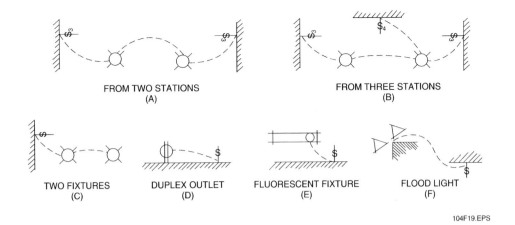

Figure 19 ◆ Electrical symbols showing control of an outlet.

Similarly, *Figure 19(B)* shows that two outlets are controlled by two three-way switches and a four-way switch.

Figures 19(C) through *19(F)* show various outlets and fixtures controlled by a single-pole switch. (A single-pole switch is a device that opens and closes one side of the circuit only.)

3.2.4 Plumbing Symbols

Plumbing symbols show all of the hardware and fixtures required for the building. Floor plans usually indicate only where the different fixtures should be located and plumbed.

Figure 20 shows some common fixture symbols as they would appear on a floor plan. It also shows some common piping symbols that would be used on special plumbing diagrams.

3.2.5 HVAC Symbols

The location and types of HVAC equipment are shown on the floor plan(s), sections, and detail drawings using symbols. For HVAC systems, symbols are also used to show the direction of movement of the hot and cold air in the system. *Figure 21* shows some common HVAC symbols.

3.2.6 Plot/Site Plan and Topographical Survey Symbols

Plot/site plans show the position and sizes of all relevant structures on the site as well as the features of the terrain. It is difficult to show the amount of information required on these drawings if symbols are not used. A good site drawing will include a legend of the symbols used, especially those that are non-standard. *Figure 22* shows some of the symbols commonly used on plot/site plan drawings.

3.2.7 Structural Member Symbols

Structural steel shapes have a system of identification by the use of symbols. Some examples of these symbols are shown in *Figure 23*.

3.2.8 Welding Symbols

Symbols for welding provide a means of conveying complete welding instructions from the designer to the welder. The symbols and their method of use are defined in the *American National Standard ANSI/AWS A2.4-79*, sponsored by the American Welding Society.

In *ANSI/AWS A2.4-79*, a distinction is made between the terms weld symbol and welding symbol. Weld symbols are used to indicate the type of welding to be performed. *Figure 24* shows some basic weld symbols.

A welding symbol is made up of as many as eight elements that are used together to provide exact welding instructions. *Figure 25* shows the standard elements of a welding symbol. As shown in *Figure 25*, the reference line is the key part of the welding symbol. All other elements are positioned with respect to this line. An arrow, which points to the location of the weld, is positioned on one end of the reference line. When necessary, a tail is positioned on the other end of the reference line. Other elements, such as the appropriate weld symbol used to show the type of weld, are placed on the reference line.

Review:
- Plumbing symbols
- HVAC symbols
- Plot/site plan symbols
- Structural steel symbols
- Welding symbols

Show Transparencies 24 through 29 (Figures 20 through 25).

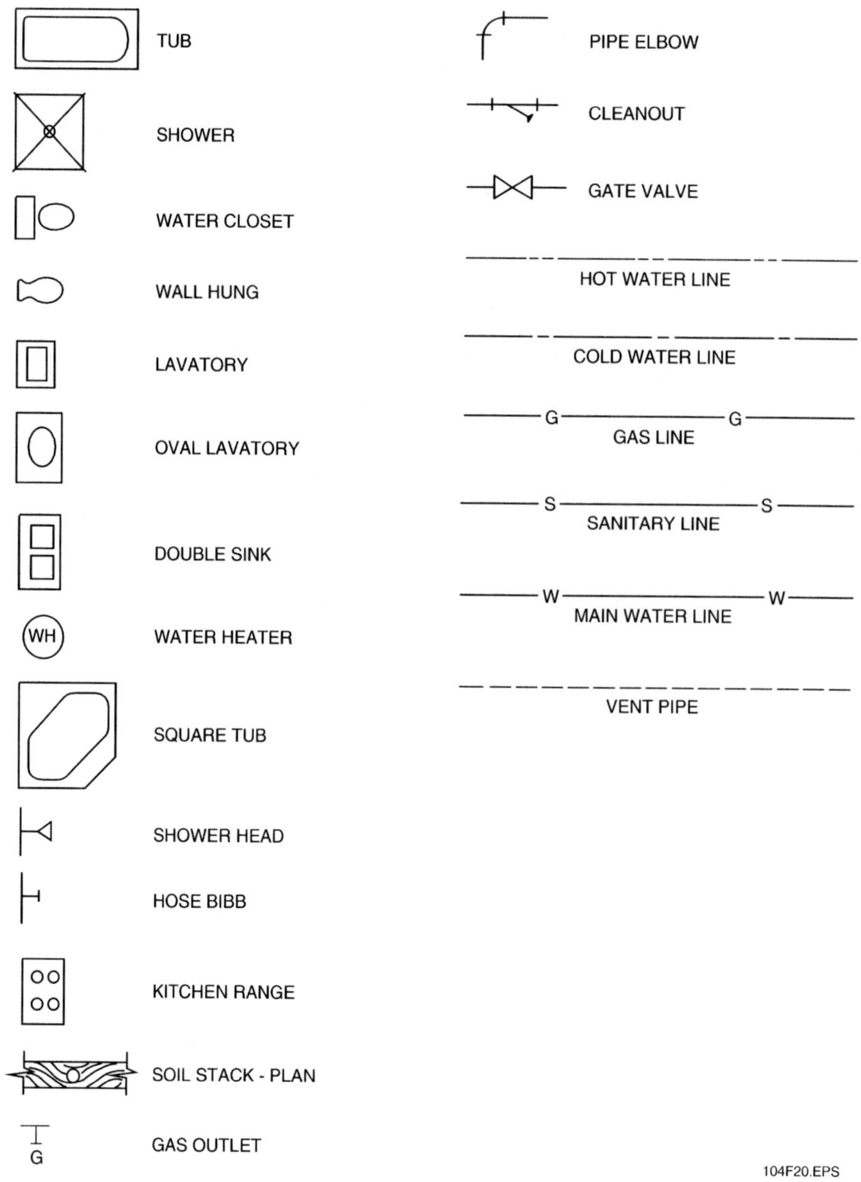
TUB
SHOWER
WATER CLOSET
WALL HUNG
LAVATORY
OVAL LAVATORY
DOUBLE SINK
WH
WATER HEATER
SQUARE TUB
SHOWER HEAD
HOSE BIBB
KITCHEN RANGE
SOIL STACK - PLAN
G
GAS OUTLET
PIPE ELBOW
CLEANOUT
GATE VALVE
HOT WATER LINE
COLD WATER LINE
G G
GAS LINE
S S
SANITARY LINE
W W
MAIN WATER LINE
VENT PIPE
104F20.EPS

Figure 20 ◆ Common plumbing symbols.

Instructor's Notes:

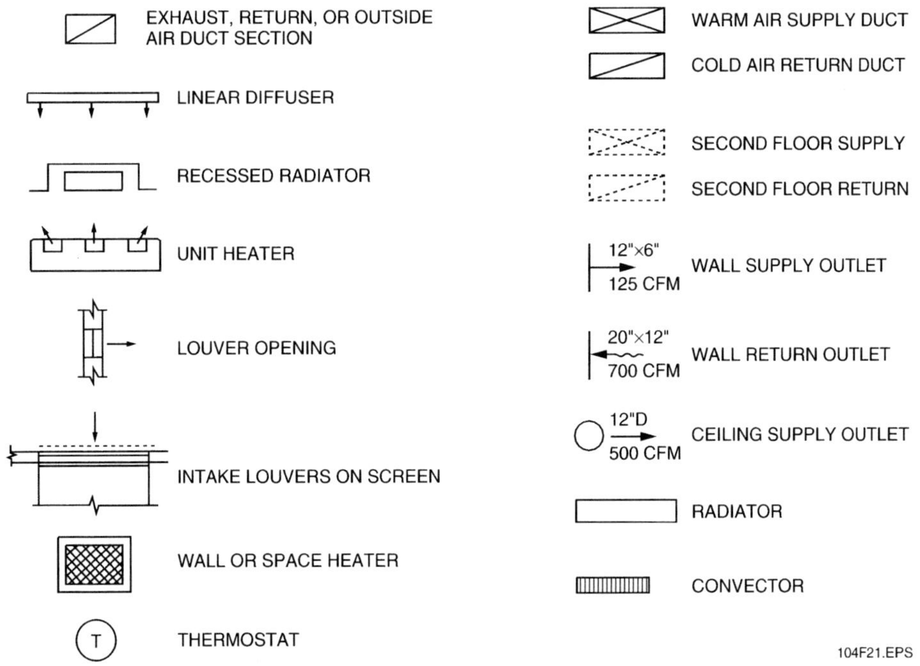

Figure 21 ◆ Common HVAC symbols.

MODULE 27104-06 ◆ READING PLANS AND ELEVATIONS 4.29

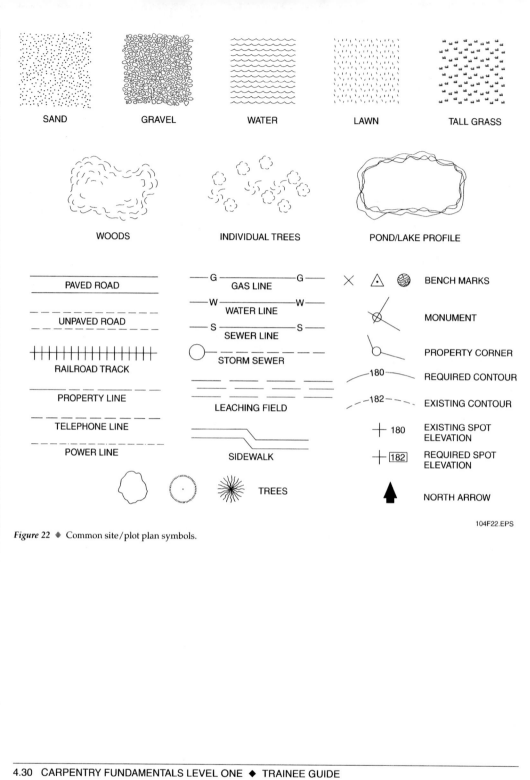

Figure 22 ♦ Common site/plot plan symbols.

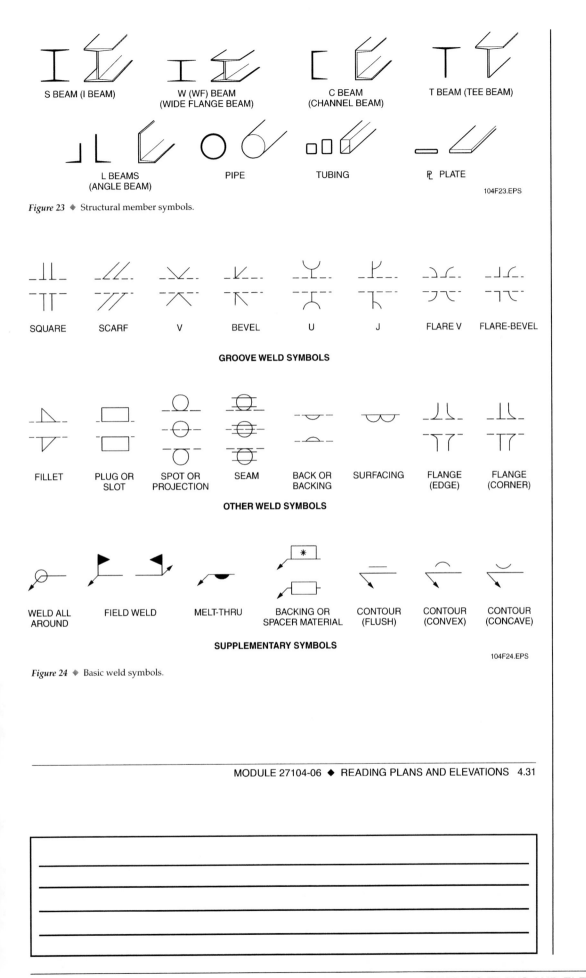

Figure 23 ◆ Structural member symbols.

Figure 24 ◆ Basic weld symbols.

Take a field trip to a building site for which you have access to the drawings. Have trainees identify various building components and locate them on the plans.

Explain that dimensions may be depicted from the outside to center, center to center, wall to wall, or outside to outside. Emphasize that dimensions shown on drawings are given in full scale.

Discuss nominal dimensions and explain how they differ from actual dimensions.

Show Transparency 30 (Figure 26).

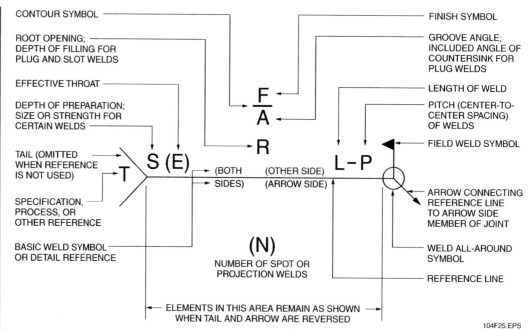

Figure 25 ♦ Elements of a welding symbol.

Information placed above the reference line indicates the weld is to be made on the other side of the joint from where the arrow points. Information placed below the reference line indicates the weld is to be made on the arrow side of the joint. Dimensions for the welds are drawn on the same side of the reference line as the weld symbol. Weld contour symbols are either flush, concave, or convex. The contour symbol is located next to the weld symbol. Finish symbols usually accompany the contour symbol. The weld groove angle is shown on the same side of the reference line as the weld symbol. The size (depth) is placed to the left of the weld symbol, and the root opening is shown inside the weld symbol.

3.3.0 Dimensioning

Dimensions given on drawings show actual sizes, distances, and heights of the objects and spaces being represented. Dimensions may be from outside to center, center to center, wall to wall, or outside to outside (*Figure 26*). In all cases, dimensions shown on drawings are given in full scale regardless of the fact that the plan shows an object or distance on a smaller scale. Note that sectional and detail views may use a **nominal size** in labeling.

Nominal sizes or dimensions are approximate or rough sizes by which lumber, block, etc., is commonly known and sold. For example, a nominal size 2 × 4 board is actually 1½" × 3½".

Sometimes it is necessary to determine dimensions not shown on a drawing by measuring them with an architect's scale. Plan drawings are created using a specified scale. Inches or fractions of an inch on the drawing are used to represent feet in the actual measurement of a building. For example, in a plan drawn to ¼" scale, ¼" on the drawing represents 1' of the building. The scale of a drawing is usually shown directly below the drawing. The same scale may not be used for all the drawings that make up a complete set of plans.

NOTE

When they are available, always use the dimensions shown on a drawing rather than ones obtained by scaling the drawing. This is because some reproduction methods used to make copies of drawings can introduce errors in the reproduced image.

4.32 CARPENTRY FUNDAMENTALS LEVEL ONE ♦ TRAINEE GUIDE

Instructor's Notes:

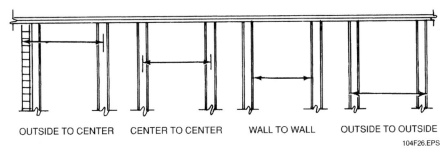

Figure 26 ♦ Common methods of indicating dimensions on drawings.

Review the metric system of measurement.

Discuss the "Think About It." Note that the answer is 24'.

Metric Measurement

In the United States, length measurements shown on architectural drawings are commonly given in feet and inches using the English system of measurement, while in Canada and other parts of the world, the metric system of measurement is used. The two common length measurements used in the metric system on architectural drawings are the meter and millimeter, the millimeter being $1/1000$ of a meter. On drawings drawn to scales between 1:1 and 1:100, the millimeter is typically used. The millimeter symbol (mm) will not be shown, but there should be a note on the drawing indicating that all dimensions are given in millimeters unless otherwise noted. On drawings drawn to scales between 1:200 and 1:2,000, the meter is generally used. Again, the meter symbol (m) will not be shown and the drawing should have a note indicating that all dimensions are given in meters unless otherwise noted. Land distances shown on plot or site plans expressed in metric units typically are given in meters or kilometers (1,000 meters). Conversion factors that can be used to convert between English and metric linear units of measure are provided below.

1 kilometer = 0.62137 mile (mi)
1 meter = 39.37 inches (in)
1 meter = 3.2808 feet (ft)
1 meter = 1.0936 yards (yd)
1 centimeter = 0.3937 inch (in)
1 millimeter = 0.03947 inch (in)

1 mile = 1.609 kilometers (km)
1 yard = 0.9144 meter (m)
1 foot = 0.3048 meter (m)
1 foot = 304.8 millimeters (mm)
1 inch = 2.54 centimeters (cm)
1 inch = 25.4 millimeters (mm)

Scaling Drawings

Measuring the length of an object or line on a drawing and converting the measurement to actual length using the given scale is called scaling a drawing. If a line on a floor plan measures 6" in actual length and the object being represented is drawn to a scale of ¼" = 1', what is the true length of the object?

MODULE 27104-06 ♦ READING PLANS AND ELEVATIONS 4.33

Describe common practices used for dimensioning drawings. Point out that these practices may vary by region.

Have the trainees determine the actual dimensions of a partition or other structure from a scaled drawing.

Architect's Scale

Measurements are usually made on architectural drawings using an architect's scale rather than a standard ruler. Architect's scales like the one shown on the left are divided into feet and inches and usually consist of several scales on one rule. Architect's scales also come in other forms such as tapes or with wheels, like the one shown on the right.

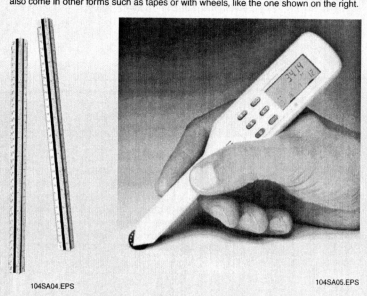

Some common practices used for dimensioning drawings are listed below. Keep in mind that these are not rules. The practices in your area may be different.

- Architectural dimension lines are unbroken lines, with the dimensions placed above and near the center of the line.
- Dimensions over one foot are shown in feet and inches (not decimals). Dimensions less than one foot are shown in inches only. The common exception to this rule is the center-to-center distances for standard construction, such as for framing 16" on center (OC) or 24" OC.
- Dimensions are placed to read from the right or from the bottom of the drawing.
- Overall building dimensions go to the outside of all other dimensions.
- Room sizes can be shown by stating width and length.
- Rooms are sometimes dimensioned from the center lines of partition walls, but wall-to-wall dimensions are more common.
- Window and door sizes are usually shown in window and door schedules.
- Dimensions that cannot be shown on the floor plan because of their size are placed at the end of leader lines.
- When stairs are dimensioned, the number of risers is placed on a line with an arrow indication in either an up or down direction.
- Architectural dimensions always refer to the actual size of the building, regardless of the scale of the drawings.

4.34 CARPENTRY FUNDAMENTALS LEVEL ONE ◆ TRAINEE GUIDE

Instructor's Notes:

3.4.0 Abbreviations

Many written instructions are needed to complete a set of construction drawings. It is impossible to print out all such references, so a system of abbreviations is used. By using standard abbreviations, such as BRK for brick or CONC for concrete, the architect ensures that the drawings will be accurately interpreted. A brief list of some commonly used abbreviations is contained in *Appendix A*. Note that some architects and engineers may use different abbreviations for the same terms. Normally, the title sheet in a drawing set contains a list of abbreviations used in the drawings. For this reason, it is important to get a complete set of drawings and specifications including the title sheet(s), so that you can better understand the exact abbreviations used.

Some practices for using abbreviations on drawings are as follows:

- Most abbreviations are capitalized.
- Periods are used when abbreviations might look like a whole word.
- Abbreviations are the same whether they are singular or plural.
- Several terms have the same abbreviations and can only be identified from the context in which they are found.
- Many abbreviations are similar.

3.5.0 Architectural Terms Used in Drawings and Specifications

There are many architectural terms used in plans and specifications. Some commonly used terms are listed in *Appendix B*. You should already be familiar with many of the terms listed.

4.0.0 ◆ GUIDELINES FOR READING A DRAWING SET

The following general procedure is suggested as a method of reading a set of drawings for maximum understanding.

Step 1 Acquire a complete set of drawings and specifications, including the title sheet(s), so that you can better understand the abbreviations, symbols, etc., used throughout the drawings.

Step 2 Read the title block. The title block tells you what the drawing is about. Take note of the critical information such as the scale, date of last revision, drawing number, and architect or engineer. After you use a sheet from a set of drawings, be sure to refold the sheet with the title block facing up.

Step 3 Find the North arrow. Always orient yourself to the structure. Knowing where north is enables you to more accurately describe the location of walls and other parts of the building.

Step 4 Always be aware that the drawings work together as a group. The reason the architect or engineer draws plans, elevations, and sections is that drawings require more than one type of view to communicate the whole project. Learn how to use more than one drawing when necessary to find the information you need.

Step 5 Check the list of drawings in your set. Note the sequence of the various plans. Some drawings have an index on the front cover. Notice that the prints in the set are of several categories:
- Architectural
- Structural
- Mechanical
- Electrical
- Plumbing

Step 6 Study the plot/site plan to determine the location of the building to be constructed, the various utilities, roadways, and any easements.

Step 7 Check the floor plan for the orientation of the building. Observe the location and features of entries, corridors, offsets, and any special features.

Step 8 Check the foundation plan for the size and types of footings, reinforcing steel, and loadbearing substructures.

Step 9 Check the floor and wall construction and other details relating to exterior and interior walls.

Step 10 Study the features that extend for more than one floor, such as plumbing and vents, stairways, elevator shafts, and heating and cooling ductwork.

Step 11 Study the plumbing and/or mechanical plans for details of any framing or other carpentry work pertaining to the installation of the heating, cooling, and plumbing equipment. You must coordinate this type of work with members of the HVAC and plumbing trades.

Explain that abbreviations are used to simplify construction drawings and may vary from drawing to drawing. Refer trainees to *Appendix A* in the Trainee Module.

Explain that carpenters must be familiar with many architectural terms, some of which are defined in *Appendix B* of the Trainee Module.

Hand out Worksheet 27104-1. Have the trainees complete the Worksheet. This laboratory corresponds to Performance Task 1.

Assign reading of Section 4.0.0 for the next class session.

Demonstrate the process of reviewing a typical set of construction drawings or plans.

Hand out Worksheets 27104-2 and 27104-3. Have the trainees complete the Worksheets. This laboratory corresponds to Performance Tasks 2 through 6.

Assign reading of Sections 5.0.0–5.1.3 for the next class session.

Explain that specifications define the quality of work to be done and the materials to be used. Provide examples for the trainees to examine.

Point out that there are two types of information contained in a set of specifications: special and general conditions and technical aspects of construction. Discuss the types of information contained in each section.

Step 12 Study the electrical plans for details of any framing or other carpentry work pertaining to the installation of the electrical system, fixtures, or special equipment. You must coordinate this type of work with members of the electrical trade.

Step 13 Check the notes on various pages and compare the specifications against the construction details.

Step 14 Thumb through the sheets of drawings until you are familiar with all the plans and structural details.

Step 15 Recognize applicable symbols and their relative locations in the plans. Note any special construction details or variations that will affect the carpentry work.

5.0.0 ◆ SPECIFICATIONS

Specifications are written instructions provided by architectural and engineering firms to the general contractor and, consequently, to the subcontractors. Specifications are just as important as the drawings in a set of plans. They furnish what the drawings cannot in that they define the quality of work to be done and the materials to be used. Specifications serve several important purposes:

- Clarify information that cannot be shown on the drawings
- Identify work standards, types of materials to be used, and the responsibility of various parties to the contract
- Provide information on details of construction
- Serve as a guide for contractors bidding on the construction job
- Serve as a standard of quality for materials and workmanship
- Serve as a guide for compliance with building codes and zoning ordinances
- Serve as the basis of agreement between the owner, architect, and contractors in settling any disputes

Specifications are legal documents. When there is a difference between the drawings and the specifications, the specifications normally take legal precedence over the working drawings. However, it is often the case that the plans are more specific to the job than the specifications. Therefore, notes on the plans may be considered by the architect/owner to be the true intent. You must be very careful to watch for discrepancies between the plans and specifications and report them to your supervisor immediately.

5.1.0 Organization and Types of Specifications

Specifications consist of various elements that may differ between construction jobs. For small projects, they may be simple; for large projects, complex. Basically, two types of information are contained in a set of specifications: special and general conditions, and technical aspects of construction.

5.1.1 Special and General Conditions

Special and general conditions cover the nontechnical aspects of the contractual agreements. Special conditions cover topics such as safety and temporary construction. General conditions cover the following points of information:

- Contract terms
- Responsibilities for examining the construction site
- Types and limits of insurance
- Permits and payments of fees
- Use and installation of utilities
- Supervision of construction
- Other pertinent items

The general conditions section is the area of the construction contract where misunderstandings often occur. Therefore, these conditions are usually much more explicit on large, complicated construction projects. Note that residential specifications often do not spell out general conditions and are basically material specifications only. An example of a typical residential material specification is shown in *Figure 27*.

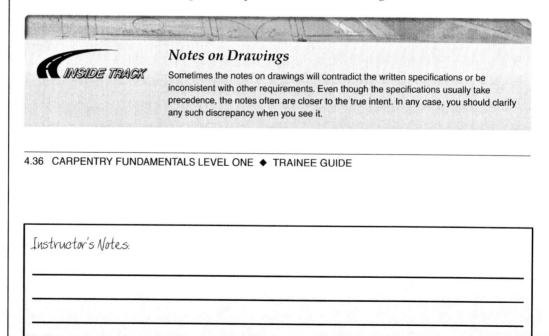

Notes on Drawings

Sometimes the notes on drawings will contradict the written specifications or be inconsistent with other requirements. Even though the specifications usually take precedence, the notes often are closer to the true intent. In any case, you should clarify any such discrepancy when you see it.

4.36 CARPENTRY FUNDAMENTALS LEVEL ONE ◆ TRAINEE GUIDE

Instructor's Notes:

Form RD 1924-2 (Rev. 7-99)	UNITED STATES DEPARTMENT OF AGRICULTURE U.S. DEPARTMENT OF HOUSING AND URBAN DEVELOPMENT FEDERAL HOUSING ADMINISTRATION U.S. DEPARTMENT OF VETERANS AFFAIRS	FORM APPROVED OMB NO. 0575-0042

☐ Proposed Construction

☐ Under Construction

DESCRIPTION OF MATERIALS

No. _____ (To be inserted by Agency)

Property address _____ City _____ State Oklahoma

Mortgagor or Sponsor _____ (Name) _____ (Address)

Contractor or Builder _____ (Name) _____ (Address)

INSTRUCTIONS

1. For additional information on how this form is to be submitted, number of copies, etc., see the instructions applicable to the FHA Application for Mortgage Insurance, VA Request for Determination of Reasonable Value, or other, as the case may be.
2. Describe all materials and equipment to be used, whether or not shown on the drawings, by marking an X in each appropriate check-box and entering the information called for in each space. If space is inadequate, enter "See misc", and describe under item 27 or on an attached sheet. THE USE OF PAINT CONTAINING MORE THAN THE PERCENT OF LEAD BY WEIGHT PERMITTED BY LAW IS PROHIBITED.
3. Work not specifically described or shown will not be considered unless required, then the minimum acceptable will be assumed. Work exceeding minimum requirements cannot be considered unless specifically described.
4. Include no alternates, "or equal" phrases, or contradictory items. (Consideration of a request for acceptance of substitute materials or equipment is not thereby precluded.)
5. Include signatures required at the end of this form.
6. The construction shall be completed in compliance with the related drawings and specifications, as amended during processing. The specifications include this Description of Materials and the applicable building code.

1. EXCAVATION:
   Bearing soil, type Firm clay; Note: Where fill is in excess of 18", concrete piers to be installed at 8' O.C. and the cost will be added to the contract.

2. FOUNDATIONS:
   Footings: concrete mix transite 14"x18" ftg ; strength psi 2500 PSI Reinforcing (4) 5/8" steel rebar
   Foundation wall: material 2500 PSI concrete concrete Reinforcing
   Interior foundation wall: material 2500 PSI concrete Party foundation wall 2500 concrete
   Columns: material and sizes _____ Piers: material and reinforcing _____
   Girders: material and sizes _____ Sills: material W.Coast Utility Douglas Fir w/sill sealer
   Basement entrance areaway _____ Window areaways _____
   Waterproofing waterproof mix in concrete Footing drains open mortar joints
   Termite protection Pretreat soil and stem wall with Chlordane and issue 5 year warranty.
   Basementless space: ground cover _____; insulation _____; foundation vents _____
   Special foundations: Thicken slab beam with (2) 5/8" steel rebars to undisturbed soil.
   Additional information: See patio details, also porch and fireplace footing.

3. CHIMNEYS:
   Material brick Prefabricated (make and size) _____
   Flue lining: material terra cotta Heater flue size 6" metalbestos Fireplace flue size 12"x12"
   Vents (material and size): gas or oil heater _____; water heater metalbestos
   Additional information: _____

4. FIREPLACES:
   Type: ☐ solid fuel☒ gas-burning; ☐ circulator (make and size) 42"vestal damper Ash dump and clean-out metal
   Fireplace: Facing brick ; lining firebrick ; hearth brick ; mantel wood
   Additional information: Note: Fireplace stubbed and keyed for gas outlet.

5. EXTERIOR WALLS: 2x4 studs @ 16" O.C.   1/2" plywood at corners
   Wood frame: wood grade, and species w.c. utility d.f. ☒ Corner bracing. Building paper or felt 15# felt
   Sheathing flintcoat ; thickness 3/4" ; width 24" ; ☒ solid; ☐ space _____ "o.c. ; ☐ diagonal; _____
   Siding polyurethane ; grade _____; type _____; size _____; exposure _____"; fastening _____
   Shingles _____; grade _____; type _____; size _____; exposure _____"; fastening _____
   Stucco _____; thickness _____"; Lath _____; weight _____ lb.
   Masonry veneer face brick Sills brick Lintels steel supporting brick Base flashing _____
   Masonry: ☐ solid ☐ faced ☐ stuccoed; total wall thickness _____"; facing thickness _____"; facing material _____
   Backup material _____; thickness _____"; bonding _____
   Door sills _____ Window sills _____ Lintels _____ Base flashing _____
   Interior surfaces: dampproofing, _____ coats of _____; furring _____
   Additional information: Two coats of exterior paint or stain as selected by owner.
   Exterior painting: material _____; number of coats _____
   Gable wall construction: ☐ same as main walls; ☒ other construction Cedar plywood or vertical siding as shown on elevations.

6. FLOOR FRAMING:
   Joists: wood, grade, and species _____; other _____; bridging _____; anchors 1/2"x8"
   Concrete slab: ☐ basement floor; ☐ first floor; ☒ ground supported; ☐ self-supporting; mix 3500 PSI concrete; thickness 4".
   reinforcing 6x6-W1.4xW1.4 WWF ; insulation 1" perimeter ; membrane waterproof concrete
   Fill under slab; material sand fill ; thickness 4 ". Additional information: _____

7. SUBFLOORING: (Describe underflooring for special floors under item 21.)
   Material: grade and species 3/4" AC exterior plywood ; size _____; type _____
   Laid: ☐ first floor; ☐ second floor ☒ attic 200 sq. ft.; ☐ diagonal; ☐ right angles. Additional information: _____

8. FINISH FLOORING: (Wood only. Describe other finish flooring under item 21.)

LOCATION	ROOMS	GRADE	SPECIES	THICK-NESS	WIDTH	BLDG. PAPER	FINISH
First floor	living area, formal dining,						
Second floor	hall and bedrooms: pad	and carpet	installed.				allowance per square foot
Attic floor	sq. ft.						

Additional information: _____

According to the Paperwork Reduction Act of 1995, an agency may not conduct or sponsor, and a person is not required to respond to a collection of information unless it displays a valid OMB control number. The valid OMB control number for this information collection is 0575-0042. The time required to complete this information collection is estimated to average 15 minutes per response, including the time for reviewing instructions, searching existing data sources, gathering and maintaining the data needed, and completing and reviewing the collection of information.

HUD-FHA 2005
VA Form 26-1852

RD 1924-2 (Rev. 7-99)

104F27A.EPS

Figure 27 ♦ Typical materials specification (1 of 4).

DESCRIPTION OF MATERIALS

9. **PARTITION FRAMING:**
   Studs: wood, grade, and species w.c. utility d.f. size and spacing 2x4 studs @ 16" oc Other _____
   Additional information: NOTE: 2x6 studs where needed for plumbing vents.

10. **CEILING FRAMING:**
    Joists: wood, grade, and species w.c. utility d.f. Other see floor plan Bridging midspan stiffener
    Additional information: See typical sheet for construction details. 2x4 flat and 2x6 up

11. **ROOF FRAMING:**
    Rafters: wood, grade, and species w.c. utility d.f. Roof trusses (see detail): grade and species _____
    Additional information: 2x6 rafters @ 24" oc, unless spans exceed allowable.

12. **ROOFING:**
    Sheathing: wood, grade, and species west coast utility Douglas fir ; ☐ solid ☒ spaced ___ " o.c.
    Roofing 240 #3 tab sq. butt ; grade #1 ; size_____ ; type_____
    Underlay asphalt shingles ; weight or thickness 5/2 ; size 16" ; fastening nail
    Built-up roofing _____ ; number of plies _____ ; surface material _____
    Flashing: material galvanized iron ; gage or weight 26 gauge ; ☐ gravel stops; ☐ snow guards
    Additional information: _____

13. **GUTTERS AND DOWNSPOUTS:**
    Gutters: material galvanized iron ; gage or weight 26 gauge; size 5" ; shape o-gee _____
    Downspouts: material galvanized iron; gage or weight 26 gauge; size 3"x4" ; shape square ; number as need
    Downspouts connected to: ☐ Storm sewer; ☐ sanitary sewer; ☐ dry-well. ☒ Splash blocks: material and size _____
    Additional information: _____

14. **LATH AND PLASTER:**
    Lath ☐ walls, ☐ ceilings: material _____ ; weight or thickness _____ Plaster: coats ___ ; finish _____
    Dry-wall ☒ walls, ☒ ceilings: material Sheetrock ; thickness 1/2" ; finish medium texture
    Joint treatment mud and tape; NOTE: Sheetrock applied according to manufacturer's specs.

15. **DECORATING:** (Paint, wallpaper, etc.)

ROOMS	WALL FINISH MATERIAL AND APPLICATION	CEILING FINISH MATERIAL AND APPLICATION
Kitchen		
Bath	ALL ROOMS: paint and/or wallpaper	all ceilings to be textured
Other	WALLPAPER ALLOWANCE: $	and painted two coats flat paint.

    Additional information: PANELING: 1/4" v-groove ash (includes cost of labor)

16. **INTERIOR DOORS AND TRIM:**
    Doors: type hollow core slab ; material ash or birch ; thickness 1-3/8"
    Door trim: type detail ; material white pine Base: type detail ; material white pine ; size 3-1/4"
    Finish: doors hand-rubbed stain ; trim hand-rubbed stain
    Other trim (item, type and location) NOTE: All headers to meet schedule as shown on detail sheet.
    Additional information: _____

17. **WINDOWS:** All Thermopane Glass - fixed and operating
    Windows: type single hung ; make Alenco or equal ; material aluminum ; sash thickness 1"
    Glass: grade DS6 ; ☐ sash weights; ☒ balances, type spring ; head flashing galvanized
    Trim: type Sheetrock return ; material Sheetrock Paint texture & paint ; number coats 2
    Weatherstripping: type with units ; material wool pile Storm sash, number _____
    Screens: ☒ full; ☐ half; type metal or aluminum ; number all ; screen cloth material 16" aluminum mesh
    Basement windows: type _____ ; material _____ ; screens, number _____ ; Storm sash, number _____
    Special windows See plans for fixed windows
    Additional information: NOTE: Sheetrock return on all windows with wood stool and apron.

18. **ENTRANCES AND EXTERIOR DETAIL:**
    Main entrance door: material ash panel ; width see plans ; thickness 1-3/4"; Frame: material w. pine ; thickness 1-3/8"
    Other entrance doors: material W.P. m.c. ; width see plans ; thickness 1-3/4"; Frame: material w. pine ; thickness 1-3/8"
    Head flashing galvanized iron Weatherstripping: type spring bronze ; Thresholds aluminum
    Screen doors: thickness ___" ; number _____ ; screen cloth material _____ Storm doors: thickness ___" ; number _____
    Combination storm and screen doors: thickness ___" ; number _____ ; screen cloth material _____
    Shutters: ☐ hinged; ☐ fixed. Railings _____ , Attic louvers galvanized back screen in soffit
    Exterior millwork: grade and species redwood or rough cedar Paint exterior paint or stain ; number coats 2
    Additional information: Thermopane glass sliding doors with aluminum frame

19. **CABINETS AND INTERIOR DETAIL:**
    Kitchen cabinets, wall units: material ash with white pine stiles ; lineal feet of shelves _____ ; shelf width 12"
    Base units: material ash with white pine ; counter top Formica ; edging self-edge Formica
    Back and end splash full Formica Finish of cabinets enamel or hand-rubbed stain ; number coats 5
    Medicine cabinets: make _____ ; model _____
    Other cabinets and built-in furniture Ash raised panel door and drawer fronts -- shop built
    Additional information: Built-in vanities with synthetic marble or Formica top and splash; built-in ash bookcases (See Plans)

20. **STAIRS:**

STAIR	TREADS		RISERS		STRINGS		HANDRAIL		BALUSTERS	
	Material	Thickness	Material	Thickness	Material	Thickness	Material	Thickness	Material	Thickness
Basement										
Main										
Attic										

    Disappearing: make and model number pull-down ladder to attic access in garage. 2040
    Additional information: _____

HUD-FHA 2005
VA Form 26-1852

Figure 27 ♦ Typical materials specification (2 of 4).

21. SPECIAL FLOORS AND WAINSCOT: (Describe carpet as listed in Certified Products Directory.)

	Location	Material, Color, Border, Sizes, Gage, Etc.	Threshold Material	Wall Base Material	Underfloor Material
Floors	Kitchen	utility: sheet vinyl			
	Bath	ceramic tile	ceramic	white pine	concrete
	Main entry	earthstone	aluminum	white pine	concrete
	Rear entry	sheet vinyl	aluminum		concrete

	Location	Material, Color, Border, Sizes, Gage, Etc.	Height (tub & shower)	Height Over Tub	Height in Showers (From Floor)
Wainscot	Bath	ceramic tile over met'l lath and grout area only)		72"	72"
	remainder of bath walls: 1/4" v-groove ash wainscoting 32" high				

Bathroom accessories: ☐ Recessed; material _chrome_ (brushed) ; number ___ ; ☐ Attached; material _chrome_ (brushed) ; number ___
Additional information: _____

22. PLUMBING:

Fixture	Number	Location	Make	Mfr's Fixture Identification No.	Size	Color
Sink	1	kitchen	Kohler	Delafield k 5950	32"x21"	color
Lavatory		baths	Kohler	Caxton K-2210	19"x15"	color
Water closet		baths	Kohler	Wellsworth K-3500 eb	reverse trap	color
Bathtub		bath	Kohler	Villager K-715	60"	color
Shower over tubΔ		bath	Kohler	Chrome divertor over tub		
Stall showerΔ		bath	ceramic stall shower with glass door (job built)			
Laundry trays						
Optional:	1	Kohler party bar sink with single lever faucet in utility room				

NOTE: Kohler faucets on all lavatories and sinks.

Δ ☒ Curtain rod   Δ ☒ Door   ☐ Shower pan: material _____
Water supply: ☒ public; ☐ community system; ☐ individual (private) system.*
Sewage disposal: ☒ public; ☐ community system; ☐ individual (private) system.*
* Show and describe individual system in complete detail in separate drawings and specifications according to requirements.
House drain (inside): ☐ cast iron; ☐ tile; ☒ other _ABS or PVC_ House sewer (outside): ☐ cast iron; ☐ tile; ☐ other _4" ABS_
Water piping: ☐ galvanized steel; ☒ copper tubing; ☐ other _typeL, hard-drawn joints_ Sill cocks, number _frostproof_
Domestic water heater: type _automatic_ ; make and model _Rheem or State_ ; heating capacity _40 gallons_
_25.2_ gph. 100° rise. Storage tank: material _glass lined_ ; capacity _40_ gallons.
Gas service: ☒ utility company; ☐ liq. pet. gas; ☐ other _____ Gas piping: ☐ cooking; ☒ house heating.
Footing drains connected to : ☐ storm sewer; ☐ sanitary sewer; ☐ dry well. Sump pump; make and model _____
_____ ; capacity _____ ; discharges into _____

23. HEATING:
☐ Hot water.   ☐ Steam.   ☐ Vapor.   ☐ One-pipe system.   ☐ Two-pipe system.
  ☐ Radiators.   ☐ Convectors.   ☐ Baseboard radiation. Make and model _____
Radiant panel: ☐ floor; ☐ wall; ☐ ceiling. Panel coil: material _____
☐ Circulator.   ☐ Return pump. Make and model _____ ; capacity _____ gpm.
Boiler: make and model _____ Output _____ Btuh; net rating _____ Btuh.
Additional information: _____
Warm air: ☐ Gravity.   ☒ Forced. Type of system _perimeter under floor_
  Duct material: supply _concrete_ ; return _sheet metal_ Insulation _fiberglass_ ; thickness _1"_ ☐ Outside air intake.
  Furnace: make and model _See heating and air conditioning_ Input _120,000_ Btuh; output _96,000_ Btuh.
Additional information: _Layout; to be prepared by mechanical contractor._
☐ Space heater; ☐ floor furnace; ☐ wall heater. Input _____ Btuh; output _____ Btuh; number units _____
  Make, model _____ Additional information: _____
Controls: make and types _____
Additional information: _____
Fuel: ☐ Coal; ☐ oil; ☐ gas; ☐ liq. pet. gas; ☐ electric; ☐ other _____ ; storage capacity _____
  Additional information: _____
Firing equipment furnished separately: ☐ Gas burner, conversion type.   ☐ Stoker: hopper feed ☐ ; bin feed ☐
  Oil burner: ☐ pressure atomizing; ☐ vaporizing _____
  Make and model _____ Control _furnace._
  Additional information: _NOTE: Gas outlet for hot water tank, furnace and fireplace jet, and garage_
Electric heating system: type _____ Input _____ watts;@ _____ volts; output _____ Btuh.
  Additional information: _____
Ventilating equipment: attic fan, make and model _____ ; capacity _7,800_ cfm.
  kitchen exhaust fan, make and model _Kitchen vent hood and exhaust fan._
Other heating, ventilating, or cooling equipment _Nutone or equal ceiling heaters in baths._

24. ELECTRIC WIRING: allow for 200 amp service   as required
Service: ☐ overhead; ☐ underground. Panel: ☐ fuse box; ☒ circuit-breaker; make _square D_ AMPs _____ No. circuits
Wiring: ☐ conduit; ☒ armored cable; ☐ nonmetallic cable; ☐ knob and tube; ☐ other _romex wiring if code allows_
Special outlets: ☒ range; ☐ water heater; ☐ other _220 volt for oven, range, dryer, and air conditioning_
☐ Doorbell.   ☒ Chimes. Push-button locations _____ Additional information: _____

25. LIGHTING FIXTURES:
Total number of fixtures _see plans_ Total allowance for fixtures, typical installation, $ _____
Nontypical installation _____
Additional information: _Optional Fluorescent grid, dropped ceiling in kitchen as shown on plans_

HUD-FHA 2005
VA Form 26-1852   3   **DESCRIPTION OF MATERIALS**

Figure 27 ◆ Typical materials specification (3 of 4).

DESCRIPTION OF MATERIALS

26. INSULATION:

Location	Thickness	Material, Type, and Method of Installation	Vapor Barrier
Roof			
Ceiling	12"	loose-fill fiberglass insulation in ceiling of house only	(blown between jsts)
Wall	4"	batts- fiberglass insulation- exterior walls of house	(stapled to studs)
Floor	1"	perimeter insulation under slab.	

27. MISCELLANEOUS: *(Describe any main dwelling materials, equipment, or construction items not shown elsewhere; or use to provide additional information where the space provided was inadequate. Always reference by item number to correspond to numbering used on this form.)*

LIGHTING FIXTURES . . . . $ as per contract
WALLPAPER . . . . $ per contract labor and material
LANDSCAPING . . . . $ as per contract
CARPET ALLOWANCE . . . . $ PER SQUARE YARD INSTALLED as per contract
Cabinet over washer and dryer.
Waterproof outlet on patio.
Ornamental fence and gate at main entrance

HARDWARE: *(make, material, and finish.)*
Dexter or equal passage sets and locks- all exterior doors keyed alike
deadbolt locks on all exterior doors.

SPECIAL EQUIPMENT: *(State material or make, model and quantity. Include only equipment and appliances which are acceptable by local law, custom and applicable FHA standards. Do not include items which, by established custom, are supplied by occupant and removed when he vacates premises or chattles prohibited by law from becoming realty.)*

APPLIANCE ALLOWANCE: $
AS SELECTED BY OWNER WITHIN SPECIFIED ALLOWANCE

PORCHES:
foundation construction: 6"x8" poured monolithic with the slab.
floor: 4" concrete slab with 6x6-W1.4xW1.4 WWF smooth trowel finish.

TERRACES:
stoops: 4" concrete slab- smooth trowel finish.
patio: 4" concrete slab- smooth trowel finish. - see plans for size.

GARAGES: automatic garage door opener
foundation: 14"x 18" concrete footing with (4) 5/8" rebars; 6" concrete stem wall
floor: 4" concrete with 6x6-W1.4xW1.4 WWF; smooth trowel finish floors.
interior: 3/8" prefinished Sheetrock on walls; texture and paint 1/2" Sheetrock on ceiling

WALKS AND DRIVEWAYS:
Driveway: width see plot ; base material tamped earth ; thickness 4 "; surfacing material concrete ; thickness 4 "
Front walk: width 36" ; material concrete ; thickness 4 ". Service walk: width ____ ; material ____ ; thickness ____ "
Steps: material ____ ; treads ____ "; risers ____ ". Check walls ____

OTHER ONSITE IMPROVEMENTS:
(Specify all exterior onsite improvements not described elsewhere, including items such as unusual grading, drainage structures, retaining walls, fence railings, and accessory structures.)
NOTE: All dimensions to be rechecked on site prior to beginning construction by builder and builder shall be responsible for the same.

LANDSCAPING, PLANTING, AND FINISH GRADING:
Topsoil ____ " thick: ☐ front yard; ☐ side yards; ☐ rear yard to ____ feet behind main building.
Lawns *(seeded, sodded, or sprigged)*: ☐ front yard ____ ; ☐ side yards ____ ; ☐ rear yard ____
Planting: ☐ as specified and shown on drawings; ☐ as follows:
____ Shade trees, decidous. ____ " caliper. ____ Evergreen trees ____ ' to ____ ', B & B.
____ Low flowering trees, decidous, ____ ' to ____ ' ____ Evergreen shrubs ____ ' to ____ ', B & B.
____ High-growing shrubs, decidous, ____ ' to ____ ' ____ Vines, 2-year ____
____ Medium-growing shrubs, decidous, ____ ' to ____ '
____ Low-growing shrubs, decidous, ____ ' to ____ '

IDENTIFICATION – This exhibit shall be identified by the signature of the builder, or sponsor, and/or the proposed mortgagor if the latter is known at the time of application.

Date ____ Signature ____

Signature ____

HUD-FHA 2005
VA Form 26-1852 4

104F27D.EPS

Figure 27 ◆ Typical materials specification (4 of 4).

4.40 CARPENTRY FUNDAMENTALS LEVEL ONE ◆ TRAINEE GUIDE

Instructor's Notes:

Addenda and Change Orders

Addenda and change orders are contractual documents used to correct or make changes to original construction drawings and/or specifications. The difference between the two documents is a matter of timing. An addenda is written before the contract is awarded, while a change order is drawn up after the award of the contract.

5.1.2 Technical Aspects

The technical aspects section includes information on materials that are specified by standard numbers and by standard national testing organizations such as the American Society of Testing Materials (ASTM). The technical data section of specifications can be of three types:

- *Outline specifications* – These specifications list the materials to be used in order of the basic parts of the job such as foundation, floors, and walls.
- *Fill-in specifications* – This is a standard form filled in with pertinent information. It is typically used on smaller jobs.
- *Complete specifications* – For ease of use, most specifications written for large construction jobs are organized in the Construction Specification Institute format called the Uniform Construction Index. This is known as the CSI format and is explained in the next section.

5.1.3 Format of Specifications

For convenience in writing, speed in estimating, and ease of reference, the most suitable organization of the specifications is a series of sections dealing with the construction requirements, products, and activities, and that is easily understandable by the different trades. Those people who use the specifications must be able to find all information needed without spending too much time looking for it.

The most commonly used specification format in North America is *MasterFormat*™. This standard was developed jointly by the Construction Specifications Institute (CSI) and Construction Specifications Canada (CSC). For many years prior to 2004, the organization of construction specifications and supplier's catalogs has been based on a standard with 16 sections, known as divisions. The divisions and their subsections were individually identified by a five-digit numbering system. The first two digits represented the division number and the next three individual numbers represented successively lower levels of breakdown. For example, the number 13213 represents division 13, subsection 2, sub-subsection 1 and sub-sub-subsection 3.

In 2004, the *MasterFormat*™ standard underwent a major change. What had been 16 divisions was expanded to four major groupings and 49 divisions with some divisions reserved for future expansion (*Figure 28*). The first 14 divisions are essentially the same as the old format. Subjects under the old Division 15—Mechanical have been relocated to new divisions 22 and 23. The basic subjects under old Division 16—Electrical have been relocated to new divisions 26 and 27.

In addition, the numbering system was changed to 6 digits to allow for more subsections in each division for finer definition. In the new

Explain that specifications are typically organized in the Construction Specification Institute (CSI) format.

Show Transparency 31 (Figure 28).

Specifications

Written specifications supplement the related working drawings in that they contain details not shown on the drawings. Specifications define and clarify the scope of the job. They describe the specific types and characteristics of the components that are to be used on the job and the methods for installing some of them. Many components are identified specifically by the manufacturer's model and part numbers. This type of information is used to purchase the various items of hardware needed to accomplish the installation in accordance with the contractual requirements.

Hand out Worksheet 27104-4. Have the trainees complete the Worksheet. This laboratory corresponds to Performance Task 7.

MasterFormat™ 2004 Edition – Numbers & Titles
6/8/04
Division Numbers & Titles

Division Numbers and Titles

PROCUREMENT AND CONTRACTING REQUIREMENTS GROUP
Division 00   Procurement and Contracting Requirements

SPECIFICATIONS GROUP

GENERAL REQUIREMENTS SUBGROUP
Division 01   General Requirements

FACILITY CONSTRUCTION SUBGROUP
Division 02   Existing Conditions
Division 03   Concrete
Division 04   Masonry
Division 05   Metals
Division 06   Wood, Plastics, and Composites
Division 07   Thermal and Moisture Protection
Division 08   Openings
Division 09   Finishes
Division 10   Specialties
Division 11   Equipment
Division 12   Furnishings
Division 13   Special Construction
Division 14   Conveying Equipment
Division 15   Reserved
Division 16   Reserved
Division 17   Reserved
Division 18   Reserved
Division 19   Reserved

FACILITY SERVICES SUBGROUP
Division 20   Reserved
Division 21   Fire Suppression
Division 22   Plumbing
Division 23   Heating, Ventilating, and Air Conditioning
Division 24   Reserved
Division 25   Integrated Automation
Division 26   Electrical
Division 27   Communications
Division 28   Electronic Safety and Security
Division 29   Reserved

SITE AND INFRASTRUCTURE SUBGROUP
Division 30   Reserved
Division 31   Earthwork
Division 32   Exterior Improvements
Division 33   Utilities
Division 34   Transportation
Division 35   Waterway and Marine Construction
Division 36   Reserved
Division 37   Reserved
Division 38   Reserved
Division 39   Reserved

PROCESS EQUIPMENT SUBGROUP
Division 40   Process Integration
Division 41   Material Processing and Handling Equipment
Division 42   Process Heating, Cooling, and Drying Equipment
Division 43   Process Gas and Liquid Handling, Purification, and Storage Equipment
Division 44   Pollution Control Equipment
Division 45   Industry-Specific Manufacturing Equipment
Division 46   Reserved
Division 47   Reserved
Division 48   Electrical Power Generation
Division 49   Reserved

Div Numbers - 1

All contents copyright © 2004, The Construction Specifications Institute and Construction Specifications Canada.
All rights reserved.

Figure 28 ◆ *2004 MasterFormat™.*

104F28.EPS

4.42   CARPENTRY FUNDAMENTALS LEVEL ONE   ◆   TRAINEE GUIDE

Instructor's Notes:

numbering system, the first two digits represent the division number. The next two digits represent subsections of the division and the two remaining digits represent the third level sub-subsection numbers. The fourth level, if required, is a decimal and number added to the end of the last two digits. For example, the number 132013.04 represents division 13, subsection 20, sub-subsection 13, and sub-sub-subsection 04.

6.0.0 ◆ BUILDING CODES

Building codes that are national in scope provide minimum standards to guard the life and safety of the public by regulating and controlling the design, construction, and quality of materials used in modern construction. They have also come to govern the use and occupancy, location of a type of building, and the on-going maintenance of all buildings and facilities. Once adopted by a local jurisdiction, these model building codes then become law. It is common for localities to change or add new requirements to any model code requirements adopted in order to meet more stringent requirements and/or local needs. The provisions of the model building codes apply to the construction, alteration, movement, demolition, repair, structural maintenance, and use of any building or structure within the local jurisdiction.

The model building codes are the legal instruments that enforce public safety in construction of human habitation and assembly structures. They are used not only in the construction industry but also by the insurance industry for compensation appraisals and claims adjustments, and by the legal industry for court litigation.

Up until 2000, there were three model building codes. The three code writing groups, Building Officials and Code Administrators (BOCA), International Conference of Building Officials (ICBO), and Southern Building Code Congress International (SBCCI), combined into one organization called the International Code Council (ICC) with the purpose of writing one nationally accepted family of building and fire codes. The first edition of the *International Building Code®* was published in 2000 and the second edition in 2003. It is intended to continue on a three-year cycle.

In 2002, the National Fire Protection Association (NFPA) published its own building code, *NFPA 5000®*. There are now two nationally recognized codes competing for adoption by the 50 states.

The format and chapter organization of the two codes differ, but the content and subjects covered are generally the same. Both codes cover all types of occupancies from single-family residences to high-rise office buildings, as well as industrial facilities. They also cover structures, building materials, and building systems, including life safety systems.

When states, counties, and cities adopt a model code as the basis for their code, they often change it to meet local conditions. They might add further restrictions, or they might only adopt part of the model code. An important general rule to remember about codes is that in almost every case the most stringent local code will apply.

The carpenter should be aware of the federal laws, building codes, and restrictions that affect the specific job being constructed. This should also include a basic understanding of the codes that pertain mainly to other trades, such as the NFPA gas and electrical codes.

7.0.0 ◆ QUANTITY TAKEOFFS

The **quantity takeoff** or takeoff procedure involves surveying, measuring, and counting all materials indicated on a set of drawings. Normally, standard takeoff sheets are available for the carpenter to use for accurate takeoffs. The sheets are useful because they provide standardization, continuity, and a permanent record. In addition, they reduce the workload as well as the potential for error. In order to make measuring, counting, and calculating tasks easier when performing a takeoff, the following materials will help:

- Colored pencils for checking off items on drawings as they are taken off. (The same color can be used to check off all similar materials.)
- Two drafting scales: one architect's scale and one engineer's scale
- Calculator
- Magnifying glass for examining details on the drawings

Following a standard procedure when doing a takeoff helps ensure accuracy and completeness.

Assign reading of Sections 6.0.0–7.0.0 for the next class session.

Discuss the purpose of building codes and explain that they may vary widely from region to region. Point out that there are several model building codes used in the United States.

Explain that codes may conflict with one another, but in most cases, the most stringent local code will apply.

Explain that the quantity takeoff procedure involves surveying, measuring, and counting all materials indicated on a set of drawings. Show examples of standard takeoff forms.

Demonstrate and discuss the overall steps involved in performing a takeoff.

MODULE 27104-06 ◆ READING PLANS AND ELEVATIONS 4.43

Using an instructor-supplied drawing set and specifications, have the trainees practice doing a material quantity takeoff for a building, one or more rooms in a building, etc. This laboratory corresponds to Performance Task 8.

Assign reading of Sections 8.0.0–10.0.0 for the next class session.

Discuss the various people involved in the design and execution of a building project.

Generally, the steps for performing a takeoff are as follows:

Step 1 Determine the scope of the work.
- Start with the plans. Leaf through the drawings to determine:
  - General location
  - Job-site conditions
  - Size of project
  - Type(s) of work to be done
  - For whom the work is being done (architect, owner, etc.)
  - All the drawings are included in your set
  - Date of the drawings and revision numbers
- Begin to fill out your takeoff sheets using these data.
- Write down a summary of important aspects of the specifications on the takeoff sheets.

Step 2 Begin a quantity takeoff.
- Review plan sheets in sequence:
  - Plot plan
  - Foundation plan
  - Floor plans
  - Exterior elevations
  - Section and detail views
  - Framing plans (if applicable)
  - Door, window, and finish schedules
- Use a systematic approach by working through the structure in a consistent order, such as from foundation to top floor, from front to rear, or from exterior roof line to ground level.
- Review related drawings:
  - Plot and landscaping
  - Mechanical, plumbing, fire sprinklers, and electrical
  - Structural steel
  - Plan sheets
  - Written notes and unusual features

Step 3 Review specifications and seek clarification, if necessary.
- Read the specifications again in light of your takeoff. Review the finish and other schedules and written notes on the drawings to see if they are in agreement with the written specifications.
- Resolve any questions with your supervisor, architect, and/or the owner, then record the answers.
- As applicable, discuss the division of responsibilities with your supervisor, architect, and/or the owner, then record these decisions.

Step 4 Determine total quantities for each category of materials.
- Make calculations and review for errors.
- Check off and summarize the total quantity needed for each item of material. Remember to adjust material quantities for normal units of purchase and any unusual waste considerations.

8.0.0 ♦ PROJECT ORGANIZATION

Many people are involved in the design, planning, and execution of a project. This section will give you some insight into the many people involved in project organization and the roles they play.

Building owner – The owner can be either a public or private entity. Public projects are funded with public money from sources such as taxes, fees, and donations. Public owners can include federal, state, and local governments, school boards, park commissions, public universities, and transit authorities.

Private projects are funded with private money from individuals, companies, and institutions.

Architect/engineer (A/E) – The architect is the person or firm who does the design work for a project. They produce the blueprints that govern every aspect of the project. In many cases, architectural firms also do engineering work because, in commercial work, a great deal of mechanical, structural, and electrical engineering, as well as other design work is required. Hence the term A/E.

Some of the engineering activities include the following:

- *Structural engineering* – The structural engineer designs the frame of the building.
- *Site engineering* – The site engineer designs everything outside the footprint (perimeter) of the building.
- *Soils engineer* – The soils engineer tests the soil and may also do concrete testing.
- *Mechanical engineering* – The mechanical engineer designs the HVAC and plumbing systems.
- *Electrical engineering* – The electrical engineer designs the lighting and electrical distribution systems.
- *Other* – Other engineering specialties may be needed to design such systems as fire protection, telecommunications, environmental (indoor air quality), and acoustical.

General contractor (GC) – In many cases, the GC manages the entire construction project, following the plans developed by the architect/engineer. The GC will schedule all the work on the project, buy the materials, and hire subcontractors to perform carpentry, plumbing, electrical, roofing, painting, and other work.

Construction manager (CM) – The CM becomes involved earlier in the project than the GC. The CM provides construction expertise throughout the design phase of the project and acts as the owner's agent for the duration of the project. The GC often acts as the CM.

Figure 29 shows the charts for several types of project organizations. The one consistent aspect of these charts is that the property owner is the ultimate authority. At first, you may have difficulty in seeing how some of the organizations differ from others. The way to figure out the subtle differences is to look at how the lines connect between the boxes on the charts. Notice, for example, that in all but one case, the owner interacts with the architect. In the case of a design/build arrangement shown in *Figure 29(D)* and *(E)*, the design/build contractor or CM is positioned between the architect and the owner. In that case, the owner has transferred responsibility and authority for design to the design/build contractor or CM.

There are also differences in the role of the GC from one project to the next. Note, for example, that in chart A, the GC is responsible for all construction work. In chart B, the GC is on a par with the CM. In chart C, the GC is subordinate to the CM.

Review various project organizations.

Show Transparency 32 (Figure 29).

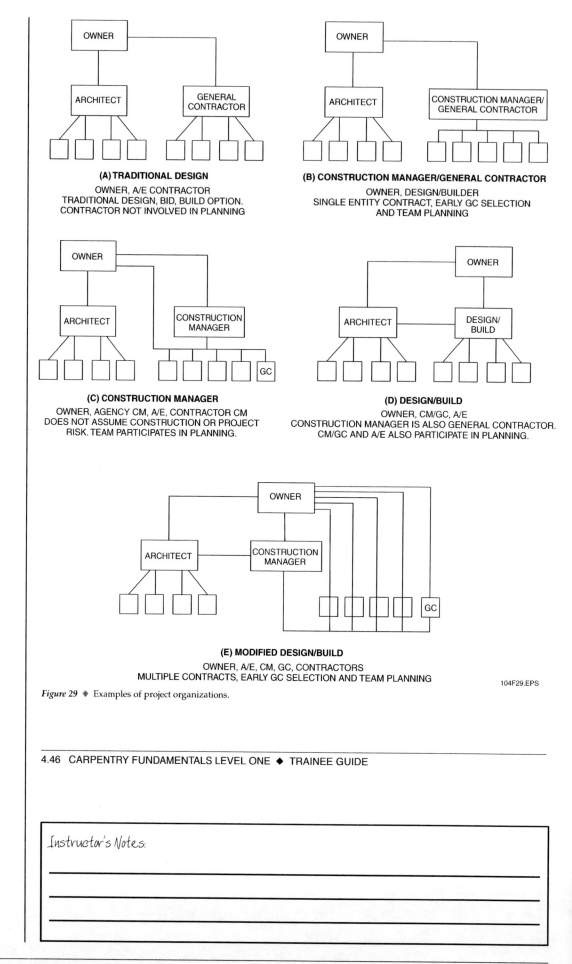

Figure 29 ♦ Examples of project organizations.

9.0.0 ♦ WORKING WITH OTHER TRADES

Cooperation among the various trades at the job site and respect for the work of other trades are essential to achieve a smooth-running project.

On many well-run jobs, a sense of togetherness (cooperation) soon develops and the trades work together in harmony. Many times, there is even a tradeoff of activities that allows the project to progress at a uniform pace. The carpenter may, for example, build a storage locker for the electricians, who may in turn provide the carpenters with additional outlets. Such practices are fine as long as the material used does not short the project of materials needed for completion and can be done without additional cost.

The point to be made here, from the craftsperson's standpoint, is that cooperation among the many trades on the job makes the work flow better and helps ensure timely completion of the project.

A job superintendent needs to coordinate the activities of the major trade groups, including:

- Mechanical
- Electrical
- Plumbing
- Concrete
- Carpentry
- Roofing
- Masonry
- Insulation
- Flooring (carpet, tile, ceramics)
- Painting
- Drywall
- Building electronic systems

The job superintendent will need to sequence activities among the different trades to avoid conflict in distribution. The foreman or job superintendent will also be responsible for seeing that the trades follow drawings or design. The job foreman or superintendent will need to coordinate special storage needs and be responsible for scheduling or coordination of work according to the schedule.

The coordination of the work is important for the efficiency of construction and timely completion. The general superintendent is the person with overall control of all the trades and will plan with each trade so that their work does not interfere with that of another trade.

Carpenters are generally on the job before most other trades. Therefore, an important part of trade coordination is ensuring that the work done by the carpenters will accommodate the other trades. For example, plans for location of heating and air conditioning duct runs, plumbing fixtures and piping, and electrical components and wiring runs must be continually reviewed by the superintendent or supervisor to ensure that carpenters are allocating space, in the correct locations, for these facilities.

The superintendent plans the routes for all chases for the sub-trades during construction and can show the other trades the chase routes as built. The superintendent tracks the scheduling for the entire job and has the start/completion time planned for each function of each trade. He or she also knows if the project is ahead of or behind schedule and can advise the other trade supervisors to either increase or cut back on work activities.

The superintendent, as the representative of the general contractor, is aware of the desires of the owner/architect and knows of changes that have come about during construction. These changes are passed on to the other trades working on the project.

10.0.0 ♦ PROJECT SCHEDULES

A construction project requires a lot of planning and scheduling because different trades, equipment, and materials are needed at different times in the process. Project planning and scheduling will be covered in more detail later in your training. For now, *Figures 30* and *31* provide an overview of where each trade fits into the commercial process for residential and commercial projects, respectively.

Emphasize the importance of coordinating building activities with other trades.

Discuss project scheduling.

Show Transparencies 33 and 34 (Figures 30 and 31).

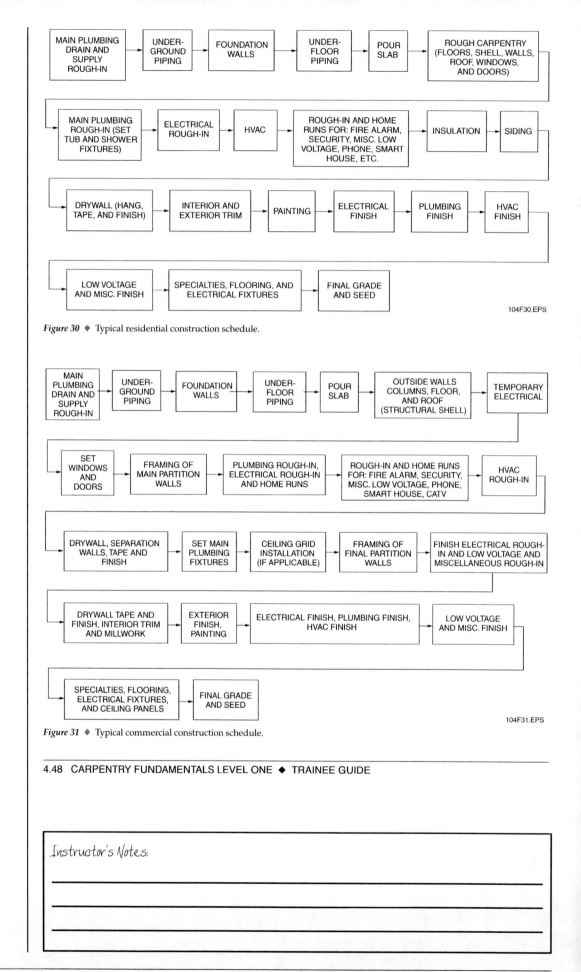

Figure 30 ♦ Typical residential construction schedule.

Figure 31 ♦ Typical commercial construction schedule.

4.48 CARPENTRY FUNDAMENTALS LEVEL ONE ♦ TRAINEE GUIDE

Summary

As a carpenter, you have to build projects that meet the needs and requirements of the owner and architect. To do this, you must be able to read and understand construction drawings and specifications.

Construction drawings are drawings that show the carpenter and other skilled tradespeople how to build a specific building or structure. Specifications are written instructions provided by architectural and engineering firms to the general contractor and, consequently, to the subcontractors. Specifications are just as important as the drawings in a set of plans. They furnish what the drawings cannot in that they define the quality of work to be done and the materials to be used. When there is a conflict between the drawings and the specifications, the specifications normally take precedence.

Summarize the major concepts presented in the module.

Notes

Have trainees complete the Review Questions, and go over the answers prior to administering the Module Examination.

Review Questions

1. What is the elevation of the point of beginning for the site plan shown in *Figure 4*?
   a. 539.05'
   b. 540.85'
   c. 551.12'
   d. 552.92'

2. How wide is the footing for the outside wall by the master bedroom shown in *Figure 7*? (Hint: Refer to the related section drawings shown in *Figure 9*.)
   a. 8"
   b. 12"
   c. 18"
   d. 22"

3. A drawing that shows the horizontal view of a building, the length and width of the building, and the floor layout of the rooms is called a(n) _____.
   a. elevation drawing
   b. foundation plan
   c. plot plan
   d. floor plan

4. What is the distance from the outside of the rear wall to the center of the left-side window in the master bedroom shown in *Figure 7*?
   a. 2'-8"
   b. 8'-5¾"
   c. 12'-2⅞"
   d. 13'-6⅞"

5. How many solid core doors are identified in the door schedule shown in *Figure 11*?
   a. 2
   b. 3
   c. 4
   d. 5

6. The catalog number 143 PW identified in the window schedule shown in *Figure 11* identifies a _____ window.
   a. 2'-0" wide × 3'-0" high
   b. 3'-6" wide × 5'-0" high
   c. 3'-8" wide × 3'-0" high
   d. 3'-8" wide × 5'-0" high

7. What type of floor is used for the entry in the finish schedule shown in *Figure 12*?
   a. Carpet
   b. Concrete
   c. Ceramic tile
   d. Rubber tile

4.50 CARPENTRY FUNDAMENTALS LEVEL ONE ♦ TRAINEE GUIDE

Instructor's Notes:

Review Questions

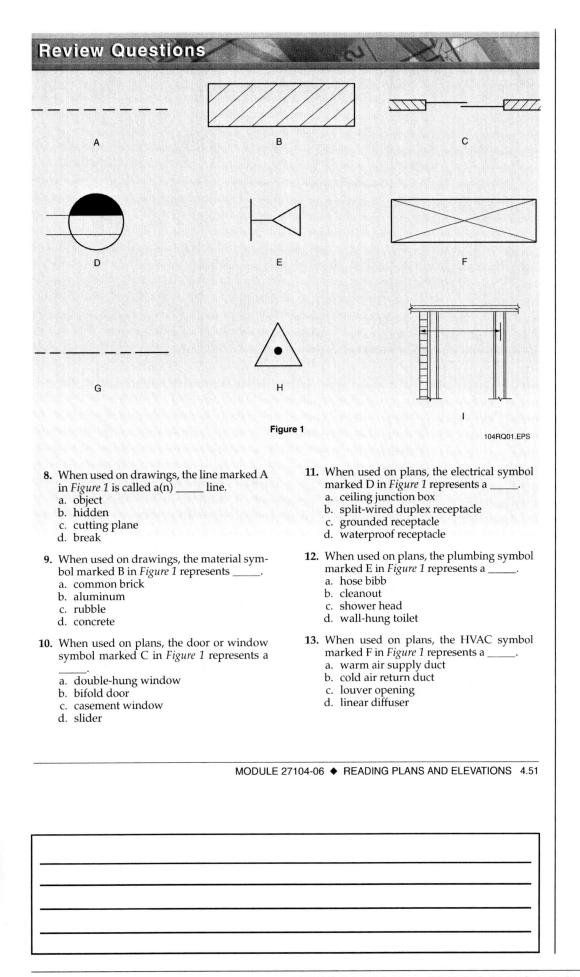

Figure 1

8. When used on drawings, the line marked A in *Figure 1* is called a(n) _____ line.
   a. object
   b. hidden
   c. cutting plane
   d. break

9. When used on drawings, the material symbol marked B in *Figure 1* represents _____.
   a. common brick
   b. aluminum
   c. rubble
   d. concrete

10. When used on plans, the door or window symbol marked C in *Figure 1* represents a _____.
    a. double-hung window
    b. bifold door
    c. casement window
    d. slider

11. When used on plans, the electrical symbol marked D in *Figure 1* represents a _____.
    a. ceiling junction box
    b. split-wired duplex receptacle
    c. grounded receptacle
    d. waterproof receptacle

12. When used on plans, the plumbing symbol marked E in *Figure 1* represents a _____.
    a. hose bibb
    b. cleanout
    c. shower head
    d. wall-hung toilet

13. When used on plans, the HVAC symbol marked F in *Figure 1* represents a _____.
    a. warm air supply duct
    b. cold air return duct
    c. louver opening
    d. linear diffuser

Review Questions

14. When used on site (plot) plans, the line marked G in *Figure 1* represents a _____ line.
    a. gas
    b. power
    c. contour
    d. property

15. When used on site (plot) plans, the symbol marked H in *Figure 1* represents a _____.
    a. property corner
    b. bench mark
    c. required spot elevation
    d. storm sewer

16. The dimension marked I in *Figure 1* represents a measurement made from _____.
    a. wall to wall
    b. center to center
    c. outside to center
    d. outside to outside

17. The special and general conditions portion of a specification covers the _____ aspects of contractual arrangements.
    a. non-technical
    b. technical
    c. equipment and material
    d. both non-technical and technical

18. The specification shown in *Figure 27* indicates the size and spacing of partition studs to be _____.
    a. 2 × 4 wood studs at 16" OC
    b. 2 × 4 wood studs at 24" OC
    c. 2 × 4 16-gauge metal studs at 16" OC
    d. 2 × 4 20-gauge metal studs at 24" OC

19. When given a specification written in *Master Format*™, you would expect to find information on doors, windows, and other building openings in Division _____.
    a. 4
    b. 6
    c. 8
    d. 9

20. When doing a quantity takeoff, the first step is to _____.
    a. perform the quantity takeoff
    b. total the quantities per category of material
    c. determine the scope of the work
    d. review the specifications and seek clarification, if necessary

Instructor's Notes:

Trade Terms Quiz

1. The different elevations on a site are shown as _____ on the site/plot plan.
2. Official survey points may be marked using _____.
3. Local building codes may include a(n) _____ requirement that determines how far back from the property line the structure must be placed.
4. A(n) _____ may also influence the position of a structure on the site.
5. The legal boundaries of a site are known as _____.
6. All the finish grade references on the drawings are keyed to a(n) _____.
7. The _____ shows the structure from either the front or the side.
8. The _____ shows the structure from above.
9. Lumber is commonly sold by its _____ rather than its actual dimensions.
10. Before purchasing any materials for a job, you should perform a(n) _____ using the information contained in the drawing set for the job.
11. A detailed drawing of a region that depicts the natural and man-made features is a(n) _____.
12. A(n) _____ depicts the connections of a piping system.

Trade Terms

Bench mark
Contour lines
Easement
Elevation view
Front setback
Monuments
Nominal size
Plan view
Property lines
Quantity takeoff
Riser diagram
Topographical survey

Have trainees complete the Trade Terms Quiz, and go over the answers.

Administer the Module Examination. Record the results on Craft Training Report Form 200, and submit the form to the Training Program Sponsor.

Administer the Performance Tests, and fill out Performance Profile Sheets for each trainee. If desired, trainee proficiency noted during laboratory sessions may be used to complete the Performance Test. Record the results on Craft Training Report Form 200, and submit the results to the Training Program Sponsor.

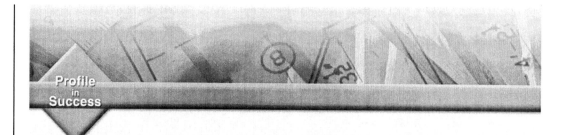

Profile in Success

Luis Alexander Gomez
Bronze Medal Winner Associated Builders and Contractors National Championships

In 2005, Luis's skills as a carpenter earned him a trip to the ABC Craft Olympics in Orlando, Florida, where he competed against some of the best talent in the trade. When the competition was over and the judges' decisions came in, Luis was awarded a bronze medal.

How did you become interested in carpentry?
I first became interested in carpentry when I was a helper at a remodeling company. I found myself associating with the carpenters more and more—asking them questions about their work. I became more interested in the trade and eventually entered an apprenticeship program at SpawGlass. My apprenticeship was four years long.

What are some of the things you do in your job?
After I graduated from my apprenticeship program, I stayed at SpawGlass. I've been with the company for seven years now. Now I'm in training as a foreman. I supervise, but I also work in the field. I make sure the work is done right. Right now, we're remodeling four floors of a hospital. We've been on this project for four months now, and it's expected to take a total of a year and four months.

What do you like most about carpentry?
I love woodworking. I love the smell of fresh-cut lumber. I really like learning about new tools. I dream of building my own house one day. Everyone has an idea of their dream home. I want to build mine.

What advice would you give someone just starting out?
Associate with knowledgeable people. If you hang around with the right persons, you'll succeed. Some people can teach you how to do things faster, better—that counts. And your employer will notice. Safety is very important, too: Your company can teach you about safety, but ultimately, you're the one who has to act safely.

4.54 CARPENTRY FUNDAMENTALS LEVEL ONE ◆ TRAINEE GUIDE

Instructor's Notes:

Trade Terms Introduced in This Module

Bench mark: A point established by the surveyor on or close to the building site. It is used as a reference for determining elevations during the construction of a building.

Contour lines: Imaginary lines on a site/plot plan that connect points of the same elevation. Contour lines never cross each other.

Easement: A legal right-of-way provision on another person's property (for example, the right of a neighbor to build a road or a public utility to install water and gas lines on the property). A property owner cannot build on an area where an easement has been identified.

Elevation view: A drawing giving a view from the front or side of a structure.

Front setback: The distance from the property line to the front of the building.

Monuments: Physical structures that mark the location of a survey point.

Nominal size: Approximate or rough size (commercial size) by which lumber, block, etc., is commonly known and sold, normally slightly larger than the actual size (for example, 2×4s).

Plan view: A drawing that represents a view looking down on an object.

Property lines: The recorded legal boundaries of a piece of property.

Quantity takeoff: A procedure that involves surveying, measuring, and counting all materials indicated on a set of drawings.

Riser diagram: A schematic drawing that depicts the layout, components, and connections of a piping system.

Topographical survey: An accurate and detailed drawing of a place or region that depicts all the natural and man-made physical features, showing their relative positions and elevations.

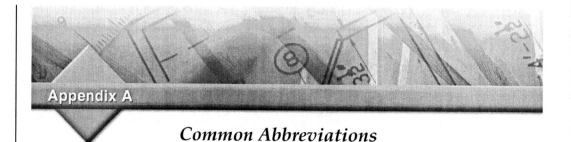

Appendix A

Common Abbreviations

COMMON ABBREVIATIONS USED ON ELEVATIONS

Aluminum	AL
Asbestos	ASB
Asphalt	ASPH
Basement	BSMT
Beveled	BEV
Brick	BRK
Building	BLDG
Cast iron	CI
Ceiling	CLG
Cement	CEM
Center	CTR
Center line	C or CL
Clear	CLR
Column	COL
Concrete	CONC
Concrete block	CONC B
Copper	COP
Corner	COR
Detail	DET
Diameter	DIA
Dimension	DIM.
Ditto	DO.
Divided	DIV
Door	DR
Double-hung window	DHW
Down	DN or D
Downspout	DS
Drawing	DWG
Drip cap	DC
Each	EA
East	E
Elevation	EL
Entrance	ENT
Excavate	EXC
Exterior	EXT
Finish	FIN.
Flashing	FL
Floor	FL
Foot or feet	' or FT
Foundation	FND
Full size	FS
Galvanized	GALV
Galvanized iron	GI
Gauge	GA
Glass	GL
Glass block	GL BL
Grade	GR
Grade line	GL
Height	HGT, H, or HT
High point	H PT
Horizontal	HOR
Hose bibb	HB
Inch or inches	" or IN.
Insulating (insulated)	INS
Length	LGTH, LG, or L
Length overall	LOA
Level	LEV
Light	LT
Line	L
Lining	LN
Long	LG
Louver	LV
Low point	LP
Masonry opening	MO
Metal	MET. or M
Molding	MLDG
Mullion	MULL
North	N
Number	NO. or #
Opening	OPNG
Outlet	OUT
Outside diameter	OD
Overhead	OVHD
Panel	PNL
Perpendicular	PERP
Plate glass	PL GL
Plate height	PL HT
Radius	R
Revision	REV
Riser	R
Roof	RF
Roof drain	RD
Roofing	RFG
Rough	RGH

4.56 CARPENTRY FUNDAMENTALS LEVEL ONE ♦ TRAINEE GUIDE

Instructor's Notes:

Saddle	SDL or S	Steel	STL
Scale	SC	Stone	STN
Schedule	SCH	Terra-cotta	TC
Section	SECT	Thick or thickness	THK or T
Sheathing	SHTHG	Typical	TYP
Sheet	SH	Vertical	VERT
Shiplap	SHLP	Waterproofing	WP
Siding	SDG	West	W
South	S	Width	W or WTH
Specifications	SPEC	Window	WDW
Square	SQ	Wire glass	W GL
Square inch	SQ. IN.	Wood	WD
Stainless steel	SST	Wrought iron	WI

COMMON ABBREVIATIONS USED ON PLAN VIEWS

Access panel	AP	Cellar	CEL
Acoustic	ACST	Cement	CEM
Acoustical tile	AT	Cement asbestos board	CEM AB
Aggregate	AGGR	Cement floor	CEM FL
Air conditioning	AIR COND	Cement mortar	CEM MORT
Aluminum	AL	Center	CTR
Anchor bolt	AB	Center to center	C to C
Angle	AN	Center line	C or CL
Apartment	APT	Center matched	CM
Approximate	APPROX	Ceramic	CER
Architectural	ARCH	Channel	CHAN
Area	A	Cinder block	CIN BL
Area drain	AD	Circuit breaker	CIR BKR
Asbestos	ASB	Cleanout	CO
Asbestos board	AB	Cleanout door	COD
Asphalt	ASPH	Clear glass	CL GL
Asphalt tile	AT	Closet	C, CL, or CLO
Basement	BSMT	Cold air	CA
Bathroom	B	Cold water	CW
Bathtub	BT	Collar beam	COL B
Beam	BM	Concrete	CONC
Bearing plate	BRG PL	Concrete block	CONC B
Bedroom	BR	Concrete floor	CONC FL
Blocking	BLKG	Conduit	CND
Blueprint	BP	Construction	CONST
Boiler	BLR	Contract	CONT
Bookshelves	BK SH	Copper	COP
Brass	BRS	Counter	CTR
Brick	BRK	Cubic feet	CU FT
Bronze	BRZ	Cutout	CO
Broom closet	BC	Detail	DET
Building	BLDG	Diagram	DIAG
Building line	BL	Dimension	DIM.
Cabinet	CAB.	Dining room	DR
Caulking	CLKG	Dishwasher	DW
Casing	CSG	Ditto	DO.
Cast iron	CI	Double acting	DA
Cast stone	CS	Double-strength glass	DSG
Catch basin	CB	Down	DN

Downspout	DS	Limestone	LS
Drain	D or DR	Linen closet	L CL
Drawing	DWG	Lining	LN
Dressed and matched	D & M	Linoleum	LINO
Dryer	D	Living room	LR
Electric panel	EP	Louver	LV
End to end	E to E	Main	MN
Excavate	EXC	Marble	MR
Expansion joint	EXP JT	Masonry opening	MO
Exterior	EXT	Material	MATL
Finish	FIN.	Maximum	MAX
Finished floor	FIN. FL	Medicine cabinet	MC
Firebrick	FBRK	Minimum	MIN
Fireplace	FP	Miscellaneous	MISC
Fireproof	FPRF	Mixture	MIX
Fixture	FIX.	Modular	MOD
Flashing	FL	Mortar	MOR
Floor	FL	Molding	MLDG
Floor drain	FD	Nosing	NOS
Flooring	FLG	Obscure glass	OBSC GL
Fluorescent	FLUOR	On center	OC
Flush	FL	Opening	OPNG
Footing	FTG	Outlet	OUT
Foundation	FND	Overall	OA
Frame	FR	Overhead	OVHD
Full size	FS	Pantry	PAN
Furring	FUR	Partition	PTN
Galvanized iron	GI	Plaster	PL or PLAS
Garage	GAR	Plastered opening	PO
Gas	G	Plate	PL
Glass	GL	Plate glass	PL GL
Glass block	GL BL	Platform	PLAT
Grille	G	Plumbing	PLBG
Gypsum	GYP	Porch	P
Hardware	HDW	Precast	PRCST
Hollow metal door	HMD	Prefabricated	PREFAB
Hose bibb	HB	Pull switch	PS
Hot air	HA	Quarry tile floor	QTF
Hot water	HW	Radiator	RAD
Hot water heater	HWH	Random	RDM
I beam	I	Range	R
Inside diameter	ID	Recessed	REC
Insulation	INS	Refrigerator	REF
Interior	INT	Register	REG
Iron	I	Reinforce or reinforcing	REINF
Jamb	JB	Revision	REV
Kitchen	K	Riser	R
Landing	LDG	Roof	RF
Lath	LTH	Roof drain	RD
Laundry	LAU	Room	RM or R
Laundry tray	LT	Rough	RGH
Lavatory	LAV	Rough opening	RGH OPNG
Leader	L	Rubber tile	R TILE
Length	L, LG, or LNG	Scale	SC
Library	LIB	Schedule	SCH
Light	LT	Screen	SCR

Instructor's Notes:

Scuttle	S	Terrazzo	TER
Section	SECT.	Thermostat	THERMO
Select	SEL	Threshold	TH
Service	SERV	Toilet	T
Sewer	SEW.	Tongue-and-groove	T & G
Sheathing	SHTHG	Tread	TR or T
Sheet	SH	Typical	TYP
Shelf and rod	SH&RD	Unexcavated	UNEXC
Shelving	SHELV	Unfinished	UNF
Shower	SH	Utility room	URM
Sill cock	SC	Vent	V
Single-strength glass	SSG	Vent stock	VS
Sink	SK or S	Vinyl tile	V TILE
Soil pipe	SP	Warm air	WA
Specifications	SPEC	Washing machine	WM
Square feet	SQ FT	Water	W
Stained	STN	Water closet	WC
Stairs	ST	Water heater	WH
Stairway	STWY	Waterproof	WP
Standard	STD	Weatherstripping	WS
Steel	ST or STL	Weep hole	WH
Steel sash	SS	White pine	WP
Storage	STG	Wide flange	WF
Switch	SW or S	Wood	WD
Telephone	TEL	Wood frame	WF
Terra-cotta	TC	Yellow pine	YP

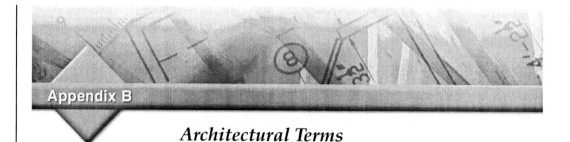

Appendix B

Architectural Terms Commonly Found on Plans

WINDOW TERMS

Apron – A plain or molded piece of finish below the stool of a window that is put on to cover the rough edge of the plastering.

Drip cap – A projection of masonry or wood on the outside top of a window to protect the window from rain.

Head jamb – Horizontal top post used in the framing of a window or doorway.

Light – A pane of glass.

Lintel – Horizontal structural member supporting a wall over a window or other opening.

Meeting rail – The horizontal center rail of a sash in a double-hung window.

Mullion – A large, vertical division of a window opening.

Muntin – A strip of wood or metal that separates and supports the panes of glass in a window sash.

Sash – The part of a window in which panes of glass are set; it is generally movable, as in double-hung windows. The two side pieces are called stiles and the upper and lower pieces are called rails.

Side jambs – Vertical sideposts used in the framing of a window or doorway.

Sill – Horizontal member at the bottom of a window or doorway.

Stool – A flat, narrow shelf forming the top member of the interior trim at the bottom of a window.

Stop bead – The strip on a window frame against which the sash slides.

PITCHED ROOF TERMS

Flashing – Sheet metal, copper, lead, or tin that is used to cover open joints to make them waterproof.

Gable – The end of a ridged roof as distinguished from the front or rear side.

Louver – An opening for ventilation that is covered by sloping slats to exclude rain.

Ridge – The top edge of a roof where the two slopes meet.

Ridge board – A board that is placed on the edge at the ridge of a roof to support the upper ends of rafters.

Saddle – A tent-shaped portion of a roof between a chimney and the main part of the roof; built to support flashing and to direct water away from the chimney.

Valley – The intersection of the bottom two inclined sides of a roof.

CORNICE TERMS

Cornice – The part of a roof that projects beyond a wall.

Cornice return – The short portion of a cornice that is carried around the corner of a structure.

Crown molding – The molding at the top of the cornice and just under the roof.

Fascia – The outside flat member of a cornice.

Frieze – A trim member used just below the cornice.

Soffit – The underside of a cornice.

STAIR TERMS

Headroom – The distance between flights of steps or between the steps and the ceiling above.

Landing – The horizontal platform in a stairway.

Nosing – The overhanging edge of a stair tread.

Rise – The vertical distance from the top of a tread to the top of the next highest tread.

Riser – The vertical portion of a step.

Run – The horizontal distance that is covered by a flight of steps. Also, the net width of a step.

4.60 CARPENTRY FUNDAMENTALS LEVEL ONE ◆ TRAINEE GUIDE

Instructor's Notes:

Stringer – The supporting timber at the sides of a staircase.

Tread – The horizontal part of a step on which the foot is placed.

STRUCTURAL TERMS

Anchor bolt – A bolt with the threaded portion projecting from a structure; generally used to hold the frame of a building secure against wind load. Anchor bolts may also be referred to as hold-down bolts, foundation bolts, and sill bolts.

Batt – A type of insulation designed to be installed between framing members.

Battens – Narrow strips of wood or metal used to cover vertical joints between boards and panels.

Beam – One of the principal horizontal members of a building.

Bridging – The process of bracing floor joists by fixing lateral members between them.

Camber – The concave or convex curvature of a surface.

Expansion joint – The separation between adjoining parts to allow for small relative movements, such as those caused by temperature changes.

Footing – The foundation for a column or the enlargement at the bottom of a wall to distribute the weight of the superstructure over a greater area to prevent settling.

Furring – Strips of wood or metal applied to a wall or other surface to make it level, form an airspace, or provide a fastening surface for a finish covering.

Girder – The main supporting beam (either timber or steel) that is used for supporting a superstructure.

Header – A wood beam that is set at a right angle to a joist to provide a seat or support.

Joist – A heavy piece of horizontal timber to which the boards of a floor or the latch of a ceiling is nailed. Joists are laid edgewise to form a floor support; they rest on the wall or on girders.

Monolithic concrete – A continuous mass of concrete that is cast as a single unit.

Plate – A structural member with a depth that is substantially smaller than its length or width.

Rake – Trim members that run parallel to the roof slope and form the finish between the roof and the wall at the gable end.

Reinforced concrete – Concrete containing metal rods, wires, or other slender members. It is designed in such a manner that the concrete and metal act together to resist forces.

Sill – A horizontal member that is supported by a foundation wall or piers, and which in turn bears the upright members of a frame.

Slab – A poured concrete floor.

Steam wall – That portion of a foundation that rests on the footing.

Studs – The vertical, slender wood or metal members that are used to support the elements in walls and partitions.

Vapor barrier – A material that is used to retard the flow of vapor or moisture into the walls or floors and thus prevent condensation within them.

Veneer – The covering layer of material for a wall or facing materials applied to the external surface of steel, reinforced concrete, or frame walls.

Additional Resources and References

This module is intended to present thorough resources for task training. The following reference works are suggested for further study. These are optional materials for continuing education rather than for task training.

Architectural Drawing and Light Construction. Upper Saddle River, NJ: Prentice Hall.

Blueprint Reading for the Building Trades. Carlsbad, CA: Craftsman Book Company.

Code Check. Newton, CT: Taunton Press.

Design Drawing. New York, NY: John Wiley & Sons.

Graphic Guide to Frame Construction. Newton, CT: Taunton Press.

International Building Code 2003. Falls Church, VA: International Code Council.

MasterFormat™ 2004 Edition. Alexandria, VA: The Construction Specifications Institute (CSI) and Construction Specifications Canada (CSC).

Measuring, Marking, and Layout. Newton, CT: Taunton Press.

Plan Reading & Material Takeoff. Kingston, MA: R.S. Means Company.

Reading Architectural Plans for Residential and Commercial Construction, Ernest R. Weidhaas. Upper Saddle River, NJ: Prentice Hall, 1998.

The Construction Specifications Institute. An organization that seeks to facilitate communication among all those involved in the building process. www.csinet.org

International Code Council. A membership organization dedicated to building safety and fire prevention through development of building codes. www.iccsafe.org

Instructor's Notes:

MODULE 27104-06 — ANSWERS TO REVIEW QUESTIONS

Answer	Section
1. b	2.2.1; Figure 4
2. c	2.2.2; 2.5.0; Figure 7; Figure 9
3. d	2.2.3
4. b	2.2.3; Figure 7
5. a	2.6.0; Figure 11
6. b	2.6.0; Figure 11
7. c	2.6.0; Figure 12
8. b	3.1.0; Figure 14
9. a	3.2.1; Figure 16
10. d	3.2.2; Figure 17
11. b	3.2.3; Figure 18
12. c	3.2.4; Figure 20
13. a	3.2.5; Figure 21
14. d	3.2.6; Figure 22
15. b	3.2.6; Figure 22
16. c	3.3.0; Figure 26
17. a	5.1.1
18. a	5.1.1; Figure 27
19. c	5.1.3
20. c	7.0.0

MODULE 27104-06 — ANSWERS TO TRADE TERMS QUIZ

1. Contour lines
2. Monuments
3. Front setback
4. Easement
5. Property lines
6. Bench mark
7. Elevation view
8. Plan view
9. Nominal size
10. Quantity takeoff
11. Topographical survey
12. Riser diagram

MODULE 27104-06 — TEACHING TIPS

The following are suggested activities or instructional methods to help you teach the material in this module.

General

When you call on someone to answer a question, the rest of the class relaxes or even tunes out because they expect that the question and answer will take place only between you and the trainee you called on. Instead, use this technique to involve more trainees in answering questions and to keep them on their toes.

1. Ask trainees to define a term or explain a concept.
2. After one trainee has answered, ask a trainee seated nearby if the answer is right. Then ask whether a trainee in the back of the room agrees.
3. Ask trainees to explain why they think an answer is right or wrong.
4. Use the session to clear up incorrect ideas and encourage trainees to learn from their mistakes.

CONTREN® LEARNING SERIES — USER UPDATE

NCCER makes every effort to keep these textbooks up-to-date and free of technical errors. We appreciate your help in this process. If you have an idea for improving this textbook, or if you find an error, a typographical mistake, or an inaccuracy in NCCER's Contren® textbooks, please write us, using this form or a photocopy. Be sure to include the exact module number, page number, a detailed description, and the correction, if applicable. Your input will be brought to the attention of the Technical Review Committee. Thank you for your assistance.

Instructors – If you found that additional materials were necessary in order to teach this module effectively, please let us know so that we may include them in the Equipment/Materials list in the Annotated Instructor's Guide.

Write: Product Development and Revision
National Center for Construction Education and Research
P.O. Box 141104, Gainesville, FL 32614-1104

Fax: 352-334-0932

E-mail: curriculum@nccer.org

Craft _____ Module Name _____

Copyright Date _____ Module Number _____ Page Number(s) _____

Description
_____
_____
_____
_____

(Optional) Correction
_____
_____
_____

(Optional) Your Name and Address
_____
_____
_____

Floor Systems
27105-06

NCCER STANDARDIZED CRAFT TRAINING PROGRAM

The National Center for Construction Education and Research (NCCER) provides a standardized national program of accredited craft training. Key features of the program include instructor certification, competency-based training, and performance testing. The program provides trainees, instructors, and companies with a standard form of recognition through the National Registry. The program is described in full in the *Guidelines for Accreditation*, published by NCCER. For more information on standardized craft training, contact NCCER by writing to P.O. Box 141104, Gainesville, FL 32614-1104; calling 352-334-0911; or e-mailing **info@nccer.org**. More information is available at **www.nccer.org**.

HOW TO USE THIS ANNOTATED INSTRUCTOR'S GUIDE

Each page presents two sections of information. The larger section displays each page exactly as it appears in the Trainee Module. The narrow column ties suggested trainee and instructor actions to each page and provides icons (detailed below) to call your attention to material, safety, audiovisual, or testing requirements. The bottom of each page includes space for your notes.

 The **Audiovisual** icon indicates an appropriate time to show a transparency or other audiovisual aid.

 The **Classroom** icon prompts you to define a term, stress a point, ask trainees to explain a concept, or give examples.

 The **Demonstration** icon directs you to show trainees how to perform tasks.

 The **Examination** icon tells you to administer the written module examination.

 The **Homework** icon is placed where you may wish to assign reading for the next class, assign a project, or advise trainees to prepare for an examination.

 The **Laboratory** icon is used when trainees are to practice performing tasks.

 The **Materials** icon is a reminder for you to gather materials needed for classes, laboratories, and testing.

 The **Performance Testing** icon tells you to administer a performance test or a portion thereof.

 The **Safety** icon is used to emphasize safety issues. It is often keyed to *Caution* and *Warning!* statements in the Trainee Module.

 The **Teaching Tip** icon indicates additional guidance is available, such as how to conduct an exercise, get the most educational value from a field trip, or encourage class participation. Teaching Tips may expand on a feature (*Think About It*, *Did You Know?*) or provide *Quick Quizzes* or similar exercises. You will be referred to the Teaching Tips section at the back of the module if there is additional material.

 The **Combination** icon indicates that the laboratory listed corresponds with a performance task. If desired, you can note the proficiency of the trainees during the laboratory, and use it to satisfy performance testing requirements.

PREPARATION

Before teaching this module, you should review the Objectives, Performance Tasks, Materials and Equipment List, and Module Outline. Be sure to allow ample time to prepare your own training or lesson plan and gather all required materials and equipment.

**Floor Systems
Annotated Instructor's Guide**

Module 27105-06

MODULE OVERVIEW

This module introduces the carpentry trainee to residential floor systems. It covers the materials and general methods used to construct floor systems, with emphasis placed on the platform method of floor framing.

PREREQUISITES

Prior to training with this module, it is suggested that the trainee shall have successfully completed *Core Curriculum*; and *Carpentry Fundamentals Level One,* Modules 27101-06 through 27104-06.

OBJECTIVES

Upon completion of this module, the trainee will be able to do the following:

1. Identify the different types of framing systems.
2. Read and interpret drawings and specifications to determine floor system requirements.
3. Identify floor and sill framing and support members.
4. Name the methods used to fasten sills to the foundation.
5. Given specific floor load and span data, select the proper girder/beam size from a list of available girders/beams.
6. List and recognize different types of floor joists.
7. Given specific floor load and span data, select the proper joist size from a list of available joists.
8. List and recognize different types of bridging.
9. List and recognize different types of flooring materials.
10. Explain the purposes of subflooring and underlayment.
11. Match selected fasteners used in floor framing to their correct uses.
12. Estimate the amount of material needed to frame a floor assembly.
13. Demonstrate the ability to:
    - Lay out and construct a floor assembly
    - Install bridging
    - Install joists for a cantilever floor
    - Install a subfloor using butt-joint plywood/OSB panels
    - Install a single floor system using tongue-and-groove plywood/OSB panels

PERFORMANCE TASKS

Under the supervision of the instructor, the trainee should be able to do the following:

1. Lay out and construct a floor assembly.
2. Install bridging.
3. Install joists for a cantilever floor.
4. Install a subfloor using butt-joint plywood/OSB panels.
5. Install a single floor system using tongue-and-groove plywood/OSB panels.
6. Estimate the amount of material needed to frame a floor assembly.
7. Given specific floor load and span data, select the proper girder/beam and joist size from a list of available girders/beams/joists.

MATERIALS AND EQUIPMENT LIST

Transparencies
Markers/chalk
Blank acetate sheets
Transparency pens
Pencils and scratch paper
Overhead projector and screen
Whiteboard/chalkboard
Appropriate personal protective equipment
Floor adhesive (optional)
Beam material
Grout
Plywood or OSB butt-joint panels to cover floor area
Plywood or OSB (tongue-and-groove, 1¼") to cover floor area
Shim materials
Sill sealer
Steel bridging and instructions
Termite shield
2 × 6s for sills
2 × 10s for joists and headers
1 × 4s or 2 × 10s for bridging
8d box nails for bridging
8d box, screw, or ring shank nails for flooring
16d box nails for joists and headers
8d doublehead box nails
Pictures, photographs, etc., showing braced, balloon, platform, and post-and-beam framing
Sets of building working drawings and specifications
Examples of several floor plans and specifications
Pictures/photos of building damage that resulted from defective floor and sill framing (optional)
Tool box consisting of standard carpenter's hand tools
Chalkline
Electric drill and assorted drill and flat bits
Framing square
Level
100' tape
Power circular saw and extension cord
Reciprocating saw
Tin snips
Copies of Worksheets 1 through 3*
Copies of Job Sheets 1 through 5*
Module Examinations**
Performance Profile Sheets**

* Packaged with this Annotated Instructor's Guide.

**Located in the Test Booklet.

SAFETY CONSIDERATIONS

Ensure that the trainees are equipped with appropriate personal protective equipment.

ADDITIONAL RESOURCES

This module is intended to present thorough resources for task training. The following reference works are suggested for both instructors and motivated trainees interested in further study. These are optional materials for continued education rather than for task training.

Builder Tips: Steps to Construct a Solid, Squeak-Free Floor System. Tacoma, WA: APA – The Engineered Wood Association.

Building with Floor Trusses. Madison, WI: Wood Truss Council of America (11-minute DVD or video).

Field Guide for Prevention and Repair of Floor Squeaks. Boise, ID: Trus Joist, a Weyerhauser business.

I-Joist Construction Details: Performance-Rated I-Joists in Floor and Roof Framing. Tacoma, WA: APA – The Engineered Wood Association.

Quality Floor Construction. Tacoma, WA: APA – The Engineered Wood Association (15-minute video).

Storage, Handling, Installation & Bracing of Wood Trusses. Madison, WI: Wood Truss Council of America (69-minute DVD or video).

American Wood Council. A trade association that develops design tools and guidelines for wood construction. **www.awc.org.**

Western Wood Products Association. A trade association representing softwood lumber manufacturers in 12 western states and Alaska. **www.wwpa.com**

Wood I-Joist Manufacturers Association. An organization representing manufacturers of prefabricated wood I-joist and structural composite lumber. **www.i-joist.org**

Wood Truss Council of America. An international trade association representing structural wood component manufacturers. **www.woodtruss.com**

TEACHING TIME FOR THIS MODULE

An outline for use in developing your lesson plan is presented below. Note that each Roman numeral in the outline equates to one session of instruction. Each session has a suggested time period of 2½ hours. This includes 10 minutes at the beginning of each session for administrative tasks and one 10-minute break during the session. Approximately 25 hours are suggested to cover *Floor Systems*. You will need to adjust the time required for hands-on activity and testing based on your class size and resources. Because laboratories often correspond to Performance Tasks, the proficiency of the trainees may be noted during these exercises for Performance Testing purposes.

Topic   **Planned Time**

Session I. Introduction; Methods of Framing Houses; Building Working Drawings and Specifications
- A. Introduction   _____
- B. Methods of Framing Houses
    1. Platform Frame   _____
    2. Braced Frame   _____
    3. Balloon Frame   _____
    4. Post-and-Beam Frame   _____
- C. Building Working Drawings and Specifications   _____
    1. Architectural Drawings   _____
    2. Plumbing, Mechanical, and Electrical Plans   _____
    3. Reading Blueprints   _____
    4. Specifications   _____

Session II. The Floor System
- A. The Floor System
    1. Sills   _____
    2. Beams/Girders and Supports   _____
    3. Floor Joists   _____
    4. Bridging   _____
    5. Subflooring   _____

Session III. Laying Out and Constructing a Platform Floor Assembly
- A. Laying Out and Constructing a Platform Floor Assembly   _____
    1. Checking the Foundation for Squareness   _____
    2. Installing the Sill   _____
    3. Installing a Beam/Girder   _____
    4. Laying Out Sills and Girders for Floor Joists   _____
    5. Laying Out Joist Locations for the Partition and Floor Openings   _____
    6. Cutting and Installing Joist Headers   _____
    7. Installing Floor Joists   _____
    8. Framing Opening(s) in the Floor   _____
    9. Installing Bridging   _____
    10. Installing Subflooring   _____
- B. Laboratory   _____

  Hand out Worksheets 27105-1 and 27105-2. Have the trainees complete the tasks on the Worksheets. Note the proficiency of each trainee.

Session IV. Laboratory
- A. Laboratory   _____

  Hand out Job Sheet 27105-1. Under your supervision, have the trainees perform the tasks on the Job Sheet. This laboratory corresponds to Performance Task 1.

Session V. Laboratory
    A. Laboratory

    Hand out Job Sheet 27105-2. Under your supervision, have the trainees perform the tasks on the Job Sheet. This laboratory corresponds to Performance Task 2.

Session VI. Laboratory
    A. Laboratory

    Hand out Job Sheet 27105-3. Under your supervision, have the trainees perform the tasks on the Job Sheet. This laboratory corresponds to Performance Task 3.

Session VII. Laboratory
    A. Laboratory

    Hand out Job Sheet 27105-4. Under your supervision, have the trainees perform the tasks on the Job Sheet. This laboratory corresponds to Performance Task 4.

Session VIII. Installing Joists for Projections and Cantilevered Floors
    A. Installing Joists for Projections and Cantilevered Floors

    B. Laboratory

    Hand out Job Sheet 27105-5. Under your supervision, have the trainees perform the tasks on the Job Sheet. This laboratory corresponds to Performance Task 5.

Session IX. Estimating the Quantity of Floor Materials
    A. Estimating the Quantity of Floor Materials
1. Sill, Sill Sealer, and Termite Shield
2. Beams/Girders
3. Joists and Joist Headers
4. Bridging
5. Flooring

    B. Laboratory

    Hand out Worksheet 27105-3. Have the trainees complete the tasks on the Worksheet. This laboratory corresponds to Performance Task 6.

Session X. Guidelines for Determining Proper Girder and Joist Sizes; Review; Module Examination and Performance Testing
    A. Guidelines for Determining Proper Girder and Joist Sizes
1. Sizing Girders
2. Sizing Joists

    B. Laboratory

    Have the trainees select the proper girder/beam and joist size from the tables in the Trainee Module for various floor plans, floor loads, and span data. This laboratory corresponds to Performance Task 7.

    C. Review

    D. Module Examination
1. Trainees must score 70 percent or higher to receive recognition from NCCER.
2. Record the testing results on Craft Training Report Form 200, and submit the results to the Training Program Sponsor.

    E. Performance Testing
1. Trainees must perform each task to the satisfaction of the instructor to receive recognition from NCCER. If applicable, proficiency noted during laboratory exercises can be used to satisfy the Performance Testing requirements.
2. Record the testing results on Craft Training Report Form 200, and submit the results to the Training Program Sponsor.

Floor Systems
27105-06

An Associated Builders and Contractors 2005 National Craft Championships contestant measures a member for metal frame. Metal framing is gaining widespread use in both commercial and residential construction.

Instructor's Notes:

Assign reading of Module 27105-06.

27105-06
Floor Systems

Topics to be presented in this module include:

1.0.0	Introduction	.5.2
2.0.0	Methods of Framing Houses	.5.2
3.0.0	Building Working Drawings and Specifications	.5.7
4.0.0	The Floor System	.5.12
5.0.0	Laying Out and Constructing a Platform Floor Assembly	.5.27
6.0.0	Installing Joists for Projections and Cantilevered Floors	.5.38
7.0.0	Estimating the Quantity of Floor Materials	.5.40
8.0.0	Guidelines for Determining Proper Girder and Joist Sizes	.5.41

Overview

The construction of a wood-frame floor begins with placement of the sill plate on the foundation. The sill plate is typically attached to the foundation with anchor bolts that are embedded into the concrete foundation.

Once the sill plate is installed, the carpenter lays out and marks the locations of the floor joists on the sill plate, and then places and attaches the joists. In many instances, the joist span is too great for a single piece of lumber, so joists are joined over a beam or girder. Once all the joists are in place and a header board has been attached to the ends of the joists, a subfloor is attached to the joists. Bridging is attached between the joists to provide stability.

Wood I-beams and floor trusses made of wood or steel are sometimes used instead of joists. These materials are stronger than comparable lengths of dimension lumber and can therefore be placed across greater spans.

Instructor's Notes:

Objectives

When you have completed this module, you will be able to do the following:

1. Identify the different types of framing systems.
2. Read and interpret drawings and specifications to determine floor system requirements.
3. Identify floor and sill framing and support members.
4. Name the methods used to fasten sills to the foundation.
5. Given specific floor load and span data, select the proper girder/beam size from a list of available girders/beams.
6. List and recognize different types of floor joists.
7. Given specific floor load and span data, select the proper joist size from a list of available joists.
8. List and recognize different types of bridging.
9. List and recognize different types of flooring materials.
10. Explain the purposes of subflooring and underlayment.
11. Match selected fasteners used in floor framing to their correct uses.
12. Estimate the amount of material needed to frame a floor assembly.
13. Demonstrate the ability to:
    - Lay out and construct a floor assembly
    - Install bridging
    - Install joists for a cantilever floor
    - Install a subfloor using butt-joint plywood/OSB panels
    - Install a single floor system using tongue-and-groove plywood/OSB panels

Required Trainee Materials

1. Pencil and paper
2. Appropriate personal protective equipment

Prerequisites

Before you begin this module, it is recommended that you successfully complete *Core Curriculum* and *Carpentry Fundamentals Level One*, Modules 27101-06 through 27104-06.

This course map shows all of the modules in *Carpentry Fundamentals Level One*. The suggested training order begins at the bottom and proceeds up. Skill levels increase as you advance on the course map. The local Training Program Sponsor may adjust the training order.

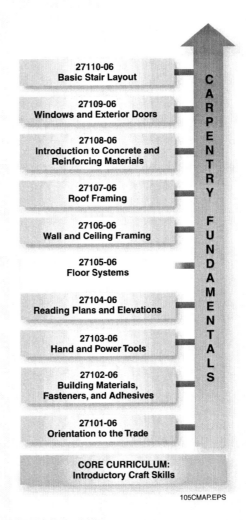

Ensure you have everything required to teach the course. Check the Materials and Equipment List at the front of this module.

Show Transparency 1, Objectives.

Show Transparency 2, Performance Tasks.

See the general Teaching Tip at the end of this module.

Explain that terms shown in bold (blue) are defined in the Glossary at the back of this module.

Identify the four basic framing methods.

Acquire photographs or drawings from architectural firms or building material manufacturers that show specific types of building construction in various stages of completion. Pass them around for the trainees to examine.

Describe platform and brace framing and identify the framing members. Point out the settling (shrinkage) problem with platform framing.

Show Transparency 3 (Figure 1).

Trade Terms

Crown
Dead load
Firestop
Foundation
Header joist
Joist hanger
Let-in
Live load
Pier

Rafter plate
Scab
Scarf
Soleplate
Span
Tail joist
Trimmer joist
Truss
Underlayment

1.0.0 ◆ INTRODUCTION

This module briefly introduces the different methods used for framing buildings. Some of the types of framing described are rarely used today. They are seen in existing installations, however, so some knowledge about them is helpful, especially when remodeling. The remainder of the module describes floor framing with an emphasis on the platform method of framing. Included are descriptions of the materials and general methods used to construct floors. Proper construction of floors is essential because, no matter how well the **foundation** of a building is constructed, a structure will not stand if the floors and sills are poorly assembled.

2.0.0 ◆ METHODS OF FRAMING HOUSES

Various areas of the country have had different methods of constructing a wood frame dwelling. This variation in methods can be attributed to economic conditions, availability of material, or climate in different parts of the country. We will discuss the various types of framing and show the approved method of construction today. Structures that are framed and constructed entirely of wood above the foundation fall into several classifications:

- Platform frame (also known as western and box frame)
- Braced frame
- Balloon frame
- Post-and-beam frame

2.1.0 Platform Frame

The platform frame, sometimes called the western frame, is used in most modern residential and light commercial construction. In this type of construction, each floor of a structure is built as an individual unit (*Figure 1*). The subfloor is laid in place prior to the exterior walls being put in place. The **soleplate** and top plate are nailed to the studs. Window openings are constructed and put in place, and studs are notched in order to let in a 1×4 diagonal brace.

In western or platform frame construction, the walls are built lying on the subfloor. Once a section or a part of a wall is complete, it is then lifted into place and nailed to the floor system and another plate is added to the top plate. This will tie in the second floor system. This method of construction allows workers to work safely on the floor while constructing the wall systems.

Platform framing is subject to settling caused by the shrinkage of a large number of horizontal, loadbearing frame members. Settling can result in various problems such as cracked plaster, cracked wallboard joints, uneven ceilings and floors, ill-fitting doors and windows, and nail pops (nails that begin to protrude through the wallboard). Settling can be minimized by using well-dried lumber.

2.2.0 Braced Frame

In early times, the brace frame method of construction was frequently used because most bearing lumber was in the vertical position. Very little shrinkage occurred in this type of construction.

Platform Framing

In platform framing, it is inevitable that the structure will begin to settle. While settling is unavoidable, the degree to which the structure settles can be minimized by not cutting any corners during the framing and construction process. Compacting the foundation soil and making sure you brace the frame properly during construction will also decrease the amount of settlement.

5.2 CARPENTRY FUNDAMENTALS LEVEL ONE ◆ TRAINEE GUIDE

Instructor's Notes:

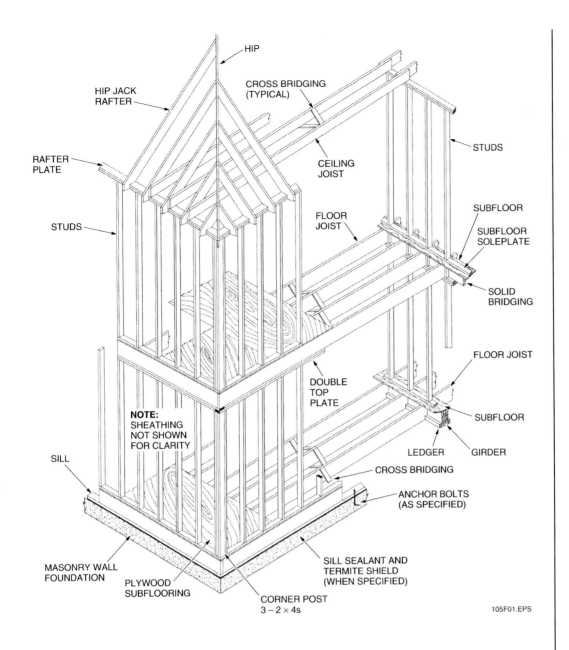

Figure 1 ♦ Platform framing.

Briefly describe older style balloon framing and identify the framing members.

Show Transparency 4 (Figure 2).

Describe post-and-beam framing and identify the framing members. Show photographs if available.

Show Transparency 5 (Figure 3).

In this method of construction, the framework does not rely on sheathing for rigidity. The framework, which in part had posts for corners and beams or girders, was mortised and tenoned together and held in place by pins and dowels. Diagonal braces were added to give the framework rigidity, thus making it stable. Therefore, the sheathing or planking could run up the sides of the structure vertically. One common type of structure built in this fashion was the barn.

2.3.0 Balloon Frame

Balloon frame construction is rarely used today because of lumber and labor costs, but a substantial number of structures built with this type of frame are still in use. In the balloon frame method of construction, the studs are continuous from the sill to the **rafter plate** (*Figure 2*). The wall studs and first floor joists rest on a solid sill (usually a 2×6), with the floor joists being nailed to the sides of the studs. The second floor joists rest on a horizontal 1×4 or 1×6 board called a ribbon that is installed in notches (**let-ins**) to the faces of the studs. Braces (usually 1×4) are notched in on a diagonal into the outside face of the studs. This method of construction makes the frame self-supporting; therefore, it does not need to rely on the sheathing for rigidity.

In the balloon frame, **firestops** must be installed in the walls in several locations. A firestop is an approved material used in the space between frame members to prevent the spread of fire for a limited period of time. Typically, 2×4 wood blocks are installed between the studs for this purpose.

One advantage of balloon framing is that the shrinkage of the wood framing members is low, thus helping to reduce settling. This is because wood shrinks across its width, but practically no shrinkage occurs lengthwise. This provides for high vertical stability, making the balloon frame adaptable for two-story structures.

2.4.0 Post-and-Beam Frame

The post-and-beam method of framing floors and roofs has been used in heavy timber buildings for many years. It uses large, widely-spaced timbers for joists, posts, and rafters. Matched planks are often used for floors and roof sheathing (*Figure 3*).

Post-and-beam homes are normally designed for this method of framing. In other words, you cannot start a house under standard framing procedures and then alter the construction to post-and-beam framing.

In post-and-beam framing, plank subfloors or roofs are usually of 2" nominal thickness, supported on beams spaced up to 8' apart. The ends of the beams are supported on posts or **piers.** Wall spaces between posts are provided with supplementary framing as required for attachment of exterior and interior finishes. This additional framing also serves to provide lateral bracing for the building. Consider this versus the conventional framing that utilizes joists, rafters, and studs placed from 16" to 24" on center. Post-and-beam framing requires fewer but larger framing members spaced further apart. The most efficient use of 2" planks occurs when the lumber is continuous over more than one **span.** When standard lengths of lumber such as 12', 14', or 16' are used, beam spacings of 6', 7', or 8' are indicated. This factor has a direct bearing on the overall dimensions of the building.

If local building codes allow end joints in the planks to fall between supports, planks of random lengths may be used and beam spacing can be adjusted to fit the house dimensions. Windows and doors are normally located between posts in the exterior walls, eliminating the need for headers over the openings. The wide spacing between posts permits ample room for large glass areas. Consideration should be given to providing an adequate amount of solid wall siding. The siding must also provide ample and adequate lateral bracing.

INSIDE TRACK

Balloon Framing

Balloon framing is frequently used in hurricane-prone areas for gable ends. In fact, this type of framing may be required by local building codes. Because balloon framing has studs that span multiple stories, alternative types of studs may be required if standard wood studs aren't available. Engineered lumber or metal studs can be used for this purpose.

Instructor's Notes:

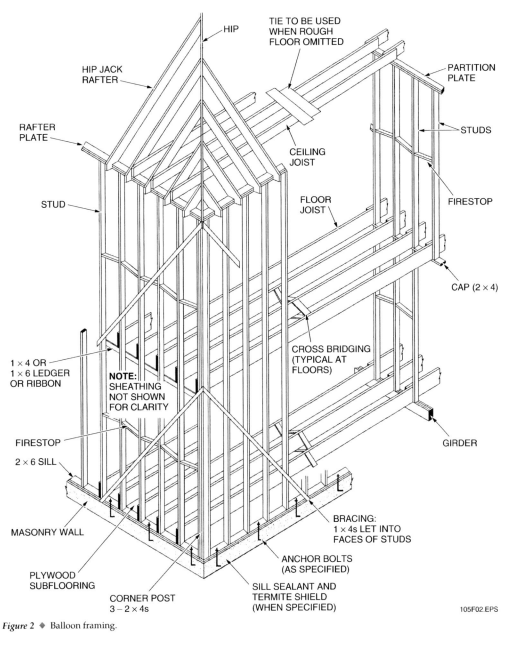

Figure 2 ♦ Balloon framing.

Using illustrations or photographs, have the trainees identify the framing members for the various types of framing.

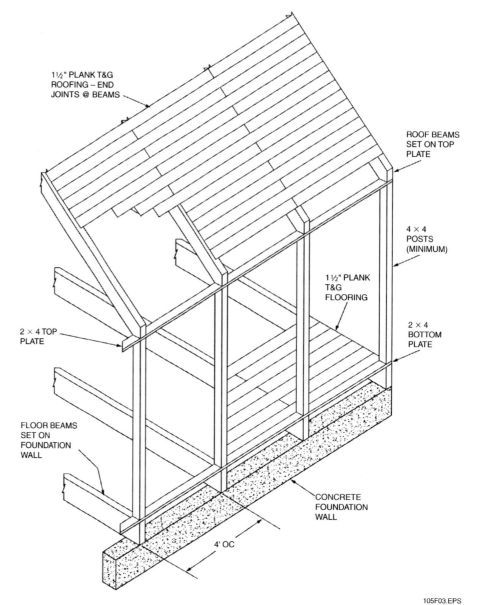

Figure 3 ♦ Post-and-beam framing.

Instructor's Notes:

Kobe, Japan

On January 17, 1995, a massive earthquake that measured 7.2 on the Richter scale leveled the city of Kobe, Japan. The quake caused an estimated $120 billion in initial damage and claimed the lives of some 5,500 people. Experts have attributed much of the residential destruction to the traditional Japanese post-and-beam construction methods. Many of the structures were poorly constructed, poorly fastened, and had few lower level partitions that would increase their ability to resist lateral movement. Most Japanese homes also had ceramic tile roofs that created very top-heavy structures. When the earthquake hit, these dwellings buckled under the pressure. New building codes have emerged in the wake of this earthquake, and the Japanese construction industry is now emphasizing platform framing. Architects and structural engineers from around the world have traveled to Kobe to research ways to improve the structural integrity of home construction. For example, they have discovered that increasing the size of wall paneling can strengthen the lateral resistance of a post-and-beam structure.

Briefly review the contents of a typical working drawing set.

Show Transparency 6 (Figure 4).

Identify the categories of architectural drawings.

A combination of conventional framing with post-and-beam framing is sometimes used where the two adjoin each other. On a side-by-side basis, no particular problems will be encountered.

Where a post-and-beam floor or roof is supported on a stud wall, a post should be placed under the end of the beam to carry the conventional load. A conventional roof can be used with post-and-beam construction by installing a header between the posts to carry the load from the rafters to the posts.

3.0.0 ◆ BUILDING WORKING DRAWINGS AND SPECIFICATIONS

Construction blueprints (working drawings) and related written specifications contain all the information and dimensions needed to build or remodel a structure. The interpretation of blueprints is critically important in the construction of floor systems, walls, and roof systems. *Figure 4* shows the contents and sequence of a typical set of working drawings. The drawings that apply when building a floor system are briefly reviewed here.

3.1.0 Architectural Drawings

The architectural drawings contain most of the detailed information needed by carpenters to build a floor system. The specific categories of architectural drawings commonly used include the following:

- Foundation plan
- Floor plan
- Section and detail drawings

3.1.1 Foundation Plan

The foundation plan is a view of the entire substructure below the first floor or frame of the building. It gives the location and dimensions of footings, grade beams, foundation walls, stem walls, piers, equipment footings, and foundations. Generally, in a detail view, it also shows the location of the anchor bolts or straps in foundation walls or concrete slabs.

3.1.2 Floor Plan

The floor plan is a cutaway view (top view) of the building, showing the length and breadth of the building and the layout of the rooms on that floor. It shows the following kinds of information:

- Outside walls, including the location and dimensions of all exterior openings
- Types of construction materials
- Location of interior walls and partitions
- Location and swing of doors
- Stairways
- Location of windows
- Location of cabinets, electrical and mechanical equipment, and fixtures
- Location of cutting plane line

Emphasize the importance of reviewing all section and detail views for construction information.

Show Transparency 7 (Figure 5).

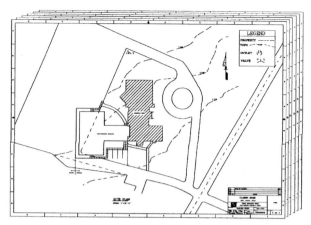

TITLE SHEET(S)
ARCHITECTURAL DRAWINGS
• SITE (PLOT) PLAN
• FOUNDATION PLAN
• FLOOR PLANS
• INTERIOR/EXTERIOR ELEVATIONS
• SECTIONS
• DETAILS
• SCHEDULES
STRUCTURAL DRAWINGS
PLUMBING PLANS
MECHANICAL PLANS
ELECTRICAL PLANS

105F04.EPS

Figure 4 ♦ Typical format of a working drawing set.

3.1.3 Section and Detail Drawings

Section drawings are cutaway vertical views through an object or wall that show its interior makeup. They are used to show the details of construction and information about walls, stairs, and other parts of construction that may not show clearly on the plan. A section view is limited to the specific portion of the building construction that the architect wishes to clarify. It may be drawn on the same sheet as an elevation or plan view or it may appear on a separate sheet. *Figure 5* shows a typical example of a section drawing. Detail views are views that are normally drawn to a larger scale. They are often used to show aspects of a design that are too small to be shown in sufficient detail on a plan or elevation drawing. Like section drawings, detail drawings may be drawn on the same sheet as an elevation or plan drawing or may appear on a separate sheet in the set of plans.

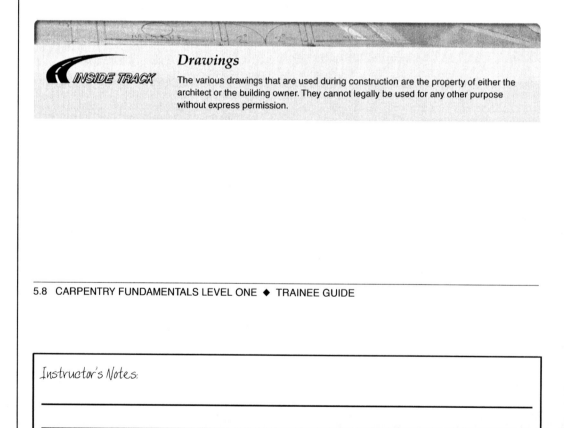

Drawings

The various drawings that are used during construction are the property of either the architect or the building owner. They cannot legally be used for any other purpose without express permission.

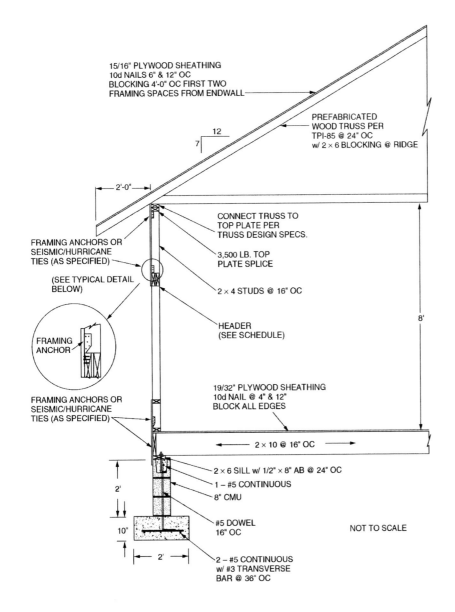

Figure 5 ♦ Typical section drawing.

Explain the purpose of structural drawings and show an example.

Floor Plan

INSIDE TRACK — A typical floor plan provides a top view that details the layout of the rooms for each floor in the building.

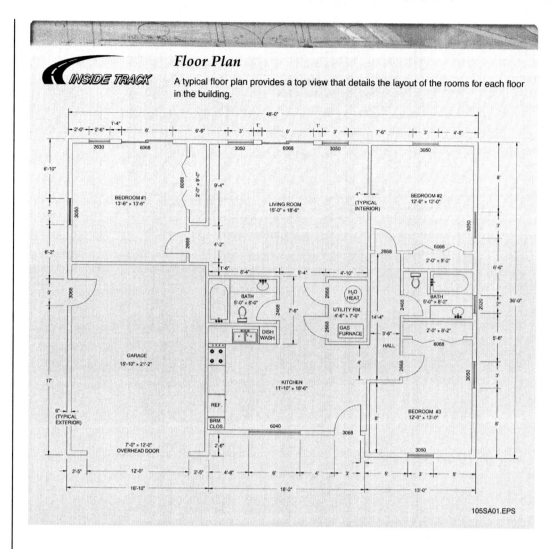

3.1.4 Structural Drawings

Structural drawings are created by a structural engineer and accompany the architect's plans. They are usually drawn for large structures such as an office building or factory. They show requirements for structural elements of the building including columns, floor and roof systems, stairs, canopies, bearing walls, etc. They include such details as:

- Height of finished floors and walls
- Height and bearing of bar joists or steel joists
- Location of bearing steel materials
- Height of steel beams, concrete planks, concrete Ts, and poured-in-place concrete
- Bearing plate locations
- Location, size, and spacing of anchor bolts
- Stairways

Instructor's Notes:

Dimension vs. Scale

The scale of a drawing refers to the amount or percentage that a document has been reduced in relation to reality (full scale). Specific dimensions on documents, however, should always take precedence over the scaled graphic representation (drawing) found on the plans. The drawing itself is meant to give a general idea of the overall layout. The dimensions found on the drawing are what you should use when planning and building.

Explain the purpose of plumbing, mechanical, and electrical plans. Show an example of each.

Demonstrate how to read a typical drawing set. Point out the information required for floor systems.

3.2.0 Plumbing, Mechanical, and Electrical Plans

Plumbing plans show the size and location of water and gas systems if they are not included in the mechanical section. Mechanical plans show temperature control and ventilation equipment including ducts, louvers, and registers. Electrical plans show all electrical equipment, lighting, outlets, etc.

It is important to note that while carpenters usually work with architectural and structural drawings, there are useful notes and views on drawings found in other sections, especially the mechanical section. For example, typical items not found on architectural drawings but that may be found on mechanical drawings are exposed heating, ventilating, and air conditioning (HVAC) ductwork; heating convectors; and fire sprinkler piping. Any one or all of these items may require the carpenter to build special framing, so make a habit of reviewing all drawings. Also, make sure to coordinate any such work with the appropriate other trades to make sure that the proper framing is done to accommodate ductwork, piping, wiring, etc.

3.3.0 Reading Blueprints

The following general procedure is suggested as a method of reading any set of blueprints for understanding:

Step 1 Read the title block. The title block tells you what the drawing is about. It contains critical information about the drawing such as the scale, date of last revision, drawing number, and architect or engineer. If you have to remove a sheet from a set of drawings, be sure to fold the sheet with the title block facing up.

Step 2 Find the north arrow. Always orient yourself to the structure. Knowing where north is enables you to more accurately describe the locations of walls and other parts of the building.

Step 3 Always be aware that blueprints work together as a team. The reason the architect or engineer draws plans, elevations, and sections is that it requires more than one type of view to communicate the whole project. Learn how to use more than one drawing, when necessary, to find the information you need.

Step 4 Check the list of blueprints in the set. Note the sequence of the various types of plans. Some blueprints have an index on the front cover. Notice that the prints are broken into several categories:
 – Architectural
 – Structural
 – Mechanical
 – Electrical
 – Plumbing

Step 5 Study the site plan (plot plan) to observe the location of the building. Notice that the geographic location of the building may be indicated on the site plan.

Step 6 Check the foundation and floor plans for the orientation of the building. Observe the location and features of entries, corridors, offsets, and any special features.

Step 7 Study the features that extend for more than one floor, such as plumbing and vents, stairways, elevator shafts, heating and cooling ductwork, and piping.

Step 8 Check the floor and wall construction and other details relating to exterior and interior walls.

MODULE 27105-06 ♦ FLOOR SYSTEMS 5.11

Explain the purpose of specifications and show examples. Point out that specifications provide detailed information not given in the drawings.

Have the trainees practice using drawing sets and specifications to determine requirements for instructor-selected floor systems.

Assign reading of Sections 4.0.0–4.5.3 for the next class session.

Explain the purpose of floor systems and identify each part of a typical platform floor system.

Show Transparencies 8 through 10 (Figures 6 through 8).

Describe sills and sill anchoring. Emphasize the need for foundation checks and the requirements for sealers, shields, and pressure-treated wood.

Step 9 Check the foundation plan for size and types of footings, reinforcing steel, and loadbearing substructures.

Step 10 Study the mechanical plans for the details of heating, cooling, and plumbing.

Step 11 Observe the electrical entrance and distribution panels, and the installation of the lighting and power supplies for special equipment.

Step 12 Check the notes on the various pages and compare the specifications against the construction details. Look for any variations.

Step 13 Thumb through the sheets of drawings until you are familiar with all the plans and structural details.

Step 14 Recognize applicable symbols and their relative locations in the plans. Note any special construction details or variations that will affect your job.

When you are building a floor system, the building plans (or specifications) should provide all the information you need to know about the floor system. Important information you should look for regarding the floor system includes:

- Type of wood or other materials used for sills, posts, girders, beams, joists, subfloors, etc.
- Size, location, and spacing of support posts or columns
- Direction of both joists and girders
- Manner in which joists connect to girders
- Location of any loadbearing interior walls that run parallel to joists
- Location of any toilet drains
- Rough opening sizes and locations of all floor openings for stairs, etc.
- Any cantilevering requirements
- Changes in floor levels
- Any special metal fasteners needed in earthquake areas
- Types of blocking or bridging
- Clearances from the ground to girder(s) and joists for floors installed over crawlspaces

3.4.0 Specifications

Written specifications are equally as important as the drawings in a set of plans. They furnish what the drawings cannot, in that they give detailed and accurate written descriptions of work to be done. They include quality and quantity of materials, methods of construction, standards of construction, and manner of conducting the work.

Specifications will be studied in detail later in your training. The basic information found in a typical specification includes:

- Contract
- Synopsis of the work
- General requirements
- Owner's name and address
- Architect's name
- Location of structure
- Completion date
- Guarantees
- Insurance requirements
- Methods of construction
- Types and quality of building materials
- Sizes

4.0.0 ◆ THE FLOOR SYSTEM

Floor systems provide a base for the remainder of the structure to rest on. They transfer the weight of people, furniture, materials, etc., from the subfloor, to the floor framing, to the foundation wall, to the footing, then finally to the earth. Floor systems are constructed over basements or crawlspaces. Single-story structures built on slabs do not have floor systems; however, multi-level structures may have both a slab and a floor system. *Figure 6* shows a typical platform floor system and identifies the various parts.

4.1.0 Sills

Sills, also called sill plates, are the lowest members of a structure's frame. They rest horizontally on the foundation and support the floor joists. The foundation is the supporting portion of a structure below the first floor construction, including the footings. Sills serve as the attachment point to the concrete or block foundation for all of the other wood framing members. The sills provide a means of leveling the top of the foundation wall and also prevent the other wood framing lumber from making contact with the concrete or masonry, which can cause the lumber to rot.

Today, sills are normally made using a single layer of 2 × 6 lumber (*Figure 7*). Local codes normally require that pressure-treated lumber and/or foundation-grade redwood lumber be used for constructing the sill whenever the sill plate comes in direct contact with any type of concrete. However, where codes allow, untreated softwood can be used.

Sills are attached to the foundation wall using either anchor bolts (*Figure 7*) or straps (*Figure 8*) embedded in the foundation. The exposed portion

5.12 CARPENTRY FUNDAMENTALS LEVEL ONE ◆ TRAINEE GUIDE

Instructor's Notes:

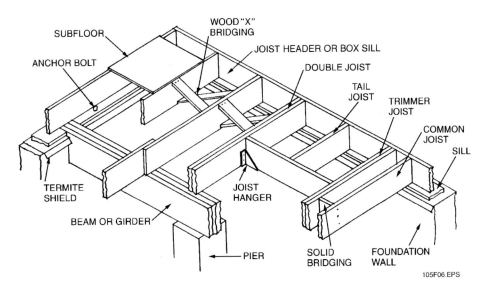

Figure 6 ◆ Typical platform frame floor system.

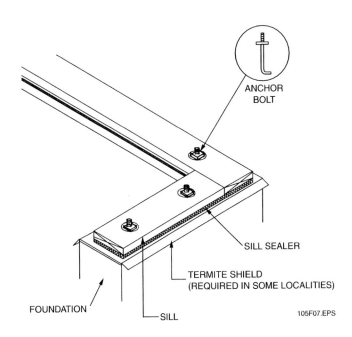

Figure 7 ◆ Typical sill installation.

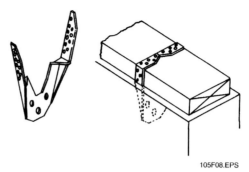

Figure 8 ♦ Typical sill anchor strap.

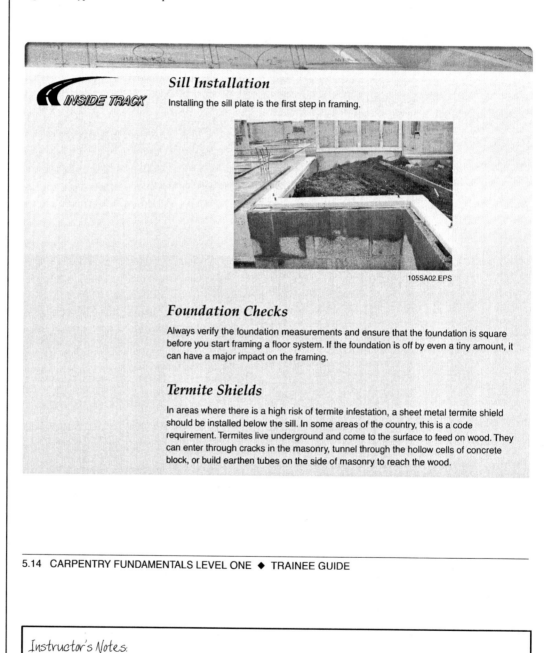

Sill Installation

Installing the sill plate is the first step in framing.

Foundation Checks

Always verify the foundation measurements and ensure that the foundation is square before you start framing a floor system. If the foundation is off by even a tiny amount, it can have a major impact on the framing.

Termite Shields

In areas where there is a high risk of termite infestation, a sheet metal termite shield should be installed below the sill. In some areas of the country, this is a code requirement. Termites live underground and come to the surface to feed on wood. They can enter through cracks in the masonry, tunnel through the hollow cells of concrete block, or build earthen tubes on the side of masonry to reach the wood.

Instructor's Notes:

of the strap-type anchor is nailed to the sill. Some types must be bent over the top of the sill, while others are nailed to the sides. The size, type, and spacing between the anchor bolts or straps must be in compliance with local building codes. Their location and other related data are normally shown on the building blueprints.

In structures where the underfloor areas are used as part of the HVAC system, for storage, or as a basement, a glass-wool insulating material, called a sill sealer (*Figure 7*), should be installed to account for irregularities between the foundation wall and the sill. It seals against drafts, dirt, and insects. Sill sealer material is made in 6" wide, 50' rolls. Uncompressed, its thickness is 1". However, it can compress to as little as ½" when the weight of the structure is upon it. The sill sealer should be installed between the sill and the foundation wall, or between the sill and a termite shield (if used).

4.2.0 Beams/Girders and Supports

The distance between two outside walls is frequently too great to be spanned by a single joist. When two or more joists are needed to cover the span, support for the inboard joist ends must be provided by one or more beams, commonly called girders. Girders carry a very large proportion of the weight of a building. They must be well designed, rigid, and properly supported at the foundation walls and on the supporting posts or columns. They must also be installed so that they will properly support the joists. Girders may be made of solid timbers, built-up lumber, engineered lumber, or steel beams (*Figure 9*). Each type has advantages and disadvantages. Note that in some instances, precast reinforced concrete girders may also be used. A general procedure for determining how to size a girder is given later in this module.

4.2.1 Solid Lumber Girders

Solid timber girder stock used for beams is available in various sizes, with 4×6, 4×8, and 6×6 being typical sizes. If straight, large timbers are available, their use can save time by not having to make built-up girders. However, solid pieces of large timber stock are often badly bowed and can create a rise in the floor unless the **crowns** are pulled down. The crowns are the high points of the crooked edges of the framing members.

4.2.2 Built-Up Lumber Girders

Built-up girders are usually made using nominal 2" stock (2×8s, 2×10s) nailed together so that they act as one piece. Built-up girders have the advantage of not warping as easily as solid wooden girders and are less likely to have decayed wood in the center. The disadvantage is that a built-up girder is not capable of carrying the same load as an equivalent size solid timber girder. When constructing a built-up girder, the individual boards must be nailed together according to code requirements. Also, it is necessary to

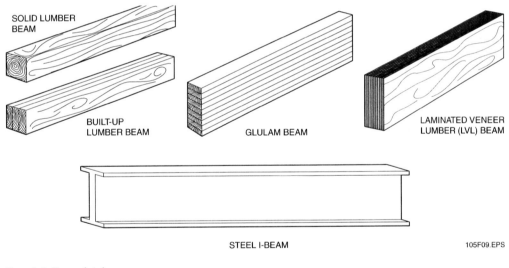

Figure 9 ♦ Types of girders.

stagger the joints at least 4' in either direction. Construction of built-up girders is covered in more detail later in this module.

4.2.3 Engineered Lumber Girders

Laminated veneer lumber (LVL) and glue-laminated lumber (glulam) are engineered lumber products that are used for girders and other framing members. Their advantage is that they are stronger than the same size structural lumber. For a given length, the greater strength of engineered lumber products allows them to span a greater distance. Another advantage is that they are very straight with no crowns or warps.

LVL girders are made from laminated wood veneer like plywood. The veneers are laid up in a staggered pattern with the veneers overlapping to increase strength. Unlike plywood, the grain of each layer runs in the same direction as the other layers. The veneers are bonded with an exterior-grade adhesive, then pressed together and heated under pressure.

Glulam girders are made from lengths of solid, kiln-dried lumber glued together. They are commonly used where the beams are to remain exposed. Glulam girders are available in three appearance grades: industrial, architectural, and premium. For floor systems like those described in this module, industrial grade would normally be used because appearance is not a priority. Architectural grade is used where beams are exposed and appearance is important. Premium grade is used where the highest-quality appearance is needed. Glulam beams are available in various widths and depths and in lengths up to 40' long.

4.2.4 Steel I-Beam Girders

Metal beams can span the greatest distances and are often used when there are few or no piers or interior supports in a basement. Also, they can span greater distances with smaller beam sizes, thereby creating greater headroom in a basement or crawlspace. For example, a 6" high steel beam may support the same load as an 8" or 10" high wooden beam. Two types of steel beams are available: standard flange (S-beam) and wide flange (W-beam). The wide beam is generally used in residential construction. Being metal, I-beams are more expensive than wood and are harder to work with. They are normally used only when the design or building code calls for it.

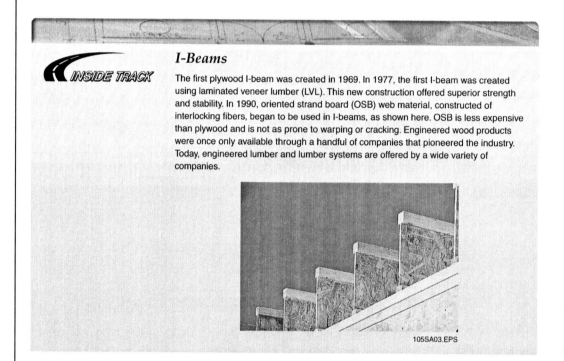

I-Beams

The first plywood I-beam was created in 1969. In 1977, the first I-beam was created using laminated veneer lumber (LVL). This new construction offered superior strength and stability. In 1990, oriented strand board (OSB) web material, constructed of interlocking fibers, began to be used in I-beams, as shown here. OSB is less expensive than plywood and is not as prone to warping or cracking. Engineered wood products were once only available through a handful of companies that pioneered the industry. Today, engineered lumber and lumber systems are offered by a wide variety of companies.

4.2.5 Beam/Girder Supports

Girders and beams must be properly supported at the foundation walls, and at the proper intervals in between, either by supporting posts, columns, or piers (*Figure 10*). Solid or built-up wood posts installed on pier blocks are commonly used to support floor girders, especially for floors built over a crawlspace. Usually, 4 × 4 or 4 × 6 posts are used. However, all posts must be as wide as the girder. Where girder stock is jointed over a post, a 4 × 6 is normally required. To secure the wood posts to their footings, pieces of ½" reinforcing rod or iron bolts are often embedded in the support footings before the concrete sets. These project into holes bored in the bottoms of the posts. The use of galvanized steel post anchors is another widely used method of fastening the bottoms of wooden posts to their footings (*Figure 11*). The

Describe the various methods of supporting beams and girders. Point out the methods of tying the supports to the beams and girders.

Show Transparencies 12 and 13 (Figures 10 and 11).

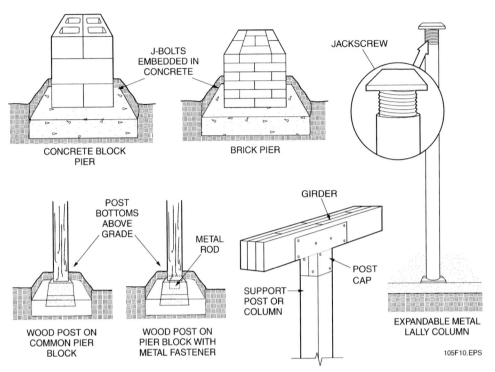

Figure 10 ◆ Typical methods of supporting girders.

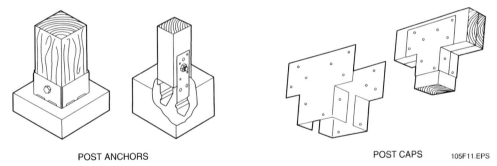

Figure 11 ◆ Typical post anchors and caps.

MODULE 27105-06 ◆ FLOOR SYSTEMS 5.17

Explain beam support spacing and size requirements. Point out that wooden beams should not rest directly on concrete and that at least 4" at the ends of each beam should rest on wall supports.

Show Transparencies 14 through 16 (Figures 12 through 14).

tops of the posts are normally fastened to the girder using galvanized steel post caps. In addition to securing the post to the girder, these caps also provide for an even bearing surface.

Four-inch round steel columns filled and reinforced with concrete (*Figure 10*), called lally columns, are commonly used as support columns in floors built over basements. Some types of lally columns must be cut to the required height, while others have a built-in jackscrew that allows the column to be adjusted to the proper height. Metal plates are installed at the top and bottom of the columns to distribute the load over a wider area. The plates normally have predrilled holes so that they may be fastened to the girder.

Support piers made of brick or concrete block (*Figure 10*) are more difficult to work with because their level cannot be adjusted. The height of the related footings must be accurate so that when using 4" thick bricks or 8" tall blocks, their tops come out at the correct height to support the girder.

The spacing or interval required between the girder posts or columns is determined by local building codes based on the stress factor of the girder beam (i.e., how much weight is put on the girder beam). It is important to note that the farther apart the support posts or columns are spaced, the heavier the girder must be in order to carry the joists over the span between them. An example of a girder and supporting columns used in a 24' × 48' building is shown in *Figure 12*. In this example, column B supports one-half of the girder load existing between the building wall A and column C. Column C supports one-half of the girder load between columns B and D. Likewise, column D will share equally the girder loads with column C and the wall E.

As shown in *Figure 13*, support of girder(s) at the foundation walls can be done by constructing posts made from solid wood or piers made of concrete block or brick. Another widely used method is to construct girder (beam) pockets into the concrete or concrete block foundation walls (*Figure 14*). Provide steel reinforcement as required by the job specifications.

The specifications for girder pockets vary with the size of the girder being used. A rule of thumb is that the pocket should be at least one inch wider than the beam and the beam must have at least 4" of bearing on the wall. Wooden girders placed in the pocket should not be allowed to come in direct contact with the concrete or masonry foundation. This is because the chemicals in the concrete or masonry can deteriorate the wood. The end of the wooden girder should sit on a steel plate that is at least ¼" thick. Some carpenters also use metal flashing to line the girder pocket to help protect the wood from the concrete. In some applications, any one of several types of galvanized steel girder hangers can be used to secure the girder to the foundation. *Figure 14* shows one common type.

It is normal to use temporary supports, such as jacks and/or 2 × 4 studs nailed together with braces, to support the girder(s) while the floor is being constructed. After the floor is assembled, but before the subflooring is installed, the permanent support posts or columns are put into place.

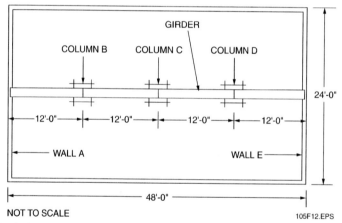

Figure 12 ◆ Example of column spacing.

5.18 CARPENTRY FUNDAMENTALS LEVEL ONE ◆ TRAINEE GUIDE

Instructor's Notes:

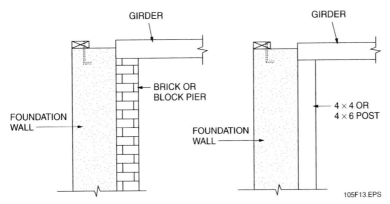

Figure 13 ♦ Post or pier support of a girder at the foundation wall.

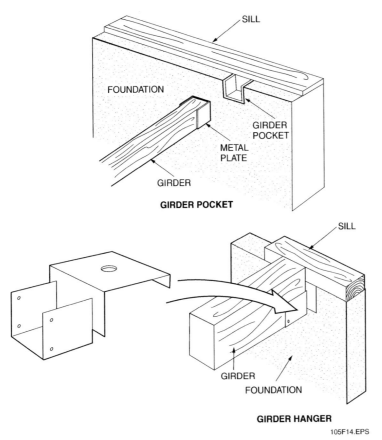

Figure 14 ♦ Girder pocket and girder hanger girder supports.

MODULE 27105-06 ♦ FLOOR SYSTEMS 5.19

Explain joist sizing, placement, and support methods. Point out joist loading and deflection requirements. Emphasize that joists are doubled under partitions, around openings, and under points of extra load.

Show Transparencies 17 through 19 (Figures 15 through 17).

4.3.0 Floor Joists

Floor joists are a series of parallel, horizontal framing members that make up the body of the floor frame (*Figure 15*). They rest on and transfer the building load to the sills and girders. The flooring or subflooring is attached to them. The span determines the length of the joist that must be used. Safe spans for joists under average loads can be found using the latest tables available from wood product manufacturers or sources such as the National Forest Products Association and the Southern Forest Products Association. For floors, this is usually figured on a basis of 50 lbs. per sq. ft. (10 lbs. **dead load** and 40 lbs. **live load**). Dead load is the weight of permanent, stationary construction and equipment included in a building. Live load is the total of all moving and variable loads that may be placed upon a building.

Joists must not only be strong enough to carry the load that rests on them; they must also be stiff enough to prevent undue bending (deflection) or vibration. Too much deflection in joists is undesirable because it can make a floor noticeably springy. Building codes typically specify that the deflection downward at the center of a joist must not exceed $\frac{1}{360}$th of the span with normal live load. For example, for a joist with a 15' span, this would equal a maximum of ½" of downward deflection (15' span $\times$ 12 = 180" $\div$ 360 = 0.5"). A general procedure for sizing joists is given later in this module.

Joists are normally placed 16" on center (OC) and are always placed crown up. However, in some applications joists can be set as close as 12" OC or as far apart as 24" OC. All these distances are used because they accommodate 4' $\times$ 8' subfloor panels and provide a nailing surface where two panels meet. Joists can be supported by the top of the girder or be framed to the side. *Figure 16* shows three methods for joist framing at the girder. Check your local code for applicability. Note that if joists are lapped over the girder, the minimum amount of lap is 4" and the maximum amount of lap is 12". *Figure 17* shows some examples of the many different types of **joist hangers** that can be used to fasten joists to girders as well as other support framing members. Joist hangers are used where the bottom of the girder must be flush with the bottoms of the joists. At the sill end of the joist, the joist should rest on at least 1½" of wood. In platform construction, the ends of all the joists are fastened to a **header joist,** also called a band joist or rim joist, to form the box sill.

Joists must be doubled where extra loads need to be supported. When a partition runs parallel to the joists, a double joist is placed underneath. Joists must also be doubled around all openings in the floor frame for stairways, chimneys, etc., to reinforce the rough opening in the floor. These additional joists used at such openings are called **trimmer joists.** They support the headers that carry short joists called **tail joists.** Double joists should spread where necessary to accommodate plumbing.

In residential construction, floors traditionally have been built using wooden joists. However, the

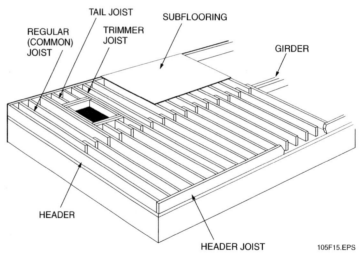

Figure 15 ◆ Floor joists.

5.20 CARPENTRY FUNDAMENTALS LEVEL ONE ◆ TRAINEE GUIDE

Instructor's Notes:

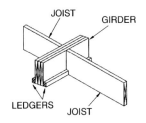 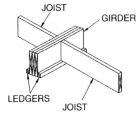 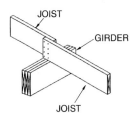

JOIST NOTCHED AROUND LEDGER   **JOIST SITS ON LEDGER**   **JOIST OVERLAP ON GIRDER**

Figure 16 ♦ Methods of joist framing at a girder.

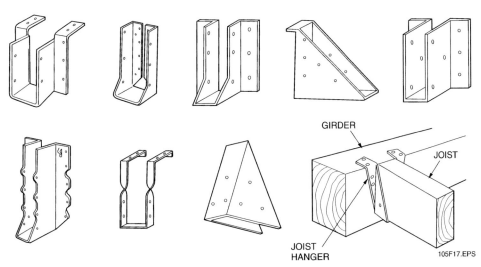

Figure 17 ♦ Typical types of joist hangers.

Maintaining a Flush Top Surface

If you are securing a joist to a girder with a ledger, you must first make sure that you maintain a flush top surface for the subflooring. Not all joists are the same size. To account for these small discrepancies, toenail the joist to the top of the girder prior to installing the ledger. Once you have established a smooth, flat surface, you can install the ledger under the joist.

Special Nails

Joist hangers require special nails to secure them to joists and girders. These nails are 1½" long and stronger than common nails. They are often referred to as joist or stub nails. The picture below shows various types of hangers as well as joist nails.

use of prefabricated engineered wood products such as wood I-beams and various types of trusses is also becoming common.

4.3.1 Notching and Drilling of Wooden Joists

When it is necessary to notch or drill through a floor joist, most building codes will stipulate how deep a notch can be made. For example, the *Standard Building Code* specifies that notches on the ends of joists shall not exceed one-fourth the depth. Therefore, in a 2 × 10 floor joist, the notch could not exceed 2½" (see *Figure 18*).

This code also states that notches for pipes in the top or bottom shall not exceed one-sixth the depth, and shall not be located in the middle third of the span. Therefore, when using a 2 × 10 floor joist, a notch cannot be deeper than 1⅜". This notch can be made either in the top or bottom of the joist, but it cannot be made in the middle third of the span. This means that if the span is 12', the middle span from 4' to 8' could not be notched.

This code further requires that holes bored for pipe or cable shall not be within 2" of the top or bottom of the joist, nor shall the diameter of any such hole exceed one-third the depth of the joist. This means that if a hole needs to be drilled, it may not exceed 3" in diameter if a 2 × 10 floor joist is used. Always check the local codes.

Some wood I-beams are manufactured with perforated knockouts in their web, approximately 12" apart. Never notch or drill through the beam flange or cut other openings in the web without checking the manufacturer's specification sheet.

Also, do not drill or notch other types of engineered lumber (e.g., LVL, PSL, and glulam) without first checking the specification sheets.

4.3.2 Wood I-Beams

Wood I-beams, sometimes called solid-web trusses, are made in various depths and with lengths up to 80'. These manufactured joists are not prone to shrinking or warping. They consist of an oriented strand board (OSB) or plywood web (*Figure 19*) bonded into grooves cut in the wood flanges on the top and bottom. This arrangement provides a joist that has a strength-to-weight ratio much greater than that of ordinary lumber. Because of their increased strength, wood I-beams can be used in greater spans than a comparable length of dimension lumber. They can be cut, hung, and nailed like ordinary lumber. Special joist hangers and strapping, similar to those used with wood joists, are used to fasten wood I-beam joists to girders and other framing members. Wood I-beam joists are typically manufactured with 1½" diameter, prestamped knockout holes in the web about 12" OC that can be used to accommodate electrical wiring. Other holes or openings can be cut into the web, but these can only be of a size and at the locations specified by the I-beam manufacturer. Under no circumstances should the flanges of I-beam joists be cut or notched. *Figure 20* shows a typical floor system constructed with wood I-beams.

Wood I-Beams

I-beams have specific guidelines and instructions to follow for cutting, blocking, and installation. It is important to always follow the manufacturer's instructions when installing these materials. Otherwise, you may create a very dangerous situation by compromising the structural integrity of the beam.

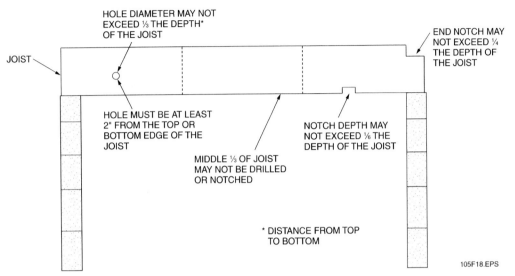

Figure 18 ♦ Notching and drilling of wooden joists.

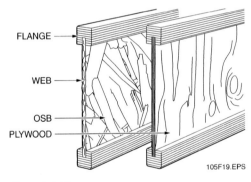

Figure 19 ♦ I-beams.

Figure 20 ♦ Typical floor system constructed with engineered I-beams (second floor shown).

Describe the various types of trusses available and their advantages over solid joists.

Show Transparencies 22 and 23 (Figures 21 and 22).

Describe the various types of bridging. Point out that subflooring should be installed before solid bridging (blocking) is installed or the bottoms of cross bridging are secured.

4.3.3 Trusses

Trusses are manufactured joist assemblies made of wood or a combination of steel and wood (*Figure 21*). Solid light-gauge steel and open-web steel trusses are also made, but these are used mainly in commercial construction. Like the wood I-beams, trusses are stronger than comparable lengths of dimension lumber, allowing them to be used over longer spans. Longer spans allow more freedom in building design because interior load-bearing walls and extra footings can often be eliminated. Trusses generally erect faster and easier with no need for trimming or cutting in the field. They also provide the additional advantage of permitting ducting, plumbing, and electrical wires to be run easily between the open webs.

Floor trusses consist of three components: chords, webs, and connector plates. The wood chords (outer members) are held rigidly apart by either wood or metal webs. The connector plates are toothed metal plates that fasten the truss web and chord components together at the intersecting points. The type of truss used most frequently in residential floor systems is the parallel-chord 4×2 truss. This name is derived from the chords being made of 2×4 lumber with the wide surfaces facing each other. Webs connect the chords. Diagonal webs positioned at 45-degree angles to the chords mainly resist the shearing stresses in the truss. Vertical webs, which are placed at right angles to the chords, are used at critical load transfer points where additional strength is required. Wood is used most frequently for webs, but galvanized steel webs are also used. Trusses made with metal webs provide greater clear spans for any given truss depth than wood-web trusses. The openings in the webs are larger too, which allows more room for HVAC ducting.

Note that there are several different kinds of parallel-chord trusses. What makes each one different is the arrangement of its webs. Typically, parallel-chord floor trusses with wood webs are available in depths ranging from 12" to 24" in 1" increments. The most common depths are 14" and 16". Some metal-web trusses are available with the same actual depth dimensions as 2×8, 2×10, and 2×12 solid wood joists, making them interchangeable with an ordinary joist-floor system. *Figure 22* shows a typical floor system constructed with trusses.

4.4.0 Bridging

Bridging is used to stiffen the floor frame to prevent unequal deflection of the joists and to enable an overloaded joist to receive some support from the joists on either side. Most building codes require that bridging be installed in rows between

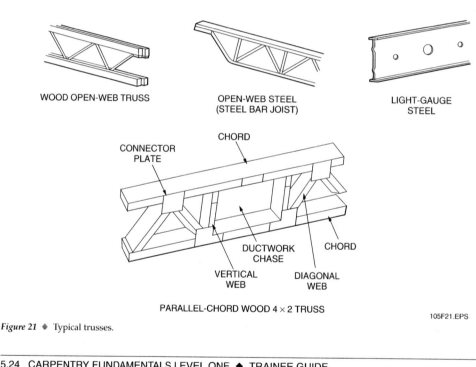

Figure 21 ♦ Typical trusses.

the floor joists at intervals of not more than 8'. For example, floor joists with spans of 8' to 16' need one row of bridging in the center of the span.

Three types of bridging are commonly used (*Figure 23*): wood cross-bridging, metal cross-bridging, and solid bridging. Wood and metal cross-bridging are composed of pieces of wood or metal set diagonally between the joists to form an X. Wood cross-bridging is typically 1 × 4 lumber placed in double rows that cross each other in the joist space.

Metal cross-bridging is installed in a similar manner. Metal cross-bridging comes in a variety of styles and different lengths for use with a particular joist size and spacing. It is usually made of 18-gauge steel and is ¾" wide. When using cross-bridging, you may nail the top, but do not nail the bottom until the subfloor is installed. Solid bridging, also called blocking, consists of solid pieces of lumber, usually the same size as the floor joists, installed between the joists. It is installed in an offset fashion to enable end nailing.

4.5.0 Subflooring

Subflooring consists of panels or boards laid directly on and fastened to floor joists (*Figure 24*) in order to provide a base for **underlayment** and/or the finish floor material. Underlayment is a material, such as particleboard or plywood, laid on top of the subfloor to provide a smoother surface for some finished flooring. This surface is normally applied after the structure is built but before the finished floor is laid. The subfloor adds rigidity to the structure and provides a surface upon which wall and other framing can be laid out and constructed. Subfloors also act as a barrier to cold and dampness, thus keeping the building

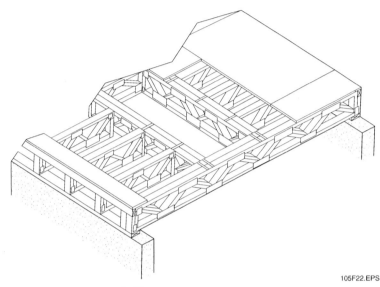

Figure 22 ◆ Typical floor system constructed with trusses.

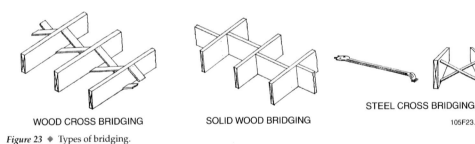

Figure 23 ◆ Types of bridging.

Explain the ratings of APA-graded plywood. Point out that subfloor panels are commonly glued as well as nailed to the joists, and that OSB is becoming a preferred subflooring material.

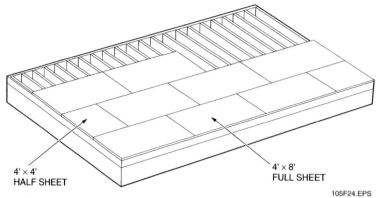

4' × 4' HALF SHEET

4' × 8' FULL SHEET

Figure 24 ♦ Subflooring installation.

warmer and drier in winter. Plywood, OSB or other manufactured board panels, or common wooden boards can be used as subflooring.

4.5.1 Plywood Subfloors

Butt-joint or tongue-and-groove plywood is widely used for residential subflooring. Used in 4' × 8' panels, typically ⅝" to ¾" thick when the joists are placed 16" OC, it goes on quickly and provides great rigidity to the floor frame. APA-rated sheathing plywood panels (*Table 1*) are generally used for subflooring in two-layer floor systems. APA-rated Sturd-I-Floor® tongue-and-groove plywood panels are commonly used in combined subfloor-underlayment (single-layer) floor systems where direct application of carpet, tile, etc., to the floor is intended.

Traditionally, plywood panels have been fastened to the floor joists using nails. Today, it is becoming more common to use a glued floor system in which the subfloor panels are both glued and nailed to the joists. This method helps stiffen the floors. It also helps eliminate squeaks and nail popping. Procedures for installing plywood subfloors, including gluing, are described later in this module.

4.5.2 Manufactured Board Panel Subfloors

Manufactured panels made of materials such as composite board, waferboard, OSB, and structural particleboard can also be used for subflooring. Detailed information on the construction and composition of these manufactured wood products is contained in an earlier module. Panels made of these materials have been rated by the American Plywood Association and meet all standards for subflooring. The method for installing these kinds of panels is basically the same as that used for plywood.

4.5.3 Board Subfloors

There are some instances when 1 × 6 or 1 × 8 boards are used as subflooring. Boards can be laid either diagonally or perpendicular to the floor joists. However, it is more common to lay them diagonally across the floor frame at a 45-degree angle. This provides for more rigidity of the floor and also assists in the bracing of the

Table 1 Guide to APA Performance-Rated Plywood Panels

Panel Grade	Thickness in Inches	Span Rating
Rated Sheathing	⅜	24/0
	⁷⁄₁₆	24/16
	¹⁵⁄₃₂	32/16
	¹⁹⁄₃₂	40/20
	²³⁄₃₂	48/24
Rated Sturd-I-Floor	¹⁹⁄₃₂	20 OC
	²³⁄₃₂	24 OC
	⅞, 1	32 OC
	1⅛	48 OC
Rated Siding	¹¹⁄₃₂	16 OC
	⁷⁄₁₆	24 OC
	¹⁵⁄₃₂	24 OC
T1-11	¹⁹⁄₃₂	16 OC
	¹⁹⁄₃₂	24 OC

Instructor's Notes:

OSB Subfloors

Today, many builders prefer to use OSB for subfloors. It offers acceptable structural strength at a reduced price.

105SA05.EPS

floor joists. Also, if laid perpendicular to the joist in a subfloor where oak flooring is to be laid over it, the oak flooring (instead of the subflooring) would have to be laid diagonally to the floor joist. This is necessary to prevent the shrinkage of the subfloor from affecting the joints in the finished oak floor, which would cause the oak flooring to pull apart. Board subflooring is nailed at each joist. Typically, two nails are used in each 1 × 6 board and three nails for wider boards. Note that a subfloor made of boards is normally not as rigid as one made of plywood or other manufactured panels.

5.0.0 ♦ LAYING OUT AND CONSTRUCTING A PLATFORM FLOOR ASSEMBLY

After the foundation is completed and the concrete or mortar has properly set up, assembly of the floor system can begin. Framing of the floors and sills is usually done before the foundation is backfilled. This is because the floor frame helps the foundation withstand the pressure placed on it by the soil. This section gives an overview of the procedures and methods used for laying out and constructing a basic platform floor assembly. When building any floor system, always coordinate your work with that of the other trades to ensure the framing is properly done to accommodate ductwork, piping, wiring, etc.

The construction of a platform floor assembly is normally done in the sequence shown below:

Step 1 Check the foundation for squareness.

Step 2 Lay out and install the sill plates.

Step 3 Build and/or install the girders and supports.

Step 4 Lay out the sills and girders for the floor joists.

Step 5 Lay out the joist locations for partitions and floor openings.

Step 6 Cut and attach the joist headers to the sill.

Step 7 Install the joists.

Step 8 Frame the openings in the floor.

Step 9 Install the bridging.

Step 10 Install the subflooring.

5.1.0 Checking the Foundation for Squareness

Before installing the sill plates, ensure that the foundation wall meets the dimensions specified on the blueprints and that the foundation is square. However, keep in mind that parallel and plumb take precedence over square. Checking the foundation for square is done by making measurements of the foundation with a 100' steel measuring tape. First, the lengths of each of the

Hand out Worksheets 27105-1 and 27105-2. Have the trainees complete the tasks on the Worksheets. Note the proficiency of each trainee.

Assign reading of Sections 5.0.0–5.10.0 for the next class session.

For floor framing, bridging, and sub-flooring demonstrations or laboratory exercises, you may wish to use half-size (half-scale) layout dimensions and lumber (for example, 1 × 2s = 2 × 4s, 1 × 4s = 2 × 8s, etc.).

Review the basic sequence of a platform floor assembly.

Emphasize the importance of checking the foundation before installing the sill plates.

Show Transparency 26 (Figure 25).

Demonstrate the diagonal-corner measurement method for checking foundation squareness.

Describe the procedure for installing the sill.

Show Transparency 27 (Figure 26).

foundation walls are measured and recorded (*Figure 25*). The measurements must be as exact as possible. Following this, the foundation is measured diagonally from one outside corner to the opposite outside corner. A second diagonal measurement is then made between the outsides of the remaining two corners. If the measured lengths of the opposite walls are equal and the diagonals are equal, the foundation is square. For buildings where the foundation is other than a simple rectangle, a good practice is to divide the area into two or more individual square or rectangular areas and measure each area as described above.

5.2.0 Installing the Sill

For floors where the sill is installed flush with the outside of the foundation walls, installation of the sill begins by snapping chalklines on the top of the foundation walls in line with the inside edge of the sill (*Figure 26*). If the sill must be set in to

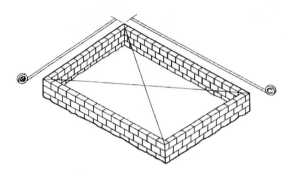

Figure 25 ◆ Checking the foundation for squareness.

Keep It Square

When it is time to install sill plates on the foundation, you may discover that the foundation wall is not exactly true and square. You shouldn't use the foundation wall as a guide. Instead, ensure that the sill plates are square with each other by using a tape measure (shown here) to measure the four plates and the diagonals. If the opposite plates and the diagonals are equal, the sill plates are square with each other. This may mean that the outside edge of the sill plates may not align exactly with the outside edge of the foundation. If necessary, some sills may overlap or underlay the wall.

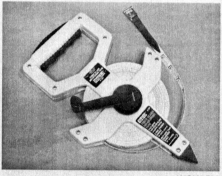

5.28 CARPENTRY FUNDAMENTALS LEVEL ONE ◆ TRAINEE GUIDE

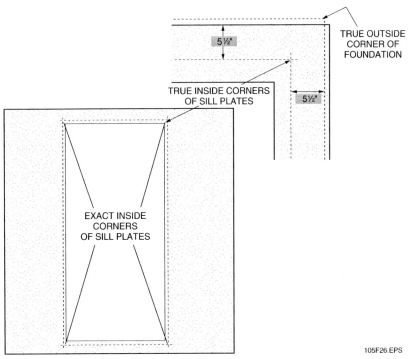

Figure 26 ♦ Inside edges of sill plates marked on the top of the foundation wall.

accommodate the thickness of wall sheathing, brick veneer, etc., the chalklines may be snapped for the outside edge of the sill plates or the inside edge (if the foundation size allows). At each corner, the true location of the outside corner of the sill is used as a reference point to mark the corresponding inside corner of the sill on the foundation wall. To do this, the exact width of the sill stock being used must be determined. For example, if using 2 × 6 sills, 5½" is the sill width measurement. After the exact inside corners are located and marked on the sill, chalklines are snapped between these points. This gives an outline on the top of the foundation wall of the exact inside edges of the sill plates. At this point, a good practice is to double-check the dimensions and squareness of these lines to make sure that they are accurate.

After the location of the sill is marked on the foundation, the sill pieces can be measured and cut. Take into consideration that there must be an anchor bolt within 12" of the end of any plate. Also, sill plates cannot butt together over any opening in the foundation wall. When selecting the lumber, choose boards that are as straight as possible for making the sills. Badly bowed pieces should not be used.

Holes must be drilled in the sill plates so that they can be installed over the anchor bolts embedded in the foundation wall. To lay out the location of these holes, hold the sill sections in place on top of the foundation wall against the anchor bolts (*Figure 27*). At each anchor bolt, use a combination square to scribe lines on the sill corresponding to both sides of the bolt. On the foundation, measure the distance between the center of each anchor bolt and the chalkline, then transfer this distance to the corresponding bolt location on the sill by measuring from the inside edge. After the sill hole layout is done, the holes in the sill are drilled. They should be drilled about ⅛" to ¼" larger than the diameter of the anchor bolt in order to allow for some adjustment of the sill plates, if necessary. Also, make sure all holes are drilled straight.

Before installing the sill plates, the termite shield (if used) and sill sealer are installed on the foundation. Following this, the sill sections are placed in position over the anchor bolts, making sure that the inside edges of the sill plate sections are aligned with the chalkline on top of the foundation wall

Demonstrate sill plate marking, measurement, cutting, and installation. Point out that pressurized lumber, sill sealer, and termite shields are required by some local codes for sill plates.

Explain the method for placement of a built-up girder.

Show Transparency 29 (Figure 28).

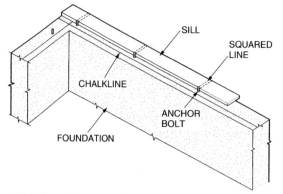
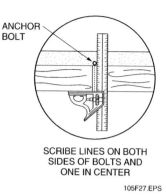

Figure 27 ◆ Square lines across the sill to locate the anchor bolt hole.

and that the inside corners are aligned with their marks. The sill plates are then loosely fastened to the foundation with the anchor bolt nuts and washers and the sill checked to make sure it is level. An 8' level can be used for this task. However, using a transit or builder's level and checking the level every 3' or 4' along the sill is more accurate. It cannot be emphasized enough how important it is that the sill be level. If it is not level, it will throw off the building's floors and walls. Low spots can be shimmed with plywood wedges or filled with grout or mortar. If the sill is too high, the high areas of the concrete foundation will need to be ground or chipped away. After the sill has been made level, the anchor bolt nuts can be fully tightened. Be careful not to overtighten the nuts, especially if the concrete is not thoroughly dry and hard, because this can crack the wall.

5.3.0 Installing a Beam/Girder

Before installing a girder, use the job specifications to determine the details related to its installation. For example, assume you are working with a structure that has a foundation that is 24' wide and 48' long (*Figure 28*). The foundation is poured concrete that is 12" thick with 6"-deep and 7"-wide girder (beam) pockets centered in the short walls. The girder is to be a built-up beam containing three thicknesses of 2 × 10. The columns used to support the girder when the floor system is complete are three 4" lally columns, concrete-filled, with ¼" steel plates top and bottom. The distance between these columns is 12'-0" OC.

To lay out the distances for each of the support columns, first use a steel tape to measure from one end of the foundation to the other in the precise location where the girder will sit. Using a plumb bob, hold the line at the 24'-0" mark. This locates the center of the middle support column. Then, measure 12'-0" to the left of center and 12'-0" to the right of center to locate the center of the other two support columns. The distance from the center of the two end columns into their girder pockets in the foundation walls is 11'-5½". This allows for ½" of space between the back of each girder pocket and the end of the girder. Given the dimensions above, the finished built-up girder for our example needs to measure 46'-11" in length. When constructing this girder, remember that the joints should fall directly over the support columns and the girder crown must face upward.

> **NOTE**
> The *Uniform Building Code* specifies that in standard framing, nails should not be spaced closer than one-half their length nor closer to the edge of a framing member than one-quarter their length. Keep this requirement in mind when nailing floor systems.

When framing floors, use the fasteners as indicated below:

- *16d nails* – Used to attach the header to joists, to install solid bridging, and to construct beams and girders.
- *Special nails* – Furnished to attach the joist hangers.
- *8d nails* – Used to install the wood cross-bridging and subfloor.
- *Pneumatic, ring shank, and screw nails, or etched galvanized staples* – Used to apply the subfloor and underlayment.
- *Construction adhesive* – Used to apply the subfloor and underlayment.

5.30 CARPENTRY FUNDAMENTALS LEVEL ONE ◆ TRAINEE GUIDE

Instructor's Notes:

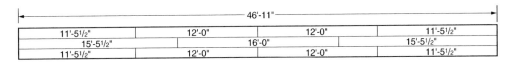
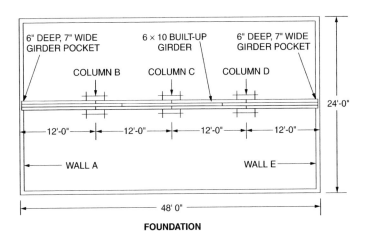

Figure 28 ♦ Example girder and support column data.

Demonstrate built-up girder construction. Point out the fasteners used for framing and emphasize that codes specify fastener spacing.

Review the procedure for laying out sills and girders.

Show Transparency 30 (Figure 29).

Demonstrate joist layout for standard spacing on the sill plate. Point out the importance of the second joist location and the method of marking the girder and opposite sill plate.

As shown in *Figure 28*, the 6 × 10 built-up girder can be constructed using eight 12' long 2 × 10s and three 16' long 2 × 10s. Four of the 12' long 2 × 10s are cut to 11'-5½" and two of the 16' long 2 × 10s are cut to 15'-5½". The 16' pieces are used in order to provide for an overlap of at least 4'-0" at the joints. To make the girder, nail the 2 × 10s together in the pattern shown using 16d nails spaced about 24" apart. Note that the nailing schedule will vary at different locations, so make sure to consult local codes for the proper schedule. Drive the nails at an angle for better holding power. Be sure to butt the joints together so they form a tight fit. Continue this process until the 46'-11" girder is finished. Once completed, the girder is put in place, supported by temporary posts or A-frames, and made level. Note that the temporary supports are removed and replaced by the permanent lally columns after the floor system joists are all installed.

5.4.0 Laying Out Sills and Girders for Floor Joists

Joists should be laid out so that the edges of standard size subfloor panels fall over the centers of the joists. There are different ways to lay out a sill plate to accomplish this. One method for laying out floor joists 16" OC is described here. Begin by using a steel tape to measure out from the end of the sill exactly 15¼" (*Figure 29*). At this point, use a speed square to square a line on the sill. To make sure of accurate spacing, drive a nail into the sill at the 15¼" line, then hook the steel tape to the nail and stretch it the length of the sill. At every point on the tape marked as a multiple of 16" (most tapes highlight these numbers), make a mark on the sill. It is important that the spacing be laid out accurately; otherwise, the subfloor panels may not fall in the center of some joists.

After the sill is marked, use your speed square to square a line at each mark. Next, mark a narrow X next to each line on the sill to show the actual position where each joist is to be placed. Note that the lines marked on the sill mark the edge of the joists, not the center. Be sure to mark the X on the proper side. If the layout has been started from the left side of the sill, as shown in *Figure 29*, the X should be placed to the right of the line. If the layout has been started from the right side, the X should be placed to the left of the line. After the locations for all common 16" OC joists have been laid out, the locations for any double joists, trimmer joists, etc., should be marked on the sill and identified with a T (or other letter) instead of an X.

After the first sill has been laid out, the process is repeated on the girder and the opposite sill. If

Explain the alternate method of marking joist locations using the box sill.

Show Transparency 31 (Figure 30).

Demonstrate the layout marking of extra joists for partitions and openings that are parallel with the joists.

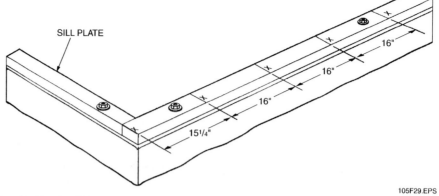

Figure 29 ◆ Marking the sill for joist locations.

Measurement Tip

The reason 15¼", not 16", is used in *Figure 29* as the first measurement is so that the first panel of flooring will come to the outside edge of the first joist, not the center of it. All flooring panels except the first and last need to fall on the center of the joists to provide a nailing surface for the adjoining piece. The first and last panels do not have adjoining panels on one side. By reducing the first measurement ¾" (half the thickness of the joist), you shift the first piece of floor paneling from the center of the first joist to the outside edge of it.

the joists are in line, Xs should be marked on the same side of the mark on both the girder and the sill plate on the opposite wall. If the joists are lapped at the girder, an X should be placed on both sides of the mark on the girder and on the opposite side of the mark on the sill plate on the opposite wall.

The location of the floor joists can be laid out directly on the sills as described above. However, in platform construction, some carpenters prefer to lay them out on the header joists rather than the sill. If done on the header, the procedure is basically the same. Also, instead of making a series of individual measurements, some carpenters make a layout rod marked with the proper measurements and use it to lay out the sills and girders.

5.5.0 Laying Out Joist Locations for the Partition and Floor Openings

After the locations of all the common 16" OC floor joists are laid out, it is necessary to determine the locations of additional joists needed to accommodate loadbearing partitions, floor openings, etc., as shown on the blueprints. Typically, these include:

- Double joists needed under loadbearing interior walls that run parallel with the joists (*Figure 30*). Depending on the structure, the double joists may need to be separated by 2 × 4 (or larger) blocks placed every 4' to allow for plumbing and electrical wires to pass into the wall from below. Loadbearing walls that run perpendicular to the joist system normally do not need extra joists added.
- Double joists needed for floor openings for stairs, chimneys, etc.

The sill plates and girder should be marked where the joists are doubled on each side of a large floor opening. Also, the sill and girder should be marked for the locations of the shorter tail joists at the ends of floor openings. They can be identified by marking their locations with a *T* instead of an X.

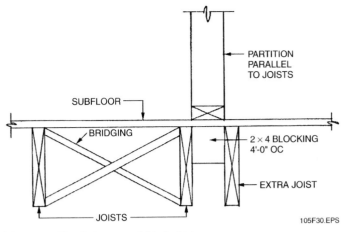

Figure 30 ◆ Double joists at a partition that runs parallel to the joists.

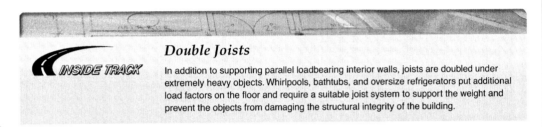

Double Joists

In addition to supporting parallel loadbearing interior walls, joists are doubled under extremely heavy objects. Whirlpools, bathtubs, and oversize refrigerators put additional load factors on the floor and require a suitable joist system to support the weight and prevent the objects from damaging the structural integrity of the building.

5.6.0 Cutting and Installing Joist Headers

After the sills and girder are laid out for the joist locations, the box sill can be built (*Figure 31*). The box sill encloses the joist system and is made of the same size stock (typically 2 × 10s). It consists of the two header joists that run perpendicular to the joists and the first and last joists (end joists) in the floor system. The headers are placed flush with the outside edges of the sill. Good straight stock should be used so that the headers do not rise above the sill plates. They need to sit flat on the sill so that they do not push up the wall. After the joist headers are cut, they are toenailed to the sill and face-nailed where the first and last joists meet the header joist. Any splices that need to be made in the header joist can meet either in the center of a joist or be joined by a **scab**. Note that some carpenters do not use header joists; they use blocks placed between the ends of the joists instead. Also, in areas subject to earthquakes, hurricanes, or tornadoes, codes may require that metal fasteners be used to further attach the header and end joists to the sill.

5.7.0 Installing Floor Joists

With the box sill in place, floor joists are placed at every spot marked on the sill. This includes the extra joists needed at the locations of partition walls and floor openings. When installing each joist, it is important to locate the crown and always point the crown up. With the joist in position at the header joist, hold the end tightly against the header and along the layout line so the sides are plumb, then end nail it to the header and toenail it to the sill (*Figure 32*). Repeat this procedure until all the joists are attached to their associated header. To facilitate framing of openings in the floor, a good practice is to leave the full-length joists out where the floor openings occur.

Following this, the joists are fastened at the girder. If they join end-to-end without overlapping at the girder, they should be joined with a **scarf** or metal fastener. Where the joists overlap at the girder, they should be face-nailed together and toenailed to the girder.

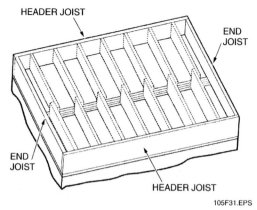

Figure 31 ◆ Box sill installed on foundation.

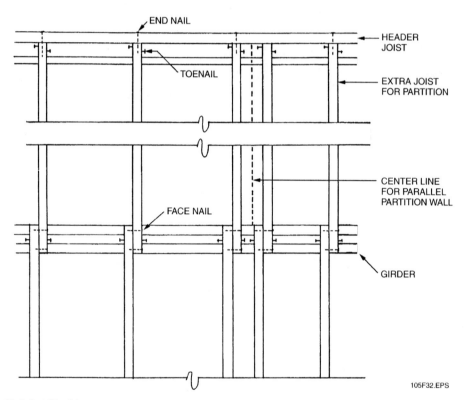

Figure 32 ◆ Installing joists.

5.34 CARPENTRY FUNDAMENTALS LEVEL ONE ◆ TRAINEE GUIDE

5.8.0 Framing Openings in the Floor

Floor openings are framed by a combination of headers and trimmer joists (*Figure 33*). Headers run perpendicular to the direction of the joists and are doubled. Full-length trimmer joists and short tail joists run parallel to the common joists. The blueprints show the location in the floor frame and the size of the rough opening (RO). This represents the dimensions from the inside edge of one trimmer joist to the inside edge of the other trimmer joist and from the inside edge of one header to the inside edge of the other header. The method used to frame openings can vary depending on the particular situation. Trimmers and headers at openings must be nailed together in a certain sequence so that there is never a need to nail through a double piece of stock.

A typical procedure for framing an opening like the one shown in *Figure 33* is given here:

Step 1 First install full-length trimmer joists A and C, then cut four header pieces with a length corresponding to the distance between the trimmer joists A and C.

Step 2 Nail two of these header pieces (headers No. 1 and No. 2) between trimmer joists A and C at the required distances.

Step 3 Following this, cut short tail joists X and Y and nail them to headers No. 1 and No. 2, as shown. Check the code to see if hangers are required.

Step 4 After headers No. 1 and No. 2 and tail joists X and Y are securely nailed, headers No. 3 and No. 4 can be installed and nailed to headers No. 1 and No. 2. Then, joists B and D can be placed next to and nailed to trimmer joists A and C, respectively.

5.9.0 Installing Bridging

Three types of bridging can be installed: wood cross-bridging, metal cross-bridging, and solid bridging. Wood cross-bridging is typically made of 1 × 4 pieces of wood installed diagonally between the joists to form an X. Normally, the bridging is installed every 8' or in the middle of the joist span. For example, joists with a 12' span would have a row of bridging installed at 6'.

A framing square can be used to lay out the bridging. First, determine the actual distance between the floor joists and the actual depth of the joist. For example, 2 × 10 joists 16" OC measure 14½" between them. The actual depth of the joist is 9¼". To lay out the bridging, position the framing square on a piece of bridging stock, as shown in *Figure 34*. This will give the proper length and angle to cut the bridging. Make sure to use the same side of the framing square in both places. Once the required length and angle of the bridging are determined, many carpenters build a jig to cut the numerous pieces of bridging.

Metal cross-bridging comes in a variety of styles and different lengths for use with a particular joist

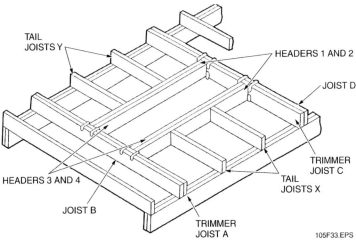

Figure 33 ♦ Floor opening construction.

Explain how to frame a floor opening. Emphasize that the correct procedure must be followed to avoid nailing through double pieces of stock.

Show Transparency 34 (Figure 33).

Demonstrate how to frame a floor opening.

Hand out Job Sheet 27105-1. Under your supervision, have the trainees perform the tasks on the Job Sheet. This laboratory corresponds to Performance Task 1.

Explain the purpose of bridging.

Show Transparency 35 (Figure 34).

Demonstrate how to construct various types of bridging.

Hand out Job Sheet 27105-2. Under your supervision, have the trainees perform the tasks on the Job Sheet. This laboratory corresponds to Performance Task 2.

Describe subflooring layout and installation. Point out that panels must be positioned to a chalkline, staggered, run lengthwise across joists, and spaced for expansion.

Show Transparency 36 (Figure 35).

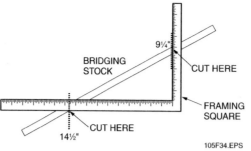

Figure 34 ♦ Framing square used to lay out cross-bridging.

size and spacing. Its installation should be done in accordance with the manufacturer's instructions. Layout of the joist system for the location of installing metal cross-bridging is done in the same way as described for wood cross-bridging.

Solid bridging consists of solid pieces of lumber joist stock installed between the joists. Layout of the joist system for the location of solid bridging is done in the same way as described for wood cross-bridging. The solid bridging is installed between pairs of joists, first on one side of the chalkline and then in the next pair of joists on the other. This staggered method of installation enables end nailing. Note that because of variations in lumber thickness and joist spacing, the length of the individual pieces of solid bridging placed between each joist pair may have to be adjusted.

5.10.0 Installing Subflooring

Installation of the subfloor begins by measuring 4' in from one side of the frame and snapping a chalkline across the tops of the floor joists from one end to the other. When installing 4' × 8' plywood, OSB, or similar floor panels, the long (8') dimension of the panels must be run across (perpendicular) to the joists (*Figure 35*). Also, the panels must be laid so that the joints are staggered in each successive course (row). Never allow an intersection of four corners. This is done by starting the first course with a full panel and continuing to lay full panels to the opposite end. Following this, start the next course using half a panel (4' × 4'), then continue to lay full panels to the opposite end. Repeat this procedure until the surface of the floor is covered. The ends of the panels that overhang the end of the building are then cut off flush with the floor frame. When butt-joint plywood panels are used,

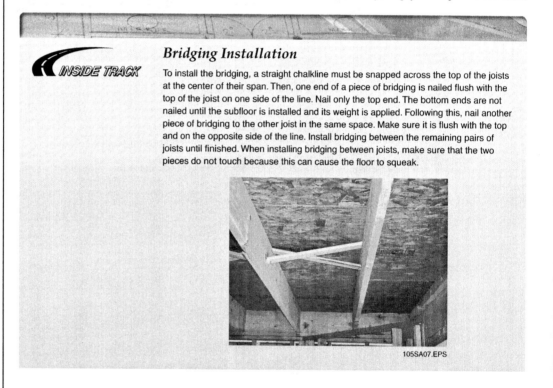

Bridging Installation

To install the bridging, a straight chalkline must be snapped across the top of the joists at the center of their span. Then, one end of a piece of bridging is nailed flush with the top of the joist on one side of the line. Nail only the top end. The bottom ends are not nailed until the subfloor is installed and its weight is applied. Following this, nail another piece of bridging to the other joist in the same space. Make sure it is flush with the top and on the opposite side of the line. Install bridging between the remaining pairs of joists until finished. When installing bridging between joists, make sure that the two pieces do not touch because this can cause the floor to squeak.

5.36 CARPENTRY FUNDAMENTALS LEVEL ONE ♦ TRAINEE GUIDE

Instructor's Notes:

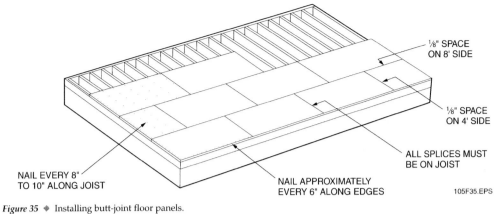

Figure 35 ◆ Installing butt-joint floor panels.

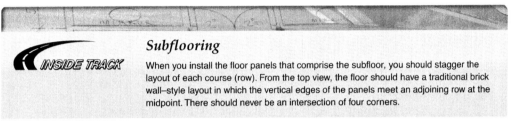

Subflooring

When you install the floor panels that comprise the subfloor, you should stagger the layout of each course (row). From the top view, the floor should have a traditional brick wall–style layout in which the vertical edges of the panels meet an adjoining row at the midpoint. There should never be an intersection of four corners.

Floor Joist Transitions

If the floor joists are overlapped at a girder, the layout for the panels will differ from one side of the floor to the other by 1½". If the lap occurs at a natural break for the panels, the next course can be slid back 1½" so that it continues to butt on a joist. However, if the panels cover the joist lap, then 2 × 4 scabs must be nailed to the joists to provide a nailing surface for the panels at this transition.

at least ⅛" of space should be left between each head joint and side joint for expansion. If installing a single-layer floor, blocking is also required under the joints of the butt-edged panels. Specifications for nailing panels to floor members vary with local codes. Typically, when nailing ⅝"-thick panels to joists on 16" centers, the nailing would be done every 6" along the edges and 8" to 10" along intermediate members using ring-shank or screw nails. To avoid fatigue, pneumatic nailers are commonly used to nail the subfloors. Some carpenters use power screw guns to screw the subfloor to the joists.

Traditionally, plywood panels have been fastened to the floor joists using nails. Today, it is more common to use a glued floor system in which the subfloor panels are both glued and nailed (or screwed) to the joists, as previously discussed. Before each of the panels is placed, a ¼" bead of subflooring adhesive is applied to the joists using a caulking gun. Two beads are applied on joists where panel ends butt together. Following this, the panel should immediately be nailed to the joists before the adhesive sets. Be sure all nails hit the floor joists.

Building a subfloor with tongue-and-groove panels is done in basically the same way as described for butt-joint panels, with the following exceptions. Begin the first course of paneling with the tongue (*Figure 36*) of the panels facing the outside of the house, not the inside. This leaves the groove as the leading edge. The next course of

Hand out Job Sheets 27105-3 and 27105-4. Under your supervision, have the trainees perform the tasks on the Job Sheets. This laboratory corresponds to Performance Tasks 3 and 4.

Assign reading of Section 6.0.0 for the next class session.

Describe joist installation for projections and cantilevered floors. Point out that these joists usually extend inward a distance that is twice the overhang unless local codes specify otherwise.

Show Transparency 38 (Figure 37).

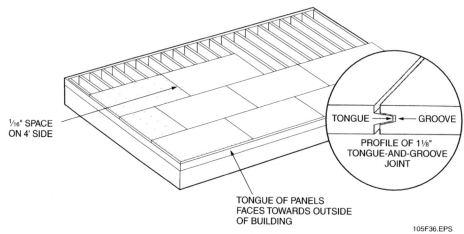

Figure 36 ◆ Installing tongue-and-groove floor panels.

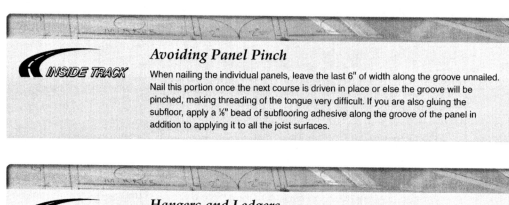

panels has to be interlocked into the previous course by driving the new sheets with a 2 × 4 block and a sledgehammer. The grooved edge of the panels can take this abuse; the tongued edge cannot.

6.0.0 ◆ INSTALLING JOISTS FOR PROJECTIONS AND CANTILEVERED FLOORS

Porches, decks, and other projections from a building present some special floor-framing situations. Projections overhang the foundation wall and are suspended without any vertical support posts. When constructing a projection, it makes a difference whether the joists run parallel or perpendicular to the common joists in the floor system. If they are parallel, longer joists are simply run out past the foundation wall (*Figure 37*). If they must be run perpendicular to the common joists, the projection must be framed with cantilevered joists. This means that you have to double up a common joist, then tie the cantilevered joists into this double joist. Regardless of whether you are using parallel or cantilevered joists, there should always be at least a ⅓ to ⅔ relationship

5.38 CARPENTRY FUNDAMENTALS LEVEL ONE ◆ TRAINEE GUIDE

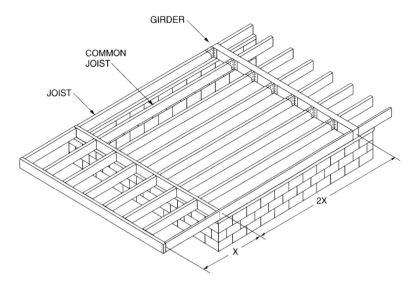

EXAMPLE OF PROJECTION JOIST RUNNING IN SAME DIRECTION AS COMMON JOISTS

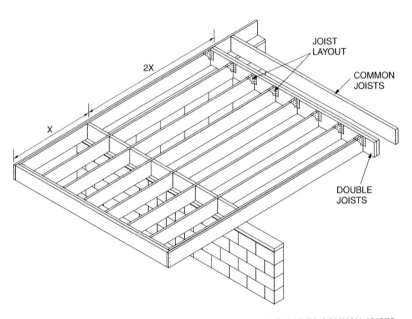

EXAMPLE OF CANTILEVERING WHEN JOISTS RUN PERPENDICULAR TO COMMON JOISTS

Figure 37 ♦ Cantilevered joists.

Hand out Job Sheet 27105-5. Under your supervision, have the trainees perform the tasks on the Job Sheet. This laboratory corresponds to Performance Task 5.

Assign reading of Sections 7.0.0–7.5.0 for the next class session.

Explain the methods of estimating flooring materials. Point out that if boards are used for the subfloor, a waste allowance must be added.

Show Transparency 39 (Figure 38).

between the length of the total joist and the distance it can project past the foundation wall. Check local codes for exact requirements. Stated another way, the joist should extend inward a distance equal to at least twice the overhang.

7.0.0 ◆ ESTIMATING THE QUANTITY OF FLOOR MATERIALS

Because of the importance of the floor in carrying the weight of the structure, it is important to correctly determine the materials required. You must be able to recognize special needs such as floor openings, cantilevers, and partition supports that affect material requirements. Once the needs are determined, you must be able to estimate the quantities of materials needed in order to construct the floor without delays or added expense from too much material. The process begins by checking the building specifications for the kinds and dimensions of materials to be used. It also requires that the blueprints be checked or scaled to determine the dimensions of the various components needed. These include:

- Sill sealer, termite shield, and sill
- Girders or beams
- Joists
- Joist headers
- Bridging
- Subflooring

For the purpose of an example in the sections below, we will use the floor system shown in *Figure 38* to determine the quantity of floor and sill framing materials needed.

7.1.0 Sill, Sill Sealer, and Termite Shield

To determine the amount of sill, sill sealer, and/or termite shield materials required, simply measure the perimeter of the foundation. For our example in *Figure 38*, the amount of material needed is 192 lineal feet [2 × (32' + 64') = 192'].

7.2.0 Beams/Girders

The quantity of girder material needed is determined by the type of girder and its length. For our example, if using a solid girder, the length of material needed would be 64 lineal feet. If a built-up beam made of three 2 × 12s is used as shown, the length of 2 × 12 material needed is 192 lineal feet (3 × 64' = 192').

7.3.0 Joists and Joist Headers

To determine the number of floor joists in a frame, divide the length of the building by the joist spacing and add one joist for the end and one joist for each partition that runs parallel to the joists. For our example, there are no partitions; therefore, the number of 2 × 8 joists is 49 [(64' × 12") ÷ 16" OC = 48 + 1 = 49]. Because there are two rows of joists (one on each side of the girder), the total number of joists needed is 98 (2 × 49 = 98). Each of these joists would be about 18' long. The amount of 2 × 8 material needed for the header joists is 128 lineal feet (2 × 64' = 128').

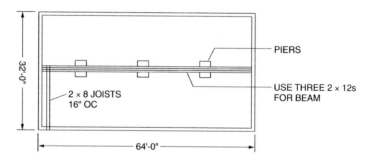

Figure 38 ◆ Determining floor system materials.

7.4.0 Bridging

Codes require one row of bridging in spans over 8' and less than 16' in length. Two rows of bridging are required in spans over 16'.

To find the amount of wood cross-bridging needed, determine the number of rows of bridging needed and the length of each row of bridging to find the total lineal footage for the bridging rows. For our example, this is 128 lineal feet (2 rows × 64' = 128'). Next, multiply the total lineal footage of bridging rows by the appropriate factor given in *Table 2* to get the total lineal feet of bridging needed. For our example, we are using 2 × 8 joists; therefore, the total amount of bridging needed is 256 lineal feet (2 × 128' = 256').

To find the amount of solid bridging needed, determine the number of rows of bridging needed and the length of each row of bridging. Then, multiply the number of rows by the length of each row to determine the total lineal feet needed. For our example, we need 128 lineal feet (2 × 64' = 128').

To determine the amount of steel bridging needed, multiply the number of rows of bridging needed by the length of each row of bridging to determine the total lineal footage of bridging rows. For our example, this is 128 lineal feet (2 × 64' = 128'). Then, multiply the total lineal footage of bridging rows by 0.75 (¾) to find the number of spaces between joists that are 16" OC. For our example, there are 96 spaces (0.75 × 128 = 96). Then, multiply the number of spaces by 2 to determine the total number of steel bridging pieces needed. For our example, we need 192 pieces (2 × 96 = 192).

7.5.0 Flooring

To determine the number of 4' × 8' plywood/OSB sheets needed to cover a floor, divide the total floor area by 32 (the area in square feet of one panel). For our example, the total floor area is 2,048 square feet (64 × 32 = 2,048). Therefore, we need 64 panels (2,048 ÷ 32 = 64). For any fractional sheets, round up to the next whole sheet.

If using lumber boards for flooring, calculate the total floor area to be covered. To this amount, add a quantity of material to allow for waste. When using 1 × 6 lumber, a rule of thumb is to add ⅕ to the total area for waste; for 1 × 8 lumber, add ⅛ for waste.

8.0.0 ◆ GUIDELINES FOR DETERMINING PROPER GIRDER AND JOIST SIZES

Normally, the sizes of the girder(s) and joists used in a building are specified by the architect or structural engineer who designs the building. However, a carpenter should be familiar with the procedures used to determine girder and joist sizes.

8.1.0 Sizing Girders

The following discussion on how to size a girder is keyed to the example first-floor plan shown in *Figure 39*. Assume that a built-up girder will be used and the ceilings are drywall (not plaster).

Step 1 Determine the distance (span) between girder supports. For our example floor, the span is 8'-0".

Step 2 Find the girder load width. The girder must be able to carry the weight of the floor on each side to the midpoint of the joist that rests upon it. Therefore, the girder load is half the length of the joist span on each side of the girder multiplied by 2. For our example floor, the girder load width is 12'-0" (6'-0" on each side of the girder).

Step 3 Find the total floor load per square foot carried by the joists and bearing partitions. This is the sum of the loads per square foot as shown in *Figure 40*, with the exception of the roof load. Roof loads are not included because these are carried on the outside walls unless braces or partitions are placed under the rafters. With a drywall ceiling and no partitions, the total load per square foot for our example floor is 50 pounds per square foot (40 pounds live load + 10 pounds dead load).

Step 4 Find the total load on the girder. This is the product of the girder span multiplied by the girder width multiplied by the total floor load. For our example floor, the total load on the girder is 4,800 pounds (8 × 12 × 50 = 4,800).

Table 2	Wood Cross Bridging Multiplication Factor	
Joist Size	Spacing (Inches OC)	Lineal Feet of Material (per Foot of Bridging Row)
2 × 6, 2 × 8, 2 × 10	16	2
2 × 12	16	2.25
2 × 14	16	2.5

Hand out Worksheet 27104-3. Have the trainees complete the tasks on the Worksheet. This laboratory corresponds to Performance Task 6.

Assign reading of Sections 8.0.0–8.2.0 for the next class session.

Demonstrate girder sizing using the example provided in the Trainee Module or another foundation plan.

Show Transparencies 40 and 41 (Figures 39 and 40).

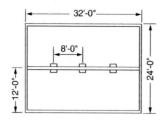

- DETERMINE LENGTH OF JOIST SPAN AND GIRDER WIDTH
- FIND TOTAL FLOOR LOAD PER SQUARE FOOT CARRIED BY JOIST AND BEARING PARTITIONS
- CALCULATE TOTAL LOAD ON GIRDER
- SELECT PROPER SIZE OF GIRDER IN ACCORDANCE WITH LOCAL CODES

Figure 39 ♦ Sizing girders.

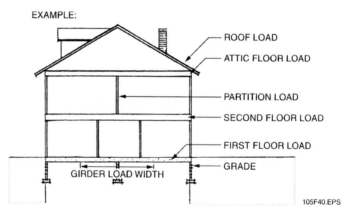

GIRDER LOADS (POUNDS PER SQUARE FOOT)		
	LIVE LOAD*	DEAD LOAD
ROOF	20	10
ATTIC FLOOR	20	20 (FLOORED)
		10 (NOT FLOORED)
SECOND FLOOR	40	20
PARTITIONS		20
FIRST FLOOR	40	20 (CEILING PLASTERED)
		10 (CEILING NOT PLASTERED)
PARTITIONS		20

*USUAL LOCAL REQUIREMENTS

Figure 40 ♦ Floor loads.

5.42 CARPENTRY FUNDAMENTALS LEVEL ONE ♦ TRAINEE GUIDE

Instructor's Notes:

Step 5 Select the proper size of girder according to local codes. *Table 3* is typical. It indicates safe loads on standard-size girders for spans from 6' to 10'. For our example floor, a 6 × 8 built-up girder is needed to carry a 4,800-pound load at an 8' span. Note that shortening the span is the most economical way to increase the load that a girder will carry.

8.2.0 Sizing Joists

The following discussion on how to size joists is keyed to the example first-floor plan shown in *Figure 41*. Assume that the plan falls into the 40-pound live load category and will be built with joists on 16" centers and drywall ceilings.

Step 1 Determine the length of the joist span. For our example, the span is 16'.

Step 2 Determine if there is a dead load on the ceiling. For our example, the ceiling is drywall; therefore, there is a dead load of 10 pounds per square foot.

Step 3 Select the proper size of joists according to local codes or by using the latest tables available from wood product manufacturers or sources such as the National Forest Products Association and the Southern Forest Products Association. *Table 4* indicates maximum safe spans for various sizes of wood joists under ordinary load conditions. For floors, this is usually figured on a basis of 50 pounds per square foot (10 pounds dead load and 40 pounds live load). For our example, *Table 4* shows that for a 40-pound live load, a 2 × 10 joist 16" OC will carry the load up to a span of 19'-2". Note that a 2 × 8 joist 16" OC will carry the required load up to 15'-3", which is not long enough for the required span.

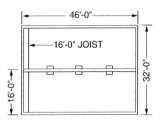

- DETERMINE LENGTH OF JOIST SPAN
- DETERMINE LIVE LOAD PER SQUARE FEET
- DETERMINE IF THERE IS A DEAD LOAD ON CEILING
- SELECT PROPER SIZE JOISTS IN ACCORDANCE WITH LOCAL CODES

105F41.EPS

Figure 41 ◆ Sizing joists.

Table 3 Typical Safe Girder Loads

Nominal Girder Size	Safe Load in Pounds for Spans Shown				
	6 ft	7 ft	8 ft	9 ft	10 ft
6 × 8 solid	8,306	7,118	6,220	5,539	4,583
6 × 8 built-up	7,359	6,306	5,511	4,908	4,062
6 × 10 solid	11,357	10,804	9,980	8,887	7,997
6 × 10 built-up	10,068	9,576	8,844	7,878	7,086
8 × 8 solid	11,326	9,706	8,482	7,553	6,250
8 × 8 built-up	9,812	8,408	7,348	6,554	5,416
8 × 10 solid	15,487	14,732	13,608	12,116	10,902
8 × 10 built-up	13,424	12,968	11,792	10,504	9,448

Table 4 Safe Joist Spans

Nominal Joist Size	Spacing	30# Live Load	40# Live Load	50# Live Load	60# Live Load
2 × 6	12"	14'-10"	13'-2"	12'-0"	11'-1"
	16"	12'-11"	11'-6"	10'-5"	9'-8"
	24"	10'-8"	9'-6"	8'-7"	7'-10"
2 × 8	12"	19'-7"	17'-5"	15'-10"	14'-8"
	16"	17'-1"	15'-3"	13'-10"	12'-9"
	24"	14'-2"	13'-6"	11'-4"	10'-6"
2 × 10	12"	24'-6"	21'-10"	19'-11"	18'-5"
	16"	21'-6"	19'-2"	17'-5"	16'-1"
	24"	17'-10"	15'-10"	14'-4"	13'-3"
2 × 12	12"	29'-4"	26'-3"	24'-0"	22'-2"
	16"	25'-10"	23'-0"	21'-0"	19'-5"
	24"	21'-5"	19'-1"	17'-4"	16'-9"
3 × 8	12"	24'-3"	21'-8"	19'-10"	18'-4"
	16"	21'-4"	19'-1"	17'-4"	16'-0"
	24"	17'-9"	15'-9"	14'-4"	13'-3"
3 × 10	12"	30'-2"	27'-1"	34'-10"	23'-0"
	16"	26'-8"	23'-10"	21'-9"	20'-2"
	24"	22'-3"	19'-10"	18'-1"	16'-8"

Demonstrate joist sizing using the example provided in the Trainee Module or another foundation plan.

Show Transparency 42 (Figure 41).

Have the trainees select the proper girder/beam and joist size from the tables in the Trainee Module for various floor plans, floor loads, and span data. This laboratory corresponds to Performance Task 7.

Summarize the major concepts presented in the module.

Summary

A great majority of a carpenter's time is devoted to building floor systems. It is important that a carpenter not only be knowledgeable about both traditional and modern floor framing techniques, but, more importantly, be able to construct modern flooring systems.

The construction of a platform floor assembly involves the tasks listed below and is normally done in the sequence listed:

- Check the foundation for squareness.
- Lay out and install the sill plates.
- Build and/or install the girders and supports.
- Lay out the joist locations for partitions and floor openings.
- Cut and attach the joist headers to the sill.
- Install the joists.
- Frame the openings in the floor.
- Install the bridging.
- Install the subflooring.

Notes

Instructor's Notes:

Review Questions

1. In braced frame construction, the corner posts and beams are _____.
   a. nailed
   b. mortised and tenoned
   c. pinned
   d. glued

2. In balloon framing, the second floor joists sit on a _____.
   a. plate nailed to the top plate of the wall assembly below
   b. 1 × 4 or 1 × 6 ribbon let into the wall studs
   c. sill attached to the top of the wall assembly below
   d. 2 × 4 let into the wall studs

3. Shrinkage in wood framing members occurs mainly _____.
   a. lengthwise
   b. both lengthwise and on the wide width
   c. on the wide width
   d. in the middle

4. The most common framing method used in modern residential and light commercial construction is _____ framing.
   a. brace
   b. balloon
   c. platform
   d. post-and-beam

5. The method of construction that experiences a relatively large amount of settling as a result of shrinkage is _____ framing.
   a. platform
   b. brace
   c. balloon
   d. post-and-beam

6. The method of construction that features widely spaced, heavy framing members is _____ framing.
   a. brace
   b. balloon
   c. platform
   d. post-and-beam

7. In a set of working drawings, the details about the floor used in a building most likely will be defined in the _____.
   a. architectural drawings
   b. structural drawings
   c. mechanical plans
   d. site plans

8. The letter *H* in *Figure 1* is pointing to the _____.
   a. sill
   b. termite shield
   c. bearing plate
   d. sill sealer

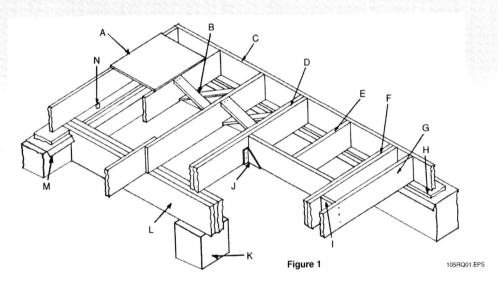

Figure 1

Review Questions

9. The letter C in *Figure 1* is pointing to the _____.
   a. tail joist
   b. trimmer joist
   c. joist header
   d. common joist

10. The letter F in *Figure 1* is pointing to the _____.
    a. tail joist
    b. trimmer joist
    c. joist header
    d. common joist

11. The letter G in *Figure 1* is pointing to the _____.
    a. tail joist
    b. trimmer joist
    c. joist header
    d. common joist

12. The letter L in *Figure 1* is pointing to the _____.
    a. joist header
    b. beam or girder
    c. column
    d. triple joist

13. For a given size, the type of girder with the least strength is the _____ girder.
    a. built-up
    b. solid lumber
    c. LVL
    d. glulam

14. For a given total load, the size of the girder _____ if the span between its support columns is increased.
    a. can be decreased
    b. must be increased
    c. can remain the same
    d. must be decreased

15. The weight of all moving and variable loads that may be placed on a building is referred to as _____ weight.
    a. dead
    b. live
    c. variable
    d. permanent

16. According to the *Standard Building Code*, notches for pipes made in the top or bottom of floor joists shall not exceed _____ of the depth.
    a. one-eighth
    b. one-sixth
    c. one-fourth
    d. one-half

17. Which of the following panels creates the least rigid subflooring?
    a. Plywood
    b. Oriented strand board (OSB)
    c. Tongue-and-groove plywood
    d. 1 × 6 boards

18. The first task that should be done when constructing a floor is to _____.
    a. lay out and install sill plates
    b. build and install girders and supports
    c. check the foundation for squareness
    d. lay out the sill plates and girders for joist locations

19. When nailing floor systems, nails should *not* be spaced closer to one another than _____ their length *nor* closer to the edge of a framing member than _____ their length.
    a. one-half; one-half
    b. one-half; one-quarter
    c. one-quarter; one-quarter
    d. one-quarter; one-half

20. When laying out the sill for joist locations 16" OC, the first measurement on the sill should be at _____.
    a. 14¼"
    b. 15⅜"
    c. 16"
    d. 15¼"

21. If building a floor to the plan shown in *Figure 28*, how much sill material is needed?
    a. 144'
    b. 128'
    c. 96'
    d. 72'

Review Questions

22. For the same structure, how many feet of 2 × 10s are needed for a built-up girder?
   a. 48'
   b. 96'
   c. 144'
   d. 192'

23. For the same structure, what is the total number of 2 × 10 joists needed if the joists are spaced 16" OC?
   a. 36
   b. 37
   c. 72
   d. 74

24. For the same structure, how many feet of wood cross-bridging are needed?
   a. 96'
   b. 192'
   c. 48'
   d. 216'

25. For the same structure, how many 4' × 8' panels of subflooring material are needed?
   a. 96
   b. 72
   c. 36
   d. 48

Have trainees complete the Trade Terms Quiz, and go over the answers.

Administer the Module Examination. Record the results on Craft Training Report Form 200, and submit the form to the Training Program Sponsor.

Administer the Performance Tests, and fill out Performance Profile Sheets for each trainee. If desired, trainee proficiency noted during laboratory sessions may be used to complete the Performance Test. Record the results on Craft Training Report Form 200, and submit the results to the Training Program Sponsor.

Trade Terms Quiz

1. The furniture placed in a building would be considered part of the _____.
2. The HVAC equipment in a building would be considered part of the _____.
3. Before installing hardwood flooring, you would most likely install _____ over the subfloor.
4. The roof of a structure is supported by a(n) _____ assembly.
5. A joist can be secured to the structure using a(n) _____.
6. The distance between girders is known as the _____.
7. The rafters of a building are attached to a horizontal piece of lumber known as the _____.
8. A(n) _____ can be used to support the girders of a structure.
9. In balloon framing, the studs must be blocked with approved _____.
10. Lumber must be installed _____ side up.
11. In platform construction, the ends of the joists are attached to a(n) _____ to form the box sill.
12. A(n) _____ can be applied over a joint to strengthen it.
13. A(n) _____ can be used to hide a joint.
14. A notch in a joist used to support another piece is known as a(n) _____.
15. A(n) _____ runs from an opening to a bearing.
16. A rough opening in a floor such as that for a stairway would be reinforced by a(n) _____.
17. Every building is supported by a(n) _____.
18. The bottom horizontal member of a wall frame is known as the _____.

Trade Terms

Crown
Dead load
Firestop
Foundation
Header joist
Joist hanger
Let-in
Live load
Pier
Rafter plate
Scab
Scarf
Soleplate
Span
Tail joist
Trimmer joist
Truss
Underlayment

5.48 CARPENTRY FUNDAMENTALS LEVEL ONE ◆ TRAINEE GUIDE

Instructor's Notes:

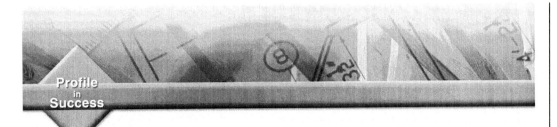

Profile in Success

Adam Deeds
SkillsUSA Carpentry Silver Medal Winner—Two Years in a Row

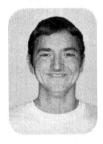

Adam grew up in Kansas. He participated in the SkillsUSA competition in his junior and senior years in high school. In his junior year, he won a gold medal at the state level and a silver at the nationals. The following year he also won the gold at the state competition and silver at the nationals. He is currently working for a residential home builder and is a freshman at Pittsburgh State University majoring in construction management with a minor in business.

How did you become interested in carpentry?
My father is a carpenter and I have three siblings. When we were younger, my dad used to tell us to go out to the garage and pound nails if we were bored. We weren't building anything in particular, I would just hammer nails into a block of wood. So I pretty much had a hammer in my hand for as long as I can remember. My father encouraged all of us to be carpenters, but I am the only one that decided to pursue it.

My father has been a carpenter for 30 years. He owned his own business and now teaches carpentry. He was a general contractor and taught me the trade as I was growing up. I built on my skills in high school. During our sophomore, junior, and senior years our class built a house. I have always enjoyed everything about construction and carpentry.

What do you think it takes to be a success in your trade?
To be successful, you have to have a strong ambition and a will to do your best. Always be willing to learn new things. Be willing to try anything because you don't know what you are good at until you try it. You can never know everything about construction—there is always more to learn. I think I have been successful because I have always enjoyed working and I am not afraid to ask questions.

What positions have you held and how did they help you get to where you are now?
I helped my father build several houses. We also built a shop in the backyard and remodeled both of our houses. It taught me a lot about residential construction. I worked for a few summers building modular homes. I learned different aspects of home construction there. Now I am working with a general contractor while I am going to school.

What do you do in your job?
Anything my boss asks me to do. I do many different tasks depending on what is needed. I have done framing, painting, fascia and soffit work, and siding, and I have only had this job for a few months.

What do you like about the work you do?
I enjoy it all. I enjoy working with my hands and building things. When you learn something you can put it to good use. You learn something and then go out and do it.

What would you say to someone entering the trade today?
Always be willing to learn a new skill. I learn best by watching people. You can learn when you are standing around watching someone. Later, you can repeat it and put those skills to work.

Trade Terms Introduced in This Module

Crown: The high point of the crooked edge of a framing member.

Dead load: The weight of permanent, stationary construction and equipment included in a building.

Firestop: An approved material used to fill air passages in a frame to retard the spread of fire.

Foundation: The supporting portion of a structure, including the footings.

Header joist: A framing member used in platform framing into which the common joists are fitted, forming the box sill. Header joists are also used to support the free ends of joists when framing openings in a floor.

Joist hanger: A metal stirrup secured to the face of a structural member, such as a girder, to support and align the ends of joists flush with the member.

Let-in: Any type of notch in a stud, joist, etc., which holds another piece. The item that is supported by the notch is said to be let in.

Live load: The total of all moving and variable loads that may be placed upon a building.

Pier: A column of masonry used to support other structural members, typically girders or beams.

Rafter plate: The top or bottom horizontal member at the top of a wall.

Scab: A length of lumber applied over a joint to strengthen it.

Scarf: To join the ends of stock together with a sloping lap joint so there appears to be a single piece.

Soleplate: The bottom horizontal member of a wall frame.

Span: The distance between structural supports such as walls, columns, piers, beams, or girders.

Tail joist: Short joists that run from an opening to a bearing.

Trimmer joist: A full-length joist that reinforces a rough opening in the floor.

Truss: An engineered assembly made of wood, or wood and metal members, that is used to support floors and roofs.

Underlayment: A material, such as particleboard or plywood, laid on top of the subfloor to provide a smoother surface for the finished flooring.

Instructor's Notes:

Additional Resources and References

This module is intended to present thorough resources for task training. The following reference works are suggested for further study. These are optional materials for continued education rather than for task training.

Builder Tips: Steps to Construct a Solid, Squeak-Free Floor System. Tacoma, WA: APA – The Engineered Wood Association.

Building with Floor Trusses. Madison, WI: Wood Truss Council of America (11-minute DVD or video).

Field Guide for Prevention and Repair of Floor Squeaks. Boise, ID: Trus Joist, a Weyerhauser business.

I-Joist Construction Details: Performance-Rated I-Joists in Floor and Roof Framing. Tacoma, WA: APA – The Engineered Wood Association.

Quality Floor Construction. Tacoma, WA: APA – The Engineered Wood Association (15-minute video).

Storage, Handling, Installation & Bracing of Wood Trusses. Madison, WI: Wood Truss Council of America (69-minute DVD or video).

American Wood Council. A trade association that develops design tools and guidelines for wood construction. www.awc.org.

Western Wood Products Association. A trade association representing softwood lumber manufacturers in 12 western states and Alaska. www.wwpa.org

Wood I-Joist Manufacturers Association. An organization representing manufacturers of prefabricated wood I-joist and structural composite lumber. www.i-joist.org

Wood Truss Council of America. An international trade association representing structural wood component manufacturers. www.woodtruss.com

MODULE 27105-06 — ANSWERS TO REVIEW QUESTIONS

Answer	Section
1. b	2.2.0
2. b	2.3.0
3. c	2.3.0
4. c	2.1.0
5. a	2.1.0
6. d	2.4.0
7. a	3.1.0
8. a	4.0.0, Figure 6
9. c	4.0.0, Figure 6
10. b	4.0.0, Figure 6
11. d	4.0.0, Figure 6
12. b	4.0.0, Figure 6
13. a	4.2.2
14. b	4.2.5
15. b	4.3.0
16. b	4.3.1
17. d	4.5.3
18. c	5.0.0
19. b	5.3.0
20. d	5.4.0
21. a*	7.1.0, Figure 28
22. c*	7.2.0, Figure 28
23. d*	7.3.0, Figure 28
24. b*	7.4.0; Table 2, Figure 28
25. c*	7.5.0, Figure 28

21. (2 × length of 48') + (2 × width of 24') = 144'

22. 3 × 48' = 144'

23. (48' × 12") ÷ 16" OC = 36 + 1 joist for end = 37 × 2 = 74

24. 2 rows x 48' = 96' × factor of 2 from *Table 2* = 192'

25. Floor area = 24' × 48' = 1,152 sq ft ÷ 32 = 36

MODULE 27105-06 — ANSWERS TO TRADE TERMS QUIZ

1. Live load
2. Dead load
3. Underlayment
4. Truss
5. Joist hanger
6. Span
7. Rafter plate
8. Pier
9. Firestops
10. Crown
11. Header joist
12. Scab
13. Scarf
14. Let-in
15. Tail joist
16. Trimmer joist
17. Foundation
18. Soleplate

MODULE 27105-06 — TEACHING TIPS

The following are suggested activities or instructional methods to help you teach the material in this module.

General

When you call on someone to answer a question, the rest of the class relaxes or even tunes out because they expect that the question and answer will take place only between you and the trainee you called on. Instead, use this technique to involve more trainees in answering questions and to keep them on their toes.

1. Ask trainees to define a term or explain a concept.
2. After one trainee has answered, ask a trainee seated nearby if the answer is right. Then ask whether a trainee in the back of the room agrees.
3. Ask trainees to explain why they think an answer is right or wrong.
4. Use the session to clear up incorrect ideas and encourage trainees to learn from their mistakes.

CONTREN® LEARNING SERIES — USER UPDATE

NCCER makes every effort to keep these textbooks up-to-date and free of technical errors. We appreciate your help in this process. If you have an idea for improving this textbook, or if you find an error, a typographical mistake, or an inaccuracy in NCCER's Contren® textbooks, please write us, using this form or a photocopy. Be sure to include the exact module number, page number, a detailed description, and the correction, if applicable. Your input will be brought to the attention of the Technical Review Committee. Thank you for your assistance.

Instructors – If you found that additional materials were necessary in order to teach this module effectively, please let us know so that we may include them in the Equipment/Materials list in the Annotated Instructor's Guide.

Write: Product Development and Revision
National Center for Construction Education and Research
P.O. Box 141104, Gainesville, FL 32614-1104

Fax: 352-334-0932

E-mail: curriculum@nccer.org

Craft _____ Module Name _____

Copyright Date _____ Module Number _____ Page Number(s) _____

Description _____

_____

_____

_____

(Optional) Correction _____

_____

_____

(Optional) Your Name and Address _____

_____

_____

Wall and Ceiling Framing
27106-06

NCCER STANDARDIZED CRAFT TRAINING PROGRAM

The National Center for Construction Education and Research (NCCER) provides a standardized national program of accredited craft training. Key features of the program include instructor certification, competency-based training, and performance testing. The program provides trainees, instructors, and companies with a standard form of recognition through the National Registry. The program is described in full in the *Guidelines for Accreditation*, published by NCCER. For more information on standardized craft training, contact NCCER by writing to P.O. Box 141104, Gainesville, FL 32614-1104; calling 352-334-0911; or e-mailing **info@nccer.org.** More information is available at **www.nccer.org.**

HOW TO USE THIS ANNOTATED INSTRUCTOR'S GUIDE

Each page presents two sections of information. The larger section displays each page exactly as it appears in the Trainee Module. The narrow column ties suggested trainee and instructor actions to each page and provides icons (detailed below) to call your attention to material, safety, audiovisual, or testing requirements. The bottom of each page includes space for your notes.

 The **Audiovisual** icon indicates an appropriate time to show a transparency or other audiovisual aid.

 The **Classroom** icon prompts you to define a term, stress a point, ask trainees to explain a concept, or give examples.

 The **Demonstration** icon directs you to show trainees how to perform tasks.

 The **Examination** icon tells you to administer the written module examination.

 The **Homework** icon is placed where you may wish to assign reading for the next class, assign a project, or advise trainees to prepare for an examination.

 The **Laboratory** icon is used when trainees are to practice performing tasks.

 The **Materials** icon is a reminder for you to gather materials needed for classes, laboratories, and testing.

 The **Performance Testing** icon tells you to administer a performance test or a portion thereof.

 The **Safety** icon is used to emphasize safety issues. It is often keyed to *Caution* and *Warning!* statements in the Trainee Module.

 The **Teaching Tip** icon indicates additional guidance is available, such as how to conduct an exercise, get the most educational value from a field trip, or encourage class participation. Teaching Tips may expand on a feature (*Think About It*, *Did You Know?*) or provide *Quick Quizzes* or similar exercises. You will be referred to the Teaching Tips section at the back of the module if there is additional material.

 The **Combination** icon indicates that the laboratory listed corresponds with a performance task. If desired, you can note the proficiency of the trainees during the laboratory, and use it to satisfy performance testing requirements.

PREPARATION

Before teaching this module, you should review the Objectives, Performance Tasks, Materials and Equipment List, and Module Outline. Be sure to allow ample time to prepare your own training or lesson plan and gather all required materials and equipment.

Wall and Ceiling Framing
Annotated Instructor's Guide

Module 27106-06

MODULE OVERVIEW

This module introduces the carpentry trainee to the materials and general procedures used in wall and ceiling framing.

PREREQUISITES

Prior to training with this module, it is suggested that the trainee shall have successfully completed *Core Curriculum*; and *Carpentry Fundamentals Level One*, Modules 27101-06 through 27105-06.

OBJECTIVES

Upon completion of this module, the trainee will be able to do the following:

1. Identify the components of a wall and ceiling layout.
2. Describe the procedure for laying out a wood frame wall, including plates, corner posts, door and window openings, partition Ts, bracing, and firestops.
3. Describe the correct procedure for assembling and erecting an exterior wall.
4. Identify the common materials and methods used for installing sheathing on walls.
5. Lay out, assemble, erect, and brace exterior walls for a frame building.
6. Describe wall framing techniques used in masonry construction.
7. Explain the use of metal studs in wall framing.
8. Describe the correct procedure for laying out ceiling joists.
9. Cut and install ceiling joists on a wood frame building.
10. Estimate the materials required to frame walls and ceilings.

PERFORMANCE TASKS

Under the supervision of the instructor, the trainee should be able to do the following:

1. Lay out, assemble, erect, and brace exterior walls.
2. Cut and install ceiling joists on a wood frame building.
3. Estimate the materials required to frame walls and ceilings.

MATERIALS AND EQUIPMENT LIST

Transparencies
Markers/chalk
Blank acetate sheets
Transparency pens
Pencils and scratch paper
Overhead projector and screen
Whiteboard/chalkboard
Appropriate personal protective equipment
8d common nails
16d box nails
Floor plan
2 × 4 or 2 × 6 framing lumber for studs and joists
2 × 12 header material
¼" CD plywood for header spacers
½" CD plywood
Stock for blocking

Metal brace material
Sheathing material
Joist lumber
Chalkline
25' tape
Steel tape
Framing hammer
Framing square or speed square
Circular saw
Extension cord
4' level
6' stepladder
Copies of Job Sheets 1 through 5*
Module Examinations**
Performance Profile Sheets**

* Packaged with this Annotated Instructor's Guide.
**Located in the Test Booklet.

SAFETY CONSIDERATIONS

Ensure that the trainees are equipped with appropriate personal protective equipment.

ADDITIONAL RESOURCES

This module is intended to present thorough resources for task training. The following reference works are suggested for both instructors and motivated trainees interested in further study. These are optional materials for continued education rather than for task training.

Builder's Essentials: Advanced Framing Methods. Kingston, MA: R.S. Means Company.

Builder's Essentials: Framing & Rough Carpentry. Kingston, MA: R.S. Means Company.

Framing Floors, Walls and Ceilings. Newton, CT: Taunton Press.

Framing Walls (DVD). Newton, CT: Taunton Press.

Graphic Guide to Frame Construction. Newton, CT: Taunton Press.

Precision Framing for Pros by Pros. Newton, CT: Taunton Press.

The Proper Construction and Inspection of Ceiling Joists and Rafters (DVD and workbook). Falls Church, VA: International Code Council.

Residential Steel Framing Handbook. New York, NY: McGraw-Hill.

International Code Council. A membership organization dedicated to building safety and fire prevention through development of building codes. **www.iccsafe.org**

National Association of Home Builders. A trade association whose mission is to enhance the climate for housing and the building industry. **www.nahb.org**

TEACHING TIME FOR THIS MODULE

An outline for use in developing your lesson plan is presented below. Note that each Roman numeral in the outline equates to one session of instruction. Each session has a suggested time period of 2½ hours. This includes 10 minutes at the beginning of each session for administrative tasks and one 10-minute break during the session. Approximately 20 hours are suggested to cover *Wall and Ceiling Framing*. You will need to adjust the time required for hands-on activity and testing based on your class size and resources. Because laboratories often correspond to Performance Tasks, the proficiency of the trainees may be noted during these exercises for Performance Testing purposes.

Topic   **Planned Time**

Session I. Introduction; Components of a Wall; Laying Out a Wall; Measuring and Cutting Studs; Assembling and Erecting Walls

   A. Introduction

   B. Components of a Wall

      1. Corners

      2. Partition Intersections

      3. Headers

   C. Laying Out a Wall

      1. Laying Out Wall Openings

   D. Measuring and Cutting Studs

   E. Assembling the Wall

      1. Firestops

   F. Erecting the Wall

      1. Plumbing and Aligning Walls

Session II. Laying Out a Wall

   A. Laying Out a Wall

   B. Laboratory

      Hand out Job Sheet 27106-1. Under your supervision, have the trainees perform the tasks on the Job Sheet. This laboratory corresponds to Performance Task 1.

Session III. Measuring and Cutting Studs
A. Measuring and Cutting Studs
B. Laboratory

Hand out Job Sheet 27106-2. Under your supervision, have the trainees perform the tasks on the Job Sheet. This laboratory corresponds to Performance Task 1.

Session IV. Assembling Walls
A. Assembling Walls
B. Laboratory

Hand out Job Sheet 27106-3. Under your supervision, have the trainees perform the tasks on the Job Sheet. This laboratory corresponds to Performance Task 1.

Session V. Erecting Walls
A. Erecting Walls
B. Laboratory

Hand out Job Sheet 27106-4. Under your supervision, have the trainees perform the tasks on the Job Sheet. This laboratory corresponds to Performance Task 1.

Session VI. Ceiling Layout and Framing
A. Ceiling Layout and Framing
  1. Cutting and Installing Ceiling Joists

Session VII. Laboratory
A. Laboratory

Hand out Job Sheet 27106-5. Under your supervision, have the trainees perform the tasks on the Job Sheet. This laboratory corresponds to Performance Task 2.

Session VIII. Estimating Materials; Wall Framing in Masonry; Steel Studs in Framing; Review; Module Examination and Performance Testing
A. Estimating Materials
B. Laboratory

Have the trainees estimate the materials required to frame example walls and ceilings. This laboratory corresponds to Performance Task 3.

C. Wall Framing in Masonry
  1. Framing Door and Window Openings in Masonry
D. Steel Studs in Framing
  1. Fabrication
E. Review
F. Module Examination
  1. Trainees must score 70 percent or higher to receive recognition from NCCER.
  2. Record the testing results on Craft Training Report Form 200, and submit the results to the Training Program Sponsor.
G. Performance Testing
  1. Trainees must perform each task to the satisfaction of the instructor to receive recognition from NCCER. If applicable, proficiency noted during laboratory exercises can be used to satisfy the Performance Testing requirements.
  2. Record the testing results on Craft Training Report Form 200, and submit the results to the Training Program Sponsor.

Wall and Ceiling Framing
27106-06

Luis Gomez, a contestant at the Associated Builders and Contractors 2005 National Craft Championships, assembles a metal frame. Metal studs are used for exterior and interior loadbearing and non-loadbearing walls.

Instructor's Notes:

Assign reading of Module 27106-06.

27106-06
Wall and Ceiling Framing

Topics to be presented in this module include:

1.0.0	Introduction	6.2
2.0.0	Components of a Wall	6.2
3.0.0	Laying Out a Wall	6.8
4.0.0	Measuring and Cutting Studs	6.12
5.0.0	Assembling the Wall	6.14
6.0.0	Erecting the Wall	6.16
7.0.0	Ceiling Layout and Framing	6.19
8.0.0	Estimating Materials	6.23
9.0.0	Wall Framing in Masonry	6.26
10.0.0	Steel Studs in Framing	6.28

Overview

The walls of most single-family dwellings are framed with 2 × 4 or 2 × 6 lumber. Exterior sheathing and siding, along with interior finishes such as drywall are then attached to the framing. There are two critical steps in the framing process. The first is accurate measuring and layout. The second is accurate leveling and plumbing of the walls. Any deviation in level or plumb will result in serious problems later in the construction process.

Once the walls are erected, the ceiling joists are placed on the walls to serve as the supporting frame for the next level. Again, proper measuring, layout, and placement of these framing members is critical.

The use of steel studs for wall framing is common in commercial construction and is becoming increasingly popular in residential construction as the cost of lumber rises. While the layout process is essentially the same as that of lumber framing, different tools and fastening methods are used.

Instructor's Notes:

Objectives

When you have completed this module, you will be able to do the following:

1. Identify the components of a wall and ceiling layout.
2. Describe the procedure for laying out a wood frame wall, including plates, corner posts, door and window openings, partition Ts, bracing, and firestops.
3. Describe the correct procedure for assembling and erecting an exterior wall.
4. Identify the common materials and methods used for installing sheathing on walls.
5. Lay out, assemble, erect, and brace exterior walls for a frame building.
6. Describe wall framing techniques used in masonry construction.
7. Explain the use of metal studs in wall framing.
8. Describe the correct procedure for laying out ceiling joists.
9. Cut and install ceiling joists on a wood frame building.
10. Estimate the materials required to frame walls and ceilings.

Trade Terms

Blocking
Cripple stud
Double top plate
Drying-in
Furring strip
Gable roof
Header
Hip roof
Jamb
Ribband
Sill
Strongback
Top plate
Trimmer stud

Required Trainee Materials

1. Pencil and paper
2. Appropriate personal protective equipment

Prerequisites

Before you begin this module, it is recommended that you successfully complete *Core Curriculum* and *Carpentry Fundamentals Level One,* Modules 27101-06 through 27105-06.

This course map shows all of the modules in *Carpentry Fundamentals Level One.* The suggested training order begins at the bottom and proceeds up. Skill levels increase as you advance on the course map. The local Training Program Sponsor may adjust the training order.

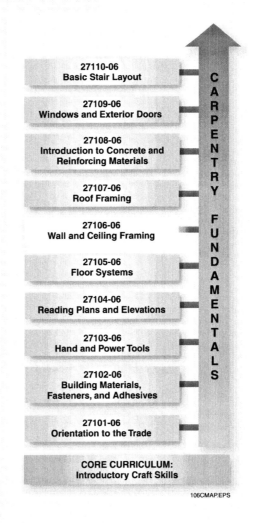

Ensure you have everything required to teach the course. Check the Materials and Equipment List at the front of this module.

Show Transparency 1, Objectives.

Show Transparency 2, Performance Tasks.

Point out that terms shown in bold (blue) are defined in the Glossary at the back of the module.

See the general Teaching Tip at the end of this module.

Identify the basic components of a wall.

Show Transparency 3 (Figure 1).

Acquire photographs or drawings from architectural firms or building material manufacturers that show specific types of building construction in various stages of completion. Pass them around for the trainees to examine.

1.0.0 ♦ INTRODUCTION

In this module, you will learn about laying out and erecting walls with openings for windows and doors. Also covered in this module are instructions for laying out and installing ceiling joists.

Precise layout of framing members is extremely important. Finish material such as sheathing, drywall, paneling, etc., is sold in 4 × 8 sheets. If the studs, rafters, and joists are not straight and evenly spaced for their entire length, you will not be able to fasten the sheet material to them. A tiny error on one end becomes a large error as you progress toward the other end. Spacings of 16" and 24" on center are used because they will divide evenly into 48".

2.0.0 ♦ COMPONENTS OF A WALL

Figure 1 identifies the structural members of a wood frame wall. Each of the members shown on the illustration is then described. You will need to know these terms as you proceed through this module.

- *Blocking* (*spacer*) – A wood block used as a filler piece and support between framing members.
- *Cripple stud* – In wall framing, a short framing stud that fills the space between a **header** and a **top plate** or between the **sill** and the soleplate.
- *Double top plate* – A plate made of two members to provide better stiffening of a wall. It is also used for connecting splices, corners, and partitions that are at right angles (perpendicular) to the wall.
- *Header* – A horizontal structural member that supports the load over an opening such as a door or window.
- *King stud* – The full-length stud next to the trimmer stud in a wall opening.
- *Partition* – A wall that subdivides space within a building. A bearing partition or wall is one that supports the floors and roof directly above in addition to its own weight.
- *Rough opening* – An opening in the framing formed by framing members, usually for a window or door.
- *Rough sill* – The lower framing member attached to the top of the lower cripple studs to form the base of a rough opening for a window.
- *Soleplate* – The lowest horizontal member of a wall or partition to which the studs are nailed. It rests on the rough floor.
- *Stud* – The main vertical framing member in a wall or partition.
- *Top plate* – The upper horizontal framing member of a wall used to carry the roof trusses or rafters.
- *Trimmer stud* – The vertical framing member that forms the sides of rough openings for doors and windows. It provides stiffening for the frame and supports the weight of the header.

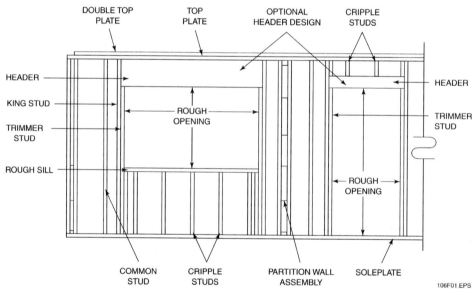

Figure 1 ♦ Wall and partition framing members.

6.2 CARPENTRY FUNDAMENTALS LEVEL ONE ♦ TRAINEE GUIDE

Instructor's Notes:

This section contains an overview of the layout and assembly requirements and procedures for walls.

2.1.0 Corners

When framing a wall, you must have solid corners that can take the weight of the structure. In addition to contributing to the strength of the structure, corners must provide a good nailing surface for sheathing and interior finish materials. Carpenters generally select the straightest, least defective studs for corner framing.

Figure 2 shows the method typically used in western platform framing. In one wall assembly, there are two common studs with blocking between them. This provides a nailing surface for the first stud in the adjoining wall. Notice the use of a double top plate at the top of the wall to provide greater strength.

Figure 3 shows a different way to construct a corner. It has several advantages:

- It doesn't require blocking, which saves time and materials.
- It results in fewer voids in the insulation.
- It promotes better coordination among trades. For example, an electrician running wiring through the corner shown in *Figure 2* would need to bore holes through two or three studs and, possibly, a piece of blocking. However, an electrician wiring through the corner shown in *Figure 3* would need to bore through only two studs.

Describe typical and alternate methods of corner construction. Point out the advantages of the alternate method. Show photographs if available.

Show Transparencies 4 and 5 (Figures 2 and 3).

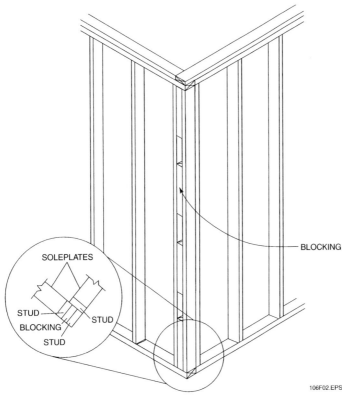

Figure 2 ◆ Corner construction typical of western platform framing.

Explain the different methods of construction for partition intersections. Point out the differences in layout plan dimensions.

Show Transparencies 6 through 8 (Figures 4 through 6).

Lay out a partition T location.

Discuss various types of header construction.

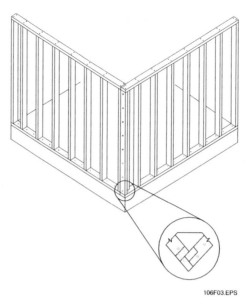

Figure 3 ♦ Alternative method of corner construction.

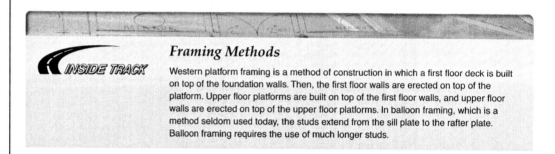

Framing Methods

Western platform framing is a method of construction in which a first floor deck is built on top of the foundation walls. Then, the first floor walls are erected on top of the platform. Upper floor platforms are built on top of the first floor walls, and upper floor walls are erected on top of the upper floor platforms. In balloon framing, which is a method seldom used today, the studs extend from the sill plate to the rafter plate. Balloon framing requires the use of much longer studs.

2.2.0 Partition Intersections

Interior partitions must be securely fastened to outside walls. For that to happen, there must be a solid nailing surface where the partition intersects the exterior frame. *Figure 4* shows a common way to construct a nailing surface for the partition intersections or Ts. The nailing surface can be a full stud nailed perpendicular between two other full studs, or it can be short pieces of 2×4 lumber, known as blocking, nailed between the two other full studs.

Figure 5 shows two other ways to prepare a nailing surface. Compare *Figures 4* and *5*. Notice that in *Figure 4* the spacing between studs differs, but in *Figure 5* the spacing between studs remains the same.

To lay out a partition location, measure from the end of the wall to the centerline of the partition opening, then mark the locations for the partition studs on either side of the centerline (*Figure 6*).

Although plans normally show stud-to-stud dimensions, they sometimes show finish-to-finish dimensions. Center-to-center dimensions are typically used in metal framing, which will be discussed later in this module.

2.3.0 Headers

When wall framing is interrupted by an opening such as a window or door, a method is needed to distribute the weight of the structure around the

6.4 CARPENTRY FUNDAMENTALS LEVEL ONE ♦ TRAINEE GUIDE

Instructor's Notes:

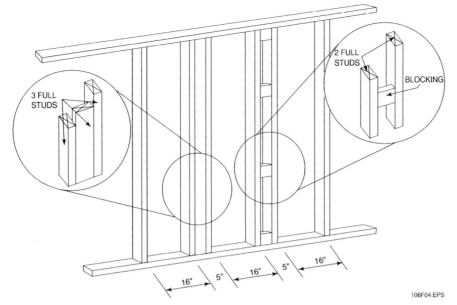

Figure 4 ♦ Constructing nailing surfaces for partitions.

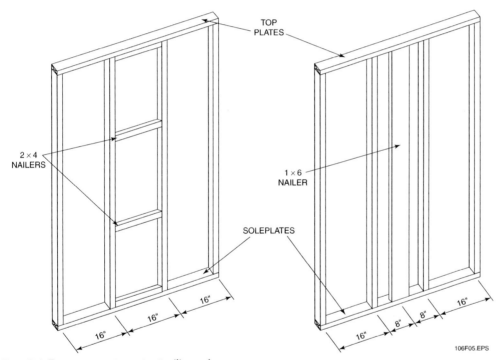

Figure 5 ♦ Two more ways to construct nailing surfaces.

Emphasize the importance of selecting the correct header type and size for the support of overhead loads.

Show Transparencies 9 and 10 (Figures 7 and 8).

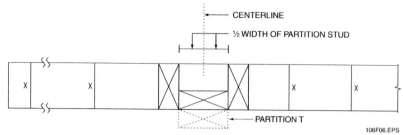

Figure 6 ◆ Partition T layout.

opening. This is done by the use of a header. The header is placed so that it rests on the trimmer studs, which transfer the weight to the soleplate or subfloor, and then to the foundation.

The width of a header should be equal to the rough opening, plus the width of the trimmer studs. For example, if the rough opening for a 3' wide window is 38", the width of the header would be 41" (1½" trimmer stud plus 38" rough opening plus the other 1½" trimmer stud, or 41" total).

2.3.1 Built-Up Headers

Headers are usually made of built-up lumber (although solid wood beams are sometimes used as headers). Built-up headers are usually made from 2 × lumber separated by ½" plywood spacers (*Figure 7*). A full header is used for large openings and fills the area from the rough opening to the bottom of the top plate. A small header with cripple studs is suitable for average-size windows and doors and is usually made from 2 × 4 or 2 × 6 lumber.

Table 1 gives the maximum span typically used for various load conditions.

Table 1 Maximum Span for Exterior Built-Up Headers

Built-Up Header Size	Single-Story Load	Two-Story Load	Three-Story Load
2 × 4	3'-6"	2'-6"	2'
2 × 6	6'	5'	4'
2 × 8	8'	7'	6'
2 × 10	10'	8'	7'
2 × 12	12'	9'	8'

2.3.2 Other Types of Headers

Figure 8 shows some other types of headers that are used in wall framing. Carpenters often use truss headers when the load is very heavy or the span is extra wide. The architect's plans usually show the design of the trusses.

Other types of headers used for heavy loads are wood or steel I-beams, box beams, and engineered wood products such as laminated veneer lumber (LVL), parallel strand lumber (PSL), and laminated lumber (glulam).

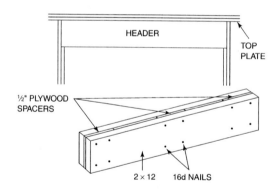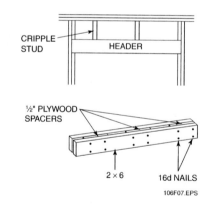

Figure 7 ◆ Two types of built-up headers.

6.6 CARPENTRY FUNDAMENTALS LEVEL ONE ◆ TRAINEE GUIDE

Instructor's Notes:

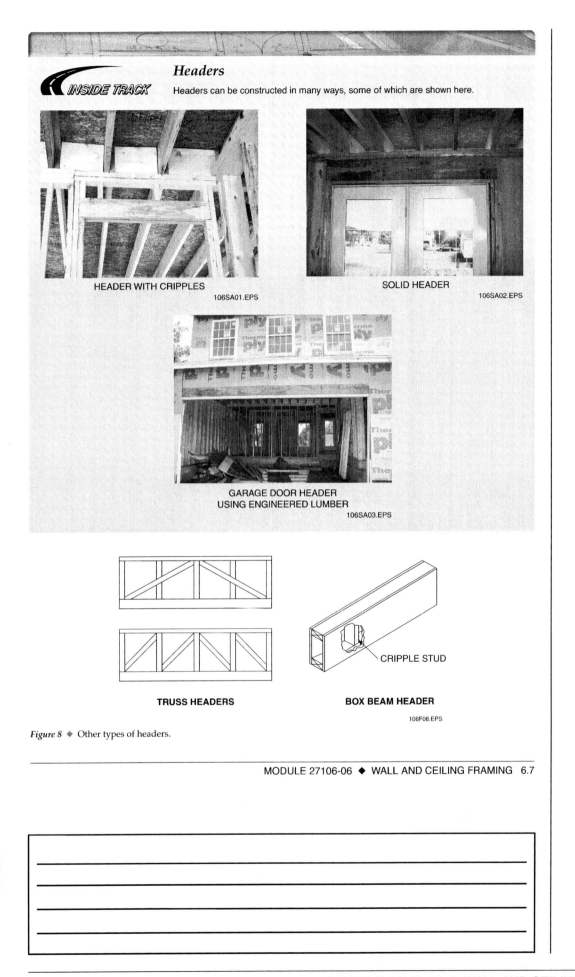

Figure 8 ◆ Other types of headers.

Describe the wall layout procedure. Emphasize the importance of properly positioning the second stud.

Show Transparencies 11 and 12 (Figures 9 and 10).

Demonstrate how to lay out a wall.

3.0.0 ◆ LAYING OUT A WALL

This section covers the basic procedures for laying out wood frame walls with correctly sized window and door openings and partition Ts. Later in this module, you will be introduced to methods for framing with metal studs, and framing window and door openings in masonry walls.

Wall framing is generally done with 2 × 4 studs spaced 16" on center. In a one-story building, 2 × 4 spacing can be 24" on center. If 24" spacing is used in a two-story building, the lower floor must be framed with 2 × 6 lumber. The following provides an overview of the procedure for laying out a wall.

Step 1 Mark the locations of the soleplates by measuring in the width of the soleplate (e.g., 3½") from the outside edge of the sill on each corner. Snap a chalkline to mark the soleplate location, then repeat this for each wall.

Step 2 The top plate and soleplate are laid out together. Start by placing the soleplate as indicated by the chalkline and tacking it in place (*Figure 9*). Lay the top plate against the soleplate so that the location of framing members can be transferred from the soleplate to the top plate. Also tack the top plate. Tacking prevents the plates from moving, which would make the critical layout lines inaccurate.

Step 3 Lay out the common stud positions. To begin, measure and square a line 15¼" from one end. Subtracting this ¾" ensures that sheathing and other panels will fall at the center of the studs because the first sheet goes to the edge rather than the center of the corner stud. Drive a nail at that point and use a continuous tape to measure and mark the stud locations every 16" (*Figure 10*). Align your framing square at each mark. Scribe a line along each side of the framing square tongue across both the soleplate and top plate. These lines will show the outside edges of each stud, centered on 16" intervals.

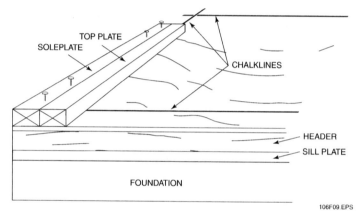

Figure 9 ◆ Soleplate and top plate positioned and tacked in place.

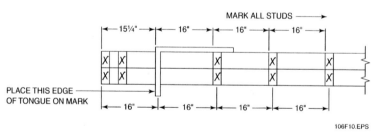

Figure 10 ◆ Marking stud locations.

6.8 CARPENTRY FUNDAMENTALS LEVEL ONE ◆ TRAINEE GUIDE

Instructor's Notes:

Chalkline Protection

If the chalkline will be exposed to the weather, spray it with a clear protective coating to keep it from being washed away.

Explain how to interpret floor plans and schedules. Point out that floor plans normally use centerline dimensions from wall centers to wall openings.

Show Transparencies 13 and 14 (Figures 11 and 12).

3.1.0 Laying Out Wall Openings

The floor plan drawings for a building (*Figure 11*) show the locations of windows and doors. Notice that each window and door on the floor plan is identified by size. In this case, the widths are shown in feet and inches, but the complete dimension information is coded. Look at the window in bedroom #1 coded 2630. This means the window is 2'-6" wide (the width is always given first) and 3'-0" high. Similar codes give widths and heights for the doors.

The window and door schedules (*Figure 12*) provided in the architect's drawings will list the dimensions of windows and doors, along with types, manufacturers, and other information.

Placement of windows is important, and is normally dealt with on the architectural drawings. A good rule of thumb is to avoid placing horizontal framework at eye level. This means you have to consider whether people will generally be sitting or standing when they look out the window. Unless they are architectural specialty

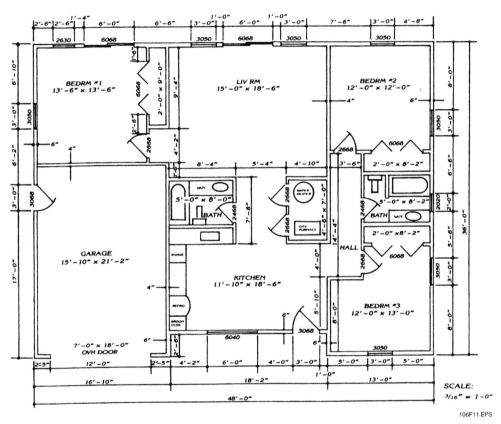

Figure 11 ◆ Sample floor plan.

Explain how to lay out a wall opening.

DOOR SCHEDULE

NO.	DOOR SIZE	MAT'L	TYPE	H.W.#	FRAME MAT'L	TYPE	HEAD	JAMB	SILL	REMARKS
104A	PR. 3'-0" x 7-8⁵⁄₁₆" x 1-³⁄₄"	W.D.	IV	11	W.D.	—	8/16	4/16	3/16	
104B	PR. 3'-0" x 8'-0" x 1-³⁄₄"	GLASS	VI	10	—	—	3/7	15/16, 16/16	4/7	W/ FULL GLASS SIDELITE
105A	3'-0" x 7'-2" x 1-³⁄₄"	W.D.	I	1	H.M.	II	5/15	6/17, 7/17	—	
105B			I	1		I	SIM. 3/15 10	SIM. 4/15 10	—	
106			I	1		I	1/15	2/15	—	
107			I	1		I	1/15	2/15	—	
108	PR. 3'-0" x 7'-2" x 1-³⁄₄"		II	4		II	8/15	9/15	—	"C" LABEL
109	3'-0" x 7'-2" x 1-³⁄₄"		I	7		I	3/15	4/15	—	
110			I	7		I	3/15	4/15	—	
112			I	13		I	SIM. 3/15 10	SIM. 4/15 10	—	
113A			I	13		III	5/15	6/17	—	"C" LABEL
113B			I	13		I	SIM. 1/15 10	SIM. 2/15 10	—	
114	2'-6" x 7'-2" x 1-³⁄₄"		I	13		I	SIM. 1/15 10	SIM. 2/15 10	—	
115A	PR. 3'-0" x 7'-0" x 1-³⁄₄"	ALUM.	V	14	ALUM.	—	28/15	33/15	38/15	
115B	PR. 3'-0" x 7'-2" x 1-³⁄₄"	W.D.	III	5	H.W.	I	10/15	11/16, 12/16	—	"C" LABEL NOTE 1
121D				17			7/15	8/10		

WINDOW SCHEDULE

SYMBOL	WIDTH	HEIGHT	MAT'L	TYPE	SCREEN & DOOR	QUANTITY	REMARKS	MANUFACTURER	CATALOG NUMBER
A	3'-8"	3'-0"	ALUM.	DOUBLE HUNG	YES	2	4 LIGHTS, 4 HIGH	LBJ WINDOW CO.	141 PW
B	3'-8"	5'-0"	ALUM.	DOUBLE HUNG	YES	1	4 LIGHTS, 4 HIGH	LBJ WINDOW CO.	145 PW
C	3'-0"	5'-0"	ALUM.	STATIONARY	STORM ONLY	2	SINGLE LIGHTS	H & J GLASS CO.	59 PY
D	2'-0"	3'-0"	ALUM.	DOUBLE HUNG	YES	1	4 LIGHTS, 4 HIGH	LBJ WINDOW CO.	142 PW
E	2'-0"	6'-0"	ALUM.	STATIONARY	STORM ONLY	2	20 LIGHTS	H & J GLASS CO.	37 TS
F	3'-6"	5'-0"	ALUM.	DOUBLE HUNG	YES	1	16 LIGHTS, 4 HIGH	LBJ WINDOW CO.	141 PW

HEADER SCHEDULE

HEADER SIZE	EXTERIOR		INTERIOR	
	26' + UNDER	26' TO 32'	26' + UNDER	26' TO 32'
(2) 2 x 4	3'-6"	3'-0"	USE (2) 2 x 6	
(2) 2 x 6	6'-6"	6'-0"	4'-0"	3'-0"
(2) 2 x 8	8'-6"	8'-0"	5'-6"	5'-0"
(2) 2 x 10	11'-0"	10'-0"	7'-0"	6'-6"
(2) 2 x 12	13'-6"	12'-0"	8'-6"	8'-0"

Figure 12 ♦ Door, window, and header schedules.

windows, the tops of all windows should be at the same height. The bottom height will vary depending on the use. For example, the bottom of a window over the kitchen sink should be higher than that of a living room or dining room window. The standard height for a residential window is 6'-8" from the floor to the bottom of the window top (head) **jamb**.

The window and door schedules will sometimes provide the rough opening dimensions for windows and doors. Another good source of information is the manufacturer's catalog. It will provide rough and finish opening dimensions, as well as the unobstructed glass dimensions.

When roughing-in a window, the rough opening width equals the width of the window plus the thickness of the jamb material; this is usually 1½" (¾" on each side), plus the shim clearance (½" on each side). Therefore, the rough opening for a 3' window would be 38½". The height of the rough opening is figured in the same way. Be sure to check the manufacturer's instructions for the dimensions of the windows you are using.

To lay out a wall opening, proceed as follows:

Step 1 Measure from the corner to the start of the opening, then add half the width of the window or door to determine the center-

Rough Opening Dimensions

The residential plan shown in *Figure 11* shows dimensions to the sides of rough openings—that is, from the corner of the building to the near side of the first rough opening, then from the far side of the first rough opening to the near side of the second rough opening. However, it is more common for plans to show centerline dimensions—that is, from the corner of the building to the center of the first rough opening, then from the center of the first rough opening to the center of the second rough opening.

Show Transparencies 15 through 17 (Figures 13 through 15).

Demonstrate how to lay out a wall opening.

line (*Figure 13*). Mark the locations of the full studs and trimmer studs by measuring in each direction from the centerline. Mark the cripples 16" on center starting with one trimmer.

Each window opening requires a common stud (king stud) on each side, plus a header, cripple studs, and a sill (*Figure 14*). If the window is more than 4' wide, local codes may require a double sill. Door openings also require trimmer studs, king studs, and a header. Cripple studs will be needed unless the door is double-wide. In that case, a full header may be called for.

Step 2 Mark the location of each common stud and king stud (X), trimmer (T), and cripple (C), as shown in *Figure 15*. (This is a suggested marking method. The only important thing is to mark the locations with codes that you and other members of your crew will recognize.)

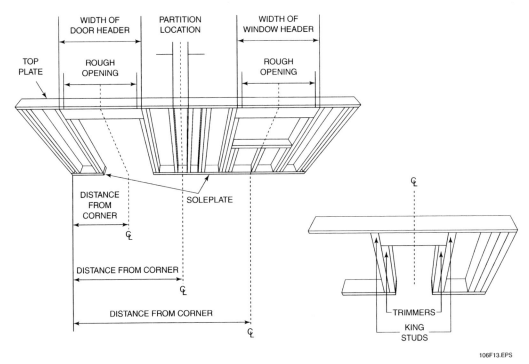

Figure 13 ◆ Laying out a wall opening.

Emphasize the importance of rechecking dimensions of the first (template) stud of each type. Point out that dimensions of precut studs must also be checked.

Show Transparency 18 (Figure 16).

Demonstrate how to measure and cut various types of studs for a wall with a window opening.

For wall and ceiling framing demonstrations or laboratory exercises, you may wish to use half-size (half-scale) layout dimensions and lumber (for example, 1 × 2s = 2 × 4s, 1 × 4s = 2 × 8s, etc.).

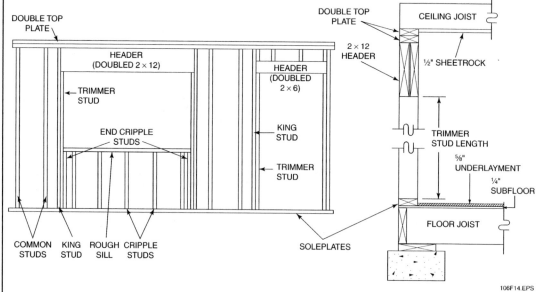

Figure 14 ♦ Window and door framing.

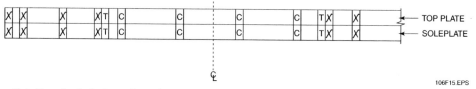

Figure 15 ♦ Example of soleplate and top plate marked for layout.

4.0.0 ♦ MEASURING AND CUTTING STUDS

It is extremely important to precisely measure the first one of each type of stud that will be used (common, trimmer, and cripple) as a template for the others.

Common and king studs – Figure 16 shows the methods for determining the exact length of a common stud for installation on a slab or wood floor.

To determine the stud length when the installation is directly on a concrete slab, simply subtract the thickness of the soleplate (1½") and double top plate (3") from the desired ceiling height and add the thickness of the ceiling material. In the case shown in *Figure 16(A)*, the length of the stud is based on the ceiling height, which is 96", plus the ½" thickness of the ceiling material, less the combined plate thicknesses of 4½", or 92".

Check Stud Lengths

Precut studs in various lengths are available from many lumberyards. These precut studs are ideal for walls on built-up wood floors for the 96" finished ceiling, a very common finished ceiling height. For example, studs precut to 92⅝" are used with exterior or loadbearing interior walls with double top plates.

Sometimes lumberyards deliver the wrong size of precut studs. Unless you look closely, a precut stud doesn't look much different from a standard 8' (96") stud. Always check the precut studs to make sure they are the right length. Taking a few seconds to measure before you start might save hours of rebuilding later on.

6.12 CARPENTRY FUNDAMENTALS LEVEL ONE ♦ TRAINEE GUIDE

Instructor's Notes:

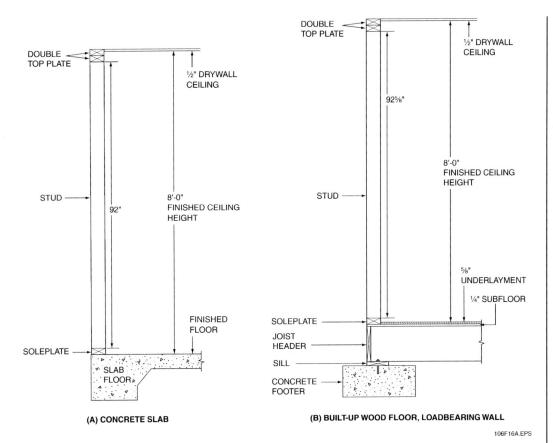

Figure 16 ◆ Calculating the length of a common stud. (1 of 2)

This example assumes that the flooring material has no appreciable thickness.

In the example shown in *Figure 16(B)*, the thickness of the underlayment must also be considered. Therefore, the length of the stud should be 92⅝"; i.e., ceiling height plus the combined thicknesses of the ceiling material and underlayment (½" + ⅝"), less the combined thicknesses of the plates (4½"). Again, this example assumes a flooring material of no appreciable thickness.)

Figure 16(C) shows an interior, nonbearing wall that does not require the use of a double top plate. Therefore, the calculated stud length is 1½" longer (94⅛").

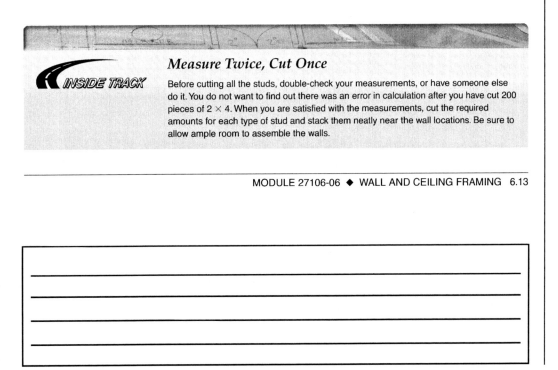

Measure Twice, Cut Once

Before cutting all the studs, double-check your measurements, or have someone else do it. You do not want to find out there was an error in calculation after you have cut 200 pieces of 2 × 4. When you are satisfied with the measurements, cut the required amounts for each type of stud and stack them neatly near the wall locations. Be sure to allow ample room to assemble the walls.

Demonstrate wall assembly. Point out that wall studs with slight crowns should all face in the same direction, and bowed studs should not be used.

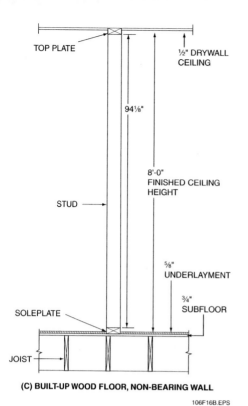

Figure 16 ♦ Calculating the length of a common stud. (2 of 2)

Trimmers – The length of a window or door trimmer stud is determined by subtracting the thickness of the soleplate from the height of the header. If the installation is on a wood floor, the thickness of the underlayment must also be subtracted.

Cripples – To determine the length of a cripple stud above a door or window, combine the height of the trimmer and the thickness of the header, then subtract that total from the length of a regular stud. To determine the length of a cripple stud below a window, determine the height of the rough opening from the floor, then subtract the combined thicknesses of the rough sill and soleplate.

5.0.0 ♦ ASSEMBLING THE WALL

The preferred procedure for assembling the wall is to lay out and assemble the wall on the floor with the inside of the wall facing down.

Step 1 Start by laying the soleplate near the edge of the floor. Then, place the top plate about a regular stud length away from the soleplate. Be sure to use treated lumber if the soleplate is in contact with a masonry floor.

Step 2 Assemble the corners and partition Ts using the straightest pieces to ensure that the corners are plumb. Also, save some of the straightest studs for placement in the wall where countertops or fixtures will hit the centers of studs (such as in kitchens, bathrooms, and laundry rooms).

Step 3 Lay a regular stud at each X mark with the crown up. If a stud is bowed, replace it and use it to make cripples.

Step 4 Assemble the window and door headers and put them in place with the crowns up.

Step 5 Lay out and assemble the rough openings, making sure that each opening is the correct size and that it is square.

Step 6 Nail the framework together. For 2 × 4 framing, use two 16d nails through the plate into the end of each stud. For 2 × 6 framing, use three nails. The use of a nail gun is recommended for this purpose; however, do not use this tool if you have not received proper training.

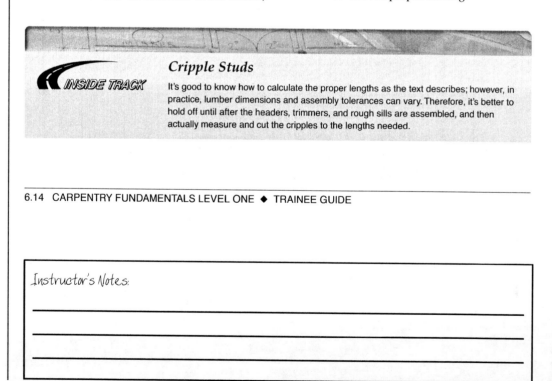

Cripple Studs

It's good to know how to calculate the proper lengths as the text describes; however, in practice, lumber dimensions and assembly tolerances can vary. Therefore, it's better to hold off until after the headers, trimmers, and rough sills are assembled, and then actually measure and cut the cripples to the lengths needed.

Instructor's Notes:

Determining Stud Length

1. If the finish ceiling height of a building with built-up wood floors is supposed to be 9'-6", how long are the studs in exterior walls (assume ½" underlayment and ⅝" drywall)?
2. If the top of the rough sill of a window is supposed to be 32" above the floor, how long are the cripple studs below the sill?

Discuss the "Think About It."

Coping with Natural Defects

Because wood is a natural material, it will have variations that can be considered defects. Lumber is almost never perfectly straight. Even when a piece of lumber is sawn straight, it will most likely curve, twist, or split as it dries.

In the old days, wood cut at a sawmill was just stacked and left to air dry. Normal changes in daily air temperature, humidity, and other conditions would result in lumber with very interesting shapes. Nowadays, lumber is dried in a kiln (a large, low-temperature oven), which will reduce the amount of twisting and curving. But even modern lumber is still somewhat distorted.

Virtually all lumber is slightly curved along its narrow side (the 1½" dimension in 2 × 4s). This is called a crown. It's normal, and, unless it is extreme, the crown doesn't prevent the lumber from being used.

When assembling a floor, you should sight down each piece of lumber to determine which side has the crown. Mark the crowned side. When using the lumber in a floor, position all the lumber crown side up. Weight on the floor will cause the crown to flatten out.

In a wall, there's no force that will flatten the crown. But, for a more uniform nailing surface for the sheathing, you should position all crowns in the same direction. In the example to follow, the crowns are placed up, so when the wall is erected the crown side will be toward the exterior of the building.

A curve along the wide side of lumber (the 3½" dimension in 2 × 4s) is called a bow. If a noticeably bowed stud was used in a wall, sheathing could not be nailed to it in a straight line. Either the nail would miss the bowed stud, or you would have to take extra time to lay out a curved line to nail through. That's not good use of your time, so discard the bowed stud and use a straight one instead.

Explain that local codes may require firestops. Mention that firestop blocking can also be used to stabilize long studs in partitions of high ceiling rooms in new construction.

Show Transparency 19 (Figure 17).

5.1.0 Firestops

In some areas, local building codes may require firestops. Firestops are short pieces of 2 × 4 blocking (or 2 × 6 pieces if the wall is framed with 2 × 6 lumber) that are nailed between studs. See *Figure 17*.

Without firestops, the space between the studs will act like a flue in a chimney. Any holes drilled through the soleplate and top plate create a draft, and air will rush through the space. In a fire, air, smoke, gases, and flames can race through the chimney-like space.

The installation of firestops has two purposes. First, it slows the flow of air, which feeds a fire through the cavity. Second, it can actually block flames (temporarily, at least) from traveling up through the cavity.

If the local code requires firestops, it may also require that holes through the soleplate and top plate (for plumbing or electrical runs) be plugged with a firestopping material to prevent airflow.

Review the procedure for erecting a wall.

Show Transparency 20 (Figure 18).

Demonstrate the erection of a wall with the help of the trainees. Emphasize that banding or cleats should be used to prevent the wall from skidding when it is raised.

Explain how to plumb/align walls using temporary bracing.

Show Transparency 21 (Figure 19).

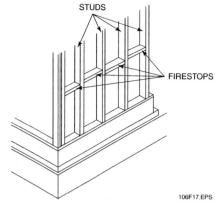

Figure 17 ♦ Firestops.

6.0.0 ♦ ERECTING THE WALL

There are four primary steps in erecting a wall:

Step 1 If the sheathing was installed with the wall lying down, or if the wall is very long, it will probably be too heavy to be lifted into place by the framing crew. In that case, use a crane or the special lifting jacks made for that purpose (*Figure 18*). Use cleats to prevent the wall from sliding.

Step 2 Raise the wall section and nail it in place using 16d nails on every other floor joist. On a concrete slab, use preset anchor bolts or powder-actuated pins. Do not use these tools if you have not received proper training and certification.

Step 3 Plumb the corners and apply temporary exterior bracing. Then erect, plumb, and brace the remaining walls. The bracing helps keep the structure square and will prevent the walls from being blown over by the wind. Generally, the braces remain in place until the roof is complete.

Step 4 As the walls are erected, straighten the walls and nail temporary interior bracing in place.

6.1.0 Plumbing and Aligning Walls

Accurate plumbing of the corners is possible only after all the walls are up. Always use a straightedge along with a hand level (*Figure 19*). The straightedge can be a piece of 2 × 4 lumber. Blocks ¾" thick are nailed to each end of the 2 × 4. The blocks make it possible to accurately plumb the wall from the bottom plate to the top plate. (If you just placed the level directly against the wall, any bow or crown in the end stud would give a false reading.)

Figure 18 ♦ Wall lifting jack.

6.16 CARPENTRY FUNDAMENTALS LEVEL ONE ♦ TRAINEE GUIDE

Instructor's Notes:

Raising a Wall

When a wall is being erected, either by hand or with a lifting jack, the bottom of the wall can slide forward. If the wall slides off the floor platform, the wall or objects on the ground below it can be damaged and workers can be injured.

Use wood blocks or cleats securely nailed to the outside of the rim joist to catch the wall as it slides forward, preventing damage or injury.

Some carpenters use metal banding (used to bundle loads of wood from the mill) to achieve the same effect. One end of a short length of banding is nailed to the floor platform. The other end is bent up 90° and nailed to the bottom of the soleplate. When the wall is raised, the flexible band straightens out horizontally, much like a hinge, but the wall cannot slide forward.

Show Transparency 22 (Figure 20).

Demonstrate how to use temporary bracing.

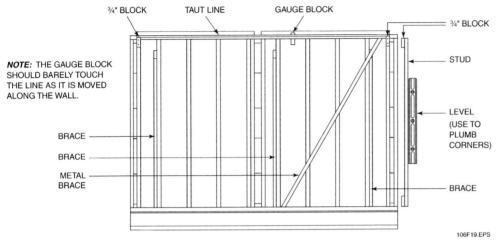

Figure 19 ◆ Plumbing and aligning a wall.

The plumbing of corners requires two people working together. One carpenter releases the nails at the bottom end of the corner brace so that the top of the wall can be moved in or out. At the same time, the second carpenter watches the level. The bottom end of the brace is renailed when the level shows a plumb wall.

Install the second plate of all the double top plates (*Figure 20*). In addition to adding strength in bearing walls, the second plate helps to straighten a bowed or curved wall. If bows are turned opposite each other, intersections of walls should be double plated after walls are erected. Overlap the corners and partition Ts. Drive two 16d nails at each end, then drive one 16d nail at each stud location.

After you have plumbed all the corners, line up the tops of the walls. This must be done before you nail the ceiling to the tops of the walls. To line up the walls, proceed as follows (refer to *Figure 19*):

Step 1 Start at the top plate at one corner of the wall. Fasten a string at that corner. Stretch the string to the top plate at the corner at the opposite end of the wall, and fasten the string.

Step 2 Cut three small blocks from 1×2 lumber. Place one block under each end of the string so that the line is clear of the wall. Use the third block as a gauge to check the wall at 6' or 8' intervals.

MODULE 27106-06 ◆ WALL AND CEILING FRAMING 6.17

Emphasize that temporary bracing should be retained until the ceiling(s) and roof are in place and permanent bracing is installed.

Describe various methods of permanent bracing and demonstrate the installation of bracing acceptable to local codes.

Show Transparency 23 (Figure 21).

Describe the various types of sheathing, and demonstrate the application of plywood sheathing.

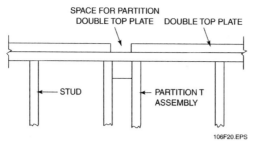

Figure 20 ◆ Double top plate layout.

Step 3 At each checkpoint, nail one end of a temporary brace near the top of a wall stud. Also attach a short 2 × 4 block to the subfloor. Adjust the wall (by moving the top of the wall in or out) so the string is barely touching the gauge block. When the wall is in the right position, nail the other end of the brace to the floor block.

 NOTE
Do not remove the temporary braces until you have completed the framing—particularly the floor or roof diaphragm that sits on top of the walls—and sheathing for the entire building.

6.1.1 Bracing

Permanent bracing is important in the construction of exterior walls. Many local building codes will require bracing when certain types of sheathing are used. In some areas where high winds are a factor, lateral bracing is a requirement even when ½" plywood is used as the sheathing.

Several methods of bracing have been used since the early days of construction. One method is to cut a notch or let-in for a 1 × 4 or 1 × 6 at a 45-degree angle on each corner of the exterior walls. Another method is to cut 2 × 4 braces at a 45-degree angle for each corner. Still another type of bracing used (where permitted by the local code) is metal wall bracing (*Figure 21*). This product is made of galvanized steel.

Metal strap bracing is easier to use than let-in wood bracing. Instead of notching out the studs for a 1 × 4 or 2 × 4, you simply use a circular saw to make a diagonal groove in the studs, top plate, and soleplate for the rib of the bracing strap and nail the strap to the framing.

With the introduction of plywood, some areas of the country have done away with corner brac-

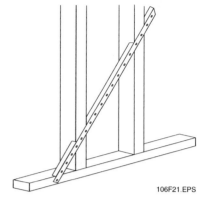

Figure 21 ◆ Use of metal bracing.

ing. However, along with plywood came different types of sheathing that are by-products of the wood industry and do not have the strength to withstand wind pressures. When these are used, permanent bracing is needed. Building codes in some areas will allow a sheet of ½" plywood to be used on each corner of the structure in lieu of diagonal bracing, when the balance of the sheathing is fiberboard. In other areas of the country, the codes require the use of bracing except where ½" plywood is used throughout for sheathing. In still other areas, the use of bracing is required regardless of the type of sheathing used.

6.1.2 Sheathing

Sheathing is the material used to close in the walls. APA-rated material, such as plywood and non-veneer panels such as OSB and other reconstituted wood products, are generally used for sheathing.

Some carpenters prefer to apply the sheathing to a squared wall while the wall frame is still lying on the subfloor. Although this helps to ensure that the wall is square, it has two drawbacks:

- It may make the wall too heavy for the framing crew to lift.
- If the floor is not perfectly straight and level, it will be a lot more difficult to square and plumb the walls once they are erected.

When plywood is used, the panels will range from 5⁄16" to ¾" thick. A minimum thickness of ⅜" is recommended when siding is to be applied. The higher end of the range is recommended when the sheathing acts as the exterior finish surface. The panels may be placed with the grain running horizontally or vertically. If they are placed hori-

---

6.18 CARPENTRY FUNDAMENTALS LEVEL ONE ◆ TRAINEE GUIDE

Instructor's Notes:

zontally, local building codes may require that blocking be used along the top edges.

Typical nailing requirements call for 6d nails for panels ½" thick or less and 8d nails for thicker panels. Nails are spaced 6" apart at the panel edges and 12" apart at intermediate studs.

Other materials that are sometimes used as sheathing are fiberboard (insulation board), exterior-rated gypsum wallboard, and rigid foam sheathing. A major disadvantage of these materials is that siding cannot be nailed to them. It must either be nailed to the studs or special fasteners must be used. If you are installing any of these materials, keep in mind that the nailing pattern is different from that of rated panels. In addition, they take roofing nails instead of common nails. Check the manufacturer's literature for more information.

When material other than rated panels is used as sheathing, rated panels can be installed vertically at the corners to eliminate the need for corner bracing in some applications. Check the plans and local codes.

6.1.3 Panelized Walls

Instead of building walls on the job site, they can be prefabricated in a shop and trucked to the job site. The walls, or panels, are either set with a small crane or by hand. The wall sections or panels will vary in length from 4' to 16'.

When working with a pre-engineered structure, the **drying-in** time is much quicker than with a field-built structure. A 1,200-square-foot residence, for example, can be dried-in within two working days. The siding would be applied at the factory and the walls erected the first day. The soffit and fascia would be installed on the morning of the second day, and the roof dried-in and ready to shingle by the morning of the third day. The residence would be ready for rough-in plumbing, electrical, heating, and cooling by the third day.

7.0.0 ◆ CEILING LAYOUT AND FRAMING

After you assemble and erect the walls, what is next? Traditionally, in a one-story building, carpenters would install ceiling joists on top of the structure, spanning the narrow dimension of the building from top plate to top plate. Then, if the carpenters intended to build a common **gable roof**, they would install rafters that extend from the ends of the ceiling joists to the peak, forming a triangle.

Nowadays, it is more likely that carpenters will install roof trusses instead of joists in a one-story building. The lowest member (or bottom chord) of the truss serves the same purpose as the ceiling joist. Alternatively, if the building is taller than one story, carpenters will install floor joists or floor trusses and build a platform and erect the walls for the second story. Again, the floor joist or truss, when viewed from below, serves the same purpose as the ceiling joist.

You will learn more about roof trusses and modern roof framing methods in later modules. For now, let's focus on the older, simpler way of framing with ceiling joists.

Ceiling joists have two important purposes:

- The joists are the top of the six-sided box structure of a building. They keep opposite walls from spreading apart.
- The joists provide a nailing surface for ceiling material, such as drywall, which is attached to the underside of joists.

As noted previously, joists extend all the way across the structure, from the outside edge of the double top plate on one wall to the outside edge of the double plate of the opposite wall. Ordinarily, carpenters lay the ceiling joists across the narrow width of a building, usually at the same positions as the wall studs.

If the spacing of the ceiling joists is the same as that of the wall studs, lay out the joists directly above the studs. This makes it easier to run ductwork, piping, flues, and wiring above the ceiling.

Laying out ceiling joists for a gable roof is very similar to laying out floor joists and wall studs. Measure along the double top plate to a point 15¼" in from the end of the building. Mark it, square the line, then use a long steel tape to mark every 16" (or 24", depending on the architect's plans). To the right of each line, mark an X for the joist location. Then mark an R for each rafter on the left side of each mark (*Figure 22*). Repeat this procedure on the opposite wall and on any bearing partitions.

NOTE
If you are installing a **hip roof**, it is also necessary to mark the end double top plates for the locations of hip rafters. Start by marking the center of the end double top plate. Then measure and place marks for joists and rafters by measuring from each corner toward the center mark.

Hand out Job Sheets 27106-1 through 27106-4. Under your supervision, have the trainees perform the tasks on the Job Sheets. These four laboratory sessions all correspond to Performance Task 1.

Assign reading of Sections 7.0.0–7.1.0 for the next class session.

Describe the various methods of ceiling framing. Point out the purpose of ceiling joists.

Show Transparency 24 (Figure 22).

Discuss the layout considerations for ceiling joists.

Show Transparency 25 (Figure 23).

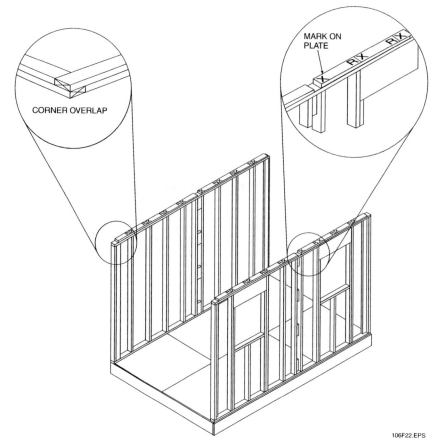

Figure 22 ♦ Marking joist and rafter locations on the double top plate.

As you learned in the module on floor layout, the actual allowable span for joists depends on the species, size, and grade of lumber, as well as the spacing and the load to be carried. If the joist exceeds the allowable span, two pieces of joist material must be spliced over a bearing wall or partition. *Figure 23* shows two ways to splice joists. In the first method, as shown in *Figure 23(A)*, two joists are overlapped above the center of the bearing partition. The overlap should be no less than 6".

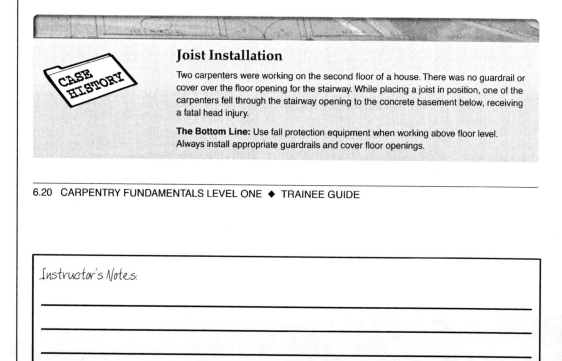

Joist Installation

Two carpenters were working on the second floor of a house. There was no guardrail or cover over the floor opening for the stairway. While placing a joist in position, one of the carpenters fell through the stairway opening to the concrete basement below, receiving a fatal head injury.

The Bottom Line: Use fall protection equipment when working above floor level. Always install appropriate guardrails and cover floor openings.

6.20 CARPENTRY FUNDAMENTALS LEVEL ONE ♦ TRAINEE GUIDE

Instructor's Notes:

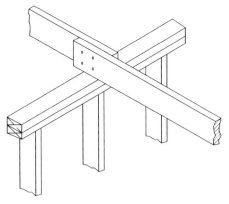
(A) CEILING JOISTS LAPPED OVER BEARING PARTITION

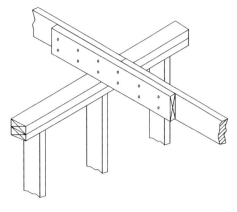
(B) CEILING JOISTS BUTTED OVER BEARING PARTITION

Figure 23 ♦ Splicing ceiling joists.

Overlapped Splices

An overlapped splice is superior to a butted splice because a butted splice reduces by one-half the critical joist-bearing surface, or the part of the joist that is supported by the bearing wall beneath it. Also, a butted splice requires more materials and labor than the simpler, stronger overlapped splice.

Figure 23(B) shows another way to splice joists. Instead of overlapping, the two joists butt together directly over the center of the bearing partition. A shorter piece of joist material is nailed to both joists to make a strong joint. Other materials used to reinforce the joist splice include plywood, 1 × lumber, steel strapping, and special anchors. The architectural plans will specify the proper method for splicing the joists.

7.1.0 Cutting and Installing Ceiling Joists

The joists must be cut to the proper length, so the ends of the joists will be flush with the outside edge of the double top plate. As you learned in the previous section, you must allow enough extra length for any overlap of the joists above bearing partitions.

You must also cut the ends of the joists at an angle matching the rafter pitch, as shown in *Figure 24*. This is so the roof sheathing will lie flush on the roof framing. You will learn more about this in the module about roof framing.

After you have cut the joists to the right length and have positioned them in the right places, toe-nail them into the double top plate. The architectural plans might also call for the installation of metal anchors.

After you install the joists, you must nail a **ribband** or **strongback** across the joists to prevent twisting or bowing (*Figure 25*). The strongback is used for longer spans. In addition to holding the joists in line, the strongback provides support for the joists at the center of the span.

Assign reading of Sections 8.0.0–10.1.0 for the next class session.

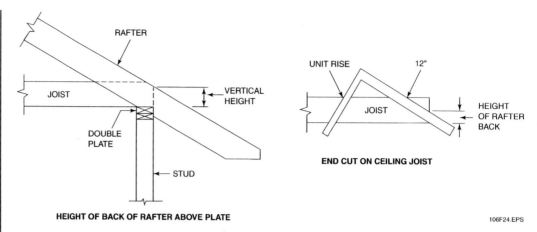

Figure 24 ♦ Cutting joist ends to match roof pitch.

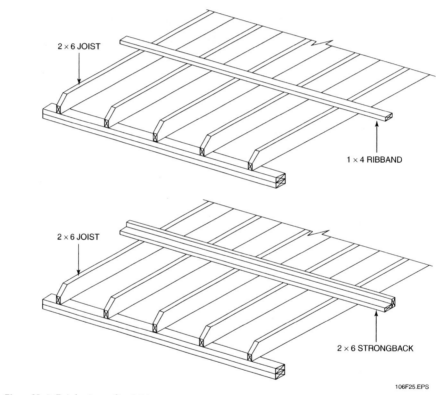

Figure 25 ♦ Reinforcing ceiling joists.

6.22 CARPENTRY FUNDAMENTALS LEVEL ONE ♦ TRAINEE GUIDE

Instructor's Notes:

8.0.0 ◆ ESTIMATING MATERIALS

In this section, you will follow the basic steps to estimate the amount of lumber you will need to frame the walls and ceilings of a building. Here's some important information that you need:

- For this example, the structure is 24' × 30'. (See *Figure 26*.)
- The walls are 8' tall.
- You will construct the building with 2 × 4s with 16" on center spacing.
- You will construct the ceiling from 2 × 6s.
- The building has two 24" wide windows on each narrow side, and two 48" wide windows and one 36" wide door on each wide side.
- Inside, the building is divided into three rooms: one is 12' × 30', two are each 11'-1½" × 12', divided by a 4'-10" wide hallway.
- The long wall that separates the large room from the two smaller rooms and hallway is a loadbearing partition.
- Each of the smaller rooms has a 36" door leading to the hallway.
- There is a 36" opening, without a door, from the large room to the hallway.

Explain the method for estimating framing materials using the example in the Trainee Module.

Show Transparency 28 (Figure 26).

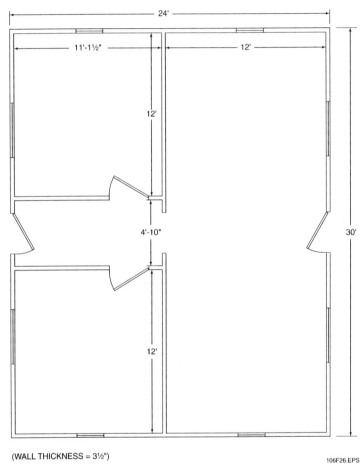

(WALL THICKNESS = 3½")

106F26.EPS

Figure 26 ◆ Sample floor plan.

Explain how to estimate soleplates, top plates, studs, and headers. Work through the examples in the Trainee Module.

Point out that only exterior walls and interior loadbearing walls require a double top plate.

8.1.0 Estimating Soleplates and Top Plates

Assume that the plate material is ordered in 12' lengths.

Step 1 Determine the length of the walls in feet. For loadbearing walls, multiply this number by 3 to account for the soleplates and double top plates (remember that a double top plate is made of two 2 × 4s stacked together).

Step 2 Divide the result by 12 to get the number of 12' pieces. Round up to the next full number, and allow for waste.

Example:

For the exterior walls, add the lengths of the two long walls, plus the two narrow walls:

$$30' + 30' + 24' + 24' = 108'$$

Also add the length of the interior partitions:

$$30' + 12' + 12' = 54'$$

Add those numbers together:

$$108' + 54' = 162'$$

Then multiply by 3 to account for the soleplate and double top plate:

$$162' \times 3' = 486'$$

Divide by 12 to get the number of 12' pieces needed:

$$486' \div 12' = 40\tfrac{1}{2} \text{ pieces}$$

Since you can buy only full pieces, you will need to round up to the next whole number, which is 41. So, for all of the soleplates, top plates, and double top plates in this building, you need 41 pieces of 12' 2 × 4s.

> **NOTE**
>
> The example estimate here for soleplate and top plate material is a very simple one and works for this small, simple structure. Actually, only the center loadbearing partition (and the exterior walls, of course) will require double top plates. But the extra amount of lumber from this simpler estimate is small, and you need to figure more lumber for waste anyway. In a larger, more complex structure, you need to calculate the different quantities of lumber for loadbearing and nonbearing partitions to get a more precise estimate.

8.2.0 Estimating Studs

Step 1 Determine the length in feet of all the walls.

Step 2 The general industry standard is to allow one stud for each foot of wall length, even when you are framing 16" on center. This should cover any additional studs that are needed for openings, corners, partition Ts, and blocking.

When you were estimating for soleplate and top plate material, you found that there were 162' of exterior walls and interior partitions. You will need one stud for every foot of wall, or 162 studs.

8.3.0 Estimating Headers

Let's assume that you'll use 2 × 12 headers with plywood spacers. This might cost a little more, but you won't need to lay out, cut, and assemble cripple studs above the headers.

Step 1 Use the architect's drawings (to be safe, check both the floor plans and the door and window schedules to make sure they agree) to find the number and size of each window and door.

Step 2 The header rests on the trimmer studs in the rough opening. Therefore, 3" must be added to the rough opening dimension to account for the trimmers. Add 3" to the finish width of each door and window.

Step 3 Double the length of each header.

Step 4 Combine the lengths obtained in Step 3 into convenient lengths for ordering.

Step 5 Order enough ½" plywood for spacers.

Example:

In *Figure 26*, there are four 4830 windows (each 48" wide):

$$48" + 3" = 51"; 51" \times 4 = 204"$$

Two 2430 windows (each 24" wide):

$$24" + 3" = 27"; 27" \times 2 = 54"$$

Four 3680 doors (each 36" wide):

$$36" + 3" = 39;" 39" \times 4 = 156"$$

One 3'-0" opening with no door:

$$36" + 3" = 39"$$

Add those numbers together:

$$204" + 54" + 156" + 39" = 453"$$

6.24 CARPENTRY FUNDAMENTALS LEVEL ONE ◆ TRAINEE GUIDE

Instructor's Notes:

How Accurate Is an Estimate?

How close does the estimate come to actual needs? Sometimes, it does not come close enough. In our example building, we estimated 162 studs based on one stud per lineal foot of wall space. However, if you calculate the number based on the actual framing spaced 16" on center, it takes:

- 17 studs to span each of the two 24' walls
- 21 studs to span each of the two 30' walls
- 21 studs for the 28'-10" interior partition (excludes the width of the two interior and two exterior stud walls, which is 1'-2" total)
- 9 studs to span each 11'-1½" interior partition

So, that subtotal is:

$$17 + 17 + 21 + 21 + 21 + 9 + 9 = 115$$

But you need to add an extra full stud and trimmer on each side of each opening. There are thirteen openings, so …

$$13 \times 4 = 52$$

…add that to the previous subtotal …

$$115 + 52 = 167$$

However, about 15 full studs will be eliminated where the openings will be, so subtract that from the subtotal …

$$167 - 15 = 152$$

Then add two extra studs for each of the six partition Ts …

$$152 + 12 = 164$$

…and two extra studs for each exterior corner …

$$164 + 8 = 172$$

Of course, you'll need studs for cripples and for extra blocking here and there. Then, some of the delivered lumber will be unusable because it is twisted or warped. Plus, measuring and cutting mistakes will be made. So, you're well over the 162 pieces calculated by allowing one stud every 12". Be smart and add a safety factor. It might vary depending on the size and complexity of the job, but for our example here, 15% (for a total of about 186 studs) would have been about right. On bigger jobs, the safety factor might be smaller or even nil. That's because the total amount of lumber is greater, and the safety margin already built into the one-stud-per-foot will cover all needs.

Explain why smaller buildings require an additional 15% added to the estimate as a safety factor.

Explain how to estimate diagonal bracing. Work through the example in the Trainee Module.

Then double that total:

$$453" \times 2 = 906"$$

Divide by 12 to get feet:

$$906" \div 12 = 75'\text{-}6"$$

Round up to 76'.

8.4.0 Estimating Diagonal Bracing

You need to install diagonal bracing at each end of all exterior walls. You can use metal strips or 1 × 4 lumber. Braces run from the top plate to the soleplate at a 45-degree angle. The walls for this building are 8' high, so they would need a 12' brace. (Review the *Core* module, *Introduction to Construction Math*, to determine how to find the length of the brace using the formula for the hypotenuse of a right triangle.)

Step 1 Determine the number of outside corners.

Step 2 Figure the length of each brace based on the height of the walls, then multiply by the number of corners.

Explain how to estimate ceiling joists. Work through the example in the Trainee Module.

Have the trainees estimate the materials required to frame example walls and ceilings. This laboratory corresponds to Performance Task 3.

Describe the methods of furring and framing for masonry walls with door and window openings. Emphasize that pressure-treated lumber must be used.

Show Transparencies 29 through 31 (Figures 27 through 29).

Take the trainees to a construction site to observe masonry wall framing in progress.

Example:

This part is pretty easy. *Figure 26* shows that this building is a rectangle with four corners. Each corner has two legs, so there are eight places where you would install diagonal braces. Since you need 12' of bracing at each location, $8 \times 12' = 96'$ of bracing altogether.

Remember that let-in or diagonal bracing is typically applied to the interior side of the exterior walls.

8.5.0 Estimating Ceiling Joists

As mentioned previously, it is more likely that carpenters will install roof trusses instead of joists in a one-story building. Roof trusses and modern roof framing methods are covered later in this book. Here the focus is on framing with ceiling joists.

Step 1 Determine the span of the building.

Step 2 Figure the number of joists based on the spacing (remember that you'll place joists 16" on center), then add one for the end joist.

Step 3 Multiply the span by the number of joists to get total length. Add 6" per joist per splice where joists will be spliced at bearing walls, partitions, and girders.

Example:

The building in *Figure 26* is 30' by 24'. Let's take the short dimension (24'). Fortunately, there is a partition down the long dimension of the building. That breaks the 24' span into two 12' spans.

Divide the long dimension of the building (30') by 16" to determine the number of joist locations. First, convert 30' to inches:

$$30' \times 12" = 360"$$

Divide that by 16":

$$360" \div 16" = 22\tfrac{1}{2} \text{ (round up to 23)}$$

Then add one for the end joist:

$$23 + 1 = 24$$

The 24' width is too big to span with one long joist, so you'll span it with two 12' joists. Multiply the last number by two:

$$24 \times 2 = 48 \text{ pieces}$$

Now calculate the total length, adding 6" for each joist at each splice.

$$48 \times 12'\text{-}6" = 600'$$

So how much lumber do you order? Twelve-foot lengths of 2×6 would be 6" short, so 14-foot lengths would be a better choice. You'd end up with 1'-6" of waste from each piece of lumber, but you can use the waste for bridging and blocking.

9.0.0 ◆ WALL FRAMING IN MASONRY

You must be aware of the methods used in furring masonry walls in order to install the interior finish. As a general rule, furring of masonry walls should be done on 16" centers. Some contractors will apply 1×2 **furring strips** 24" OC. This may save material, but it does not provide the same quality as a wall constructed on 16" centers. Carpenters have little control over the quality of the masonry structure. It is important that the walls are put up square and plumb. If a structure starts off level and plumb, little difficulty will be encountered with measuring, cutting, and fitting. If the walls or floor are not plumb and level, problems can be expected throughout the structure.

Furring is applied to a masonry wall with masonry nails measuring 1¼" to 1¾" long. In addition to the furring strips, 1×4 and 1×6 stock is used. Remember, all material that comes in contact with concrete or masonry must be pressure-treated.

Backing for partitions against a masonry block wall is done using one of the following methods. In the first method, locate where the partition is going to be, then nail a 1×6 board, centered on the center of the partition (i.e., locate the center of the partition at the bottom corner of the wall, move back 2¾", and mark the wall by plumbing with a straightedge and level). Attach the 1×6 to the wall with one edge on the mark. This will allow an even space on either side of the partition to receive the drywall, as shown in *Figure 27*.

In the second method, secure the partition to the block wall and install a furring strip on each side of the partition. This will cause a problem when the drywall is installed; i.e., the partition stud is 1½", and by adding a ¾" furring strip to the wall we have only ¾" of stud left to work with. Nailing the ½" drywall to the furring strip on the wall will leave only ¼" of nailing on the stud. This does not allow enough room to nail the ½" drywall, so it may become necessary to install a 16" 2×4 block to the bottom, center, and top of the partition, as shown in *Figure 28*.

The proper nailing surface for drywall is ¾" to 1". In preparing the corners of the block wall to receive the furring strips, enough space should be allowed for the drywall to slip by the furring strips. Come away from the corner ⅝" in either direction with the strips (*Figure 29*).

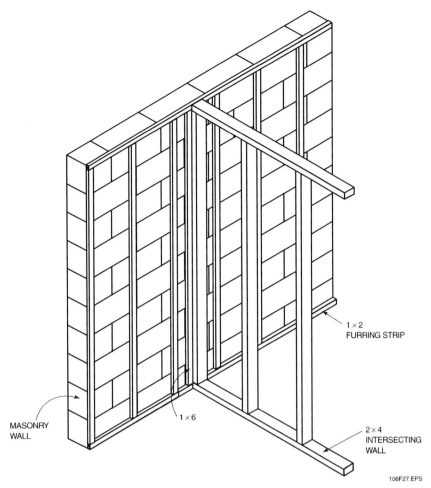

Figure 27 ◆ Partition backing using a 1 × 6.

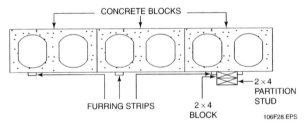

Figure 28 ◆ Partition backing using 2 × 4 blocks.

Point out the special precautions required when wood framing is used with masonry walls.

Discuss the applications and advantages of steel studs. Describe their layout and mounting methods.

Show Transparency 32 (Figure 30).

Take the trainees to a construction site to observe steel stud framing in progress.

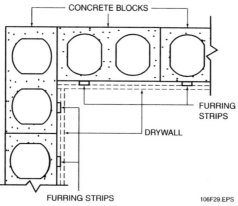

Figure 29 ◆ Placement of furring strips at corners of masonry walls.

A 1 × 4 is used at floor level to receive the baseboard. Either a narrow or wide baseboard can be used. Some carpenters will install a simple furring strip at floor level and depend on the vertical strips for baseboard nailing. Once the drywall has been installed, it is difficult to find the strips when nailing the baseboard.

A sequence of installation should be established for nailing the furring strips to the wall. Either start from the right and work to the left or work from left to right. This sequence will allow the person doing the trim work to locate the furring strips. When applying the furring strips to the wall, remember that the strips must be placed 16" OC for the drywall to be nailed properly. Start the first strip at 15¼" and then lay out the second strip 16" from the first mark. Laying out furring strips is done in the same way as laying out wall studs.

9.1.0 Framing Door and Window Openings in Masonry

In modern construction, metal door frames and metal window frames or channels are used almost exclusively with masonry walls. However, in rare instances, wood framing is used. Each installation is unique and a complete examination of this subject is beyond the scope of this module. There are two general rules, however:

- Use treated wood in contact with masonry.
- Use cut nails, expansion bolts, anchor bolts, or other fasteners to attach the wood to the masonry securely.

10.0.0 ◆ STEEL STUDS IN FRAMING

Depending on the gauge, steel studs are typically stronger, lighter, and easier to handle than wood. Unlike wood studs, steel studs will not split, warp, swell, or twist. Furthermore, steel studs will not burn as wood studs would. Steel studs are currently more expensive than wood studs, but it is expected that as lumber prices continue to rise, the costs will eventually equalize.

Steel studs are prepunched to permit quick installation of piping, wiring, and bracing members.

Steel studs have become popular in residential, commercial, and industrial construction. Steel studs may be spaced 16" or 24" OC. On a nonbearing wall, spacing is determined by drywall type and thickness. Unlike wood (which has defects), steel studs are consistent in material composition.

There are three types of steel studs. The first is used for nonbearing walls that have facings to accept drywall. The second will accept lath and plaster on both interior and exterior walls. The third type is a wide-flange steel stud, which is used for both loadbearing and nonbearing walls.

A wide variety of accessories are available for steel studs. There are tracks for floors and ceilings to which the studs are fastened. Tracks are also available for sills, fascia, and joint-end enclosures. Other accessories include channels, angles, and clips. For residential construction, steel trusses are also available.

10.1.0 Fabrication

For layout, steel studs are marked to the centerline rather than the edge. The open side of the stud should always face the beginning point of the layout. The bottom channel is fastened to the concrete floor with small powder-actuated fasteners (*Figure 30*). Note that the T-channel is held back slightly to allow room for the drywall to slide between the two channel sections. A metal self-tapping screw is used to fasten the studs to the track (*Figure 31*). The electric screwgun takes the place of the hammer. The studs may also be welded instead of screwed.

When constructing a rough opening, two studs are put back to back and screwed or welded together. The stud that will act as the trimmer stud will be cut to the height specified to receive the header (*Figure 32*). A section of floor track can be used for the bottom part of the header, with short pieces of studs put in place over the header and secured in place. Blocking may be required to fasten millwork.

6.28 CARPENTRY FUNDAMENTALS LEVEL ONE ◆ TRAINEE GUIDE

Figure 30 ◆ Steel stud channel.

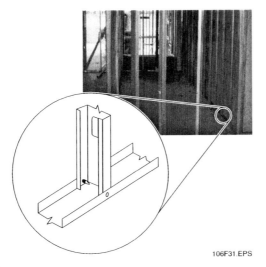

Figure 31 ◆ Stud wall section.

Figure 32 ◆ Window opening framed with steel studs.

Summarize the major concepts presented in the module.

Summary

This module covered how to identify and use all the wall and ceiling components; how to lay out walls and ceilings; how to measure and cut lumber; how to assemble, erect, plumb, brace, and sheath walls; how to lay out and build doors, windows, and other wall openings; how to install ceilings; how to estimate materials; how to work with masonry walls; how to construct walls with steel studs; and much more.

The most important skill in wall and ceiling framing is accuracy. Walls must be straight, plumb, and square. Precise layout and measuring of studs and headers is also critical.

Even a very small error can cause big problems. For example, over a span of 20', an error of only $\frac{1}{16}$" at one end of the span could become $1\frac{1}{4}$" at the other end. The closer you get to the end of the wall, the more patching and fitting you will have to do to get the sheathing, windows, and exterior doors to fit.

In other words, saving a little time at the beginning can cost you a lot of time at the end. Remember this as you move forward in your training.

Notes

Instructor's Notes:

Review Questions

1. A short framing member that fills the space between the rough sill and the soleplate is a _____.
   a. spacer
   b. cripple stud
   c. trimmer stud
   d. top plate

2. The framing member that forms the side of a rough opening for a window or door is the _____.
   a. trimmer stud
   b. header
   c. soleplate
   d. cripple

3. The straightest, least defective studs are normally used for _____.
   a. window and door trimmers
   b. common studs
   c. corners
   d. partition Ts

4. The framing member that distributes the weight of the structure around a door or window opening is the _____.
   a. trimmer stud
   b. cripple stud
   c. top plate
   d. header

5. The width of a header is equal to the _____.
   a. width of the rough opening
   b. width of the opening plus the width of two trimmer studs plus the width of two common studs
   c. width of the opening plus the width of two trimmer studs
   d. width of the trimmer plus the width of two cripple studs

6. Fabricated headers normally use _____ as spacers.
   a. furring strips
   b. ½" plywood
   c. ¾" plywood
   d. 2 × 4 blocks

7. The length of a cripple stud above a window opening equals the length of a common stud less the height of the trimmer, combined with the thickness of the _____.
   a. soleplate
   b. rough sill
   c. top plate
   d. header

8. When calculating the length of a common stud for a wood frame floor, the thickness of the _____ is not considered.
   a. soleplate
   b. subfloor
   c. underlayment
   d. double top plate

9. Pieces of 2 × 4 placed horizontally between each pair of studs are used to _____.
   a. brace the frame
   b. retard the spread of fire
   c. provide nailers for siding
   d. provide a place for carpenters to stand when installing ceiling joists

10. A long 2 × 4 with a ¾" standoff block nailed to each end is used as a _____.
    a. wall brace
    b. straightedge
    c. gauge block
    d. trimmer stud

11. When the sheathing acts as the finish surface, you should use _____ sheets.
    a. ⅜"
    b. ½"
    c. ¾"
    d. 1"

12. When you are installing APA-rated sheathing material, you should space the nails _____ apart at the panel edges.
    a. 6"
    b. 12"
    c. 16"
    d. 24"

Review Questions

13. A ribband is a(n) _____.
    a. brace placed between studs
    b. L-shaped joist brace
    c. type of joist hanger
    d. 1 × 4 strip used to prevent joists from twisting or bowing

14. The purpose of prepunched holes in steel studs is to _____.
    a. make them lighter
    b. save material
    c. provide runs for piping and wiring
    d. make it easier to attach drywall

15. Steel studs are attached to the bottom channel using _____.
    a. wood screws
    b. powder-actuated fasteners
    c. construction adhesive
    d. self-tapping metal screws

Instructor's Notes:

Trade Terms Quiz

1. The part of a door frame that comes in contact with the door is the _____.
2. A roof with four sides running toward the center is a(n) _____.
3. A roof with two slopes that meet at a center ridge is a(n) _____.
4. A(n) _____ is an L-shaped support member used to strengthen and align the ceiling joists.
5. _____ is used to provide filler and support between framing members.
6. The framing at a wall corner is stiffened using a(n) _____.
7. A(n) _____ is a short framing member that fills the space between longer framing members.
8. A(n) _____ is a framing member that forms the sides of the rough opening for a door.
9. The top support member of a rough door or window opening is known as the _____.
10. The bottom framing member of a rough door or window opening is known as the _____.
11. A(n) _____ provides a nailing surface when a wall is attached to masonry.
12. The roof trusses or rafters are supported by the _____.
13. A(n) _____ is installed between ceiling joists to prevent twisting or bowing.
14. The installation of sheathing, doors, and windows makes up the _____ stage of building construction.

Trade Terms

Blocking
Cripple stud
Double top plate
Drying-in
Furring strip
Gable roof
Header
Hip roof
Jamb
Ribband
Sill
Strongback
Top plate
Trimmer stud

Have trainees complete the Trade Terms Quiz, and go over the answers.

Administer the Module Examination. Record the results on Craft Training Report Form 200, and submit the form to the Training Program Sponsor.

Administer the Performance Tests, and fill out Performance Profile Sheets for each trainee. If desired, trainee proficiency noted during laboratory sessions may be used to complete the Performance Test. Record the results on Craft Training Report Form 200, and submit the results to the Training Program Sponsor.

Profile in Success

Barry Caldwell
Silver Medal Winner Associated Builders and Contractors National Championships

Barry Caldwell won the silver medal in carpentry at the 2005 ABC Craft Olympics. He remembers the tough competition and the testing areas. In a mock building, contestants had to build a concrete form, construct sub-flooring, a metal wall, and a wooden wall. The competition was timed over six hours, and Barry finished all task areas with just five minutes remaining. The judge's criteria were safety, workmanship, job-site etiquette, cleanliness, efficient use of materials, and of course—accuracy. There was an eighth-of-an-inch tolerance, Barry says.

How did you become interested in carpentry?
I've been working around the trade basically my whole life. I started helping my father (who's also a carpenter) on job sites when I was five years old. After high school, I went to a vocational school for two years, and then I entered an apprenticeship program with Cleveland Construction. I've been there for the past three years.

What are some of the things you do in your job?
I work for the Interior Division of Cleveland Construction, which specializes in interior and exterior framing, and drywall and ceiling tile installation. The Interior Division is where people start out. As they progress, they can move on to other areas in the company, or they can stay with that division. Eventually, I'd like to move into a management position with the company.

What do you like most about carpentry?
I like it because there's a real sense of accomplishment. Fifty years later, you can go back and see what you've done. I love being able to go down the road and say "I did that" or "I helped with that." I love that new technologies, new tools, are always coming out. For example, right now there's a move toward using metal studs in framing instead of wood studs. I also like working outside. And there's always something new—a new project, a new site.

You're still in your apprenticeship now. What separates a good apprentice from the rest?
Successful apprentices want to stay busy—they want to work. A good apprentice is motivated. The trade is not for everybody. If you don't want to do it, then don't do it. The best apprentices are those people who want to do the work. I love my job. I have fun at work every day.

6.34 CARPENTRY FUNDAMENTALS LEVEL ONE ♦ TRAINEE GUIDE

Instructor's Notes:

Trade Terms Introduced in This Module

Blocking: A wood block used as a filler piece and a support between framing members.

Cripple stud: In wall framing, a short framing stud that fills the space between a header and a top plate or between the sill and the soleplate.

Double top plate: A plate made of two members to provide better stiffening of a wall. It is also used for connecting splices, corners, and partitions that are at right angles (perpendicular) to the wall.

Drying-in: Applying sheathing, windows, and exterior doors to a framed building.

Furring strip: Narrow wood strips nailed to a wall or ceiling as a nailing base for finish material.

Gable roof: A roof with two slopes that meet at a center ridge.

Header: A horizontal structural member that supports the load over a window or door.

Hip roof: A roof with four sides or slopes running toward the center.

Jamb: The top (head jamb) and side members of a door or window frame that come into contact with the door or window.

Ribband: A 1×4 nailed to the ceiling joists at the center of the span to prevent twisting and bowing of the joists.

Sill: The lower framing member attached to the top of the lower cripple studs to form the base of a rough opening for a window.

Strongback: An L-shaped arrangement of lumber used to support ceiling joists and keep them in alignment.

Top plate: The upper horizontal framing member of a wall used to carry the roof trusses or rafters.

Trimmer stud: A vertical framing member that forms the sides of rough openings for doors and windows. It provides stiffening for the frame and supports the weight of the header.

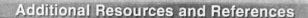

Additional Resources and References

This module is intended to present thorough resources for task training. The following reference works are suggested for further study. These are optional materials for continuing education rather than for task training.

Builder's Essentials: Advanced Framing Methods. Kingston, MA: R.S. Means Company.

Builder's Essentials: Framing & Rough Carpentry. Kingston, MA: R.S. Means Company.

Framing Floors, Walls and Ceilings. Newton, CT: Taunton Press.

Framing Walls (DVD). Newton, CT: Taunton Press.

Graphic Guide to Frame Construction. Newton, CT: Taunton Press.

Precision Framing for Pros by Pros. Newton, CT: Taunton Press.

The Proper Construction and Inspection of Ceiling Joists and Rafters (DVD and workbook). Falls Church, VA: International Code Council.

Residential Steel Framing Handbook. New York, NY: McGraw-Hill.

International Code Council. A membership organization dedicated to building safety and fire prevention through development of building codes. www.iccsafe.org

National Association of Home Builders. A trade association whose mission is to enhance the climate for housing and the building industry. www.nahb.org

---

Instructor's Notes:

MODULE 27106-06 — ANSWERS TO REVIEW QUESTIONS

Answer	Section
1. b	2.0.0
2. a	2.0.0
3. c	2.1.0
4. d	2.3.0
5. c	2.3.0
6. b	2.3.1
7. d	4.0.0
8. b	4.0.0
9. b	5.1.0
10. b	6.1.0
11. c	6.1.2
12. a	6.1.2
13. d	7.1.0; Figure 25
14. c	10.0.0
15. d	10.1.0

MODULE 27106-06 — ANSWERS TO TRADE TERMS QUIZ

1. Jamb
2. Hip roof
3. Gable roof
4. Strongback
5. Blocking
6. Double top plate
7. Cripple stud
8. Trimmer stud
9. Header
10. Sill
11. Furring strip
12. Top plate
13. Ribband
14. Drying-in

MODULE 27106-06 — TEACHING TIPS

The following are suggested activities or instructional methods to help you teach the material in this module.

General

When you call on someone to answer a question, the rest of the class relaxes or even tunes out because they expect that the question and answer will take place only between you and the trainee you called on. Instead, use this technique to involve more trainees in answering questions and to keep them on their toes.

1. Ask trainees to define a term or explain a concept.
2. After one trainee has answered, ask a trainee seated nearby if the answer is right. Then ask whether a trainee in the back of the room agrees.
3. Ask trainees to explain why they think an answer is right or wrong.
4. Use the session to clear up incorrect ideas and encourage trainees to learn from their mistakes.

MODULE 27106-06 — "THINK ABOUT IT" ANSWERS

Section 5.0.0 — Think About It—Determining Stud Length

The answers to these questions depend on the ceiling thickness and, if used, the floor underlayment thickness. Assuming the same ceiling and underlayment thicknesses shown in Figure 16, the answers are:

1. Stud length = 9'-6" + (underlayment and ceiling thickness) − (sole and double top plate thickness).

   Substituting a ceiling thickness of ½", an underlayment thickness of ⅝", and 1½" for the sole and each plate thickness yields:

   Stud length = 9'-6" + (½" + ⅝") − (1½" + 3") = 9'-2⅝".

2. Cripple length = 32" + (underlayment thickness) − (sole and sill plate thickness).

   Substituting an underlayment thickness of ⅝" and 1½" for the sole and sill thicknesses yields:

   Cripple length = 32" + ⅝" − (1½" + 1½") = 29⅝".

If desired, have the trainees determine appropriate stud and cripple lengths by substituting the values used locally for underlayment and finish ceiling thicknesses.

CONTREN® LEARNING SERIES — USER UPDATE

NCCER makes every effort to keep these textbooks up-to-date and free of technical errors. We appreciate your help in this process. If you have an idea for improving this textbook, or if you find an error, a typographical mistake, or an inaccuracy in NCCER's Contren® textbooks, please write us, using this form or a photocopy. Be sure to include the exact module number, page number, a detailed description, and the correction, if applicable. Your input will be brought to the attention of the Technical Review Committee. Thank you for your assistance.

Instructors – If you found that additional materials were necessary in order to teach this module effectively, please let us know so that we may include them in the Equipment/Materials list in the Annotated Instructor's Guide.

Write: Product Development and Revision
National Center for Construction Education and Research
P.O. Box 141104, Gainesville, FL 32614-1104

Fax: 352-334-0932

E-mail: curriculum@nccer.org

Craft

Module Name

Copyright Date

Module Number

Page Number(s)

Description

(Optional) Correction

(Optional) Your Name and Address

Roof Framing
27107-06

NCCER STANDARDIZED CRAFT TRAINING PROGRAM

The National Center for Construction Education and Research (NCCER) provides a standardized national program of accredited craft training. Key features of the program include instructor certification, competency-based training, and performance testing. The program provides trainees, instructors, and companies with a standard form of recognition through the National Registry. The program is described in full in the *Guidelines for Accreditation*, published by NCCER. For more information on standardized craft training, contact NCCER by writing to P.O. Box 141104, Gainesville, FL 32614-1104; calling 352-334-0911; or e-mailing **info@nccer.org**. More information is available at **www.nccer.org**.

HOW TO USE THIS ANNOTATED INSTRUCTOR'S GUIDE

Each page presents two sections of information. The larger section displays each page exactly as it appears in the Trainee Module. The narrow column ties suggested trainee and instructor actions to each page and provides icons (detailed below) to call your attention to material, safety, audiovisual, or testing requirements. The bottom of each page includes space for your notes.

The **Audiovisual** icon indicates an appropriate time to show a transparency or other audiovisual aid.

The **Classroom** icon prompts you to define a term, stress a point, ask trainees to explain a concept, or give examples.

The **Demonstration** icon directs you to show trainees how to perform tasks.

The **Examination** icon tells you to administer the written module examination.

The **Homework** icon is placed where you may wish to assign reading for the next class, assign a project, or advise trainees to prepare for an examination.

The **Laboratory** icon is used when trainees are to practice performing tasks.

The **Materials** icon is a reminder for you to gather materials needed for classes, laboratories, and testing.

The **Performance Testing** icon tells you to administer a performance test or a portion thereof.

The **Safety** icon is used to emphasize safety issues. It is often keyed to *Caution* and *Warning!* statements in the Trainee Module.

The **Teaching Tip** icon indicates additional guidance is available, such as how to conduct an exercise, get the most educational value from a field trip, or encourage class participation. Teaching Tips may expand on a feature (*Think About It, Did You Know?*) or provide *Quick Quizzes* or similar exercises. You will be referred to the Teaching Tips section at the back of the module if there is additional material.

The **Combination** icon indicates that the laboratory listed corresponds with a performance task. If desired, you can note the proficiency of the trainees during the laboratory, and use it to satisfy performance testing requirements.

PREPARATION

Before teaching this module, you should review the Objectives, Performance Tasks, Materials and Equipment List, and Module Outline. Be sure to allow ample time to prepare your own training or lesson plan and gather all required materials and equipment.

**Roof Framing
Annotated Instructor's Guide**

Module 27107-06

MODULE OVERVIEW

This module introduces the carpentry trainee to the methods and procedures used in roof framing.

PREREQUISITES

Prior to training with this module, it is suggested that the trainee shall have successfully completed *Core Curriculum*; and *Carpentry Fundamentals Level One*, Modules 27101-06 through 27106-06.

OBJECTIVES

Upon completion of this module, the trainee will be able to do the following:

1. Understand the terms associated with roof framing.
2. Identify the roof framing members used in gable and hip roofs.
3. Identify the methods used to calculate the length of a rafter.
4. Identify the various types of trusses used in roof framing.
5. Use a rafter framing square, speed square, and calculator in laying out a roof.
6. Identify various types of sheathing used in roof construction.
7. Frame a gable roof with vent openings.
8. Frame a roof opening.
9. Erect a gable roof using trusses.
10. Estimate the materials used in framing and sheathing a roof.

PERFORMANCE TASKS

Under the supervision of the instructor, the trainee should be able to do the following:

1. Use a framing square and speed square in laying out a roof.
2. Frame and sheathe a gable roof with an opening.
3. Erect a gable roof using trusses.
4. Estimate the materials used in framing and sheathing a roof.

MATERIALS AND EQUIPMENT LIST

Transparencies
Markers/chalk
Blank acetate sheets
Transparency pens
Pencils and scratch paper
Overhead projector and screen
Whiteboard/chalkboard
Appropriate personal protective equipment
Scientific calculator
8d common nails
8d box nails
16d box nails
16d common nails
Roof framing plan
2 × 4 or 2 × 6 framing lumber for rafters and ridgeboards
Joist and header material for roof opening
½" CD plywood or other sheathing material
Nails for sheathing
H-clips
Roof trusses
1 × 6 lumber or plywood for catwalk
2 × 4 lumber for braces and stakes

Sample blueprints
Chalkline
String line
Steel tape with markings at 16" OC
Framing hammer
Claw hammer
Spreader for lifting trusses (if applicable)
Crane for lifting trusses (if applicable)
Rafter framing square
Sawhorses
Speed square and booklet
Circular saw
Extension cord
Handsaw
4' level
6' stepladders
Plumb bob and line
Copies of Job Sheets 1 through 6*
Module Examinations**
Performance Profile Sheets**

* Packaged with this Annotated Instructor's Guide.
** Located in the Test Booklet.

SAFETY CONSIDERATIONS

Ensure that the trainees are equipped with appropriate personal protective equipment.

ADDITIONAL RESOURCES

This module is intended to present thorough resources for task training. The following reference works are suggested for both instructors and motivated trainees interested in further study. These are optional materials for continued education rather than for task training.

Advanced Framing Methods. Kingston, MA: R.S. Means Company.

Build a Better Home: Roofs. Tacoma, WA: APA – The Engineered Wood Association.

Framing Roofs. Newton, CT: Taunton Press.

Graphic Guide to Frame Construction. Newton, CT: Taunton Press.

Miller's Guide to Framing and Roofing. New York: McGraw-Hill Professional.

New Roof Construction. Sumas, WA: Cedar Shake and Shingle Bureau (15-minute video).

Quality Roof Construction. Tacoma, WA: APA – The Engineered Wood Association (15-minute video).

Roof Framer's Bible: The Complete Pocket Reference to Roof Framing. Jenkintown, PA: M.E.I. Publishing.

Wood Frame Construction Manual. Washington, D.C.: American Wood Council.

American Wood Council. A trade association that develops design tools and guidelines for wood construction. **www.awc.org.**

Cedar Shake and Shingle Bureau. A trade organization that promotes the common interests of members involved in quality cedar shake and shingle roofing. **www.cedarbureau.org.**

Western Wood Products Association. A trade association representing softwood lumber manufacturers in 12 western states and Alaska. **www.wwpa.com.**

Wood I-Joist Manufacturers Association. An organization representing manufacturers of prefabricated wood I-joist and structural composite lumber. **www.i-joist.org.**

Wood Truss Council of America. An international trade association representing structural wood component manufacturers. **www.woodtruss.com.**

TEACHING TIME FOR THIS MODULE

An outline for use in developing your lesson plan is presented below. Note that each Roman numeral in the outline equates to one session of instruction. Each session has a suggested time period of 2½ hours. This includes 10 minutes at the beginning of each session for administrative tasks and one 10-minute break during the session. Approximately 37½ hours are suggested to cover *Roof Framing*. You will need to adjust the time required for hands-on activity and testing based on your class size and resources. Because laboratories often correspond to Performance Tasks, the proficiency of the trainees may be noted during these exercises for Performance Testing purposes.

Topic **Planned Time**

Session I. Introduction; Types of Roofs; Basic Roof Layout
    A. Introduction   _____
    B. Types of Roofs   _____
    C. Basic Roof Layout   _____
        1. Rafter Framing Square   _____
        2. Basic Rafter Layout   _____

Session II. Laboratory
- A. Laboratory
  Hand out Job Sheet 27107-1. Under your supervision, have the trainees perform the tasks on the Job Sheet. This laboratory corresponds to Performance Task 1.

Session III. Erecting a Gable Roof
- A. Erecting a Gable Roof
  1. Installing Rafters

Session IV. Laboratory
- A. Laboratory
  Hand out Job Sheet 27107-2. Under your supervision, have the trainees perform the tasks on the Job Sheet. This laboratory corresponds to Performance Task 2.

Session V. Framing the Gable Ends; Framing a Gable Overhang
- A. Framing the Gable Ends
- B. Framing a Gable Overhang

Session VI. Laboratory
- A. Laboratory
  Hand out Job Sheet 27107-3. Under your supervision, have the trainees perform the tasks on the Job Sheet. This laboratory corresponds to Performance Task 2.

Session VII. Framing an Opening in the Roof
- A. Framing an Opening in the Roof

Session VIII. Laboratory
- A. Laboratory
  Hand out Job Sheet 27107-4. Under your supervision, have the trainees perform the tasks on the Job Sheet. This laboratory corresponds to Performance Task 2.

Session IX. Installing Sheathing
- A. Installing Sheathing
- B. Laboratory
  Hand out Job Sheet 27107-5. Under your supervision, have the trainees perform the tasks on the Job Sheet. This laboratory corresponds to Performance Task 2.

Session X. Rafter Layout Using a Speed Square
- A. Rafter Layout Using a Speed Square
  1. Procedure for Laying Out Common Rafters

Session XI. Truss Construction
- A. Truss Construction
  1. Truss Installation
  2. Bracing of Roof Trusses

Session XII. Laboratory
- A. Laboratory
  Hand out Job Sheet 27107-6. Under your supervision, have the trainees perform the tasks on the Job Sheet. This laboratory corresponds to Performance Task 3.

Session XIII. Determining Quantities of Materials
- A. Determining Quantities of Materials
  1. Determine Materials Needed for a Gable Roof
- B. Laboratory
  Have the trainees estimate the materials used in framing and sheathing a roof. This laboratory corresponds to Performance Task 4.

Session XIV. Dormers; Plank-and-Beam Framing
   A. Dormers
   B. Plank-and-Beam Framing

Session XV. Metal Roof Framing; Review; Module Examination and Performance Testing
   A. Metal Roof Framing
   B. Review
   C. Module Examination
      1. Trainees must score 70 percent or higher to receive recognition from NCCER.
      2. Record the testing results on Craft Training Report Form 200, and submit the results to the Training Program Sponsor.
   D. Performance Testing
      1. Trainees must perform each task to the satisfaction of the instructor to receive recognition from the NCCER. If applicable, proficiency noted during laboratory exercises can be used to satisfy the Performance Testing requirements.
      2. Record the testing results on Craft Training Report Form 200, and submit the results to the Training Program Sponsor.

Roof Framing
27107-06

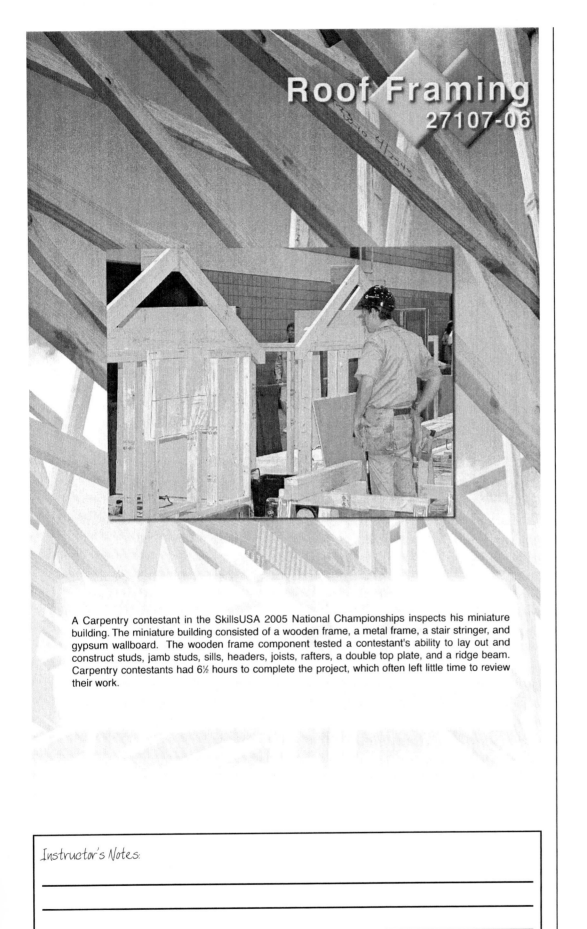

A Carpentry contestant in the SkillsUSA 2005 National Championships inspects his miniature building. The miniature building consisted of a wooden frame, a metal frame, a stair stringer, and gypsum wallboard. The wooden frame component tested a contestant's ability to lay out and construct studs, jamb studs, sills, headers, joists, rafters, a double top plate, and a ridge beam. Carpentry contestants had 6½ hours to complete the project, which often left little time to review their work.

Instructor's Notes:

Assign reading of Module 27107-06.

27107-06
Roof Framing

Topics to be presented in this module include:

1.0.0	Introduction	7.2
2.0.0	Types of Roofs	7.2
3.0.0	Basic Roof Layout	7.3
4.0.0	Installing Sheathing	7.15
5.0.0	Rafter Layout Using a Speed Square	7.18
6.0.0	Truss Construction	7.20
7.0.0	Determining Quantities of Materials	7.25
8.0.0	Dormers	7.26
9.0.0	Plank-and-Beam Framing	7.27
10.0.0	Metal Roof Framing	7.28

Overview

Before the invention of truss systems, all the rafters of a framed roof had to be measured and cut by carpenters. This was time-consuming work involving precise calculations and measurement, and often resulted in a lot of rework.

Today, roof frame members are generally integrated into truss systems that are designed and fabricated off site by companies that specialize in manufacturing trusses. However, there are occasions when a carpenter will have to stick-frame all or part of a roof. In such cases, it is essential to be able to perform the length and angle calculations needed to correctly size and cut each type of rafter.

Instructor's Notes:

Objectives

When you have completed this module, you will be able to do the following:

1. Understand the terms associated with roof framing.
2. Identify the roof framing members used in gable and hip roofs.
3. Identify the methods used to calculate the length of a rafter.
4. Identify the various types of trusses used in roof framing.
5. Use a rafter framing square, speed square, and calculator in laying out a roof.
6. Identify various types of sheathing used in roof construction.
7. Frame a gable roof with vent openings.
8. Frame a roof opening.
9. Erect a gable roof using trusses.
10. Estimate the materials used in framing and sheathing a roof.

Trade Terms

Barge rafter
False fascia
Gable
Lookout
Purlin

Required Trainee Materials

1. Pencil and paper
2. Appropriate personal protective equipment
3. Framing square

Prerequisites

Before you begin this module, it is recommended that you successfully complete *Core Curriculum* and *Carpentry Fundamentals Level One*, Modules 27101-06 through 27106-06.

This course map shows all of the modules in *Carpentry Fundamentals Level One*. The suggested training order begins at the bottom and proceeds up. Skill levels increase as you advance on the course map. The local Training Program Sponsor may adjust the training order.

Ensure you have everything required to teach the course. Check the Materials and Equipment List at the front of this module.

Show Transparency 1, Objectives.

Show Transparency 2, Performance Tasks.

Explain that terms shown in bold (blue) are defined in the Glossary at the back of this module.

See the general Teaching Tip at the end of this module.

Discuss the importance of proper roof framing.

Describe the six types of roofs. Point out that roofs can be stick-built or truss-built. Stick-built roofs usually require interior loadbearing walls for the ceiling joists.

Show Transparency 3 (Figure 1).

Take the trainees to local sites to observe various types of completed roofs as well as stick-built and truss-built roofs under construction.

1.0.0 ◆ INTRODUCTION

Roof framing is the most demanding of the framing tasks. Floor and wall framing generally involves working with straight lines. Residential roofs are usually sloped in order to shed water from rain or melting snow. In areas where there is heavy snowfall, the roof must be constructed to bear extra weight. Because a roof is sloped, laying out a roof involves working with precise angles in addition to straight lines.

In this module, you will learn about the different types of roofs used in residential construction. You will also learn how to lay out and frame a roof.

2.0.0 ◆ TYPES OF ROOFS

The most common types of roofs used in residential construction are shown in *Figure 1* and described below.

- *Gable roof* – A **gable** roof has two slopes that meet at the center (ridge) to form a gable at each end of the building. It is the most common type of roof because it is simple, economical, and can be used on any type of structure.
- *Hip roof* – A hip roof has four sides or slopes running toward the center of the building. Rafters at the corners extend diagonally to meet at the ridge. Additional rafters are framed into these rafters.
- *Mansard roof* – The mansard roof has four sloping sides, each of which has a double slope. As compared with a gable roof, this design provides more available space in the upper level of the building. The upper slope is typically not visible from the ground.
- *Gable and valley roof* – This roof consists of two intersecting gable roofs. The part where the two roofs meet is called a valley.
- *Hip and valley roof* – This roof consists of two intersecting hip roofs.
- *Gambrel roof* – The gambrel roof is a variation on the gable roof in which each side has a break, usually near the ridge. The gambrel roof provides more available space in the upper level.
- *Shed roof* – Also known as a lean-to roof, the shed roof is a flat, sloped construction. It is common on high-ceiling contemporary construction, and is often used on additions.

There are two basic roof framing systems. In stick-built framing, ceiling joists and rafters are laid out and cut by carpenters on site and the frame is constructed one stick at a time.

In truss-built construction, the roof framework is prefabricated off site. The truss contains both the rafters and the ceiling joist.

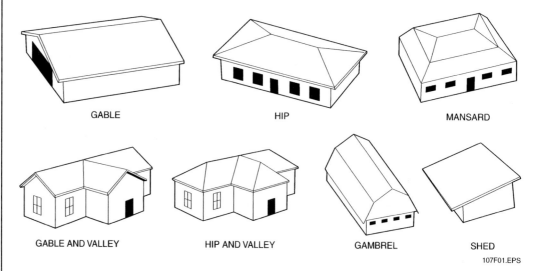

Figure 1 ◆ Types of roofs.

7.2 CARPENTRY FUNDAMENTALS LEVEL ONE ◆ TRAINEE GUIDE

Instructor's Notes:

Inside Track

Mansard Roof

Mansard roofs are popular in the design of many fast-food restaurants and small office buildings.

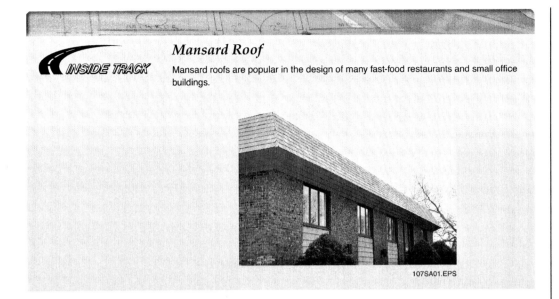

3.0.0 ◆ BASIC ROOF LAYOUT

Rafters and ceiling joists provide the framework for all roofs. The main components of a roof are shown in *Figure 2* and described below.

- *Ridge (ridgeboard)* – The highest horizontal roof member. It helps to align the rafters and tie them together at the upper end. The ridgeboard is one size larger than the rafters.
- *Common rafter* – A structural member that extends from the top plate to the ridge in a direction perpendicular to the wall plate and ridge. Rafters often extend beyond the roof plate to form the overhang (eaves) that protect the side of the building.
- *Hip rafter* – A roof member that extends diagonally from the outside corner of the plate to the ridge.
- *Valley rafter* – A roof member that extends from the inside corner of the top plate to the ridge along the lines where two roofs intersect.
- *Jack rafter* – A roof member that does not extend the entire distance from the ridge to the top plate of a wall. The hip jack and valley jack are shown in *Figure 2*. A rafter fitted between a hip rafter and a valley rafter is called a cripple jack. It touches neither the ridge nor the plate.
- *Plate* – The wall framing member that rests on top of the wall studs. It is sometimes called the rafter plate because the rafters rest on it. It is also referred to as the top plate.

As you can see in *Figure 3*, on any pitched roof, rafters rise at an angle to the ridgeboard. Therefore, the length of the rafter is greater than the horizontal distance from the plate to the ridge. In order to calculate the correct rafter length, the carpenter must factor in the slope of the roof. Here are some additional terms you will need to know in order to lay out rafters:

- *Span* – The horizontal distance from the outside of one exterior wall to the outside of the other exterior wall.
- *Run* – The horizontal distance from the outside of the top plate to the center line of the ridgeboard (usually equal to half of the span).
- *Rise* – The total height of the rafter from the top plate to the ridge. This is stated in inches per foot of run.
- *Pitch* – The angle or degree of slope of the roof in relation to the span. Pitch is expressed as a fraction; e.g., if the total rise is 6' and the span is 24', the pitch would be ¼ (6 over 24).
- *Slope* – The inclination of the roof surface expressed as the relationship of rise to run. It is stated as a unit of rise to so many horizontal units; e.g., a roof that has a rise of 5" for each foot of run is said to have a 5 in 12 slope (*Figure 3*). The roof slope is sometimes referred to as the roof cut.

The first step in determining the correct length of a rafter is to find the unit rise, which is usually

Identify the components of a typical stick-built roof.

Define the terms encountered in roof construction.

Show Transparencies 4 and 5 (Figures 2 and 3).

Explain roof slope and pitch. Point out that they are two different concepts used to define the same thing.

Explain the purpose of a rafter framing square. The trainees should have a framing square for their own use.

Show Transparency 6 (Figure 4).

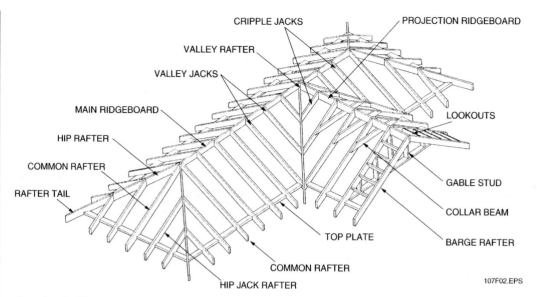

Figure 2 ♦ Roof framing members.

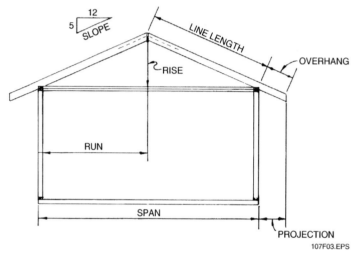

Figure 3 ♦ Roof layout factors.

shown on the building's elevation drawing. The unit rise is the number of inches the rafter rises vertically for each foot of run. The greater the rise per foot of run, the greater the slope of the roof.

There are several ways to calculate the length of a rafter. It can be done with a framing (rafter) square or speed square, or it can be done using a calculator. We will show the framing square method now, then discuss the speed square.

3.1.0 Rafter Framing Square

The rafter framing square is a special carpenter's square that is calibrated to show the length per foot of run for each type of rafter (*Figure 4*). Note that the tongue is the short (16") section of the square and the blade (or body) is the long (24") section. The corner is known as the heel. Rafter tables are normally provided on the back of the

7.4 CARPENTRY FUNDAMENTALS LEVEL ONE ♦ TRAINEE GUIDE

Instructor's Notes:

Pitch and Slope

Carpenters may use the terms pitch and slope interchangeably on the job site, but the two terms actually refer to two different concepts. Slope is the amount of rise per foot of run and is always referred to as a number in 12. For example, a roof that rises 6" for every foot of run has a 6 in 12 slope (the 12 simply refers to the number of inches in a foot). Pitch, on the other hand, is the ratio of rise to the span of the roof and is expressed as a fraction. For example, a roof that rises 8' over a 32' span is said to have a pitch of ¼ (⁸⁄₃₂ = ¼).

Show Transparency 7 (Figure 5).

Discuss the tasks involved in basic rafter layout.

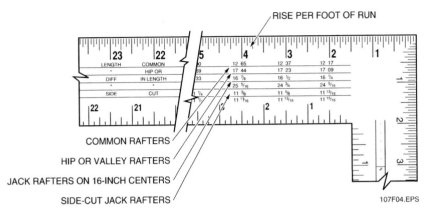

Figure 4 ♦ Rafter tables on a framing square.

square. The rafter tables usually give the rafter dimensions in length per foot of run, but some give length per given run.

The framing square is used to determine the rafter length and to measure and mark the cuts that must be made in the rafter (*Figure 5*). As you can see, you can relate the pitch and slope to the rise per foot of run. The rise per foot of run is always the same for a given pitch or slope. For example, a pitch of ½, which is the same as a 12 in 12 slope, equals 12" of rise per foot of run.

3.2.0 Basic Rafter Layout

Laying out the framing for a roof involves four tasks:

- Marking off the rafter locations on the top plate
- Determining the length of each rafter
- Making the plumb cuts at the ridge end and tail end of each rafter
- Making the bird's mouth cut in each rafter

Rafters must be laid out and cut so that the ridge end will fit squarely on the ridge and the tail end

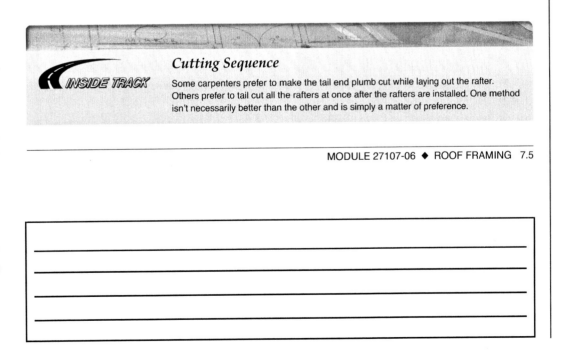

Cutting Sequence

Some carpenters prefer to make the tail end plumb cut while laying out the rafter. Others prefer to tail cut all the rafters at once after the rafters are installed. One method isn't necessarily better than the other and is simply a matter of preference.

MODULE 27107-06 ♦ ROOF FRAMING 7.5

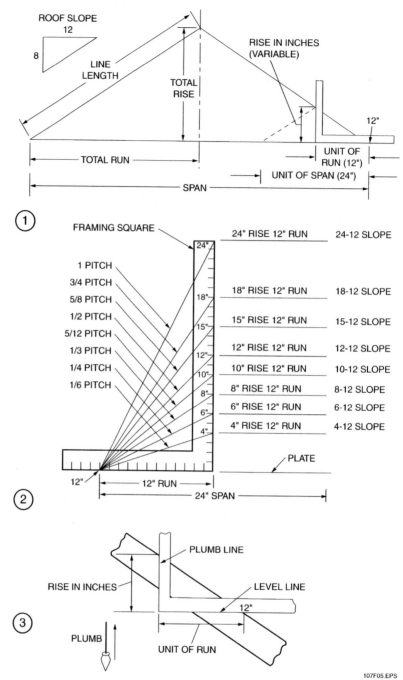

Figure 5 ♦ Application of the rafter square.

7.6 CARPENTRY FUNDAMENTALS LEVEL ONE ♦ TRAINEE GUIDE

will present a square surface for the **false fascia** board. In addition, a bird's mouth cut (*Figure 6*) must be made at the correct location and angle for the rafter to rest squarely on the plate.

3.2.1 Laying Out Rafter Locations

The following is a basic procedure for marking the locations of the rafters on the top plate (*Figure 7*) for 24" on center (OC) construction. Keep in mind

Show Transparency 8 (Figure 6).

Explain how to mark the location of the rafters on the top plate.

Show Transparency 9 (Figure 7).

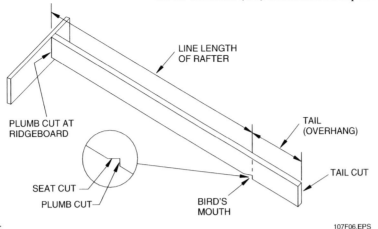

Figure 6 ♦ Parts of a rafter.

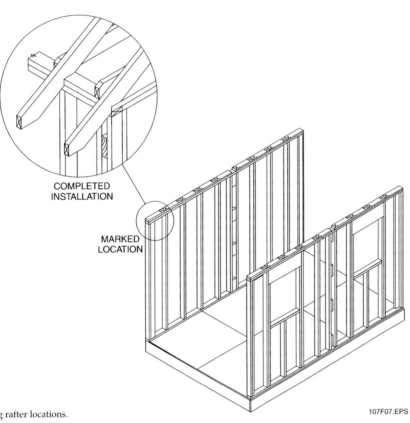

Figure 7 ♦ Marking rafter locations.

Demonstrate the layout of rafter locations. Point out that rafters can also be positioned on 16" centers to support high roof loads caused by ice or snow accumulation.

For roof framing demonstrations or laboratory exercises, you may wish to use half-size (half-scale) layout dimensions and lumber (for example, 1 × 2s = 2 × 4s, 1 × 4s = 2 × 8s, etc.).

Discuss the "Think About It."

Hand out Job Sheet 27107-1. Under your supervision, have the trainees perform the tasks on the Job Sheet. This laboratory corresponds to Performance Task 1.

that in most cases, the ceiling joists would be in place and the rafter locations would have already been marked.

Step 1 Locate the first rafter flush with one end of the top plate.

Step 2 Measure the width of a rafter and mark and square a line the width of a rafter in from the end.

NOTE
The distance between the last two rafters may be less than 24", but not more than 24".

Step 3 Use a measuring tape to space off and mark the rafter locations every 24" all the way to the end. Square the lines.

Step 4 Repeat the process on the opposite top plate, starting from the same end as before.

Step 5 Cut the ridgeboard to length, allowing for a **barge rafter** at each end, if required.

Step 6 Place the ridgeboard on the top plate and mark it for correct position. Then measure and mark the rafter locations with an R on the ridgeboard.

To determine the number of rafters you need, count the marks on both sides of the ridge or top plate.

Calculating Rise and Pitch

Now that you are familiar with the concepts of pitch, rise per foot of run, and total rise, practice your skills by completing the following example problems.

Find the pitch for the following roofs:

Total Rise	Span	Pitch
8'	24'	
9'	36'	
6'	24'	
12'	36'	
8'	32'	

Find the total rise for the following roofs:

Pitch	Span	Rise per Foot of Run	Total Rise
½	24'	12"	
⅙	32'	4"	
⅓	24'	8"	
5/12	30'	10"	
¾	24'	18"	

Find the rise per foot of run for the following roofs:

Span	Total Rise	Rise per Foot of Run
18'	6'	
24'	8'	
28'	7'	
16'	4'	
32'	12'	

7.8 CARPENTRY FUNDAMENTALS LEVEL ONE ♦ TRAINEE GUIDE

Instructor's Notes:

3.2.2 Determining the Length of a Common Rafter

The following is an overview of the procedure for laying out the rafters and joists of a gable roof.

Here is an easy way of determining the required length of a rafter:

Step 1 Start by measuring the building span, then divide that in half to determine the run.

Step 2 Determine the rise. This can be done in either of the following ways:
- Calculate the total rise by multiplying the span by the pitch (e.g., 40' span × ¼ pitch = 10' rise).
- Look for it on the slope diagram on the roof plan as discussed previously.

Step 3 Divide the total rise (in inches) by the run (in feet) to obtain the rise per foot of run.

Step 4 Look up the required length on the rafter tables on the framing square.

For example, if the roof has a span of 20', the run would be 10'. Then, assuming that the blueprint shows the rise per foot of run to be 8", the correct rafter length would be 14.42" (14 5⁄12") per foot of run. Since you have 10' of run, the rafter length would be 144⅛" (*Figure 8*). If an overhang is used, the overhang length must be added to the rafter length.

Another way to calculate rafter length is to measure the distance between 8 on the framing square blade and 12 on the tongue, as shown in *Figure 5*.

Yet another method of determining the approximate length of a rafter is the framing square step-off method shown in *Figure 9*. In this procedure,

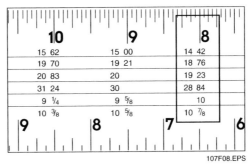

Figure 8 ♦ Determining rafter length with the framing square.

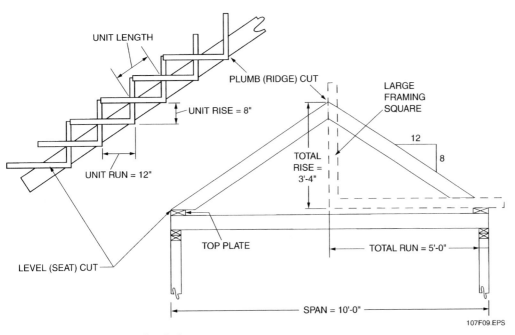

Figure 9 ♦ Framing square step-off method.

Describe the alternate methods of determining rafter length.

Provide an overview of the procedure for laying out a common rafter.

Show Transparency 12 (Figure 10).

Demonstrate the layout of a common rafter.

Discuss the "Think About It."

Assign reading of Sections 3.3.0–3.3.1 for the next class session.

the ridge plumb cut is determined, then the framing square is stepped once for every foot of run. The final step marks the plumb cut for the bird's mouth. This is not the preferred method because it is not as precise as other methods.

3.2.3 Laying Out a Common Rafter

The following is an overview of the procedure for laying out and cutting a common rafter:

Step 1 Start with a piece of lumber a little longer than the required length of the rafter, including the tail. If the lumber has a crown or bow, it should be at the top of the rafter. Lay the rafter on sawhorses with the crown (if any) at the top.

Step 2 Start by marking the ridge plumb cut using the framing square (*Figure 10*). Be sure to subtract half the thickness of the ridgeboard. Make the cut.

Step 3 Measure the length of the rafter from the plumb cut mark to the end (excluding the tail) and mark another plumb cut for the bird's mouth. Reposition the framing square and mark the bird's mouth seat.

Step 4 Make the end plumb cut, then cut out the bird's mouth. Cut the bird's mouth partway with a circular saw, then use a hand saw to finish the cuts.

Step 5 Use the first rafter as a template for marking the remaining rafters. As the rafters are cut, stand them against the building at the joist locations.

Calculating Common Rafter Lengths

Now that you are familiar with the method of arriving at the length of a rafter using the square, find the lengths of common rafters for the following spans:

Span	Run	Rise per Foot of Run	Rafter Length
26'	13'	6"	
24'	12'	4"	
28'	14'	8"	
32'	16'	12"	
30'	15'	7"	

NOTE
12 is a factor used to obtain a value in feet. Be sure to reduce or convert to the lowest terms.

Bird's Mouth

When making the cut for the bird's mouth, the depth should not exceed ⅓ the width of the rafter.

7.10 CARPENTRY FUNDAMENTALS LEVEL ONE ◆ TRAINEE GUIDE

Instructor's Notes:

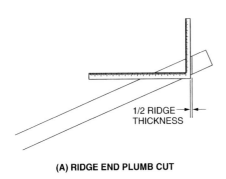

Figure 10 ♦ Marking the rafter cuts.

3.3.0 Erecting a Gable Roof

This section contains an overview of the procedure for erecting a gable roof. The layout and construction of hips and valleys is covered in *Appendix A*.

 WARNING!
Be sure to follow applicable fall protection procedures.

3.3.1 Installing Rafters

Rafters are installed using the following procedure.

 NOTE
It is a good idea to mark a 2 × 4 in advance with the total rise and use it as a guide for the height of the ridgeboard.

Step 1 Start by placing boards over the joists to walk on. Nail a rafter at each end of the ridgeboard, then lift the ridgeboard to a temporary position, secure it, and nail the bird's mouth of each rafter to the joists (*Figure 11*). Nail the rafters in pairs.

Step 2 On the opposite side, start by nailing the bird's mouth to the joist, then toenail the plumb cut into the ridgeboard. Once this is done, use a temporary brace to hold the ridgeboard in place while installing the remaining rafters. Remember to keep the ridgeboard straight and the rafters plumb.

Step 3 Run a line and trim the rafter tails.

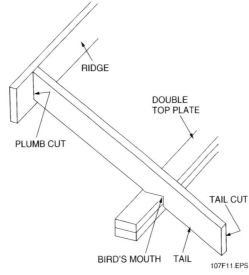

Figure 11 ♦ Rafter installation.

If the rafter span is long, additional support will be required. *Figure 12* shows the use of strongbacks, **purlins**, braces, and collar beams (collar ties) for this purpose. Two by six collar ties are installed at every second rafter. Two by four diagonal braces are notched into the purlins. Strongbacks are L-shaped members that run the length of the roof. They are used to straighten and strengthen the ceiling joists.

Discuss gable venting and framing.

Show Transparencies 15 and 16 (Figures 13 and 14).

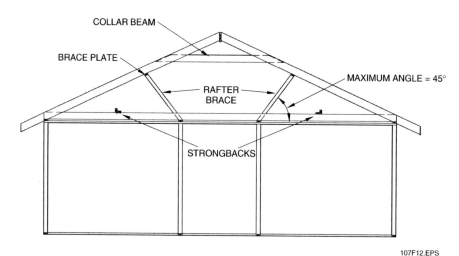

Figure 12 ♦ Bracing a long roof span.

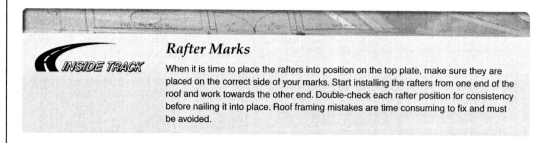

Rafter Marks

When it is time to place the rafters into position on the top plate, make sure they are placed on the correct side of your marks. Start installing the rafters from one end of the roof and work towards the other end. Double-check each rafter position for consistency before nailing it into place. Roof framing mistakes are time consuming to fix and must be avoided.

3.3.2 Framing the Gable Ends

Attics must be vented to allow heat that rises from the lower floors of the building to escape to the outdoors. There is also a need to vent moisture that accumulates in an attic due to condensation that occurs when rising heat from below meets the cooler air in the unheated attic space. Several methods are used to vent roofs. *Figure 13* shows two types of gable end vents.

Notice that the lengths of the studs decrease as they approach the sides. Each pair of studs must be measured and cut to fit.

Step 1 Start by plumbing down from the center of the rafter to the top plate, then mark and square a line on the top plate at this point.

Step 2 Lay out the header and sill for the vent opening by measuring 8" (in this case) on either side of the plumb line.

Step 3 Mark the stud locations above the wall studs, then stand a stud upright at the first position, plumb it, and mark the diagonal cut for the top of the stud.

Step 4 Measure and mark the next stud in the sequence. The difference in length between the first and second studs is the common difference, which can be applied to all remaining studs.

Step 5 Cut and install the studs as shown in *Figure 14*. Toenail or straight nail the gable stud to the double top plate. The notch should be as deep as the rafter.

Step 6 When studs are in place, cut and install the header and sill for the vent opening. Then lay out, cut, and install cripple studs above and below the vent using the same method as before.

7.12 CARPENTRY FUNDAMENTALS LEVEL ONE ♦ TRAINEE GUIDE

Instructor's Notes:

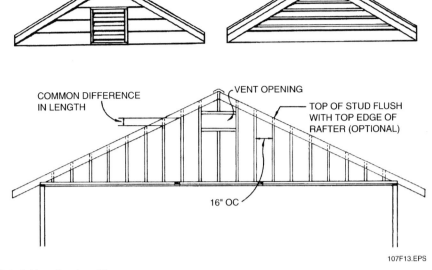

Figure 13 ◆ Gable end vents and frame.

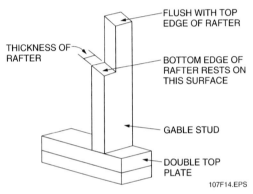

Figure 14 ◆ Gable stud.

3.3.3 Framing a Gable Overhang

If an extended gable overhang (rake) is required, the framing must be done before the roof framing is complete. *Figure 15* shows two methods of framing overhangs. In the view on the right, 2 × 4 **lookouts** are laid into notched rafters. In the view on the left, which can be used for a small overhang, a 2 × 4 barge rafter and short lookouts are used.

NOTE

Install a brace across the rafters to be cut. It will temporarily hold the framing in place until the headers are installed.

3.3.4 Framing an Opening in the Roof

It is sometimes necessary to make an opening in a roof for a chimney, skylight, or roof window (*Figure 16*). The following is a general procedure for framing such an opening.

Step 1 Lay out the opening on the floor beneath the opening, then use a plumb bob to transfer the layout to the roof. If you are framing a chimney, be sure to leave adequate clearance. If the opening is large, allow for double headers.

Step 2 Cut the rafters per the layout. Install the headers, then install a double trimmer rafter on either side of the opening, as shown in *Figure 16*.

Show Transparency 17 (Figure 15).

Demonstrate how to frame a portion of a gable end and gable overhang.

Hand out Job Sheet 27107-3. Under your supervision, have the trainees perform the tasks on the Job Sheet. This laboratory corresponds to Performance Task 2.

Assign reading of Section 3.3.4 for the next class session.

Provide an overview of the procedure for framing a roof opening.

Show Transparency 18 (Figure 16).

Demonstrate how to frame a roof opening. Point out that a brace should be nailed across all rafters to be cut, and to adjacent uncut rafters to hold the cut rafters in place until the headers are installed.

Hand out Job Sheet 27107-4. Under your supervision, have the trainees perform the tasks on the Job Sheet. This laboratory corresponds to Performance Task 2.

Assign reading of Section 4.0.0 for the next class session.

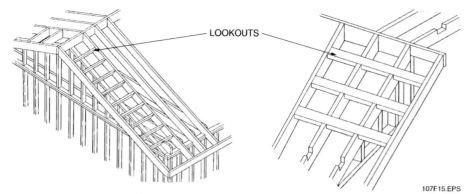

Figure 15 ♦ Framing a gable overhang.

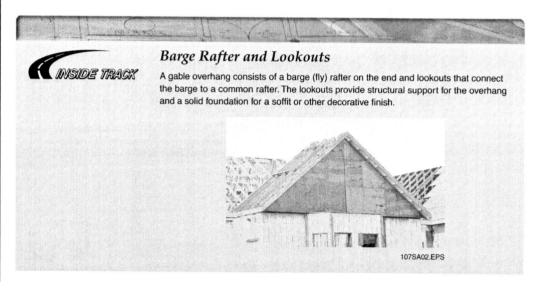

Barge Rafter and Lookouts

A gable overhang consists of a barge (fly) rafter on the end and lookouts that connect the barge to a common rafter. The lookouts provide structural support for the overhang and a solid foundation for a soffit or other decorative finish.

Framing Roof Openings

Before you cut an opening in a roof, check with the site engineer to make sure that you are proceeding according to specifications. If you don't double-check with the engineer, you may cut an incorrect opening and new rafters or trusses may be required. This type of error is very time-consuming and expensive to fix. Also, you should always be sure to check local fire codes for the proper clearance around a chimney and other roof openings. If the framing for the opening is too close to a chimney, it could cause a fire.

7.14 CARPENTRY FUNDAMENTALS LEVEL ONE ♦ TRAINEE GUIDE

Instructor's Notes:

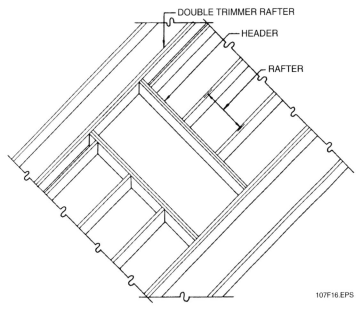

Figure 16 ♦ Framing an opening in a roof.

4.0.0 ♦ INSTALLING SHEATHING

The sheathing should be applied as soon as the roof framing is finished. The sheathing provides additional strength to the structure, and provides a base for the roofing material.

Some of the materials commonly used for sheathing are plywood, OSB, waferboard, shiplap, and common boards. When composition shingles are used, the sheathing must be solid. If wood shakes are used, the sheathing boards may be spaced. When solid sheathing is used, leave a ⅛" space between panels to allow for expansion.

The following is an overview of roof sheathing requirements using plywood or other 4 × 8 sheet material.

Step 1 Start by measuring up 48¼" from where the finish fascia will be installed. Chalk a line at that point, then lay the first sheet down and nail it. Install H-clips midway between the rafters or trusses before starting the next course (*Figure 17*). These clips eliminate the need for tongue-in-groove panels.

Step 2 Apply the remaining sheets. Stagger the panels by starting the next course with a half sheet. Let the edges extend over the hip, ridge, and gable end. Cut the extra sheathing off with a circular saw.

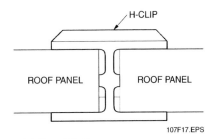

Figure 17 ♦ Use of H-clips.

Once the sheathing has been installed, an underlayment of asphalt-saturated felt or other specified material must be installed to keep moisture out until the shingles are laid. For roofs with a slope of 4" or more, 15-pound roofer's felt is commonly used.

Material such as coated sheets or heavy felt that could act as a vapor barrier should not be used. They can allow moisture to accumulate between the sheathing and the underlayment. The underlayment is applied horizontally with a 2" top lap and a 4" side lap, as shown in *Figure 18(A)*. A 6" lap should be used on each side of the centerline of hips and valleys. A metal drip edge is installed along the rakes and eaves to keep out wind-driven moisture.

Discuss the importance of sheathing.

Provide an overview of the procedure for applying roof sheathing.

Show Transparency 19 (Figure 17).

Emphasize the use of proper fall protection and footwear when applying roof sheathing.

Describe the application of various types of roofing underlayment. Point out that underlayment must be applied to dry sheathing as soon as possible.

Show Transparency 20 (Figure 18).

Demonstrate the application of roof sheathing. Point out that sheathing must be properly stacked on a roof, sheathing thickness may be specified by local codes, and sheathing joints must be staggered and spaced for expansion.

Hand out Job Sheet 27107-5. Under your supervision, have the trainees perform the tasks on the Job Sheet. This laboratory corresponds to Performance Task 2.

Rafter Tails

Prior to installing the sheathing, check the rafter tails to make sure they form a straight line. If you made accurate measurements and cuts, the rafters should all be the same length. If they are not, identify and measure the shortest rafter tail. Then mark the same distance on the rafter tails at each end of the roof span. Use these marks to snap a chalkline across the entire span of rafter tails, then trim each end to the same length. This will avoid an unsightly, crooked roof edge and provide a solid nailing base for the fascia.

Crane Delivery

If a crane is used to place a stack of sheathing on a roof, a special platform must be in place to provide a level surface. The platform must be placed over a loadbearing wall so that the weight of the sheathing doesn't cause structural damage to the framing.

Roof Openings

A 21-year-old apprentice was installing roofing on a building with six unguarded skylights. During a break, he sat down on one of the skylights. The plastic dome shattered under his weight and he fell to a concrete floor 16 feet below, suffering fatal head injuries.

The Bottom Line: Never sit or lean on a skylight. Always provide appropriate guarding and fall protection for work around skylights and other roof openings.

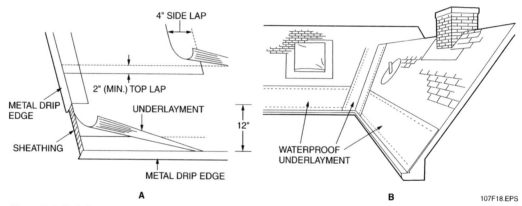

Figure 18 ♦ Underlayment installation.

In climates where snow accumulates, a waterproof underlayment, as shown in *Figure 18(B)*, should be used at roof edges and around chimneys, skylights, and vents. This underlayment has an adhesive backing that adheres to the sheathing. It protects against water damage that can result from melting ice and snow that backs up under the shingles.

Instructor's Notes:

Sheathing Safety

When installing sheathing on an inclined roof, always be sure to use a toeboard and safety harness. You can gain additional footing by wearing skid-resistant shoes. Falling from a roof can result in serious injury or death. Taking the time to follow the proper precautions will prevent you from slipping or falling. The picture shows a safety harness anchor and toeboard.

Assign reading of Sections 5.0.0–5.1.0 for the next class session.

107SA03.EPS

Sheathing

If the trusses or rafters are on 24" centers, use ⅝" sheathing. With 16" centers, ½" sheathing may be used. This may vary depending on the local codes. Before starting construction, be sure to check the sheathing and fastener code requirements.

Felt Installation

The felt underlayment should be applied to the installed sheathing as soon as possible. It is important that both the sheathing and felt be dry and smooth at the time of installation. If the roof is damp or wet, wait a couple of days for it to dry completely before installing the felt. Moisture will cause long-term damage to both the sheathing and underlayment.

Discuss the scales shown on a speed square.

Show Transparency 21 (Figure 20).

Demonstrate common rafter layout using a speed square.

Under your supervision, have the trainees perform common rafter layout using a speed square. This laboratory corresponds to Performance Task 1.

5.0.0 ♦ RAFTER LAYOUT USING A SPEED SQUARE

The speed square, also known as a super square or quick square, is a combination tool consisting of a protractor, try miter, and framing square. A standard speed square is a 6" triangular tool with a large outer triangle and a smaller inner triangle (*Figure 19*). The large triangle has a 6" scale on one edge, a full 90-degree scale on another edge, and a T-bar on the third edge. The inner triangle has a 2" square on one side. The speed square is the same on both sides. A 12" speed square is used for stair layout.

To use a square, you need to know the pitch of the roof. When you buy a speed square, it usually comes with an instruction booklet. This booklet normally contains (among other information) tables that show the required rafter length for every pitch.

5.1.0 Procedure for Laying Out Common Rafters

Step 1 Choose a piece of lumber that is slightly longer than that needed for the rafter. Remember, the eaves or overhang are not included in the measurements found in the information booklet. The length of the overhang is added after you have determined the length of the rafter. Place the lumber on a pair of sawhorses with the top edge of the rafter stock facing away from you. Be sure the crown is facing away from you (the crown, if any, will be the top edge of the rafter). The rafter will be cut for a building that is 24' wide with a 4" rise per foot of run. Therefore, the length of the rafter from the center of the ridge to the seat cut (bird's mouth) will be 12'-7¾". Any overhang must be added to this.

107F19.EPS

Figure 19 ♦ Speed square.

NOTE

When working with the speed square, there is no measuring line. The length is measured on the top edge of the rafter (this will be shown later in this module). Assume that the right side of the lumber will be the top of the rafter. Place the speed square against the top edge of the lumber and set the square with the 4 on the common scale.

Step 2 Draw a line along the edge of the speed square, as shown in *Figure 20(A)*. From the mark just made, measure with a steel tape along the top edge of the rafter the length required for the total length of the rafter (12'-7¾") and make a mark, as shown in *Figure 20(B)*.

Step 3 Place the speed square with the pivot point against that mark, and move the square so the 4 on the common scale is even with the edge. Draw a line along the edge of the square, as shown in *Figure 20(C)*. This will establish the vertical seat cut (bird's mouth) or plumb cut.

Step 4 To lay out the horizontal cut, reverse the speed square so that the short line at the edge of the square is even with the line previously drawn. Place the line so that the edge of the square is even with the lower edge of the rafter, as shown in *Figure 20(D)*.

Step 5 Draw a second line at a right angle to the plumb line. This line will establish the completed seat cut (bird's mouth), as shown in *Figure 20(E)*.

Step 6 The top and bottom cuts have been established for the total length of the rafter. If a ridgeboard is being used, be sure to deduct half of the thickness of the ridge from the top plumb cut. If any overhang is needed, it should be measured at right angles to the plumb cut of the seat cut (bird's mouth), as shown in *Figure 20(F)*.

Step 7 Once the measurement has been established, place the square against the top edge of the rafter, lining up the 4 on the common scale and marking it, as shown in *Figure 20(G)*.

Step 8 If a bottom or vertical cut is required, follow the procedure for cutting the horizontal cut of the seat cut (bird's mouth). In following the procedure described above,

7.18 CARPENTRY FUNDAMENTALS LEVEL ONE ♦ TRAINEE GUIDE

Instructor's Notes:

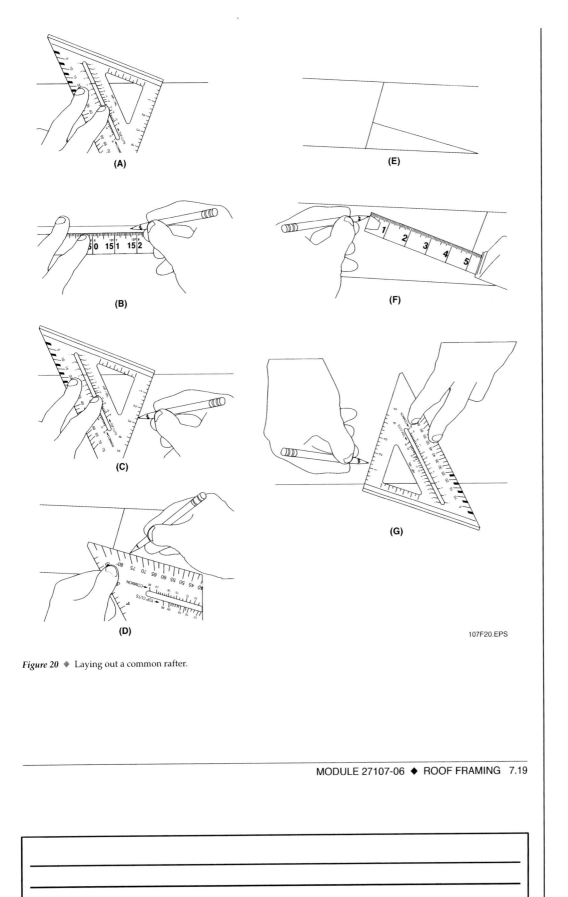

Figure 20 ◆ Laying out a common rafter.

MODULE 27107-06 ◆ ROOF FRAMING 7.19

Assign reading of Sections 6.0.0–6.2.0 for the next class session.

Discuss the advantages of truss construction.

Show Transparencies 22 and 23 (Figures 21 and 22).

a pattern for a common rafter has been established. Two pieces of scrap material (1 × 2 or 1 × 4) should be nailed to the rafter, one about 6" from the top or plumb cut and one at the seat cut (bird's mouth).

6.0.0 ◆ TRUSS CONSTRUCTION

In most cases, it is much faster and more economical to use prefabricated trusses in place of rafters and joists. Even if a truss costs more to buy than the comparable framing lumber (not always the case), it takes significantly less labor than stick framing. Another advantage is that a truss will span a greater distance without a bearing wall than stick framing. Just about any type of roof can be framed with trusses.

There are special terms used to identify the members of a truss (see *Figure 21*).

A truss is a framed or jointed structure. It is designed so that when a load is applied at any intersection, the stress in any member is in the direction of its length. *Figure 22* shows some of the many kinds of trusses. Even though some trusses

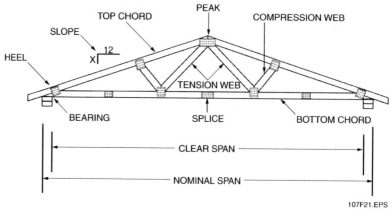

Figure 21 ◆ Components of a truss.

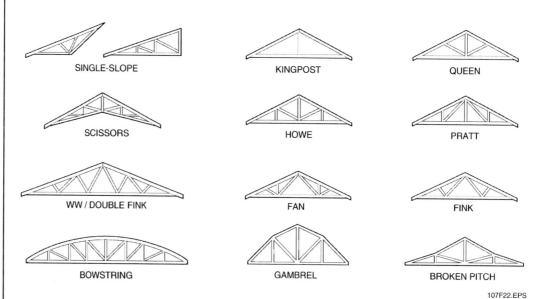

Figure 22 ◆ Types of trusses.

7.20 CARPENTRY FUNDAMENTALS LEVEL ONE ◆ TRAINEE GUIDE

Instructor's Notes:

look nearly identical, there is some variation in the interior (web) pattern. Each web pattern distributes weight and stress a little differently, so different web patterns are used to deal with different loads and spans. The decision of which truss to use for a particular application will be made by the architect or engineer and will be shown on the blueprints. Do not substitute or modify trusses on site, because it could affect their weight and stress-bearing capabilities. Also, be careful how you handle trusses. They are more delicate than stick lumber and cannot be thrown around or stored in a way that applies uneven stress to them.

Trusses are stored or carried on trucks either lying on their sides or upright in cradles and protected from water. It is a good idea to use a crane to lift the trusses from the truck, and unless the trusses are small and light, it is usually recommended that a crane be used to lift them into place on the building frame. *Figure 23* shows examples of erecting methods used for trusses. Note the tag lines. These lines are held by someone on the ground to stabilize the truss while it is being lifted.

A single lift line can be used with trusses with a span of less than 20'. From 20' to 40', two chokers are needed. If the span exceeds 40', a spreader bar is required.

 WARNING!
Installing trusses can be extremely dangerous. It is very important to follow the manufacturer's instructions for bracing and to follow all applicable safety procedures.

6.1.0 Truss Installation

Before installing roof trusses, refer to the framing plans for the proper locations. *Figure 24* is an example of a truss placement diagram. If a truss is damaged before erection, obtain a replacement or instructions from a qualified individual to repair the truss. Remember, repairs made on the ground are usually a lot better and are always easier. Never alter any part of a truss without consulting the job superintendent, the architect/engineer, or the manufacturer of the roof truss. Cutting, drilling, or notching any member without the proper approval could destroy the structural integrity of the truss and void any warranties given by the truss manufacturer.

Girders are trusses that carry other trusses or a relatively large area of roof framing. A common truss or even a double common truss will rarely serve as a girder. If there is a question about

Discuss the importance of proper truss storage, handling, and rigging.

Emphasize the safety precautions necessary when working with trusses.

Provide an overview of truss installation and bracing.

Show Transparency 24 (Figure 23).

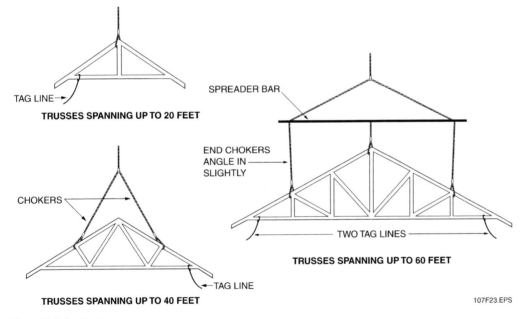

Figure 23 ◆ Erecting trusses.

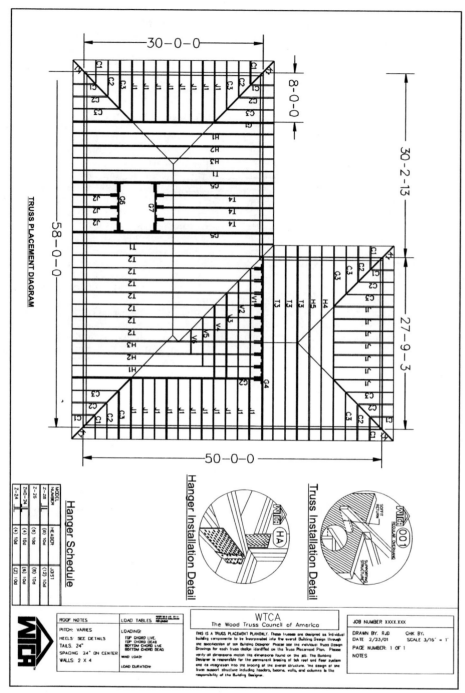

Figure 24 ◆ Example of a truss placement diagram.

Truss Storage

Trusses should only be stored on site for a short length of time. They are very large and take up a lot of storage area. In addition, they are more likely to get wet when stored on site. When a truss is saturated with moisture, excess shrinkage will occur, causing structural damage to the webs and chords. If on-site storage of trusses is unavoidable, make sure they are well covered with a waterproof tarp and raised off the ground on pallets. During the planning phase of a project, trusses should be figured into the schedule so that they can be delivered and erected in a timely manner.

Truss Rigging

Trusses are very large and can be difficult to position. Proper rigging technique is important to avoid seriously injuring workers and damaging equipment. The rigging setup varies depending on the type of truss being used. Enlist the help of experienced carpenters to ensure that the truss is securely fastened and balanced properly. Never stand directly below a truss that has been hoisted into the air.

whether a girder is needed, your job superintendent should confer with the truss manufacturer. Double girders are commonplace. Triple girders are sometimes required to ensure the proper load-carrying capacity. Always be sure that multiple-member girders are properly laminated together. Spacing the trusses should be done in accordance with the truss design. In some cases, very small deviations from the proper spacing can create a big problem. Always seek the advice of the job superintendent whenever you need to alter any spacing. Make certain that the proper temporary bracing is installed as the trusses are being set.

 WARNING!
Never leave a job at night until all appropriate bracing is in place and secured well.

Light trusses (under 30' wide, for example) can be installed by having them lifted up and anchored to the top plate, then pushed into place with Y-shaped poles by crew members on the ground. Larger trusses require a crane.

When the bottom chord is in place, the truss is secured to the top plate. This can be done by toe-nailing with 10d nails. In some cases, however, metal tiedowns are required (*Figure 25*). An example is a location where high winds occur.

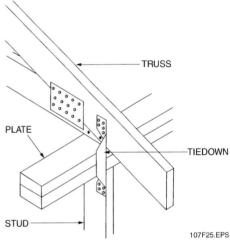

Figure 25 ♦ Use of a tiedown to secure a truss.

Explain that some trusses require permanent bracing.

Show Transparencies 26 and 27 (Figures 26 and 29).

Demonstrate how to install and brace several trusses.

Hand out Job Sheet 27107-6. Under your supervision, have the trainees perform the tasks on the Job Sheet. This laboratory corresponds to Performance Task 3.

Assign reading of Sections 7.0.0–7.1.0 for the next class session.

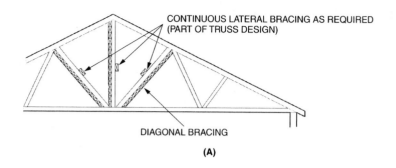

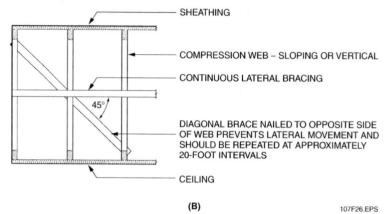

Figure 26 ◆ Example of permanent bracing specification.

6.2.0 Bracing of Roof Trusses

In some circumstances, it may be necessary to add permanent bracing to the trusses. This requirement would be established by the architect and would appear on the blueprints. *Figure 26* shows an example of such a requirement.

Temporary bracing of trusses is required until the sheathing is in place. *Figure 27* shows an example of lateral bracing across the tops of the trusses. Gable ends are braced from the ground using lengths of 2×4 or similar lumber anchored to stakes driven into the ground.

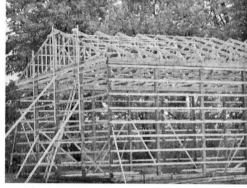

Figure 27 ◆ Example of temporary bracing of trusses.

7.24 CARPENTRY FUNDAMENTALS LEVEL ONE ◆ TRAINEE GUIDE

Instructor's Notes:

Inside Track

Temporary Truss Bracing

The leading cause of truss collapses during construction is insufficient temporary bracing. One crucial aspect of bracing that is often overlooked is diagonal bracing (shown in red in the figure below). The truss industry provides guides for safe and efficient handling, installation, and bracing of metal plate connected wood trusses called the Building Component Safety Information (BCSI) series. These documents cover recommendations for correct hoisting and bracing procedures based on the span of the truss, installation tolerances, and limits on construction loading. Following BCSI recommendations will minimize truss damage during construction, lead to better long-term performance of the truss system, and create a safer work site.

Text courtesy of the Wood Truss Council of America (WTCA). For more information, visit www.woodtruss.com.

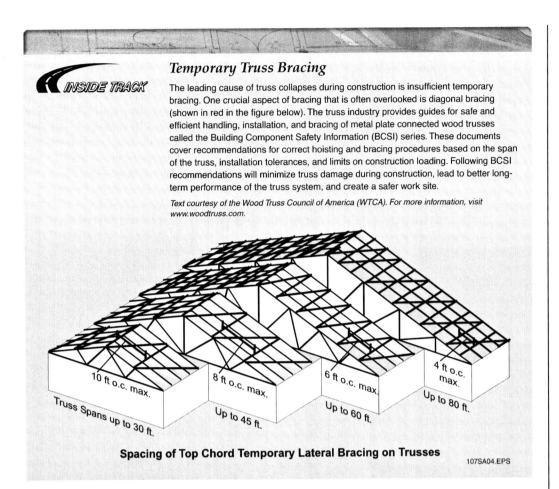

Spacing of Top Chord Temporary Lateral Bracing on Trusses

7.0.0 ♦ DETERMINING QUANTITIES OF MATERIALS

Estimating the material you will need for a roof depends first of all on the type of roof you are planning to construct, the size of the roof, the spacing of the framing members, and the load characteristics. Local building codes will usually dictate these factors and they will be disclosed on the building plans. Lumber for conventional framing may be from 2×4 to 2×10, depending on the span and the load. For example, 2×4 framing on 16" centers might support a 9' span. By comparison, 2×10 framing on 16" centers would support a span of more than 25'.

Explain how to estimate the materials required for an example roof.

Provide a number of different roof configurations and have the trainees estimate the materials required.

Assign reading of Sections 8.0.0–9.0.0 for the next class session.

Describe shed and gable dormers. Point out that shed dormers sometimes extend almost entirely across the length of a roof.

Show Transparency 27 (Figure 29).

7.1.0 Determine Materials Needed for a Gable Roof

Determining the rafter material:

Step 1 To determine how much lumber you will need for rafters on a gable roof, first determine the length of each common rafter (including the overhang) using the framing square or another method.

Step 2 Figure out the number of rafters based on the spacing (16", 24", etc.). Remember that you will need one rafter for each gable end. You will also need barge rafters in the gable overhang. Note that these are usually one size smaller than the common rafters.

Step 3 Multiply the result by two to account for the two sides of the ridge.

Step 4 Convert the result to board feet.

Estimating the ridgeboard:

The ridgeboard is usually one dimension thicker than the rafters.

Step 1 Determine the length of the plate on one side of the structure and add as needed to account for gable overhang.

Step 2 Convert the result to board feet.

Estimating the sheathing:

Step 1 Multiply the length of the roof including overhangs by the length of a common rafter. This yields half the area of the roof.

Step 2 Divide the roof area by 32 (the number of square feet in a 4 × 8 sheet) to get the approximate number of sheets of sheathing you will need. Round up if you get a fractional number.

Step 3 Multiply by 2 to obtain the number of sheets needed for the full roof area.

8.0.0 ◆ DORMERS

A dormer is a framed structure that projects out from a sloped roof. A dormer provides additional space and is often used in a Cape Cod–style home, which is a single-story dwelling in which the attic is often used for sleeping rooms. A shed dormer (*Figure 28*) is a good way to obtain a large amount of additional living space. If it is added to the rear of the house, it can be done without affecting the appearance of the house from the front.

A gable dormer (*Figure 29*) serves as an attractive addition to a house, in addition to providing a little extra space as well as some light and ventilation. They are sometimes used over garages to provide a small living area or studio.

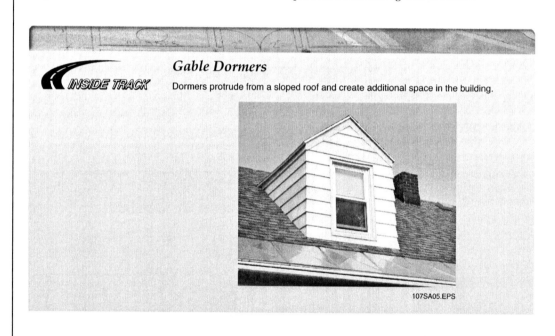

INSIDE TRACK

Gable Dormers

Dormers protrude from a sloped roof and create additional space in the building.

107SA05.EPS

Instructor's Notes:

Figure 28 ◆ Shed dormer.

9.0.0 ◆ PLANK-AND-BEAM FRAMING

Plank-and-beam framing, also known as post-and-beam framing (*Figure 30*), employs much sturdier framing members than common framing. It is often used in framing roofs for luxury residences, churches, and lodges, as well as other public buildings where a striking architectural effect is desired.

Because the beams used in this type of construction are very sturdy, wider spacing may be used. Vertical supports are typically spaced 48" OC, as compared with 16" OC used in conventional framing. When plank-and-beam framing is used for a roof, the beams and planking can be finished and left exposed. The underside of the planking takes the place of an installed ceiling.

In light construction, solid posts or beams such as 4 × 4s are used. In heavy construction, laminated beams made of glulam, LVL, and PSL are used.

Discuss plank-and-beam roof construction.

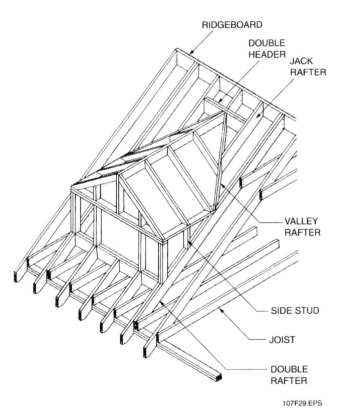

Figure 29 ◆ Gable dormer framing.

Assign reading of Section 10.0.0 for the next class session.

Discuss metal roof framing.

Figure 30 ◆ Example of post-and-beam construction.

10.0.0 ◆ METAL ROOF FRAMING

When steel framing is used, the roof framing is often done with prefabricated metal trusses. If the trusses are fabricated on site, the chords and webs are cut with a portable electric saw and placed into a jig. They are then welded and/or connected with self-tapping screws.

Trusses that look like wood are used in some applications (*Figure 31*). Commercial projects may use open web joists, known as bar joists, to support the roof (*Figure 32*).

Figure 31 ◆ Metal roof trusses.

Figure 32 ◆ Bar joists.

In post-and-beam framing, plank subfloors or roofs are usually of 2" nominal thickness, supported on beams spaced up to 8' apart. The ends of the beams are supported on posts or piers. Wall spaces between posts are provided with supplementary framing as required for attachment of exterior and interior finishes. This additional framing also provides lateral bracing for the building.

If local building codes allow end joints in the planks to fall between supports, planks of random lengths may be used and the beam spacing adjusted to fit the house dimensions. Windows and doors are normally located between posts in the exterior walls, eliminating the need for headers over the openings. The wide spacing between posts permits ample room for large glass areas.

A combination of conventional framing with post-and-beam framing is sometimes used where the two adjoin each other. Where a post-and-beam floor or roof is supported on a stud wall, a post is usually placed under the end of the beam to carry a conventional load. A conventional roof can be used with post-and-beam construction by installing a header between posts to carry the load from the rafters to the posts.

Summary

The correct layout and framing of a roof requires patience and skill. If the measurement, cutting, and installation work are not done carefully and precisely, the end result will never look right. Fortunately, there are many tools and reference tables that help to simplify the process. The important thing is to be careful and precise with the layout and cutting of the first rafter of each type. This is the rafter that is used as a pattern for the others.

Summarize the major concepts presented in the module.

Notes

Have trainees complete the Review Questions, and go over the answers prior to administering the Module Examination.

Review Questions

1. The type of roof that has four sides running toward the center of the building is the _____ roof.
   a. gable
   b. hip
   c. shed
   d. gable and valley

2. The letter A in *Figure 1* is pointing to a _____.
   a. hip jack
   b. valley jack
   c. hip rafter
   d. cripple jack

3. The letter B in *Figure 1* is pointing to the _____.
   a. main ridgeboard
   b. projection ridgeboard
   c. top plate
   d. collar beam

4. The letter D in *Figure 1* is pointing to a _____.
   a. valley cripple
   b. hip jack
   c. valley rafter
   d. valley jack

5. The letter E in *Figure 1* is pointing to the _____.
   a. top plate
   b. projection ridgeboard
   c. main ridgeboard
   d. gable stud

6. The letter G in *Figure 1* is pointing to a _____.
   a. gable end
   b. common rafter
   c. collar beam
   d. gable stud

7. The letter M in *Figure 1* is pointing to a _____ rafter.
   a. hip jack
   b. valley jack
   c. cripple jack
   d. hip

8. The roof run measurement is usually equal to _____.
   a. twice the span
   b. half the span
   c. the distance from the top plate to the ridgeboard
   d. the length of a common rafter

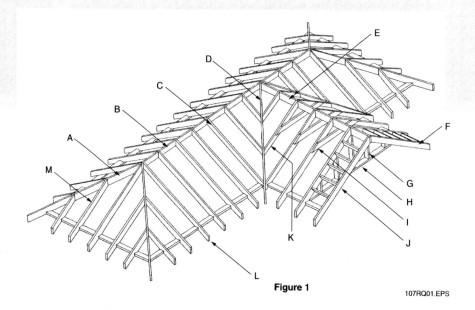

Figure 1

107RQ01.EPS

7.30 CARPENTRY FUNDAMENTALS LEVEL ONE ♦ TRAINEE GUIDE

Instructor's Notes:

7.30 CARPENTRY FUNDAMENTALS LEVEL ONE ♦ ANNOTATED INSTRUCTOR'S GUIDE

Review Questions

9. The horizontal distance from the outside of the top plate to the center of the ridgeboard is called the _____.
   a. slope
   b. span
   c. pitch
   d. run

Refer to *Figure 2* to answer Questions 10 and 11.

10. The length of a common rafter for a 20' wide building with a 10" rise per foot of run is _____.
    a. 19.7'
    b. 20'
    c. 13'
    d. 14'-8"

11. The rise per foot of run of a roof with a pitch of ½ is _____.
    a. 6"
    b. 12"
    c. 18"
    d. 24"

12. A roof with a pitch of ½ has a(n) _____ rise per foot of run.
    a. 6"
    b. 8"
    c. 10"
    d. 12"

13. When laying out a common rafter, it is necessary to deduct _____ in order to arrive at the final measurement.
    a. half the thickness of the ridgeboard
    b. half the thickness of the rafter
    c. half the thickness of the top plate
    d. nothing

14. An L-shaped brace used to strengthen and straighten ceiling joists is known as a _____.
    a. purlin
    b. strongback
    c. collar tie
    d. ridgeboard

15. Double trimmers are used when framing _____.
    a. roof openings
    b. gable ends
    c. overhangs
    d. valleys

16. When placing underlayment over roof sheathing, it is best to use the heaviest felt you can find.
    a. True
    b. False

17. Immediately after nailing the plywood sheathing on a roof frame, you should install _____.
    a. a vapor barrier
    b. a felt underlayment
    c. a drip edge
    d. plywood clips

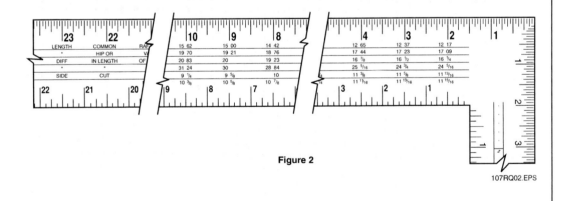

Figure 2

Review Questions

18. When you are using a speed square, the length of the rafter is determined by _____.
    a. stepping the square
    b. reading it directly from the square
    c. using rafter tables in the instruction manual
    d. multiplying the run by the rise

19. When you are laying out rafters with a speed square, you don't have to add the length of the overhang because it is included in the rafter tables provided with the square.
    a. True
    b. False

Refer to the following illustration to answer Questions 21 and 22.

20. Item 6 in *Figure 3* is pointing to the _____.
    a. center chord
    b. slope
    c. tension web
    d. compression web

21. Item 4 in *Figure 3* is pointing to the _____.
    a. top chord
    b. clear span
    c. peak
    d. slope

22. When rigging a truss with a span of 45', you should use a _____.
    a. spreader bar
    b. single choker
    c. double choker
    d. triple choker

23. The number of common rafters needed to frame a 20' long gable roof for a house framed on 16" centers with no overhang is _____.
    a. 15
    b. 32
    c. 30
    d. 16

24. The number of sheets of plywood sheathing needed for a house 30' long with a span of 20' and ¼ pitch is _____. (Assume no overhang.)
    a. 14
    b. 18
    c. 26
    d. 28

25. Vertical supports for plan-and-beam framing are typically spaced _____ on center.
    a. 16"
    b. 24"
    c. 36"
    d. 48"

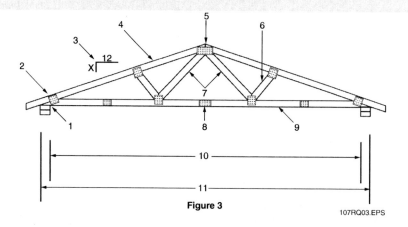

Figure 3

7.32 CARPENTRY FUNDAMENTALS LEVEL ONE ♦ TRAINEE GUIDE

Trade Terms Quiz

1. A _____ roof has two slopes that meet at the center to form a point.
2. The board nailed to the rafter tails in order to space the rafters is the _____.
3. A gable with an overhang uses a _____ to support the extension and provide a nailing surface for a decorative edge.
4. A _____ is a horizontal support member used to strengthen a long rafter span.
5. An overhang is framed using short lengths of boards known as a _____.

Trade Terms

Barge rafter
False fascia
Gable
Lookout
Purlin

Have trainees complete the Trade Terms Quiz, and go over the answers.

Administer the Module Examination. Record the results on Craft Training Report Form 200, and submit the form to the Training Program Sponsor.

Administer the Performance Tests, and fill out Performance Profile Sheets for each trainee. If desired, trainee proficiency noted during laboratory sessions may be used to complete the Performance Test. Record the results on Craft Training Report Form 200, and submit the results to the Training Program Sponsor.

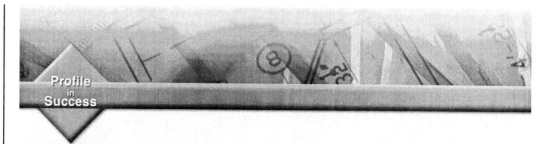

Profile in Success

Alan Van Holten
Gold Medal Winner, Associated Builders and Contractors National Championships

Before becoming a carpenter, Alan was a welder for 12 years. Although he enjoyed welding, he believes that carpentry offers a wider range of job opportunities. He won a gold medal at both the regional and national ABC championships. He is currently working as a foreman for a building company that specializes in tilt-up construction.

How did you become interested in carpentry?
I learned to weld in high school when I was 13. I enjoyed it and worked as a welder for 12 years. But I found that I was often doing the same types of work and I wanted a change. I went to work as a laborer for a building company. They sent me to school for carpentry. I really enjoyed carpentry because you are always doing different kinds of jobs. Each day has new challenges and new things to figure out.

What positions have you held and how did they help you get to where you are now?
In carpentry I started as a laborer, then I was promoted to lead man. Then I became a journeyman and a foreman. Each position was a stepping-stone to achieving my goals. Eventually, I want to become a project manager and superintendent, but I have a lot to learn before I can achieve that goal.

Did you enjoy the Associated Builders and Contractors National Championships?
I think that Associated Builders and Contractors is a good organization. They have been very good to me. The competition itself was pretty intense but it was also really gratifying. The first day is a written test. On the second day, you have six hours to complete a project. At first I was pretty nervous. When you start, you are in a big room and have no idea what you will be doing. Then they give you the drawings. Once I had the drawings, I relaxed and did what I had to do. It was like going to work. Just figure out what to do with the materials you have on hand and build the project.

Later on in my career, I was a judge for the SkillsUSA Championship. I have to say that it was pretty nice being on the other side of the competition.

What are some of the things you do in your job?
As a foreman I do a lot of paperwork. We do tilt-up construction for commercial buildings, retail stores, and office buildings. Everything is done on site. We build the forms for the concrete to make the panels. Then we assemble the building.

What do you think it takes to be a success in your trade?
Frankly, it takes lots of hard work. You need to have a desire to achieve success and grow as a person. You have to want to become the best at what you do. That's what drives me. I always want to be the best at whatever I am doing.

What do you like about the work you do?
I like that I get to do many different things. It's not the same tasks every day. There is much more variety in carpentry than there was in welding.

What advice would you give someone just starting out?
Stay focused on where you want to be in your life. There will be rough spots and you will stumble. But don't give up in the hard times. Keep striving for what you want.

Work hard for your company. The harder you work, the more a company will appreciate you and the more opportunities you will have. Foremen notice the people who work really hard every day.

Instructor's Notes:

Trade Terms Introduced in This Module

Barge rafter: A gable end roof member that extends beyond the gable to support a decorative end piece. Also known as a fly rafter.

False fascia: The board that is attached to the tails of the rafters to straighten and space the rafters and provide a nailer for the fascia. Also called sub fascia and rough fascia.

Gable: The triangular wall enclosed by the sloping ends of a ridged roof.

Lookout: A structural member used to frame an overhang.

Purlin: A horizontal roof support member parallel to the plate and installed between the plate and the ridgeboard.

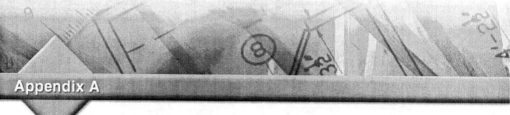

Appendix A

Laying Out and Erecting Hips and Valleys

An intersecting roof contains two or more sections sloping in different directions. Examples are the connection of two gable sections or a gable and hip combination such as that shown in *Figure A-1*. *Figure A-2* shows an overhead view of the same layout.

A valley occurs wherever two gable or hip roof sections intersect. Valley rafters run at a 45-degree angle to the outside walls of the building.

The material that follows provides an overview of the procedures for laying out hip and valley sections and the various types of rafters used in framing these sections. The first step, as always, is to lay out the rafter locations on the top plate (*Figure A-3*). The layout of the common rafters for a hip roof is the same as that for a gable roof. The next step would be to lay out, cut, and install the common rafters and the main and projection ridgeboards, as described earlier in this module. At that point, you are ready to lay out the hip or valley section.

HIP RAFTERS

A hip rafter is the diagonal of a square formed by the walls and two common rafters (*Figure A-2*). Because they travel on a diagonal to reach the ridge, the hip rafters are longer than the common rafters. The unit run is 17" (16.97" rounded up), which is the length of the diagonal of a 12" square. You can see this on the top line of the rafter tables on the framing square. There are two hip rafters in every hip section.

For every hip roof of equal pitch, for every foot of run of common rafter, the hip rafter has a run of 17". This is a very important fact to remember. It can be said that the run of a hip rafter is 17 divided by 12 times the run of a common rafter. The total rise of the hip rafter is the same as that of a common rafter. For example, if a common rafter has an 8" rise per foot of run, the hip rafter would also have a rise of 8" per 17" run. Therefore, the rise of a hip rafter would be the same as a common rafter at any given corresponding point.

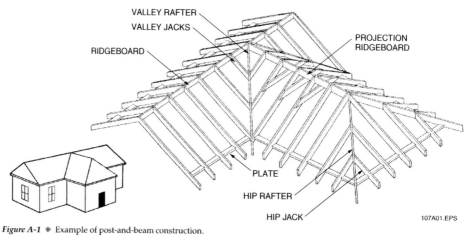

Figure A-1 ◆ Example of post-and-beam construction.

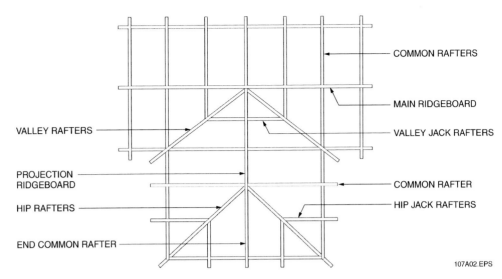

Figure A-2 ◆ Overhead view of a roof layout.

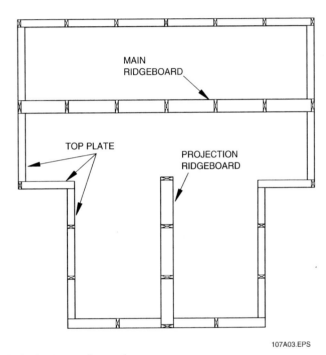

Figure A-3 ◆ Layout of rafter locations on the top plate.

To find the length of a hip rafter, first find the rise per foot of run of the roof, then locate that specific number on the inch scale line of the framing square. Find the corresponding numbers on the second line of the framing square. For example, assume a roof has a rise per foot of run of 8" (8 in 12). The span of the building is 10'. The second line under the 8 on the blade of the rafter (framing) square is 18.76. You have now established that the unit length of the hip rafter is 18.76" or 18¾" for every foot of the common rafter. This unit length must be multiplied by the run of the structure, which is 5'. The sum of the two factors multiplied together must then be divided to find the length of the hip rafter. For example, 18.76" × 5 = 93.8". 93.8" ÷ 12 = 7.81" or 7'-9¾". Therefore, the total length of the hip rafter would be 7'-9¾".

No matter what the rise per foot of run is, the length of the common and hip rafters is determined in the same fashion. The method of laying out and marking a hip rafter is very similar to the layout of a common rafter, with one exception. A common rafter is laid out using 12 on the tongue and the rise per foot of run on the blade. A hip rafter is laid out by using 17 on the blade and the rise per foot of run on the tongue (*Figure A-4*).

The basic procedure for laying out a hip rafter is as follows:

Step 1 Determine the length of the hip rafters using the framing square (*Figure A-4*) or rafter tables and add the overhang.

Step 2 Mark the plumb cuts on the hip rafters. The hip rafters must be shortened by half the diagonal thickness of the ridgeboard. Also, they must be cut on two sides to fit snugly between the common rafter and the ridgeboard (*Figure A-5*). These cuts are known as side cuts or cheek cuts.

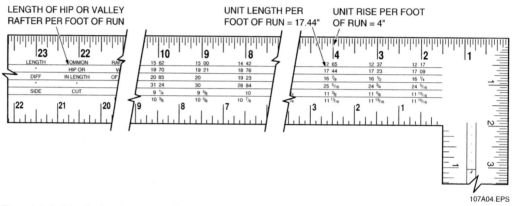

Figure A-4 ♦ Using the framing square to determine the length of a hip rafter.

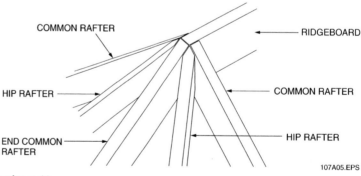

Figure A-5 ♦ Hip rafter position.

Step 3 The bird's mouth plumb and seat cuts are determined in the same way as for a common rafter. However, the seat cut is dropped half the thickness of the rafter to align it with the top plane of the common rafter (distance A, *Figure A-6*). This drop is necessary because the corners of the rafters would otherwise be higher than the plane on which the roof surface is laid. Another way to accomplish this is to chamfer the top edges of the rafter. This procedure is known as backing. Most carpenters use the dropping method because it is faster.

VALLEY LAYOUT

Each valley requires a valley rafter and some number of valley jack rafters (*Figure A-7*). The layout of a valley rafter is basically the same as that of the hip rafter, with 17" used for the unit run. The only difference in layout between the hip and valley rafters is in the seat and tail cuts.

For a valley rafter, the bird's mouth plumb cut must be angled to allow the rafter to drop down into the inside corner of the building (*Figure A-8*). In addition, the tail end cuts must be made so that corner made by the valley will line up with the rest of the roof overhang. Like the hip rafter, the valley rafter must be aligned with the plane of the common rafter. Cheek cuts are also required on valley rafters to allow them to fit between the two ridgeboards.

The layout of valley jacks is the same as that for hip jacks, with the exception that the valley jacks usually run in the opposite direction; i.e., toward the ridge.

JACK RAFTER LAYOUT

Hip jack rafters run from the top plate to the hip rafters. Valley jacks run from the valley rafter to the ridge (see *Figure A-1*). Notice that layout of valley jack rafters usually starts at the building line and moves toward the ridge. As with gable

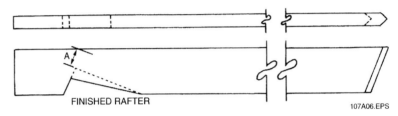

Figure A-6 ◆ Dropped bird's mouth cut.

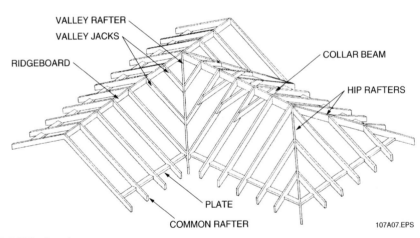

Figure A-7 ◆ Valley layout.

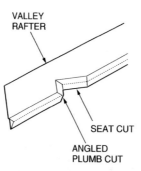

Figure A-8 ♦ Valley rafter layout.

end studs, there is a common difference from one jack rafter to the next (*Figure A-9*).

Here is an overview of the jack rafter layout process.

Step 1 The third and fourth lines of the framing square have the information you need to determine the lengths of jack rafters for 16" and 24" OC construction, respectively. The number you read from the framing square is the difference in calculated length between the common rafter and the jack rafter (*Figure A-10*). The longest jack rafter is referred to on the plans as the #1 jack rafter. Jack rafters must be cut and installed in pairs to prevent the hip or valley from bowing.

Step 2 Use the common rafter to mark the bird's mouth.

Step 3 To lay out the additional jack rafters, subtract the common difference from the last jack rafter you laid out. For each jack rafter with a cheek cut on one side, there must be one of equal length with the cheek cut on the opposite side.

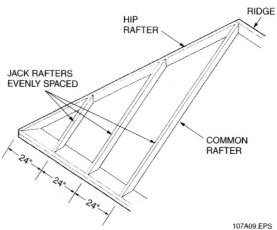

Figure A-9 ♦ Jack rafter locations.

7.40 CARPENTRY FUNDAMENTALS LEVEL ONE ♦ TRAINEE GUIDE

Instructor's Notes:

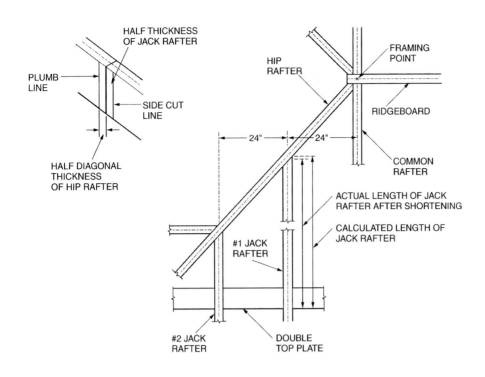
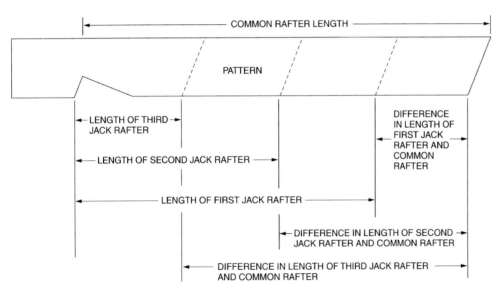

Figure A-10 ◆ Hip and jack rafter layout.

MODULE 27107-06 ◆ ROOF FRAMING 7.41

Calculating Jack Rafter Lengths

Use the rafter tables to find the lengths of the following jack rafters at 16" OC when a common rafter is 14'-6" long.

Rise per Foot of Run	Jack Rafter	Length
6"	2nd	
4"	4th	
9"	5th	
12"	3rd	
5"	4th	

LAYING OUT A HIP RAFTER WITH A SPEED SQUARE

Assume the hip rafter to be cut is on the same building as the common rafter (24' wide with a 4" rise per foot of run). The hip rafter must be cut with a different plumb or top cut because it is at a 45° angle to the common rafter (that is, sitting diagonally at the intersection of the two wall plates).

Because of the additional length of run in the hip rafter (17" of the run to 12" of run for a common rafter), the plumb or top cut must be figured at a different angle on the speed square. The speed square has a set pattern to establish the cuts for a hip or valley rafter, as shown in *Figure A-11(A)*.

The booklet that comes with the speed square indicates that the total length of a hip or valley rafter for a building 24' wide and with a 4 in 12 rise is 17'-5¼". Remember, this is the total length with no overhang figured.

The eaves or overhang must be added to the length. Assume that the hip rafter being cut has no overhang.

Step 1 Choose a piece of lumber slightly longer than necessary. Place the lumber on a pair of sawhorses with the top edge and crown facing away from you. Assume that the right end of the lumber will be the top of the hip rafter. Place the speed square a few inches away from the top. Move the speed square so that the pivot point and the 4 on the hip valley scale are even with the outside edge, and draw a line along the edge as shown in *Figure A-11(B)*. This will establish the top or plumb cut of the hip rafter.

Step 2 Measure with a tape along the top edge of the rafter the length required for the hip rafter and mark this point. Place the speed square with the pivot point on the edge of the mark and move the square until the 4 on the hip valley scale is even with the edge of the lumber; then draw a line along the edge of the square. This will establish the vertical cut of the seat cut (bird's mouth).

Step 3 To obtain the horizontal cut, reverse the square so that the small line at the pointed end of the speed square is covering the line previously drawn. Draw a line along the top edge of the speed square. This will establish the seat cut (bird's mouth), as shown in *Figure A-11(C)*.

Step 4 The next step is to turn the lumber on edge and establish the center of the rafter. Draw a line on the center of the rafter so that it is over the ridge cut, as shown in *Figure A-11(D)*. This is done in order to establish the double 45° angle that must be cut so that the hip rafter will sit properly at the intersection of the two common rafters.

Step 5 Deduct ¾" plus half the diagonal measure of the ridge (1¹⁄₁₆"). Add the two together and mark that distance back from the original plumb cut (ridge cut).

NOTE

If the ridge is 2" stock, the total measure deducted is 1¹³⁄₁₆".

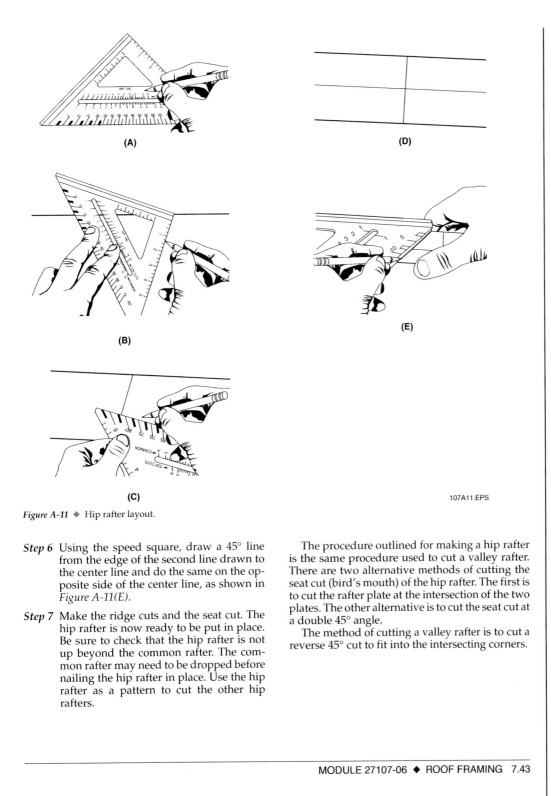

Figure A-11 ♦ Hip rafter layout.

Step 6 Using the speed square, draw a 45° line from the edge of the second line drawn to the center line and do the same on the opposite side of the center line, as shown in *Figure A-11(E)*.

Step 7 Make the ridge cuts and the seat cut. The hip rafter is now ready to be put in place. Be sure to check that the hip rafter is not up beyond the common rafter. The common rafter may need to be dropped before nailing the hip rafter in place. Use the hip rafter as a pattern to cut the other hip rafters.

The procedure outlined for making a hip rafter is the same procedure used to cut a valley rafter. There are two alternative methods of cutting the seat cut (bird's mouth) of the hip rafter. The first is to cut the rafter plate at the intersection of the two plates. The other alternative is to cut the seat cut at a double 45° angle.

The method of cutting a valley rafter is to cut a reverse 45° cut to fit into the intersecting corners.

Additional Resources and References

This module is intended to present thorough resources for task training. The following reference works are suggested for further study. These are optional materials for continuing education rather than for task training.

Advanced Framing Methods. Kingston, MA: R.S. Means Company.

Build a Better Home: Roofs. Tacoma, WA: APA – The Engineered Wood Association.

Framing Roofs. Newton, CT: Taunton Press.

Graphic Guide to Frame Construction. Newton, CT: Taunton Press.

Miller's Guide to Framing and Roofing. New York: McGraw-Hill Professional.

New Roof Construction. Sumas, WA: Cedar Shake and Shingle Bureau (15-minute video).

Quality Roof Construction. Tacoma, WA: APA – The Engineered Wood Association (15-minute video).

Roof Framer's Bible: The Complete Pocket Reference to Roof Framing. Jenkintown, PA: M.E.I. Publishing.

Wood Frame Construction Manual. Washington, D.C.: American Wood Council.

American Wood Council. A trade association that develops design tools and guidelines for wood construction. www.awc.org

Cedar Shake and Shingle Bureau. A trade organization that promotes the common interests of members involved in quality cedar shake and shingle roofing. www.cedarbureau.org

Western Wood Products Association. A trade association representing softwood lumber manufacturers in 12 western states and Alaska. www.wwpa.org

Wood I-Joist Manufacturers Association. An organization representing manufacturers of prefabricated wood I-joist and structural composite lumber. www.i-joist.org

Wood Truss Council of America. An international trade association representing structural wood component manufacturers. www.woodtruss.com

---

7.44 CARPENTRY FUNDAMENTALS LEVEL ONE ◆ TRAINEE GUIDE

Instructor's Notes:

MODULE 27107-06 — ANSWERS TO REVIEW QUESTIONS

Answer	Section
1. b	2.0.0
2. c	3.0.0; Figure 2
3. a	3.0.0; Figure 2
4. c	3.0.0; Figure 2
5. b	3.0.0; Figure 2
6. d	3.0.0; Figure 2
7. a	3.0.0; Figure 2
8. b	3.0.0
9. d	3.0.0
10. c*	3.1.0; 3.2.2; Figures 4 and 8
11. b	3.1.0; Figure 5
12. d	3.1.0; Figure 5
13. a	3.2.3
14. b	3.3.1
15. a	3.3.4; Figure 16
16. b	4.0.0
17. b	4.0.0
18. c	5.0.0
19. b	5.1.0
20. d	6.0.0; Figure 21
21. a	6.0.0; Figure 21
22. a	6.0.0
23. b*	7.1.0
24. c*	7.1.0
25. d	9.0.0

*10. $20' \div 2 = 10'$ run
  $15.62" \times 10' = 156\frac{1}{5}" \div 12 = 13$

*23. $16" = 1.3'$
  $20 \div 1.3 = 15.38$
  15 plus one end rafter = 16 × 2 sides = 32

*24. Pitch = ¼
  By the 6" mark on the framing square, we see that the length of the common rafter is 13.42.
  13.42 × 30 (length of house) = 402.6
  402.6 ÷ 32 (# of sq ft in a 4 × 8 sheet of plywood) = 12.58 (round up to 13 because plywood is purchased in whole sheets)
  13 × 2 = 26
  26 sheets of 4 × 8 plywood are needed to sheathe the roof.

MODULE 27107-06 — ANSWERS TO TRADE TERMS QUIZ

1. Gable
2. False fascia
3. Barge rafter
4. Purlin
5. Lookouts

MODULE 27107-06 — TEACHING TIPS

The following are suggested activities or instructional methods to help you teach the material in this module.

General

When you call on someone to answer a question, the rest of the class relaxes or even tunes out because they expect that the question and answer will take place only between you and the trainee you called on. Instead, use this technique to involve more trainees in answering questions and to keep them on their toes.

1. Ask trainees to define a term or explain a concept.
2. After one trainee has answered, ask a trainee seated nearby if the answer is right. Then ask whether a trainee in the back of the room agrees.
3. Ask trainees to explain why they think an answer is right or wrong.
4. Use the session to clear up incorrect ideas and encourage trainees to learn from their mistakes.

MODULE 27107-06 — "THINK ABOUT IT" ANSWERS

Section 3.2.1 *Think About It—Calculating Rise and Pitch*

Find the pitch for the following roofs:

Total Rise	Span	Pitch
8'	24'	1/3
9'	36'	1/4
6'	24'	1/4
12'	36'	1/3
8'	32'	1/4

Find the total rise for the following roofs:

Pitch	Span	Rise per Foot of Run	Total Rise
1/2	24'	12"	12'
1/6	32'	4"	5'-4"
1/3	24'	8"	8'
5/12	30'	10"	12'-6"
3/4	24'	18"	18'

Find the rise per foot of run for the following roofs:

Span	Total Rise	Rise per Foot of Run
18'	6'	8"
24'	8'	8"
28'	7'	6"
16'	4'	6"
32'	12'	9"

Section 3.2.3 *Think About It—Calculating Common Rafter Lengths*

Find the lengths of the common rafters for the following spans:

Span	Run	Rise per Foot of Run	Rafter Length
26'	13'	6"	14.539
24'	12'	4"	12.65
28'	14'	8"	16.823
32'	16'	12"	22.262
30'	15'	7"	17.362

CONTREN® LEARNING SERIES — USER UPDATE

NCCER makes every effort to keep these textbooks up-to-date and free of technical errors. We appreciate your help in this process. If you have an idea for improving this textbook, or if you find an error, a typographical mistake, or an inaccuracy in NCCER's Contren® textbooks, please write us, using this form or a photocopy. Be sure to include the exact module number, page number, a detailed description, and the correction, if applicable. Your input will be brought to the attention of the Technical Review Committee. Thank you for your assistance.

Instructors – If you found that additional materials were necessary in order to teach this module effectively, please let us know so that we may include them in the Equipment/Materials list in the Annotated Instructor's Guide.

Write: Product Development and Revision
National Center for Construction Education and Research
P.O. Box 141104, Gainesville, FL 32614-1104

Fax: 352-334-0932

E-mail: curriculum@nccer.org

Craft _____ Module Name _____

Copyright Date _____ Module Number _____ Page Number(s) _____

Description _____

_____

_____

_____

_____

(Optional) Correction _____

_____

_____

(Optional) Your Name and Address _____

_____

_____

Introduction to Concrete, Reinforcing Materials, and Forms
27108-06

NCCER STANDARDIZED CRAFT TRAINING PROGRAM

The National Center for Construction Education and Research (NCCER) provides a standardized national program of accredited craft training. Key features of the program include instructor certification, competency-based training, and performance testing. The program provides trainees, instructors, and companies with a standard form of recognition through the National Registry. The program is described in full in the *Guidelines for Accreditation*, published by NCCER. For more information on standardized craft training, contact NCCER by writing to P.O. Box 141104, Gainesville, FL 32614-1104; calling 352-334-0911; or e-mailing **info@nccer.org**. More information is available at **www.nccer.org**.

HOW TO USE THIS ANNOTATED INSTRUCTOR'S GUIDE

Each page presents two sections of information. The larger section displays each page exactly as it appears in the Trainee Module. The narrow column ties suggested trainee and instructor actions to each page and provides icons (detailed below) to call your attention to material, safety, audiovisual, or testing requirements. The bottom of each page includes space for your notes.

 The **Audiovisual** icon indicates an appropriate time to show a transparency or other audiovisual aid.

 The **Classroom** icon prompts you to define a term, stress a point, ask trainees to explain a concept, or give examples.

 The **Demonstration** icon directs you to show trainees how to perform tasks.

 The **Examination** icon tells you to administer the written module examination.

 The **Homework** icon is placed where you may wish to assign reading for the next class, assign a project, or advise trainees to prepare for an examination.

 The **Laboratory** icon is used when trainees are to practice performing tasks.

 The **Materials** icon is a reminder for you to gather materials needed for classes, laboratories, and testing.

 The **Performance Testing** icon tells you to administer a performance test or a portion thereof.

 The **Safety** icon is used to emphasize safety issues. It is often keyed to *Caution* and *Warning!* statements in the Trainee Module.

 The **Teaching Tip** icon indicates additional guidance is available, such as how to conduct an exercise, get the most educational value from a field trip, or encourage class participation. Teaching Tips may expand on a feature (*Think About It, Did You Know?*) or provide *Quick Quizzes* or similar exercises. You will be referred to the Teaching Tips section at the back of the module if there is additional material.

 The **Combination** icon indicates that the laboratory listed corresponds with a performance task. If desired, you can note the proficiency of the trainees during the laboratory, and use it to satisfy performance testing requirements.

PREPARATION

Before teaching this module, you should review the Objectives, Performance Tasks, Materials and Equipment List, and Module Outline. Be sure to allow ample time to prepare your own training or lesson plan and gather all required materials and equipment.

Introduction to Concrete, Reinforcing Materials, and Forms
Annotated Instructor's Guide

Module 27108-06

MODULE OVERVIEW

This module introduces the carpentry trainee to various cements and other materials that, when mixed together, form various types of concrete. Concrete volume estimates and concrete forms are also covered. In addition, reinforcement materials such as reinforcement bars, bar supports, and welded-wire fabric are discussed.

PREREQUISITES

Prior to training with this module, it is recommended that the trainee shall have successfully completed *Core Curriculum*; and *Carpentry Fundamentals Level One*, Modules 27101-06 through 27107-06.

OBJECTIVES

Upon completion of this module, the trainee will be able to do the following:

1. Identify the properties of cement.
2. Describe the composition of concrete.
3. Perform volume estimates for concrete quantity requirements.
4. Identify types of concrete reinforcement materials and describe their uses.
5. Identify various types of footings and explain their uses.
6. Identify the parts of various types of forms.
7. Explain the safety procedures associated with the construction and use of concrete forms.
8. Erect, plumb, and brace a simple concrete form with reinforcement.

PERFORMANCE TASKS

Under supervision of the instructor, the trainee should be able to do the following:

1. Perform volume estimates for concrete quantity requirements.
2. Construct a simple concrete form with reinforcement.

MATERIALS AND EQUIPMENT LIST

Transparencies
Markers/chalk
Blank acetate sheets
Transparency pens
Pencils and scratch paper
Overhead projector and screen
Whiteboard/chalkboard
Appropriate personal protective equipment
Hand calculator
Concrete calculator
Copies of a concrete table
Form boards, stakes, braces, ties, and spreaders
16-gauge tying wire
Samples of various aggregates
Samples of concrete mix
Various bar supports and accessories
Various mechanical splices for reinforcement steel
Various sizes, types, and grades of reinforcement materials
Samples of various types and sizes of wire fabric
Exterior plywood or plyform
Steel tape or rule
Basic carpenter's toolbox
Level
Plumb bob
String line
Duplex nails
Plan for simple form
Circular saw and extension cord
Copies of Worksheet 1*
Module Examinations**
Performance Profile Sheets**

* Packaged with this Annotated Instructor's Guide.
**Located in the Test Booklet.

SAFETY CONSIDERATIONS

Ensure that the trainees are equipped with appropriate personal protective equipment.

ADDITIONAL RESOURCES

This module is intended to present thorough resources for task training. The following reference works are suggested for both instructors and motivated trainees interested in further study. These are optional materials for continued education rather than for task training.

Concrete Masonry Handbook for Architects, Engineers, and Builders, Fifth Edition. W.C. Panarese, S.H. Kosmatka, and F.A. Randall, Jr. Portland Cement Association.

The Homeowner's Guide to Building with Concrete, Brick, and Stone. The Portland Cement Association.

TEACHING TIME FOR THIS MODULE

An outline for use in developing your lesson plan is presented below. Note that each Roman numeral in the outline equates to one session of instruction. Each session has a suggested time period of 2½ hours. This includes 10 minutes at the beginning of each session for administrative tasks and one 10-minute break during the session. Approximately 10 hours are suggested to cover *Introduction to Concrete, Reinforcing Materials, and Forms*. You will need to adjust the time required for hands-on activity and testing based on your class size and resources. Because laboratories often correspond to Performance Tasks, the proficiency of the trainees may be noted during these exercises for Performance Testing purposes.

Topic **Planned Time**

Session I. Introduction; Concrete and Concrete Materials; Normal Concrete Mix Proportions and Measurements; Special Types of Concrete; Curing Methods and Materials; Concrete Slump Testing

    A. Introduction

    B. Concrete and Concrete Materials

        1. Portland Cement

        2. Aggregates for Concrete

        3. Water for Concrete

        4. Admixtures for Concrete

    C. Normal Concrete Mix Proportions and Measurements

    D. Special Types of Concrete

    E. Curing Methods and Materials

    F. Concrete Slump Testing

Session II. Estimating Concrete Volume; Concrete Reinforcement Materials

    A. Estimating Concrete Volume

        1. Rectangular Volume Calculations

        2. Circular Volume Calculations

    B. Laboratory

       Hand out Worksheet 27108-1. Have the trainees complete the Worksheet. This laboratory corresponds to Performance Task 1.

    C. Concrete Reinforcement Materials

        1. Reinforcing Bars

        2. Bar Supports

        3. Splicing Reinforcing Bars

        4. Welded-Wire Fabric

Session III. Concrete Forms
- A. Concrete Forms
    1. Form Safety
    2. Footings
    3. Wall Forms
    4. Edge Forms
    5. Removing Forms
- B. Laboratory

    Under your supervision, have the trainees erect, plumb, and brace a simple concrete form. This laboratory corresponds to Performance Task 2.

Session IV. Review; Module Examination and Performance Testing
- A. Review
- B. Module Examination
    1. Trainees must score 70 percent or higher to receive recognition from NCCER.
    2. Record the testing results on Craft Training Report Form 200, and submit the results to the Training Program Sponsor.
- C. Performance Testing
    1. Trainees must perform each task to the satisfaction of the instructor to receive recognition from NCCER. If applicable, proficiency noted during laboratory exercises can be used to satisfy the Performance Testing requirements.
    2. Record the testing results on Craft Training Report Form 200, and submit the results to the Training Program Sponsor.

Introduction to Concrete, Reinforcing Materials, and Forms
27108-06

In the Associated Builders and Contractors 2005 National Craft Championships, Carpentry contestants are expected to be able to perform several masonry and concrete finishing tasks, such as mixing concrete; building footings, edges, and wall forms; and using concrete reinforcing materials.

Instructor's Notes:

Assign reading of Module 27108-06.

27108-06
Introduction to Concrete, Reinforcing Materials, and Forms

Topics to be presented in this module include:

1.0.0	Introduction	.8.2
2.0.0	Concrete and Concrete Materials	.8.2
3.0.0	Normal Concrete Mix Proportions and Measurements	.8.5
4.0.0	Special Types of Concrete	.8.6
5.0.0	Curing Methods and Materials	.8.8
6.0.0	Concrete Slump Testing	.8.8
7.0.0	Estimating Concrete Volume	.8.9
8.0.0	Concrete Reinforcement Materials	.8.12
9.0.0	Concrete Forms	.8.17

Overview

Though the finishing of footings, foundations, and floor slabs is typically done by subcontractors, there will be occasions when the residential carpenter is called on to build basic forms and place reinforcing materials. A residential carpenter might also have to mix a batch of concrete simply to pour into a hole to support deck pillars and other vertical supports.

Carpenters doing commercial and industrial construction can expect to spend a lot of time building, bracing, and stripping concrete forms for walls, columns, slabs, beams, and other structures. Such work requires knowledge of form ties and form support structures. In this module, you will learn about concrete, reinforcing materials, and basic formwork.

Instructor's Notes:

Objectives

When you have completed this module, you will be able to do the following:

1. Identify the properties of cement.
2. Describe the composition of concrete.
3. Perform volume estimates for concrete quantity requirements.
4. Identify types of concrete reinforcement materials and describe their uses.
5. Identify various types of footings and explain their uses.
6. Identify the parts of various types of forms.
7. Explain the safety procedures associated with the construction and use of concrete forms.
8. Erect, plumb, and brace a simple concrete form with reinforcement.

Trade Terms

ACI
Admixtures
Aggregates
Axle-steel
Brace
Cured concrete
Flatwork
Footing
Forms
Green concrete
Hydration
Kip
Monolithic slab (monolithic pour)
Pascal
Piles
Plastic concrete
Plyform
Pozzolan
Rail-steel
Rebars
Screeding
Shoring
Slab-on-grade (slab-at-grade)
Slag
Slump
Studs
Subgrade
Walers
Water-cement ratio

Required Trainee Materials

1. Pencil and paper
2. Appropriate personal protective equipment

Prerequisites

Before you begin this module, it is recommended that you successfully complete *Core Curriculum* and *Carpentry Fundamentals Level One,* Modules 27101-06 through 27107-06.

This course map shows all of the modules in *Carpentry Fundamentals Level One.* The suggested training order begins at the bottom and proceeds up. Skill levels increase as you advance on the course map. The local Training Program Sponsor may adjust the training order.

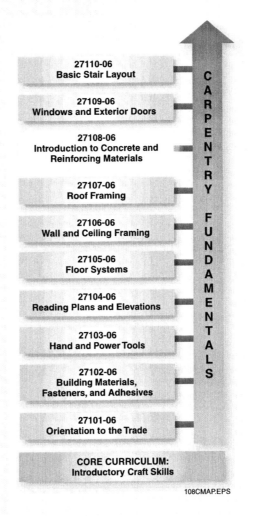

Ensure you have everything required to teach the course. Check the Materials and Equipment List at the front of this module.

Show Transparency 1, Objectives.

Show Transparency 2, Performance Tasks.

Explain that terms shown in bold (blue) are defined in the Glossary at the back of this module.

See the general Teaching Tip at the end of this module.

Emphasize the importance of understanding how to properly place and reinforce concrete.

Discuss the basic ingredients in concrete. Provide samples of concrete mix for the trainees to examine.

Define the three states of concrete.

List the ingredients in portland cement and explain the process by which it hardens.

Explain the function of concrete aggregates. Obtain different types and sizes of aggregates and pass them around for the trainees to examine.

Discuss the safety precautions associated with the use of concrete.

1.0.0 ♦ INTRODUCTION

This module introduces the trainee to the types of cement, **aggregates,** and additives used in concrete. It describes general types of concrete, concrete mixing information, and various concrete tests. It also covers concrete quantity estimating procedures for various job applications as well as compression specimen casting and **slump** (consistency) testing of freshly mixed concrete.

This module also introduces the trainee to some types of concrete reinforcement material such as steel reinforcement bars, welded-wire mesh, and various reinforcement bar supports.

Because it is in a semiliquid state, concrete must be contained in **forms** until it hardens. Forms can be built on the job using lumber and plywood, or they can be prefabricated form systems that the contractor buys or rents. In this module, we will introduce the basic job-built forms for concrete walls, **footings,** and floor slabs.

2.0.0 ♦ CONCRETE AND CONCRETE MATERIALS

Concrete is a mixture of four basic materials: portland cement, fine aggregates, coarse aggregates, and water. When first mixed, concrete is in a semiliquid state and is referred to as **plastic concrete.** When the concrete hardens, but has not yet gained structural strength, it is called **green concrete.** After concrete has hardened and gained its structural strength, it is called **cured concrete.** Various types of concrete can be obtained by varying the basic materials and/or by adding other materials to the mix. These added materials are called **admixtures.**

2.1.0 Portland Cement

Portland cement is a finely ground powder consisting of varying amounts of lime, silica, alumina, iron, and other trace components. While dry, it may be moved in bulk or can be bagged in moisture-resistant sacks and stored for relatively long periods of time. Portland cement is a hydraulic cement because it will set and harden by reacting with water with or without the presence of air. This chemical reaction is called **hydration,** and it can occur even when the concrete is submerged in water. The reaction creates a calcium silicate hydrate gel and releases heat. Hydration begins when water is mixed with the cement and continues as the mixture hardens and cures. The reaction occurs rapidly at first, depending on how finely the cement is ground and what admixtures are present. Then, after its initial cure and strength are achieved, the cement mixture continues to slowly cure over a longer period of time until its ultimate strength is attained.

Today, portland cement is manufactured by heating lime mixed with clay, shale, or **slag** to about 3000°F. The material that is produced is pulverized, and gypsum is added to regulate the hydration process. Manufacturers use their own brand names, but nearly all portland cements are manufactured to meet American Society for Testing and Materials (ASTM) or American National Standards Institute (ANSI) specifications.

 WARNING!
Those working with dry cement or wet concrete should be aware that it is harmful. Dry cement dust can enter open wounds and cause blood poisoning. The cement dust, when it comes in contact with body fluids, can cause chemical burns to the membranes of the eyes, nose, mouth, throat, or lungs. It can also cause a fatal lung disease known as silicosis. Wet cement or concrete can also cause chemical burns to the eyes and skin. Make sure that appropriate personal protective equipment is worn when working with dry cement or wet concrete. If wet concrete enters waterproof boots from the top, remove the boots and rinse your legs, feet, boots, and clothing with clear water as soon as possible. Repeated contact with cement or wet concrete can also cause an allergic skin reaction known as cement dermatitis.

2.2.0 Aggregates for Concrete

Natural sand, gravel, crushed stone, blast furnace slag, and manufactured sand (from crushed stone, gravel, or slag) are the most commonly used aggregates in concrete. In some cases, recycled crushed concrete is used for low-strength concrete applications. There are other types of special aggregates for concrete.

Aggregates make up 60 percent to 80 percent of the volume of concrete and function not only as filler material, but also provide rigidity to greatly restrain volume shrinkage as the concrete mass

---

8.2 CARPENTRY FUNDAMENTALS LEVEL ONE ♦ TRAINEE GUIDE

Instructor's Notes:

The History of Portland Cement

Portland cement is a successor to a hydraulic lime that was first developed by John Smeaton in 1756 when he built a lighthouse in the English Channel. He used a burned mixture of limestone and clay capable of setting and hardening under water as well as in air. The next developments took place around 1800 in England and France where a cement material was made by burning nodules of clayey limestone. About the same time in the United States, a similar material was obtained by burning a naturally-occurring substance called cement rock. These materials belong to a class known as natural cement, similar to portland cement but more lightly burned and not of a controlled composition.

In 1824, Joseph Aspdin, another Englishman, patented a manufacturing process for making a hydraulic cement out of limestone and clay that he called portland cement because when set it resembled portland stone, a type of limestone from the Isle of Portland, England. In 1850, this was followed by a more heavily burned portland cement product developed by Isaac Charles Johnson in southeastern England. Since then, the manufacture of portland cement has spread rapidly throughout the world accompanied by numerous new developments and improvements in its uses and manufacture.

Care of Portland Cement

In the United States, portland cement is normally sold in paper bags that hold one cubic foot of cement by volume and weigh 94 pounds. As shown here, these bags of cement should always be stored off the ground and in a dry place to prevent the cement from absorbing moisture that causes lumps to form in the cement powder. Cement powder should be free flowing. Any cement with lumps that cannot be broken up easily into powder by squeezing them in your hand should not be used.

108SA01.EPS

Discuss the "Think About It."

Concrete, Mortar, and Grout

What is the difference between concrete, mortar, and grout?

Explain how aggregate strength, size, shape, and cleanliness affect concrete quality.

Explain that using water with high concentrations of contaminants may adversely affect the quality of the concrete.

Discuss the use of various admixtures.

Sand

Desert sand is the most plentiful commodity in Middle Eastern countries such as Egypt and Saudi Arabia. However, grains of desert sand are round and smooth, whereas concrete requires irregularly shaped grains. If desert sand is used, the concrete might disintegrate. The irony is that countries with plentiful supplies of sand must import sand for use in making concrete.

Strength of Concrete

The compression strength required for a specific concrete structure, such as a footing or slab, is normally specified on the construction drawings and/or specifications. Compression strength is expressed in pounds per square inch (psi).

cures. At the usual aggregate content of about 75 percent by volume, shrinkage of concrete is only one-tenth that of pure cement paste. The presence of aggregates in concrete provides an enormous contact area for an intimate bond between the paste and aggregate surfaces.

Since aggregates are such a large portion of the concrete volume, they have a substantial effect on the quality of the finished product. In addition to concrete volume stability, they influence concrete weight, strength, and resistance to environmental destruction. Therefore, aggregates must be strong, clean, free of chemicals and coatings, and of the proper size and weight.

2.3.0 Water for Concrete

Generally, any drinkable water (unless it is extremely hard or contains too many sulfates) may be used to make concrete. If the quality of the water is unknown or questionable, the water should be analyzed, or mortar cubes should be made and tested against control cubes made with drinkable water. If the water is satisfactory, the test cubes should show the same compressive strength as the control cubes after a 28-day cure time.

High concentrations of chlorides, sulfates, alkalis, salts, and other contaminants have corrosive effects on concrete and/or metal reinforcing rods, mesh, and cables. Sulfates can cause disintegration of concrete, while alkalis, sodium carbonate, and bicarbonates may affect the hardening times as well as the strength of concrete.

2.4.0 Admixtures for Concrete

Admixtures are materials added to a concrete mix before or during mixing to modify the characteristics of the final concrete. They may be used to improve workability during placement, increase strength, retard or accelerate strength development, and increase frost resistance. Usually, an admixture will affect more than one characteristic of concrete. Therefore, its effect on all the properties of the concrete must be considered. Admixtures may increase or decrease the cost of concrete work by lowering cement requirements, changing the volume of the mixture, or lowering the cost of handling and placement.

Control of concrete setting time may reduce costs by decreasing waiting time for floor finishing and form removal. Extending the setting time may reduce costs by keeping the concrete plastic,

Spacing

The spacing between steel reinforcing bars must be taken into consideration when selecting the size of the coarse aggregate to use in a concrete mixture.

108SA02.EPS

8.4 CARPENTRY FUNDAMENTALS LEVEL ONE ♦ TRAINEE GUIDE

Instructor's Notes:

thereby eliminating bulkheads and construction joints. In practice, it is desirable to fully pretest all admixtures with the specific concrete mix to be used since the mix may affect the admixture efficiency and, ultimately, the final concrete.

There are a number of other admixtures available that are not in general use, such as gas-forming, air-entraining, grouting, expansion-producing or expansion-reducing, bonding, corrosion-inhibiting, fungicidal, germicidal, and insecticidal. These admixtures are for very specialized work, generally under the supervision of the manufacturer or a specialist.

3.0.0 ◆ NORMAL CONCRETE MIX PROPORTIONS AND MEASUREMENTS

Project specifications for concrete can generally be divided into two classes: prescription specifications and performance specifications, with some including features of both. Typical of the prescription type is the outmoded volume-proportion requirement such as 1:2:3, meaning one part cement, two parts sand, and three parts aggregate by volume with water only being limited by the maximum amount of slump allowed. Most prescription requirements are now stated by weight, such as kilograms per cubic meter or pounds per cubic yard, and often cover only minimum cement content along with a **water-cement ratio**, usually by weight. The water-cement ratio includes all cementitious components of the concrete, including fly ash and **pozzolan**, as well as portland cement. *Figure 1* is an example of the typical strength range of a Type I portland cement at various water-cement ratios. In these types of specifications, the strength of the concrete will vary over a certain range due to variations in the cementitious materials, along with additional variations that can be caused by the quality of the aggregates. Numerous prescription-type tables have been prepared and presented in various publications by the **ACI** (American Concrete Institute). The recommendations of these tables can be combined to produce a calculated concrete mix

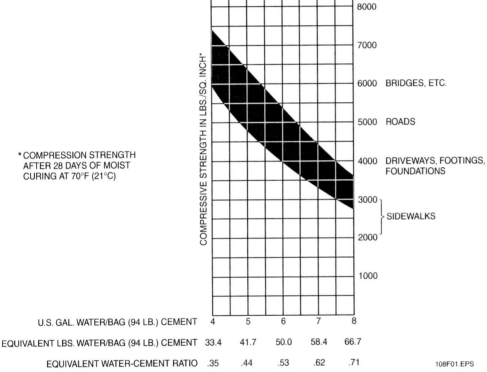

Figure 1 ◆ Typical Type I portland cement strength for various water-cement ratios.

Discuss special types of concrete and their applications.

Show Transparency 4 (Figure 2).

American Concrete Institute

The American Concrete Institute (ACI) is an international organization established in 1905 whose purpose is to further engineering and technical education, scientific investigation and research, and the development of standards for design and construction incorporating concrete and related materials. Its mission is to be the world's premier developer and disseminator of information on the use of concrete structures and facilities. It does this through the work of thousands of professional volunteers at local, national, and international levels, assisted and guided by a professional staff located at the ACI headquarters in Farmington Hills, Michigan.

design that will be on the side of safety in both strength and durability.

Concrete performance specifications usually state the strength required, the minimum cement content or maximum water-cement ratio, consistency as measured by slump, air-entrainment (if required), and usually some limitation on aggregate sizes and properties. By keeping the limitations to a minimum, concrete producers or contractors can use prescription tables as a starting point to design (with laboratory assistance) the most effective concrete mix based on local conditions. These types of specifications usually require that the concrete producer or contractor test a number of cured samples of the proposed concrete mixes to determine and demonstrate strength and durability.

4.0.0 ♦ SPECIAL TYPES OF CONCRETE

In addition to normal concrete mixes, there are a number of specialized types of concrete for specific applications. A few of these special types of concrete do not look anything like normal concrete during placement.

- *Mass concrete* – This concrete requires reduction of hydration heat to prevent shrinking and cracking. It is used in dams, spillways, and other large structures. It is basically the same as normal concrete except that aggregates up to 6" or more in diameter are used. Admixtures such as water-reducing/water-retarding agents and pozzolans are widely used to decrease heat and cement content and to increase strength.
- *High-strength and/or high-performance concrete* – High-performance concrete meets special requirements for high elasticity values, high density, watertightness, high resistance to sulfate attack, etc. High-strength concrete is concrete with a compressive strength over 6,000 psi (92 mega**pascals** or MPa).
- *Roller-compacted concrete (RCC)* – This is used for concreting large areas that contain little or no reinforcement. One type is a low-cement content type used for dams or other very large structures. The other type is a high-cement content type used for pavements. Both are very dry, stiff mixes that are placed without forms and spread with bulldozers or pavers. After spreading, the concrete is immediately compacted by 10-ton vibratory rollers.
- *Lightweight concrete* – As shown in *Figure 2*, this concrete is classified as either low-density, structural lightweight, or moderate-strength lightweight concrete and uses various density aggregates depending on its classification. Low-density concrete is used primarily for insulation. Structural lightweight concrete is used as a substitute for normal concrete and has reduced weight as well as reduced strength. Moderate-strength lightweight concrete falls between low-density and structural lightweight concrete and has enough strength for use in roofs.
- *Preplaced aggregate (prepacked) concrete* – This concrete is widely used for repair of large concrete structures or in the placement of heavyweight concrete. It involves placing dry, coarse aggregate in forms and then pumping a highly fluidized cement mixture from the bottom to the top of the forms to make the concrete. Because the coarse aggregates are in contact with each other, the concrete has very low shrinkage and very good strength.
- *Heavyweight (high-density) concrete* – A concrete containing very heavy aggregates such as iron ore, iron shot or punchings, iron slag, and other minerals. It is used primarily for radiation shielding or in construction of counterweights or deadweights. This concrete can be placed by the preplaced-aggregate method or by conventional methods if heavy-duty equipment is used. Forms for this concrete require increased strength.

8.6 CARPENTRY FUNDAMENTALS LEVEL ONE ♦ TRAINEE GUIDE

Instructor's Notes:

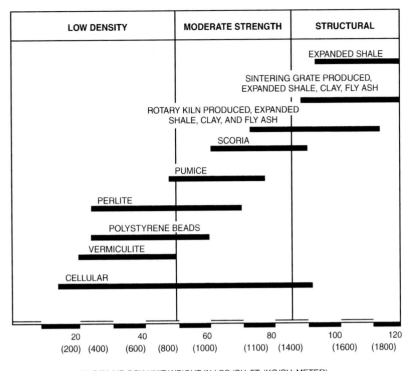

Figure 2 ◆ Lightweight concrete aggregates and unit weights.

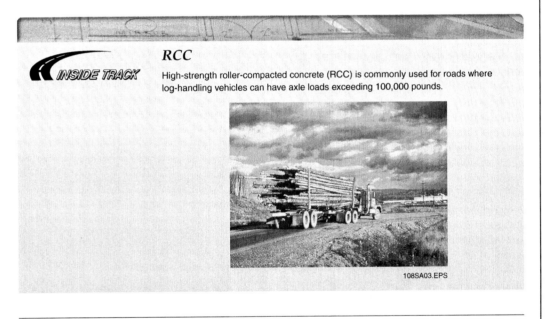

RCC

High-strength roller-compacted concrete (RCC) is commonly used for roads where log-handling vehicles can have axle loads exceeding 100,000 pounds.

Emphasize the relationship between concrete strength and proper cure time.

Describe concrete curing materials and methods.

Explain the purpose of concrete slump testing.

Assign reading of Sections 7.0.0–8.4.0 for the next class session.

Colored and Patterned Concrete

Concrete is sometimes colored and/or patterned to enhance its appearance. Concrete can be colored by adding a mineral-oxide pigment prepared especially for use with concrete to the dry cement powder prior to mixing the concrete. Plastic cement can be colored by working a commercially prepared color dye into the concrete surface after first floating, edging, and grooving the surface.

Geometric patterns can also be stamped, sawed, or scored into concrete surfaces to enhance their appearance. An imprint roller was used to simulate a stone pattern in the concrete shown here.

108SA04.EPS

5.0.0 ♦ CURING METHODS AND MATERIALS

Concrete gains strength by the process of hydration, which is a cement-water reaction. If concrete is allowed to dry prematurely, the reaction ceases, and the strength is severely reduced and cannot be recovered even if the concrete is re-wetted. This is especially true on the surface of concrete. Re-wetting dried concrete also causes cracking, especially on the surface.

Since any loss of moisture from the concrete will result in a loss of strength or surface cracking, proper concrete curing, regardless of the method used, should commence as soon as forms are removed or as soon as the surface will not be damaged by the curing method.

- *Water* – Moist curing with water is ideal and is accomplished by keeping the concrete surface covered in water by the use of sprinklers, fog sprayers, or dikes. Water used for making concrete is suitable for curing use.
- *Curing compounds* – Liquid membrane-forming compounds of various types can be applied to fresh concrete to help ensure that the concrete will not dry out due to neglect. Some of these compounds will discolor the concrete, and others will seal the surface for subsequent application of paint or floor coverings.
- *Curing paper or plastic sheeting* – These products perform the same function as curing compounds and do not discolor the concrete. However, they must be weighted down and sealed at the edges to prevent water evaporation and wind damage.
- *Mats and blankets* – Wet curing with cotton curing mats or moving blankets eliminates the need for continuous spraying. For winter concreting, commercial insulating blankets, fiberglass batts, or expanded plastic cell slabs can be used to protect the concrete from freezing; however, they generally must be protected from snow and water penetration.

6.0.0 ♦ CONCRETE SLUMP TESTING

A slump test is used to determine the consistency of concrete, referred to as its flowability or workability. The test is the distance, in inches, that the top of a conical pile of fresh concrete will sag after a standardized cone mold is removed (*Figure 3*). Concrete slump must conform to the structure specifications.

8.8 CARPENTRY FUNDAMENTALS LEVEL ONE ♦ TRAINEE GUIDE

Instructor's Notes:

Figure 3 ◆ Measurement of concrete slump.

7.0.0 ◆ ESTIMATING CONCRETE VOLUME

Accurate measuring and estimating of concrete quantities is required for concrete work. Fortunately, most concrete structures can be divided into rectangular or circular shapes, individually estimated, and the results added together to obtain the required volume of concrete. For instance, a floor is a rectangular horizontal slab. A footing and a wall can be divided into a long rectangular (nearly square) shape representing the footing and a vertical slab representing the wall. The volume of each can be calculated and the results added together for the total volume. Blueprints for a project will provide the dimensions for the various portions of a concrete structure, and these dimensions are used to calculate the volume of concrete required for the structure. For reference purposes, the *Appendix* provides area or volume formulas for various geometric shapes and a table for conversion of inches to fractions of a foot or a decimal equivalent.

7.1.0 Rectangular Volume Calculations

A number of methods can be used to determine the volume of concrete required for a rectangular object. One method uses the following formula:

$$\frac{\text{Cubic yards of concrete}}{\text{(rounded up to next } \frac{1}{4} \text{ yard)}} = \frac{\text{width or heigth (ft)} \times \text{length (ft)} \times \text{thickness (ft)}}{27 \text{(cubic ft/yard)}}$$

To use the formula, all dimensions in inches must be converted to feet and/or fractions of a foot and then into a decimal equivalent. For example:

$$7" = \tfrac{7}{12}' \text{ or } 7 \div 12 = 0.58'$$
$$8" = \tfrac{8}{12}' \text{ or } \tfrac{2}{3}' \text{ or } 2 \div 3 = 0.66'$$
$$23" = \tfrac{23}{12}' \text{ or } 23 \div 12 = 1.92'$$

The other two methods involve knowing the width or height and length of an area in feet, along with the thickness in inches, and then using a concrete calculator (*Figure 4*) or a concrete table (*Figure 5*) and a simple formula to determine the volume. The concrete calculator cannot be used when the volume of concrete required is less than ½ yard. Also, the results of the calculator must be estimated by sight between rule marks, which makes the results subject to some variation.

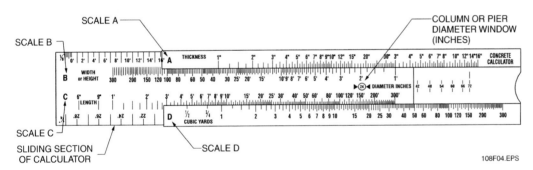

Figure 4 ◆ Typical concrete calculator.

Explain how a volume formula can be used to determine the amount of concrete required for a rectangular structure.

Show Transparency 7 (Figure 6).

ONE CUBIC YARD OF CONCRETE WILL PLACE:					
THICKNESS	SQ FT	THICKNESS	SQ FT	THICKNESS	SQ FT
1"	324	5"	65	9"	36
1¼"	259	5¼"	62	9¼"	35
1½"	216	5½"	59	9½"	34
1¾"	185	5¾"	56	9¾"	33
2"	162	6"	54	10"	32.5
2¼"	144	6¼"	52	10¼"	31.5
2½"	130	6½"	50	10½"	31
2¾"	118	6¾"	48	10¾"	30
3"	108	7"	46	11"	29.5
3¼"	100	7¼"	45	11¼"	29
3½"	93	7½"	43	11½"	28
3¾"	86	7¾"	42	11¾"	27.5
4"	81	8"	40	12"	27
4¼"	76	8¼"	39	15"	21.5
4½"	72	8½"	38	18"	18
4¾"	68	8¾"	37	24"	13.5

Figure 5 ♦ Portion of typical concrete table.

7.1.1 Example Calculation Using the Formula

Using the calculation formula, determine the amount of concrete required for the partial wall, footing, and floor slab plan shown in *Figure 6*.

Step 1 The entire footing and wall length must be determined. Since the wall is centered on the footing, the wall length is the same as the footing length:

Footing/wall length = 20' + (15' − 2') = 33'

Step 2 Determine the floor slab length and width:

Length = 20' − 16" = 20' − 1.33' = 18.67'
Width = 15' − 16" = 15' − 1.33' = 13.67'

Step 3 Using the formula, determine the volume of the wall, footing, and slab:

$$\text{Volume} = \frac{\text{width or height (ft)} \times \text{length (ft)} \times \text{thickness (ft)}}{27 (\text{cubic ft / yard})}$$

$$\text{Wall} = \frac{3 \times 33 \times 0.67}{27} = \frac{66.33}{27} = 2.46 \text{ cubic yards}$$

$$\text{Footing} = \frac{2 \times 33 \times 0.67}{27} = \frac{44.22}{27} = 1.64 \text{ cubic yards}$$

$$\text{Slab} = \frac{13.67 \times 18.67 \times 0.5}{27} = \frac{127.6}{27} = 4.73 \text{ cubic yards}$$

Step 4 Add the wall, footing, and slab volumes:

2.46 + 1.64 + 4.73 = 8.83 rounded up to 9 cubic yards

For the plan shown, 9 cubic yards of concrete will be required.

Instructor's Notes:

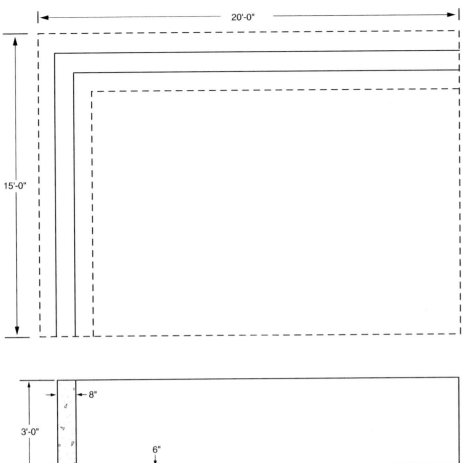

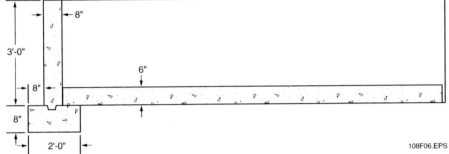

Figure 6 ♦ Partial wall, footing, and floor slab plan.

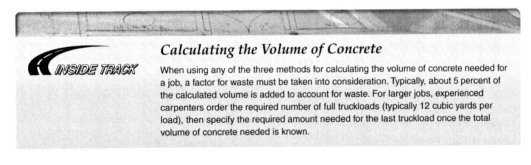

Calculating the Volume of Concrete

When using any of the three methods for calculating the volume of concrete needed for a job, a factor for waste must be taken into consideration. Typically, about 5 percent of the calculated volume is added to account for waste. For larger jobs, experienced carpenters order the required number of full truckloads (typically 12 cubic yards per load), then specify the required amount needed for the last truckload once the total volume of concrete needed is known.

Show how a volume formula can be used to determine the amount of concrete required for a circular structure.

Show Transparency 8 (Figure 7).

Take the trainees to a site that has been prepared for a concrete pour and have them estimate the amount of concrete required. After the pour is complete, compare the actual amount poured with the trainee estimates.

Hand out Worksheet 27108-1. Have the trainees complete the Worksheet. This laboratory corresponds to Performance Task 1.

Discuss common concrete reinforcement materials. Pass around examples for the trainees to examine.

7.2.0 Circular Volume Calculations

The volume of concrete required for a circular column or pier can be calculated by using the following formula:

$$\text{Cubic yards of concrete} = \frac{\pi \times \text{radius}^2(\text{sq ft}) \times \text{height (ft)}}{27 (\text{cubic ft/yd})}$$

Where:
- $\pi = 3.14$
- Radius = diameter (ft) of column ÷ 2
- Inches are expressed as fractions or decimal equivalents

As with rectangular volume calculations, a certain percentage of waste must be added to the circular volume estimate based on the job site conditions.

7.2.1 Example Calculation of Circular Volume

Using the circular volume calculation, determine the volume of concrete required for the circular column plan shown in *Figure 7*.

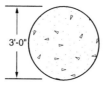

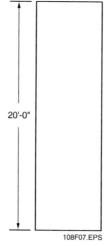

Figure 7 ♦ Typical circular column plan.

Step 1 Determine the radius of the column:

$$3' \div 2 = 1.5'$$

Step 2 Calculate the volume using the formula:

$$\text{Cubic yards of concrete} =$$

$$\frac{\pi \times \text{radius}^2(\text{sq ft}) \times \text{height (ft)}}{27 (\text{cubic ft/yd})}$$

$$\text{Column} = \frac{3.14 \times (1.5)^2 \times 20}{27} = \frac{141.3}{27}$$

= 5.23 rounded up to 5¼ cubic yards

For the plan shown, 5¼ cubic yards of concrete will be required.

The concrete calculator can also be used for circular column/pier volume estimates by setting the sliding section of the calculator so that the diameter of the column or pier (in inches) is approximately centered in the view window. Then the length (or height) of the pier or column is located on Scale C and the number of cubic yards of concrete required for the structure is read from Scale D opposite the length value on Scale C.

8.0.0 ♦ CONCRETE REINFORCEMENT MATERIALS

Concrete has good compressive strength, but is relatively weak in tension or if it is subjected to lateral or shearing forces. Many kinds of proprietary reinforcement have been used for concrete in the past. Today, steel is the material that is generally used. This is because it has nearly the same temperature expansion and contraction rate as concrete. Additionally, modern reinforcement conforms to ASTM standards that govern both its form and the types of steel used. As an alternative to steel reinforcement, fibers made from steel, fiberglass, or plastics, such as nylon, are sometimes added to concrete mixes to provide reinforcement.

8.1.0 Reinforcing Bars

Reinforcing bars (**rebars**), sometimes referred to as rerods, are available in several grades. These grades vary in yield strength, ultimate strength, percentage of elongation, bend-test requirements, and chemical composition. Furthermore, reinforcing bars can be coated with different compounds, such as epoxy, for use in concrete when corrosion could be a problem. In order to obtain uniformity throughout the United States, the ASTM has established standard specifications for these bars. These grades will appear on bar-bundle tags, in

---

8.12 CARPENTRY FUNDAMENTALS LEVEL ONE ♦ TRAINEE GUIDE

Instructor's Notes:

color coding, in rolled-on markings on the bars, and/or on bills of materials.

The specifications are:

- *A615–Standard Specification for Deformed and Plain Carbon-Steel Bars for Concrete Reinforcement*
- *A996–Standard Specification for **Rail-Steel** and **Axle-Steel** Deformed Bars for Concrete Reinforcement* (This standard replaces A616 and A617.)
- *A706–Standard Specification for Low-Alloy Steel Deformed Bars and Plain Bars for Concrete Reinforcement*

The standard configuration for reinforcing bars is the deformed bar. Different patterns may be impressed upon the bars depending on which mill manufactured them, but all are rolled to conform to ASTM specifications. The deformation improves the bond between the concrete and the bar and prevents the bar from moving in the concrete.

Plain bars are smooth and round without deformations on them and are used for special purposes, such as for dowels at expansion joints where the bars must slide in a sleeve, for expansion and contraction joints in highway pavement, and for column spirals.

Deformed bars are designated by a number in eleven standard sizes (metric or inch-pound), as shown in *Table 1*. The number denotes the approximate diameter of the bar in eighths of an inch or metric (mm). For example, a #5 bar has an approximate diameter of ⅝". The nominal dimension of a deformed bar (nominal does not include the deformation) is equivalent to that of a plain bar having the same weight per foot.

As shown in *Figure 8*, bar identification is accomplished by ASTM specifications, which require that each bar manufacturer roll the following information onto the bar:

- A letter or symbol to indicate the manufacturer's mill
- A number corresponding to the size number of the bar (*Table 1*)
- A symbol or marking to indicate the type of steel (*Table 2*)
- A marking to designate the grade (*Table 3*)

The grade represents the minimum yield (tension strength) measured in **kips** per square inch (ksi) or megapascals (MPa) that the type of steel used will withstand before it permanently stretches (elongates) and will not return to its original length. Today, Grade 420 is the most commonly used rebar. Bars are normally supplied from the mill bundled in 60' lengths.

Bar fabrication is accomplished for straight bars by cutting them to specified lengths from the 60' stock. Bent bars are cut to length the same as straight bars and then they are assigned to a bending machine that is best suited for the type of bend and size of the bar. *Table 1* provides size and weight information for various bars so that

Describe the ASTM specifications for reinforcing bars.

Show Transparency 9 (Figure 8).

Explain the markings on a typical reinforcing bar. Demonstrate how to identify various sizes, types, and grades of reinforcement steel based on markings.

Table 2   Reinforcement Bar Steel Types

Symbol/Marking	Type of Steel
A	Axle (ASTM A996)
S or N	Billet (ASTM A615)
I or IR	Rail (ASTM A996)
W	Low-alloy (ASTM A706) (for welded lap, butt joints, etc.)

Table 1   ASTM Standard Metric and Inch-Pound Reinforcing Bars

Bar Size		Nominal Characteristics*					
		Diameter		Cross-Sectional Area		Weight	
Metric	[in-lb]	mm	[in]	mm	[in]	kg/m	[lbs/ft]
#10	[#3]	9.5	[0.375]	71	[0.11]	0.560	[0.376]
#13	[#4]	12.7	[0.500]	129	[0.20]	0.944	[0.668]
#16	[#5]	15.9	[0.625]	199	[0.31]	1.552	[1.043]
#19	[#6]	19.1	[0.750]	284	[0.44]	2.235	[1.502]
#22	[#7]	22.2	[0.875]	387	[0.60]	3.042	[2.044]
#25	[#8]	25.4	[1.000]	510	[0.79]	3.973	[2.670]
#29	[#9]	28.7	[1.128]	645	[1.00]	5.060	[3.400]
#32	[#10]	32.3	[1.270]	819	[1.27]	6.404	[4.303]
#36	[#11]	35.8	[1.410]	1006	[1.56]	7.907	[5.313]
#43	[#14]	43.0	[1.693]	1452	[2.25]	11.380	[7.650]
#57	[#18]	57.3	[2.257]	2581	[4.00]	20.240	[13.600]

*The equivalent nominal characteristics of inch-pound bars are the values enclosed within the brackets.

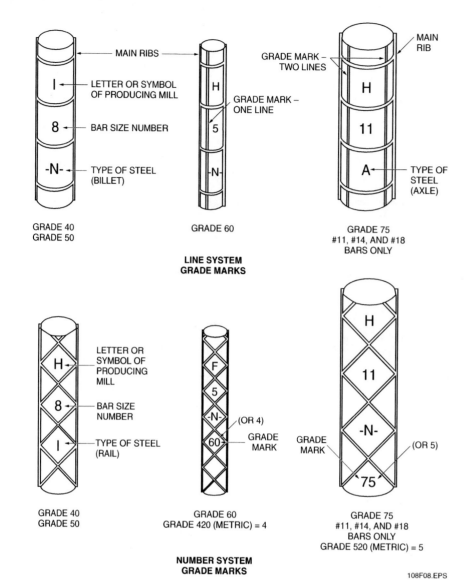

Figure 8 ♦ Reinforcement bar identification.

Table 3	Reinforcement Bar Grades	
Grade	**Identification**	**Minimum Yield Strength**
40 and 50	None	40,000 to 50,000 psi (40 to 50 ksi)
60	One line or the number 60	60,000 psi (60 ksi)
70	Two lines or the number 70	70,000 psi (70 ksi)
420	The number 4	420 MPa (60,000 psi or 60 ksi)
520	The number 5	520 MPa (75,000 psi or 70 ksi)

Instructor's Notes:

proper handling and bending equipment can be selected.

Uncoated reinforcement steel that has not been contaminated by oil, grease, or preservatives will normally rust when stored—even for short lengths of time under cover. A number of studies, some conducted over 70 years ago, have shown that rust and tight mill scale actually improve the bond between the steel and the concrete. Other studies have shown that normal handling (moving, bending, etc.) of extremely rusted reinforcement steel prepares it sufficiently for proper bonding with concrete without additional effort to remove the rust.

Explain that reinforcing bars must be sufficiently covered to prevent the rust on the bars from damaging the concrete.

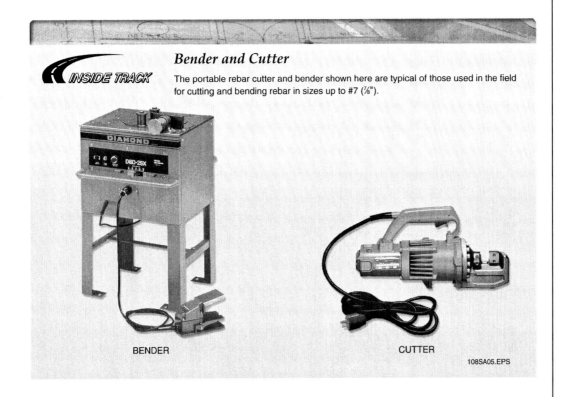

Bender and Cutter

The portable rebar cutter and bender shown here are typical of those used in the field for cutting and bending rebar in sizes up to #7 (⅞").

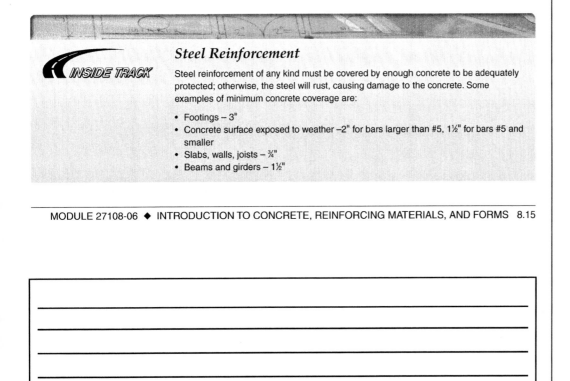

Steel Reinforcement

Steel reinforcement of any kind must be covered by enough concrete to be adequately protected; otherwise, the steel will rust, causing damage to the concrete. Some examples of minimum concrete coverage are:

- Footings – 3"
- Concrete surface exposed to weather –2" for bars larger than #5, 1½" for bars #5 and smaller
- Slabs, walls, joists – ¾"
- Beams and girders – 1½"

Discuss common types of bar supports and their applications. Pass around examples for the trainees to examine.

Show Transparency 10 (Figure 9).

Point out that the placing drawings will show the location and type of reinforcing bar splices.

Discuss the use of welded-wire fabric. Provide examples for the trainees to examine.

Show Transparency 11 (Figure 10).

Explain that OSHA requires exposed bars to be guarded to eliminate the hazard of impalement.

Assign reading of Sections 9.0.0–9.5.0 for the next class session.

8.2.0 Bar Supports

Bar supports are used to support, hold, and space reinforcing bars and mats or wire fabric before and during concrete placement. Bar supports are made from steel, concrete, or plastic. When used with coated reinforcement steel, the supports should also be coated with the same material or made of concrete or plastic to prevent corrosion. It is also important that a sufficient number of supports be used to prevent the reinforcement from shifting out of position or deforming when the concrete is placed.

Figure 9 shows a standard roll of tie wire and a wire tie (pigtail) used to secure lengths of reinforcing steel to each other or to various supports.

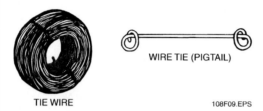

Figure 9 ◆ Tie wire and pigtail.

8.3.0 Splicing Reinforcing Bars

Because in most situations it is impossible to provide full-length bars that run continuously throughout a structure, making splices in reinforcing bars is a common occurrence. The placing drawings will show the location and type of splice to use. Sometimes several methods of splicing will be listed by the placing drawings, and the best method may be chosen from among them. No other type of splice should be used without consulting the proper authority.

There are three basic types of splices used in reinforcing steel work. These are the lap splice, the welded splice, and the mechanical-coupling splice.

8.4.0 Welded-Wire Fabric

When reinforcement is required for concrete pavement, parking lots, driveways, floor slabs, etc., welded-wire fabric (WWF) can be used instead of individual rebars. WWF consists of longitudinal and transverse steel wires electrically welded together to form a square or rectangular mesh or mat. Depending on the wire diameter, which can range up to ¾" or more, WWF is available in roll form or in 4' × 8' flat mats. *Figure 10* illustrates some standard sizes of plain wire WWF in roll form. Bar supports can be used to space and secure WWF as well as rebar.

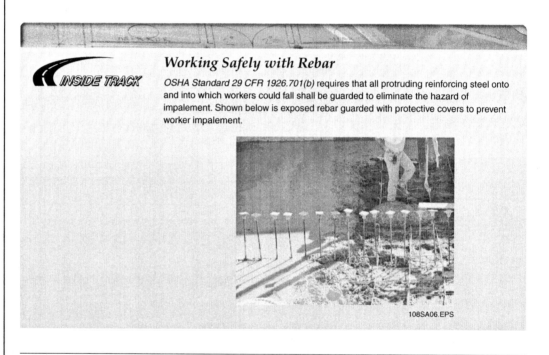

Working Safely with Rebar

OSHA Standard 29 CFR 1926.701(b) requires that all protruding reinforcing steel onto and into which workers could fall shall be guarded to eliminate the hazard of impalement. Shown below is exposed rebar guarded with protective covers to prevent worker impalement.

8.16 CARPENTRY FUNDAMENTALS LEVEL ONE ◆ TRAINEE GUIDE

Instructor's Notes:

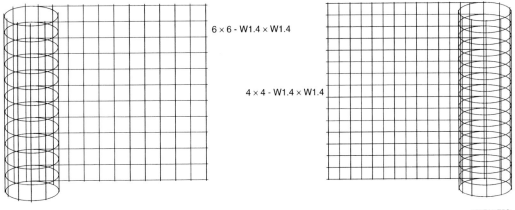

Figure 10 ◆ Typical rolled, plain wire welded-wire fabric.

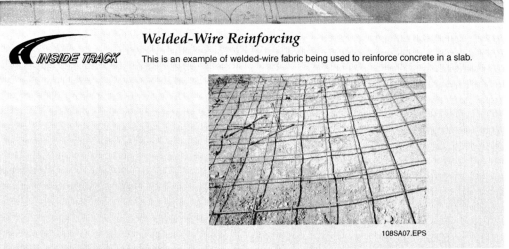

Welded-Wire Reinforcing

This is an example of welded-wire fabric being used to reinforce concrete in a slab.

9.0.0 ◆ CONCRETE FORMS

A concrete form is a temporary structure or mold used to support concrete until the hardening process gives it sufficient strength to be self-supporting. Carpenters generally assemble the job-built concrete forms that are used to construct foundations, walls, columns, and other concrete structures.

9.1.0 Form Safety

A concrete form is subjected to tremendous pressure from wet concrete. The pressure increases as the height of the wall increases. The rate at which the concrete is poured also affects the pressure on the form. The faster the rate of pour, the greater the pressure.

If the forms are not properly constructed and supported, or if the concrete is poured too quickly, the forms will not be able to withstand the pressure of the wet concrete. They will fracture, releasing the concrete. This can be very dangerous to workers, notwithstanding the mess it will make at the job site. Therefore, it is very important that the form builder construct the form in accordance with established standards and

Discuss the purpose of concrete forms.

Review the safety guidelines for basic formwork.

Discuss the use of footings as the basis for a foundation system.

Discuss the factors that determine the size and shape of footings, such as soil conditions, climate, and building codes.

Form Pressures

Concrete forms must be strong enough to resist the pressures created by plastic (liquid) concrete. These pressures greatly increase as the height of the wall or column form increases. This is because the amount of pressure at any point within the form is determined by the height and weight of the liquid concrete above it. For example, regular liquid concrete weighs about 150 pounds per cubic foot. If instantly placed in an 8' high wall form, this liquid concrete would create a pressure of about 1,200 pounds per square foot (150 × 8) at the bottom of the form. This amount of lateral pressure can be enough to cause the form to rupture. Pressures within a form increase as the placement rate of the concrete increases. For this reason, concrete is not placed in a wall form at a rapid rate, especially in high forms.

Concrete normally is placed in a form in layers about 12" to 18" deep using a slower, consistent placement rate. Use of a slower rate allows the concrete placed in a lower layer within the form enough time to begin to harden before more liquid concrete is placed on top of it. As the lower layer sets up and starts to harden, it becomes stable and ceases to exert pressure on the form even though liquid concrete continues to be placed above it.

safety practices. Basic formwork safety guidelines are:

- The structure must be capable of supporting the load created by the concrete, forms, machines, and people. Forms and **shoring** must be designed by qualified designers.
- Exposed reinforcing steel on which people could fall must be guarded.
- People are not permitted to work under concrete buckets. Buckets must be routed so as to avoid workers.
- Appropriate personal protective equipment must be worn.
- Equipment and tools used for the placement and finishing of concrete must be equipped with safety features as specified by *OSHA Standard 1926.702*.
- Formwork must be erected and braced in a manner that ensures it will support all vertical and lateral loads.
- Shoring equipment must be inspected and properly maintained. Damaged or defective shoring equipment must be repaired or replaced.
- Forms and shoring equipment may not be removed until the concrete has been verified as strong enough to support its weight and that of its loads.
- Reinforcing steel must be adequately supported to prevent it from falling or collapsing.
- Measures must be taken to prevent uncoiled wire mesh from recoiling.
- Measures must be taken to avoid skin contact with concrete, which is a hazardous material.

9.2.0 Footings

Footings provide the base of a foundation system for walls, columns, and chimneys. A footing bears directly on the soil and is made wider than the object it supports to distribute the weight over a greater area. Footings must be located accurately and must be built to the specified dimensions.

9.2.1 Factors That Determine the Size and Shape of Footings

The size and shape of footings are determined by:

- **Subgrade** loadbearing capacity
- Climate
- Building codes
- Size of the structure
- Type of material used

When designing the foundation for a structure, an engineer must consider the soil conditions of the building site. Weak or unstable soils may require larger footings or engineered foundations. Some soils are so weak that a slab may be required to spread the weight over the greatest possible area. Building a structure on soil where the soil conditions can cause a large amount of uneven settlement to occur can result in cracks in the foundation and structural damage to the rest of the building. Typically, the architect consults a soil engineer who makes test bores of the soil on the building site and analyzes these samples. Based on the results, the architect designs suitable footings and a foundation for the structure.

Instructor's Notes:

Footings

This is an example of concrete placed in a continuous footing form.

Reinforced Footings

Steel reinforcing rods, commonly called rebar, are embedded in footings to make them stronger. Additional reinforcement of footings is typically done over weak spots in the soil, such as where there have been excavations for sewer, gas, or other connections.

On sites where the soil is firm and not too porous, the construction of a footing form is often unnecessary. A trench is dug into the soil to the width and depth of the footing and the concrete is placed into the trench. Note that some problems can occur when using this method, such as the soil being porous enough to absorb too much water from the concrete, or the soil from the trench walls falling into the concrete.

WARNING

Any time you are working in or near a trench, you must follow applicable safety precautions.

Frost in the ground also affects foundation design. During the winter months in colder regions, moisture from the rain and snow penetrates the ground and freezes. The depth to which the soil freezes in a region is called the frost line. Freezing and thawing of the soil above the frost line causes the soil to expand and contract. This would cause foundations with footings placed above the frost line to heave and buckle. For this reason, foundation footings must always be installed below the frost line. Local building codes normally specify how deep the foundation footings should be in the area. In extreme northern climates, footings may be as much as 6' to 8' below the surface. In tropical climates, footings usually only need to reach solid soil, with no regard for a frost line.

Building codes can be restrictive when it comes to foundations. Codes regulate the size and thickness of footings and foundation walls, the depth of the footing, and other construction factors. In areas subject to earthquakes (seismic risk zones), the foundation will receive the forces of the quake and therefore needs to be built much stronger than foundations in non-seismic risk zones. In seismic risk zones, local building codes normally require that a foundation be designed to withstand greater stress. This is accomplished by embedding reinforcing steel rods (rebar) in all concrete or masonry foundation walls constructed in these zones.

Name the four basic types of footings.

Describe continuous footings, including the types of foundation walls they support.

Describe stepped continuous footings, emphasizing the importance of maintaining the proper distances between one part of the footing and another.

Show Transparencies 12 and 13 (Figures 11 and 12).

Discuss the use of piers and explain how their size is determined. List various pier shapes.

Show Transparency 14 (Figure 13).

9.2.2 Types of Footings

There are four basic types of footings:

- Continuous or spread
- Stepped continuous
- Pier
- Grade beam

Continuous footings – Continuous footings (*Figure 11*), also called spread footings, are commonly used to support poured concrete or concrete block foundation walls. They can be formed on the ground or partially in or out of the ground. The method used depends on the soil conditions and specifications defined in the working drawings. For a residential building with normal soil conditions, the width of a continuous footing is typically equal to twice the thickness of the foundation wall, and its height or depth is equal to the wall thickness. Both sides of the footing project out from the wall by one-half of the wall thickness. For commercial and industrial buildings, the dimensions of the footings vary depending on the type of structure and the building loads. Also, the foundation wall is not always centered on the footing.

When a foundation is to be built on hilly or steeply sloped land, it is necessary to step the footing at intervals (*Figure 12*) to achieve grade levels. When building stepped footing forms, the thickness of the footing must be maintained. Building codes normally govern the construction of stepped footings. Many building codes specify that for residential buildings the distance between one horizontal step and another must be no less than 2'. The vertical footing must be no higher than ¾ of the distance between one horizontal step and another and it must be at least 6" thick. Vertical boards are placed between the forms to retain the concrete at each step.

Piers – Piers are isolated footings set in soil to directly support posts, columns, or grade beams. The pier size, which is normally specified in the working drawings, is determined by soil conditions, the weight of the structure, and the loads supported by the structure. Piers are made in different shapes (*Figure 13*) including square, rectangular, tapered (battered), and round. Regardless of pier type, the ground must be dug so that the base of the pier rests on firm soil and is below the frost line, or is supported by a proper footing.

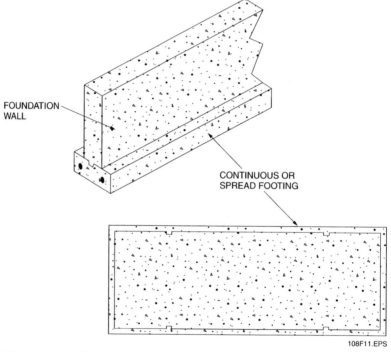

Figure 11 ♦ Continuous or spread footing.

8.20 CARPENTRY FUNDAMENTALS LEVEL ONE ♦ TRAINEE GUIDE

Instructor's Notes:

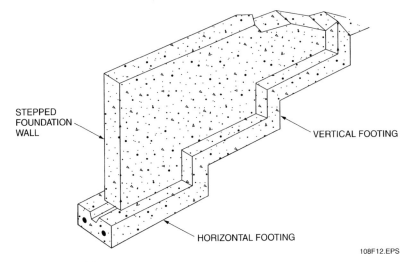

Figure 12 ♦ Continuous stepped footing.

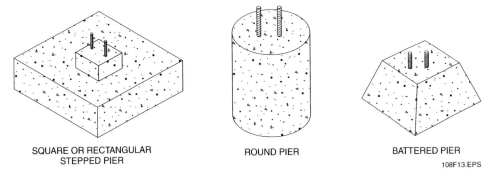

Figure 13 ♦ Typical piers.

Grade beams – Grade beams are foundation walls that receive their main support from concrete piers that extend deep into the ground (*Figure 14*). They are often used with stepped or ramped foundations on steep hillsides. Grade beams are also used on level sites where unstable soil conditions exist. In heavy construction, it may not be practical to excavate deep enough to reach loadbearing soil. Instead, **piles** are driven and capped with a grade beam.

9.2.3 Components of Footing and Pier Forms

The components of footing and pier forms are basically the same with some minor exceptions. Both types are described here.

Footing forms – Continuous or spread footing forms (*Figure 15*) are typically made from construction lumber (1×4s, 2×4s, 2×6s, etc.) selected to match the approximate width of the footing thickness. The use of thicker boards allows the form stakes to be placed farther apart and requires less bracing. The side (edge form) boards are placed on edge and held in place by the form stakes, braces, and cross spreaders. The form stakes are driven into the soil to provide primary support for the form.

The form members are fastened together with double-headed (duplex) nails or screws. The use of duplex nails allows for easy nail removal when the form is taken apart after use. The stakes, spreaders, and braces may be made of wood or metal. Wood stakes are typically made from 1×4

Discuss the applications of grade beams.

Show Transparency 15 (Figure 14).

Describe how footing forms are constructed.

Show Transparency 16 (Figure 15).

Explain the purpose of keyways and describe how they are formed.

Describe how pier forms are constructed. Explain that they can be disassembled for reuse.

Show Transparency 17 (Figure 16).

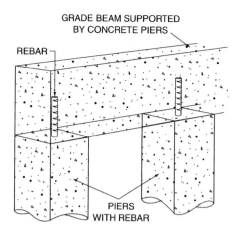

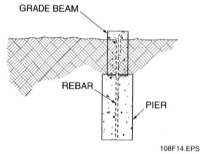

Figure 14 ♦ Grade beams.

or 2 × 4 lumber cut to length and pointed on one end. The length of the stakes used depends on the height of the form and the condition of the ground. Soft ground requires the use of longer stakes. Normally, a stake should be driven into the ground to a depth of 8" to 10". Metal stakes are commonly used when the ground is especially hard. These are made with a series of predrilled holes used for fastening them to the edge form boards with nails or screws.

A groove in the footing, called a keyway (*Figure 15*), shapes the footing concrete for interlocking with the future foundation wall pour. This joint acts to lock or secure the bottom of the foundation wall to the footing. It helps the foundation wall resist the pressure of the back-filled earth against it and also helps to prevent seepage of water into the basement or crawlspace.

A keyway strip is not always hung from the form spreaders, as shown in *Figure 15(A)*. Instead, after the concrete has been placed in the form and before it sets, pieces of beveled 2 × 4s (or metal or plastic strips) are pressed into the concrete toward the center of the footing. After the concrete has hardened, these pieces are removed, leaving the keyway groove, as shown in *Figure 15(B)*.

In seismic risk areas, horizontal and vertical rebars are positioned in the footing forms before the concrete is placed. The vertical rebars sticking out of the footing forms are later tied to the rebars placed in the foundation wall formwork.

Pier forms – Forms for piers in residential and light commercial construction typically are built by nailing plywood together in square, rectangular, or tapered (battered) shapes to the specified size. The form is built in such a way that it can be easily removed after the concrete is cured, and then reset for another pier footing. For battered forms, the angle involved should be specified in the working drawings. Note that when concrete is placed in a pier or column form, it exerts pressure that tends to push the form upward. Therefore, the form must be weighted down or staked to hold it securely in place. Forms for circular piers are generally manufactured as patented steel forms or as treated, waterproof, fiber tube forms that are cut to length on site. Steel forms can be reused, whereas fiber forms can only be used once. Fiber forms are set to grade height and staked. Pier forms for piers in commercial and industrial applications are typically built in a manner similar to the one shown in *Figure 16*. The sides of the form can be made from almost any exterior-type plywood sheathing.

The **walers** shown are horizontal members placed on the outside of the form to brace or stiffen the form and to which the form ties are fastened. Form ties are wire or metal cross ties used to hold the pressure of wet concrete and hold the form together to maintain its thickness. There are many types of form ties available, including washer-type, strap-type, cone-type, and she-bolt.

8.22 CARPENTRY FUNDAMENTALS LEVEL ONE ♦ TRAINEE GUIDE

Instructor's Notes:

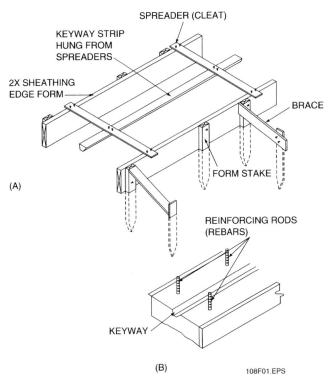

Figure 15 ♦ Parts of a continuous or spread footing form.

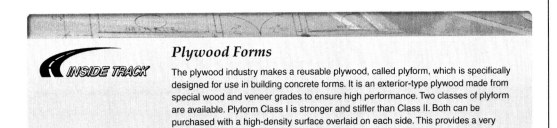

Plywood Forms

The plywood industry makes a reusable plywood, called plyform, which is specifically designed for use in building concrete forms. It is an exterior-type plywood made from special wood and veneer grades to ensure high performance. Two classes of plyform are available. Plyform Class I is stronger and stiffer than Class II. Both can be purchased with a high-density surface overlaid on each side. This provides a very smooth, grainless surface that resists abrasion and moisture penetration.

List the tasks involved in laying out and constructing forms for a continuous footing.

Discuss the different types of wall forms.

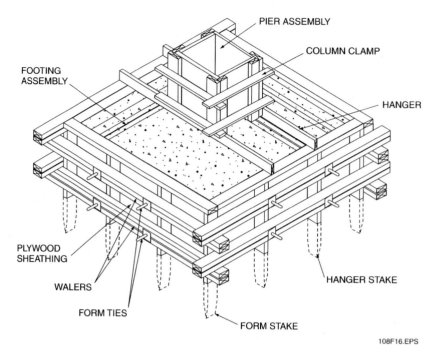

Figure 16 ♦ Parts of a stepped pier footing form.

9.2.4 Laying Out and Constructing Forms for a Continuous Footing

Laying out and constructing forms for a continuous footing involves the following tasks:

- Establishing the exact location of the building on site
- Establishing building lines
- Performing required excavation and/or trenching, observing applicable safety precautions
- Laying out and constructing footing forms

Before any footing or foundation layout can be done, the dimensions of the building and its location on the site must be determined from the site (plot) plan. Following this, the exact location of the four corners of the building must be determined and staked out. This task can be accomplished in different ways. One method involves taping and the use of a builder's level or transit level. You will learn this method later in your training.

9.3.0 Wall Forms

A wall form consists of two sets of panels separated by the thickness of the concrete wall. The following are different types of wall forms:

- Forms that are constructed primarily from wood panels, lumber, and nails, with some specialized hardware
- Forms that are constructed using wood panels and lumber support members secured with patented attaching hardware
- Patented form systems that use metal panels and factory-built attaching hardware

While there may be variations of the above, these represent the basic types. In this module, we will cover the forms that are constructed primarily from wood panels, lumber, and nails, with some specialized hardware.

9.3.1 Components of a Wall Form

See *Figure 17*. The main components of a wall form panel include the following:

- *Top and bottom plates* – The horizontal members of a wall frame panel.
- *Studs* – The upright members that support the sheathing. They are nailed to the top and bottom plates. Studs are usually 2 × 4 or 2 × 6 lumber. Heavier studs may be needed, depending on the pressure that will be applied to the forms by the concrete. In job-built wooden forms, studs are required. Many patented forms have the upright supports built into the frame and therefore do not require separate studs.
- *Sheathing* – Gives the wall surface its shape and texture. It keeps the concrete in place.
- *Walers (wales)* – Horizontal members used to support the form and the studs. Like the studs, walers usually are made of 2 × 4 or 2 × 6 lumber, but heavier lumber may be required by the particular job.
- *Strongbacks* – Upright members used to stiffen or reinforce the form.
- *Form ties or spreaders* – Used to keep the form from being spread apart by the weight of the wet concrete and to keep the walls from shifting while the concrete is being poured. Spreaders must be installed above the level of the concrete.
- *Braces* – Fastened to one side of the form and nailed to stakes driven into the ground. Braces are usually placed 8' to 10' apart. They support the form and are also used to plumb the form. Adjustable metal braces, which make it easier to plumb the form, are available (*Figure 18*). Forms must be properly braced to ensure that they do not collapse under the weight of the plastic concrete.

Sheathing can be made of exterior plywood sheets. Just about any type of exterior plywood may be used. However, a special product called **plyform** is popular because it is specifically made for use in concrete forms. Plyform is an exterior plywood that is made only from certain grades and types of wood. Plyform panels are sanded on both sides and are usually oiled at the mill. The oil reduces moisture penetration and aids in releasing the form from the concrete. Plyform edges exposed during construction must be treated with oil or another release agent before use. There are two basic grades of plyform (Class I and Class II), plus a structural grade that is designed to withstand greater pressures.

Planks may also be used to form walls, in some cases. The planks are nailed to studs.

Quarter-inch tempered hardboard that is specially treated and coated may also be used as

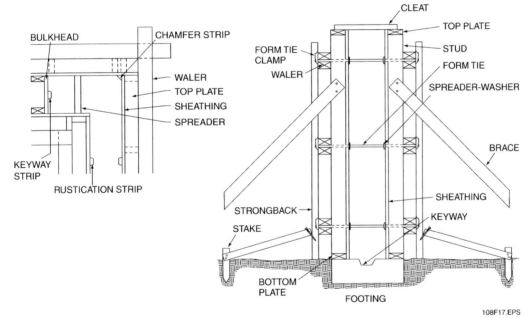

Figure 17 ◆ Parts of a wall form panel.

Explain the purpose of form ties and spreaders. Provide examples for the trainees to examine.

Show Transparency 20 (Figure 19).

Integrity of Wall Forms

Always make sure to use lumber that is free of defects when building wall forms, especially lumber used for braces and supports. Before placing concrete in a wall form or any other type of form, the form must be inspected for structural adequacy and compliance with the job specifications by a competent, authorized person.

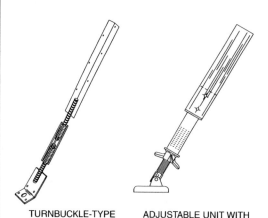

Figure 18 ◆ Adjustable form braces.

sheathing. If hardboard is used, however, it must be backed with lumber. Hardboard used in this manner acts as a liner to obtain a smooth concrete finish.

Form ties are metal rods or metal straps that hold the two sides of a wall form in position (*Figure 19*). Spreaders, which serve the same purpose, are often made from pieces of wood cut to the thickness of the wall and nailed in place between the two wall form sections above the concrete. Ties generally remain in the concrete; wood spreaders are removed as the concrete is poured to their level. Some ties are equipped with cones or other devices that allow them to act as spreaders.

When the forms are stripped, ties will project out from the hardened concrete. They are broken off, twisted off, disconnected, or removed, depending on the type of tie. The tapered tie is removable. With the other types of ties, a portion of the tie remains embedded in the concrete.

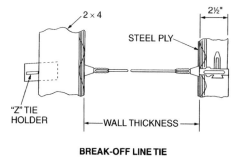
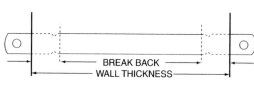

Figure 19 ◆ Ties and spreaders.

Ties go through both panels and either through a single waler or between double walers. There are several different systems for securing the ties to the form (*Figure 20*). In some systems, slotted metal wedges are driven behind the buttons at the ends of the ties. In other cases, bolts and washers are used.

Additional support is provided by horizontal members known as walers and vertical members known as strongbacks. Diagonal braces are also used to support the forms.

9.4.0 Edge Forms

Edge forms are low-height perimeter forms constructed to contain concrete poured for flat surfaces (**flatwork**) such as on-grade building slabs with or without a foundation, or for outdoor slabs such as parking lots, driveways, streets, sidewalks, and approaches. A slab is defined as a section of concrete that is larger in its horizontal dimensions than in its thickness. A **slab-on-grade** (**slab-at-grade**) is a concrete slab supported by the ground.

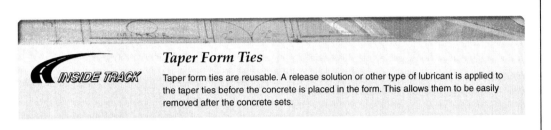

Taper Form Ties

Taper form ties are reusable. A release solution or other type of lubricant is applied to the taper ties before the concrete is placed in the form. This allows them to be easily removed after the concrete sets.

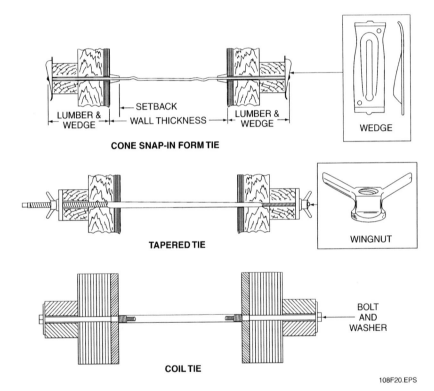

Figure 20 ♦ Methods of securing wall ties.

List the general requirements for the construction of on-grade slabs.

Show Transparency 22 (Figure 21).

Explain that slabs can be constructed either with a foundation or with a thickened edge.

Discuss the parts of an edge form.

Show Transparencies 23 and 24 (Figures 22 and 23).

Built-in-Place Form

This is an example of a partially constructed built-in-place form for a foundation wall.

9.4.1 General Requirements for On-Grade Slabs

Some general requirements associated with the construction of on-grade slabs are as follows:

- The finished level of the slab must be high enough so that the finished grade around the slab can be sloped for good drainage.
- The slab should be placed on a solid (compacted) base consisting of 4" to 6" of gravel, crushed stone, or other material to help prevent ground water from collecting under the slab. Never lay a slab without removing the top soil.
- The slab should be reinforced with fiber mesh, wire mesh, or rebar. Note that fiber mesh is placed in the concrete during mixing.
- For floor slabs, a vapor barrier must be placed under the concrete slab to prevent moisture from rising through the slab. Joints in the vapor barrier must be lapped at least 4" and sealed. The barrier material must be strong enough to resist puncturing during the placing of the concrete. Typically, 6 mil or greater polyethylene sheeting is used.
- In colder climates, a rigid waterproof insulation is typically installed around the perimeter of the slab to prevent heat loss through the floor and foundation walls.
- All required piping and ductwork to be run under the slab must be in position before the gravel and concrete are placed.

9.4.2 Types of Slabs

There are two basic kinds of slabs: slabs with a foundation and slabs with a thickened edge (*Figure 21*). For slabs with a foundation, the foundation walls are poured first, then the slab is poured separately after the walls have set up. For slabs with a thickened edge, the foundation and slab are poured as one unit. This is called a **monolithic slab (monolithic pour)**.

9.4.3 Parts of Edge Forms

Figure 22 shows the parts of a typical edge form constructed of wood, used for pouring concrete slabs without a wall and for pouring driveways, sidewalks, etc. As shown, the construction of this type of edge form is basically the same as for forms built for footings. *Figure 23* shows the construction of a typical edge form made of wood used for pouring concrete slabs with a foundation. Edge forms of both types are commonly made from construction lumber (1×4s, 2×4s, 2×6s, etc.) or plyform sheathing supported by wood or steel stakes and bracing. The edge form or sheathing contains the perimeter of the slab. The edge form material should be smooth, straight, and free of knotholes and other surface defects. Edge forms for large projects are commonly made from metal or a combination of wood and metal. Metal forms save time, can withstand rough handling, and can be used over and over again.

Instructor's Notes:

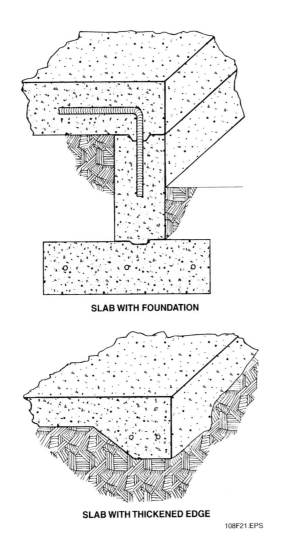

Figure 21 ♦ Types of slabs.

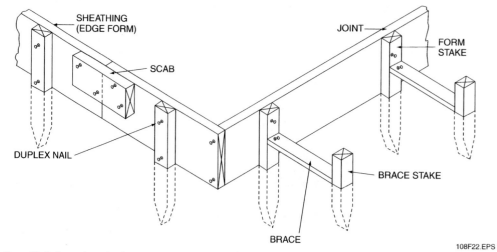

Figure 22 ◆ Parts of an edge form for a slab without a wall.

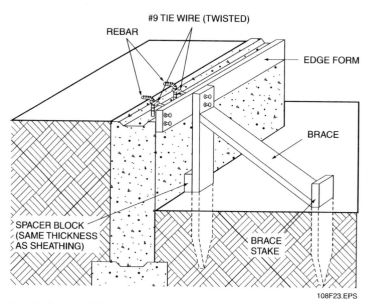

Figure 23 ◆ Parts of an edge form for a slab with a wall.

The form stakes driven into the soil provide primary support for the edge form. When used, brace stakes and the related braces provide additional form support. Braces and brace stakes should be used when the subgrade is too loose or when the form stakes cannot be driven deep enough. Wooden stakes can be made from 1×2, 1×4, 2×2, or 2×4 material. Stakes are normally spaced about 4' apart when using 2" edge form lumber. When using 1" edge form lumber, they are placed about 2' apart. Steel stakes are also made for use with wood or metal forms. They are easier to drive and last longer. Stakes should be driven straight and plumb so that the form is true. They should be driven slightly below the top of the forms to help in **screeding** and finishing the concrete slab. The edge form is attached to the stakes using duplex nails driven through the stakes, then into the edge form. Scabs are pieces of wood or other appropriate material fastened across a splice or joint in the form edge to strengthen and hold the pieces together. Another method of reinforcing a joint in the form edge is to make the joint at a form stake.

9.5.0 Removing Forms

Depending on strength and deflection considerations, forms for floor and similar kinds of slabs may have to be left in place for several days before removal. The time of removal should be determined by a structural engineer. On large construction jobs, the recommended times for removing various types of forms are often spelled out in the specifications.

Forms should be constructed in a manner that allows them to be easily taken apart and removed. Form surfaces that come in contact with the concrete should be coated with form oil before the concrete is placed to prevent the concrete from adhering to the forms. Generally, forms are taken apart in reverse order of the way in which they were built. Removal of forms must be done carefully to avoid damaging the concrete. If the forms are to be used again, they should be carefully removed using tools and equipment that will not damage either the concrete or the forms. Wooden wedges, not steel tools, should be used to separate the forms from the concrete.

Once removed, reusable forms should be cleaned, repaired, handled, and stored in such a way that they are not damaged. Wooden forms can be cleaned using a wood scraper and stiff fiber brush. Holes in forms can be patched with metal plates, corks, or plastic inserts. When the forms are coated with form oil, they should be allowed to dry before stacking or storing them. Reusable form hardware (such as form ties, nut washers, and wedges) should be sorted, cleaned, and stored for reuse. All damaged hardware should be thrown away.

Discuss the tasks associated with form removal. Cover:
- **Setting time before removal**
- **Order of removal**
- **Proper cleaning, handling, and storage**

Demonstrate how to erect, plumb, and brace a simple concrete form.

Under your supervision, have the trainees erect, plumb, and brace a simple concrete form. This laboratory corresponds to Performance Task 2.

Summarize the major concepts presented in the module.

Summary

In some instances, a carpenter will be called upon to mix concrete and build concrete forms. Concrete is a mixture of portland cement, sand, and water in specified proportions. When mixed, concrete is in a semiliquid (plastic) state and must be contained in forms until it hardens. Concrete will support a great deal of weight on its own, but must usually be reinforced with steel bars or steel mesh so it can resist lateral or shearing forces. Job-built concrete forms can be made entirely out of lumber and wood sheathing. In practice, however, many of the attaching components, such as form ties, are made of metal. Plyform is the most common sheathing used in forms because it is specially made for that purpose. Plyform is sanded on both sides and treated with oil, which makes it easier to strip. Only certain kinds of wood are used to make plyform.

The most important thing to remember about building or assembling forms is safety. If forms are improperly built or are not adequate for the concrete pressure they must handle, people can be killed or injured as a result of structural failure. Therefore it is very important to follow the plans and specifications established by the building designers, especially when it comes to supporting and bracing the forms.

Notes

Instructor's Notes:

Review Questions

1. The basic ingredients of concrete are _____.
   a. sand, water, and aggregates
   b. aggregates, water, and portland cement
   c. admixtures, aggregates, and water
   d. portland cement, water, and admixtures

2. Aggregates make up about _____ percent of the volume of concrete.
   a. 15
   b. 30
   c. 50
   d. 75

3. High-strength concrete is defined as concrete with a compressive strength over _____ psi.
   a. 4,000
   b. 6,000
   c. 8,000
   d. 10,000

4. A concrete calculator *cannot* be used when the volume required is under _____ yard.
   a. ¼
   b. ½
   c. ¾
   d. ⅞

5. Steel is used as concrete reinforcement because of its _____.
   a. weight
   b. strength
   c. expansion/contraction rate
   d. resistance to sulfate attack

6. Reinforcement bars are identified by a line or number system that designates their _____.
   a. weight
   b. grade
   c. manufacturer
   d. type of steel

7. Footings provide the base of a foundation system for _____.
   a. patios
   b. walkways
   c. driveways
   d. chimneys

8. Construction of forms for a footing is sometimes *not* needed at sites _____
   a. in tropical climates with firm surface topsoil conditions
   b. where the soil is firm and not too porous
   c. where the frost line is less than 2" to 3" below the surface
   d. where the frost line is more than 6" to 8" below the surface

9. A _____ is a type of footing that is commonly used at level sites where unstable soil conditions exist.
   a. continuous footing
   b. stepped continuous footing
   c. pier
   d. grade beam

Have trainees complete the Review Questions, and go over the answers prior to administering the Module Examination.

Review Questions

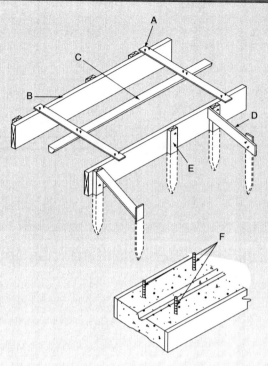

Figure 1

10. The letter D in *Figure 1* is pointing to _____.
    a. a stake
    b. a spreader
    c. a brace
    d. sheathing

11. The letter C in *Figure 1* is pointing to _____.
    a. a keyway strip
    b. a spreader
    c. a brace
    d. rebar

12. Which of the following requirements applies when using tempered hardboard as form sheathing?
    a. It must be sanded on both sides.
    b. It must be backed with lumber.
    c. It may not have any special treatment or coating.
    d. It may only be used as a form liner in combination with plywood or plyform sheets.

Instructor's Notes:

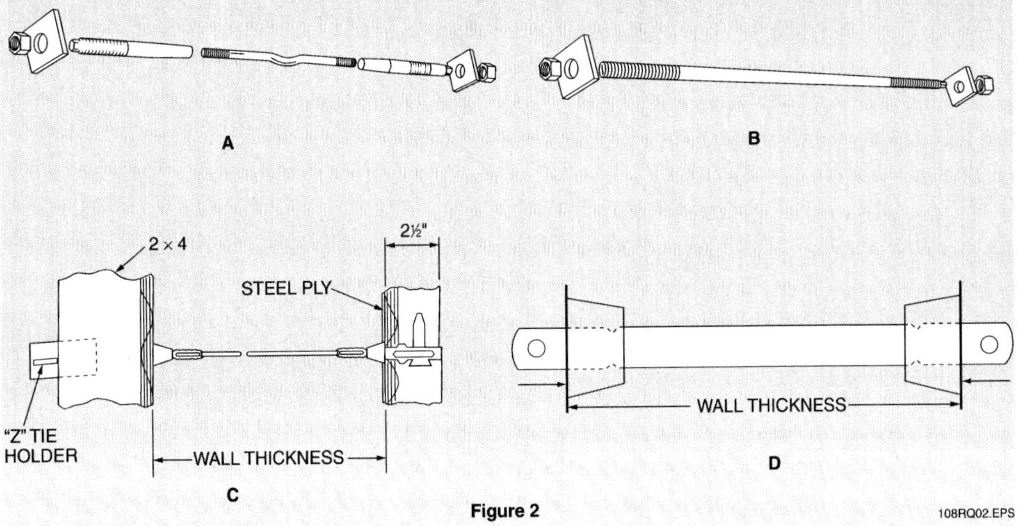

Review Questions

Figure 2

13. Which item in *Figure 2* shows a form tie that can be completely removed after the concrete has hardened?
    a. A
    b. B
    c. C
    d. D

14. A piece of wood placed inside a form to hold the sides of the form the proper distance apart is known as a ____.
    a. spandrel
    b. cleat
    c. ledger
    d. spreader

15. A slab poured as a single unit of concrete is called a(n) ____.
    a. above-grade slab
    b. monolithic slab
    c. slab with a foundation
    d. slab without a foundation

Have trainees complete the Trade Terms Quiz, and go over the answers.

Administer the Module Examination. Record the results on Craft Training Report Form 200, and submit the form to the Training Program Sponsor.

Administer the Performance Tests, and fill out Performance Profile Sheets for each trainee. If desired, trainee proficiency noted during laboratory sessions may be used to complete the Performance Test. Record the results on Craft Training Report Form 200, and submit the results to the Training Program Sponsor.

Trade Terms Quiz

1. The metric unit of pressure is the _____.
2. A diagonal form support is a(n) _____.
3. Vertical form supports are called _____.
4. _____ is used to support above-grade slabs until they harden.
5. The reinforcing bars used in concrete are also called _____.
6. The materials used as filler in concrete are _____.
7. The materials used to alter the properties of concrete, such as the setting time, are _____.
8. When concrete is first mixed, it is called _____.
9. When concrete has hardened, but has not yet cured, it is called _____.
10. When concrete has gained its full structural strength, it is called _____.
11. _____ provide support for grade beams or columns that carry the structural load of a building.
12. Freshly mixed concrete should be tested for the proper _____ before it is placed.
13. _____ is the chemical reaction that causes cement to harden.
14. Reinforcing bars that are rolled from carbon-steel axles used on railroad cars are known as _____.
15. Reinforcing bars that are rolled from used railroad rails are known as _____.
16. Because concrete is a semiliquid, it must be contained in _____ until it hardens.
17. _____ is a byproduct of the ironmaking process.
18. Reusable concrete forms are often constructed of _____.
19. The building of slabs for walkways, patios, and driveways is known as _____.
20. A concrete slab that is supported by the ground is a(n) _____.
21. One thousand pounds is also referred to as one _____.
22. The strength of cured concrete is determined by the _____.
23. _____ is an international organization that develops standards for the design and construction of concrete structures.
24. A wall rests on a foundation known as a(n) _____.
25. The ancient Romans used a type of volcanic ash cement known as _____.
26. Form ties are attached to horizontal form supports known as _____.
27. After the concrete is placed, it is leveled to an established grade using a process known as _____.
28. Building a foundation directly on an unstable _____ can result in cracks and structural damage.
29. A(n) _____ is a continuous pour without construction joints.

Trade Terms

ACI
Admixtures
Aggregates
Axle-steel
Brace
Cured concrete
Flatwork
Footing
Forms
Green concrete
Hydration
Kip
Monolithic slab (monolithic pour)
Pascal
Piles
Plastic concrete
Plyform
Pozzolan
Rail-steel
Rebars
Screeding
Shoring
Slab-on-grade (slab-at-grade)
Slag
Slump
Studs
Subgrade
Walers
Water-cement ratio

8.36 CARPENTRY FUNDAMENTALS LEVEL ONE ♦ TRAINEE GUIDE

Instructor's Notes:

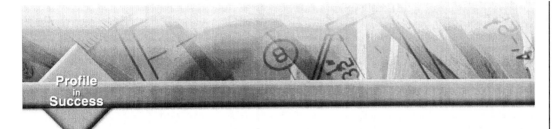

Profile in Success

John Payne
Craft Training Adminstrator
Brasfield and Gorrie

When he was a young man, John worked with his dad on odd jobs around the house. In high school, he had a summer job as a builder's assistant. His main occupation is as a field foreman or field engineer. Now he works for Brasfield and Gorrie, which was ranked by *Engineering News-Record* among the top 400 contractors in the United States.

What kinds of work have you done in your career?
I went to college for industrial education, but that only lasted one semester. I preferred working with my hands. I got a job as a surveyor's rodman. Through the years, I always worked on carpentry crews. For eight years, I was a superintendent doing mid-rise commercial buildings. Currently, I am in craft training administration and will be moving up to become a general superintendent soon.

My most mentally challenging project was a nine-story domed structure. I spent many late nights figuring out how to lay it out. It was all radial math. The most physically challenging project I did was a seven-story office building. We poured 248,000 square feet of concrete in four months.

What are some of the things you do in your job?
I run our in-house training program for carpenters, surveyors, and foremen. Our 19 operating divisions maintain a combined staff of more than 2,600 skilled professionals. In my company, our greatest strength is our people and their performance. This job gave me the opportunity to be in one place and not travel as much as I did doing field work.

What do you like about the work you do?
The construction business is a tangible occupation. At the end of the day, you can see and touch what you have built. In my present job, I like to talk about what I do. I support our organization through longevity, training the new people, building the future.

What do you think it takes to be a success in your trade?
It's the old adage, show up on time and be enthusiastic. This industry does not lend itself to the wishy-washy. You have to stay after it, keep your eyes wide open, and learn every day. There are many opportunities to learn new things. They are always developing new products, new materials, and new methods of doing things.

What advice would you give someone just starting out?
Explore all areas of the trade and the construction industry. What you enjoy doing will become evident. It may be carpentry or it could be another trade. You only know if you try them out. If you enjoy your work, you will put in the time to learn your trade.

Study your math skills. There is a direct correlation between your ability to think rationally and your ability to do simple math. It is impossible to advance without a good math background and reasoning skills. I see my students in their late teens and early twenties who realize that they should have paid more attention in math class. Anyone can cut to the mark, but the one who can tell where to make the mark gets paid more.

Trade Terms Introduced in This Module

ACI: American Concrete Institute.

Admixtures: Materials that are added to a concrete mix to change certain properties of the concrete such as retarding setting time, reducing water requirements, or making the concrete easier to work with.

Aggregates: Materials used as filler in concrete; may include mixtures of sand, gravel, crushed stone, crushed gravel, or blast-furnace slag.

Axle-steel: Deformed reinforcing bars that are rolled from carbon-steel axles used on railroad cars.

Brace: A diagonal supporting member used to reinforce a form against the weight of the concrete.

Cured concrete: Concrete that has hardened and gained its structural strength.

Flatwork: Work connected with concrete slabs used for walks, driveways, patios, and floors.

Footing: The base of a foundation system for a wall, column, and chimney. It bears directly on the undisturbed soil and is made wider than the object it supports to distribute the weight over a greater area.

Forms: Wood or metal structures built to contain plastic concrete until it hardens.

Green concrete: Concrete that has hardened but has not yet gained its structural strength.

Hydration: The catalytic action water has in transforming the chemicals in portland cement into a hard solid. The water interacts with the chemicals to form calcium silicate hydrate gel.

Kip: An informal unit of force that equals one thousand (kilo) pounds.

Monolithic slab (monolithic pour): Concrete placed in forms in a continuous pour without construction joints.

Pascal: A metric measurement of pressure.

Piles: Column-like structural members that penetrate through unstable, nonbearing soil to lower levels of loadbearing soil. They provide support for grade beams or columns that carry the structural load of a building.

Plastic concrete: Concrete when it is first mixed and is in a semiliquid and moldable state.

Plyform: American Plywood Association's tradename for a reusable material for constructing concrete forms.

Pozzolan: The name given by the ancient Romans to describe the volcanic ash they used as a type of cement. Today, the term is used for natural or calcined materials (including fly ash and silica fume) or air-cooled blast furnace slag.

Rail-steel: Deformed reinforcing bars that are rolled from selected used railroad rails.

Rebars: Abbreviation for reinforcing bars. Also called rerod.

Screeding: Leveling newly placed concrete to an established grade. Also called striking off.

Shoring: Temporary bracing used to support above-grade concrete slabs while they harden.

Slab-on-grade (slab-at-grade): A ground-supported concrete slab 3½" or thicker that is used as a foundation system. It combines concrete foundation walls with a concrete floor slab that rests directly on an approved base that has been placed over the ground.

Slag: The ash produced during the reduction of iron ore to iron in a blast furnace.

Slump: The distance a standard-sized cone made of freshly mixed concrete will sag. This is known as a slump test.

Studs: Vertical members of a form panel used to support sheathing.

Subgrade: Soil prepared and compacted to support a structure or pavement system.

Walers: Horizontal pieces placed on the outsides of the form walls to strengthen and stiffen the walls. The form ties are also fastened to the walers.

Water-cement ratio: The ratio of water to cement, usually by weight (water weight divided by cement weight), in a concrete mix. The water-cement ratio includes all cementitious components of the concrete, including fly ash and pozzolans, as well as portland cement.

Instructor's Notes:

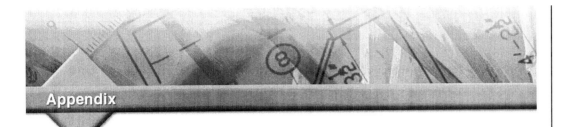

Appendix

Formulas for Geometric Shapes and Conversion Tables

Conversion Table for Changing Measurements
(in Inches) to Fractions and Decimal Parts of a Foot

Inches	Fractional Part of Foot	Decimal Part of Foot
1	1/12	0.08
2	1/6	0.17
3	1/4	0.25
4	1/3	0.33
5	5/12	0.42
6	1/2	0.50
7	7/12	0.58
8	2/3	0.67
9	3/4	0.75
10	5/6	0.83
11	11/12	0.92
12	1	1.00

Metric Conversion of Pounds to Grams or Kilograms

1 pound (16 ounces) = 453.6 grams or
0.4536 kilogram (kg)

AREAS OF PLANE FIGURES

NAME / FORMULA (A = Area)	SHAPE
Parallelogram $A = B \times h$	parallelogram with base B and height h
Trapezoid $A = \dfrac{B + C}{2} \times h$	trapezoid with parallel sides C (top) and B (bottom), height h
Triangle $A = \dfrac{B \times h}{2}$	triangle with base B and height h
Trapezium (Divide into 2 triangles) A = Sum of the 2 triangles (See above)	irregular quadrilateral
Regular Polygon $A = \dfrac{\text{Sum of sides (s)}}{2} \times \text{inside Radius (R)}$	hexagon with side s and inside radius R
Circle $\pi = 3.14$ (1) πR^2 $A =$ (2) $.784 \times D^2$	circle with circumference C, diameter D, radius R
Sector (1) $\dfrac{a^2}{360°} \times \pi R^2$ $A =$ (2) Length of arc $\times \dfrac{R}{2}$ ($\pi = 3.14$)	circle with sector marked
Segment A = Area of sector minus triangle (see above)	circle with segment marked
Ellipse $A = M \times m \times .7854$	ellipse with major axis M and minor axis m
Parabola $A = B \times \dfrac{2h}{3}$	parabola with base B and height h

VOLUMES OF SOLID FIGURES

NAME / FORMULA (V - volume)	SHAPE
Cube $V = a^3$ (in cubic units)	cube with side a
Rectangular Solids $V = L \times W \times h$	rectangular solid with length L, width W, height h
Prisms $V(1) = \dfrac{B \times A}{2} \times h$ $V(2) = \dfrac{s \times R}{2} \times 6 \times h$ V = Area of end × h	triangular prism and hexagonal prism
Cylinder $V = \pi R^2 \times h$ ($\pi = 3.14$)	cylinder with radius R and height h
Cone $V = \dfrac{\pi R^2 \times h}{3}$ ($\pi = 3.14$)	cone with radius R and height h
Pyramids $V(1) = L \times W \times \dfrac{h}{3}$ $V(2) = \dfrac{B \times A}{2} \times \dfrac{h}{3}$ V = Area of Base $\times \dfrac{h}{3}$	rectangular pyramid and triangular pyramid
Sphere $V = \dfrac{1}{6} \pi D^3$	sphere with diameter D
Circular Ring (Torus) $V = 2\pi^2 \times R r^2$ V = Area of section $\times 2\pi R$	torus with radii R and r

METRIC CONVERSION CHART

inches (fractions)	inches (decimals)	mm	inches (fractions)	inches (decimals)	mm	inches (fractions)	inches (decimals)	mm	inches (fractions)	inches (decimals)	mm
–	.0004	.01	25/32	.781	19.844	–	2.165	55.	3-11/16	3.6875	93.663
–	.004	.10	–	.7874	20.	2-3/16	2.1875	55.563	–	3.7008	94.
–	.01	.25	51/64	.797	20.241	–	2.2047	56.	3-23/32	3.719	94.456
1/64	.0156	.397	13/16	.8125	20.638	2-7/32	2.219	56.356	–	3.7401	95.
–	.0197	.50	–	.8268	21.	–	2.244	57.	3-3/4	3.750	95.250
–	.0295	.75	53/64	.828	21.034	2-1/4	2.250	57.150	–	3.7795	96.
1/32	.03125	.794	27/32	.844	21.431	2-9/32	2.281	57.944	3-25/32	3.781	96.044
–	.0394	1.	55/64	.859	21.828	–	2.2835	58.	3-13/16	3.8125	96.838
3/64	.0469	1.191	–	.8661	22.	2-5/16	2.312	58.738	–	3.8189	97.
–	.059	1.5	7/8	.875	22.225	–	2.3228	59.	3-27/32	3.844	97.631
1/16	.062	1.588	57/64	.8906	22.622	2-11/32	2.344	59.531	–	3.8583	98.
5/64	.0781	1.984	–	.9055	23.	–	2.3622	60.	3-7/8	3.875	98.425
–	.0787	2.	29/32	.9062	23.019	2-3/8	2.375	60.325	–	3.8976	99.
3/32	.094	2.381	59/64	.922	23.416	–	2.4016	61.	3-29/32	3.9062	99.219
–	.0984	2.5	15/16	.9375	23.813	2-13/32	2.406	61.119	–	3.9370	100.
7/64	.109	2.778	–	.9449	24.	2-7/16	2.438	61.913	3-15/16	3.9375	100.013
–	.1181	3.	61/64	.953	24.209	–	2.4409	62.	3-31/32	3.969	100.806
1/8	.125	3.175	31/32	.969	24.606	2-15/32	2.469	62.706	–	3.9764	101.
–	.1378	3.5	–	.9843	25.	–	2.4803	63.	4	4.000	101.600
9/64	.141	3.572	63/64	.9844	25.003	2-1/2	2.500	63.500	–	4.0551	103.188
5/32	.156	3.969	1	1.000	25.400	–	2.5197	64.	4-1/16	4.062	104.775
–	.1575	4.	–	1.0236	26.	2-17/32	2.531	64.294	4-1/8	4.125	105.
11/64	.172	4.366	1-1/32	1.0312	26.194	–	2.559	65.	–	4.1338	106.363
–	.177	4.5	–	1.062	26.988	2-9/16	2.562	65.088	4-3/16	4.1875	107.950
3/16	.1875	4.763	1-1/16	1.063	27.	2-19/32	2.594	65.881	4-1/4	4.250	109.538
–	.1969	5.	1-3/32	1.094	27.781	–	2.5984	66.	4-5/16	4.312	110.
13/64	.203	5.159	–	1.1024	28.	2-5/8	2.625	66.675	–	4.3307	111.125
–	.2165	5.5	1-1/8	1.125	28.575	–	2.638	67.	4-3/8	4.375	112.713
7/32	.219	5.556	–	1.1417	29.	2-21/32	2.656	67.469	4-7/16	4.438	114.300
15/64	.234	5.953	1-5/32	1.156	29.369	–	2.6772	68.	4-1/2	4.500	115.
–	.2362	6.	–	1.1811	30.	2-11/16	2.6875	68.263	–	4.5275	115.888
1/4	.250	6.350	1-3/16	1.1875	30.163	–	2.7165	69.	4-9/16	4.562	117.475
–	.2559	6.5	1-7/32	1.219	30.956	2-23/32	2.719	69.056	4-5/8	4.625	119.063
17/64	.2656	6.747	–	1.2205	31.	2-3/4	2.750	69.850	4-11/16	4.6875	120.
–	.2756	7.	1-1/4	1.250	31.750	–	2.7559	70.	–	4.7244	120.650
9/32	.281	7.144	1-9/32	1.2598	32.	2-25/32	2.781	70.6439	4-3/4	4.750	120.650
–	.2953	7.5	–	1.281	32.544	–	2.7953	71.	4-13/16	4.8125	122.238
19/64	.297	7.541	1-5/16	1.2992	33.	2-13/16	2.8125	71.4376	4-7/8	4.875	123.825
5/16	.312	7.938	–	1.312	33.338	–	2.8346	72.	–	4.9212	125.
–	.315	8.	–	1.3386	34.	2-27/32	2.844	72.2314	4-15/16	4.9375	125.413
21/64	.328	8.334	1-11/32	1.344	34.131	–	2.8740	73.	5	5.000	127.000
–	.335	8.5	1-3/8	1.375	34.925	2-7/8	2.875	73.025	–	5.1181	130.
11/32	.344	8.731	–	1.3779	35.	2-29/32	2.9062	73.819	5-1/4	5.250	133.350
–	.3543	9.	1-13/32	1.406	35.719	–	2.9134	74.	5-1/2	5.500	139.700
23/64	.359	9.128	–	1.4173	36.	2-15/16	2.9375	74.613	–	5.5118	140.
–	.374	9.5	1-7/16	1.438	36.513	–	2.9527	75.	5-3/4	5.750	146.050
3/8	.375	9.525	–	1.4567	37.	2-31/32	2.969	75.406	–	5.9055	150.
25/64	.391	9.922	1-15/32	1.469	37.306	–	2.9921	76.	6	6.000	152.400
–	.3937	10.	–	1.4961	38.	3	3.000	76.200	6-1/4	6.250	158.750
13/32	.406	10.319	1-1/2	1.500	38.100	3-1/32	3.0312	76.994	–	6.2992	160.
–	.413	10.5	1-17/32	1.531	38.894	–	3.0315	77.	6-1/2	6.500	165.100
27/64	.422	10.716	–	1.5354	39.	3-1/16	3.062	77.788	–	6.6929	170.
–	.4331	11.	1-9/16	1.562	39.688	–	3.0709	78.	6-3/4	6.750	171.450
7/16	.438	11.113	–	1.5748	40.	3-3/32	3.094	78.581	7	7.000	177.800
29/64	.453	11.509	1-19/32	1.594	40.481	–	3.1102	79.	–	7.0866	180.
15/32	.469	11.906	–	1.6142	41.	3-1/8	3.125	79.375	–	7.4803	190.
–	.4724	12.	1-5/8	1.625	41.275	–	3.1496	80.	7-1/2	7.500	190.500
31/64	.484	12.303	–	1.6535	42.	3-5/32	3.156	80.169	–	7.8740	200.
–	.492	12.5	1-21/32	1.6562	42.069	3-3/16	3.1875	80.963	8	8.000	203.200
1/2	.500	12.700	1-11/16	1.6875	42.863	–	3.1890	81.	–	8.2677	210.
–	.5118	13.	–	1.6929	43.	3-7/32	3.219	81.756	8-1/2	8.500	215.900
33/64	.5156	13.097	1-23/32	1.719	43.656	–	3.2283	82.	–	8.6614	220.
17/32	.531	13.494	–	1.7323	44.	3-1/4	3.250	82.550	9	9.000	228.600
35/64	.547	13.891	1-3/4	1.750	44.450	–	3.2677	83.	–	9.0551	230.
–	.5512	14.	–	1.7717	45.	3-9/32	3.281	83.344	–	9.4488	240.
9/16	.563	14.288	1-25/32	1.781	45.244	–	3.3071	84.	9-1/2	9.500	241.300
–	.571	14.5	–	1.8110	46.	3-5/16	3.312	84.1377	–	9.8425	250.
37/64	.578	14.684	1-13/16	1.8125	46.038	3-11/32	3.344	84.9314	10	10.000	254.001
–	.5906	15.	1-27/32	1.844	46.831	–	3.3464	85.	–	10.2362	260.
19/32	.594	15.081	–	1.8504	47.	3-3/8	3.375	85.725	–	10.6299	270.
39/64	.609	15.478	1-7/8	1.875	47.625	–	3.3858	86.	11	11.000	279.401
5/8	.625	15.875	–	1.8898	48.	3-13/32	3.406	86.519	–	11.0236	280.
–	.6299	16.	1-29/32	1.9062	48.419	–	3.4252	87.	–	11.4173	290.
41/64	.6406	16.272	–	1.9291	49.	3-7/16	3.438	87.313	–	11.8110	300.
–	.6496	16.5	1-15/16	1.9375	49.213	–	3.4646	88.	12	12.000	304.801
21/32	.656	16.669	–	1.9685	50.	3-15/32	3.469	88.106	13	13.000	330.201
–	.6693	17.	1-31/32	1.969	50.006	3-1/2	3.500	88.900	–	13.7795	350.
43/64	.672	17.066	2	2.000	50.800	–	3.5039	89.	14	14.000	355.601
11/16	.6875	17.463	–	2.0079	51.	3-17/32	3.531	89.694	15	15.000	381.001
45/64	.703	17.859	2-1/32	2.03125	51.594	–	3.5433	90.	–	15.7480	400.
–	.7087	18.	–	2.0472	52.	3-9/16	3.562	90.4877	16	16.000	406.401
23/32	.719	18.256	2-1/16	2.062	52.388	–	3.5827	91.	17	17.000	431.801
–	.7283	18.5	–	2.0866	53.	3-19/32	3.594	91.281	–	17.7165	450.
47/64	.734	18.653	2-3/32	2.094	53.181	–	3.622	92.	18	18.000	457.201
–	.7480	19.	2-1/8	2.125	53.975	3-5/8	3.625	92.075	19	19.000	482.601
3/4	.750	19.050	–	2.126	54.	3-21/32	3.656	92.869	–	19.6850	500.
49/64	.7656	19.447	2-5/32	2.156	54.769	–	3.6614	93.	20	20.000	508.001

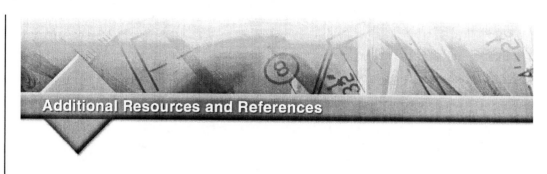

Additional Resources and References

This module is intended to present thorough resources for task training. The following reference works are suggested for further study. These are optional materials for continuing education rather than for task training.

Concrete Masonry Handbook for Architects, Engineers, and Builders, Fifth Edition. W.C. Panarese, S.H. Kosmatka, and F.A. Randall, Jr. Portland Cement Association.

The Homeowner's Guide to Building with Concrete, Brick, and Stone. The Portland Cement Association.

Instructor's Notes:

MODULE 27108-06 — ANSWERS TO REVIEW QUESTIONS

Answer	Section
1. b	2.0.0
2. d	2.2.0
3. b	4.0.0
4. b	7.1.0
5. c	8.0.0
6. b	8.1.0
7. d	9.2.0
8. b	9.2.1
9. d	9.2.2
10. c	9.2.3; Figure 15
11. a	9.2.3; Figure 15
12. b	9.3.1
13. b	9.3.1; Figure 19
14. d	9.3.1
15. b	9.4.2

MODULE 27108-06 — ANSWERS TO TRADE TERMS QUIZ

1. Pascal
2. Brace
3. Studs
4. Shoring
5. Rebars
6. Aggregates
7. Admixtures
8. Plastic concrete
9. Green concrete
10. Cured concrete
11. Piles
12. Slump
13. Hydration
14. Billet-steel
15. Rail-steel
16. Forms
17. Slag
18. Plyform
19. Flatwork
20. Slab-on-grade (slab-at-grade)
21. Kip
22. Water-cement ratio
23. ACI
24. Footing
25. Pozzolan
26. Walers
27. Screeding
28. Subgrade
29. Monolithic slab (monolithic pour)

MODULE 27108-06 — TEACHING TIPS

The following are suggested activities or instructional methods to help you teach the material in this module.

General

When you call on someone to answer a question, the rest of the class relaxes or even tunes out because they expect that the question and answer will take place only between you and the trainee you called on. Instead, use this technique to involve more trainees in answering questions and to keep them on their toes.

1. Ask trainees to define a term or explain a concept.
2. After one trainee has answered, ask a trainee seated nearby if the answer is right. Then ask whether a trainee in the back of the room agrees.
3. Ask trainees to explain why they think an answer is right or wrong.
4. Use the session to clear up incorrect ideas and encourage trainees to learn from their mistakes.

MODULE 27108-06—"THINK ABOUT IT" ANSWERS

Section 2.2.0 *Think About It—Concrete, Mortar, and Grout*

Concrete is made by mixing portland cement, fine aggregate, coarse aggregate, and water in proper proportions. All four ingredients must be used in the mixture for it to be called concrete. A mixture consisting of cement, fine aggregate, and water (but no coarse aggregate) is called mortar. When this same mixture of cement, fine aggregate, and water has a high water-to-cement ratio, it is called grout.

CONTREN® LEARNING SERIES — USER UPDATE

NCCER makes every effort to keep these textbooks up-to-date and free of technical errors. We appreciate your help in this process. If you have an idea for improving this textbook, or if you find an error, a typographical mistake, or an inaccuracy in NCCER's Contren® textbooks, please write us, using this form or a photocopy. Be sure to include the exact module number, page number, a detailed description, and the correction, if applicable. Your input will be brought to the attention of the Technical Review Committee. Thank you for your assistance.

Instructors – If you found that additional materials were necessary in order to teach this module effectively, please let us know so that we may include them in the Equipment/Materials list in the Annotated Instructor's Guide.

Write: Product Development and Revision
National Center for Construction Education and Research
P.O. Box 141104, Gainesville, FL 32614-1104

Fax: 352-334-0932

E-mail: curriculum@nccer.org

Craft _____ Module Name _____

Copyright Date _____ Module Number _____ Page Number(s) _____

Description _____

_____

_____

_____

_____

(Optional) Correction _____

_____

_____

_____

(Optional) Your Name and Address _____

_____

_____

_____

Windows and Exterior Doors
27109-06

NCCER STANDARDIZED CRAFT TRAINING PROGRAM

The National Center for Construction Education and Research (NCCER) provides a standardized national program of accredited craft training. Key features of the program include instructor certification, competency-based training, and performance testing. The program provides trainees, instructors, and companies with a standard form of recognition through the National Registry. The program is described in full in the *Guidelines for Accreditation*, published by NCCER. For more information on standardized craft training, contact NCCER by writing to P.O. Box 141104, Gainesville, FL 32614-1104; calling 352-334-0911; or e-mailing **info@nccer.org**. More information is available at **www.nccer.org**.

HOW TO USE THIS ANNOTATED INSTRUCTOR'S GUIDE

Each page presents two sections of information. The larger section displays each page exactly as it appears in the Trainee Module. The narrow column ties suggested trainee and instructor actions to each page and provides icons (detailed below) to call your attention to material, safety, audiovisual, or testing requirements. The bottom of each page includes space for your notes.

 The **Audiovisual** icon indicates an appropriate time to show a transparency or other audiovisual aid.

 The **Classroom** icon prompts you to define a term, stress a point, ask trainees to explain a concept, or give examples.

 The **Demonstration** icon directs you to show trainees how to perform tasks.

 The **Examination** icon tells you to administer the written module examination.

 The **Homework** icon is placed where you may wish to assign reading for the next class, assign a project, or advise trainees to prepare for an examination.

 The **Laboratory** icon is used when trainees are to practice performing tasks.

 The **Materials** icon is a reminder for you to gather materials needed for classes, laboratories, and testing.

 The **Performance Testing** icon tells you to administer a performance test or a portion thereof.

 The **Safety** icon is used to emphasize safety issues. It is often keyed to *Caution* and *Warning!* statements in the Trainee Module.

 The **Teaching Tip** icon indicates additional guidance is available, such as how to conduct an exercise, get the most educational value from a field trip, or encourage class participation. Teaching Tips may expand on a feature (*Think About It, Did You Know?*) or provide *Quick Quizzes* or similar exercises. You will be referred to the Teaching Tips section at the back of the module if there is additional material.

 The **Combination** icon indicates that the laboratory listed corresponds with a performance task. If desired, you can note the proficiency of the trainees during the laboratory, and use it to satisfy performance testing requirements.

PREPARATION

Before teaching this module, you should review the Objectives, Performance Tasks, Materials and Equipment List, and Module Outline. Be sure to allow ample time to prepare your own training or lesson plan and gather all required materials and equipment.

Windows and Exterior Doors
Annotated Instructor's Guide

Module 27109-06

MODULE OVERVIEW

This module introduces the carpentry trainee to methods and procedures used in the selection and installation of residential windows and exterior doors.

PREREQUISITES

Prior to training with this module, it is suggested that the trainee shall have successfully completed *Core Curriculum*; and *Carpentry Fundamentals Level One*, Modules 27101-06 through 27108-06.

OBJECTIVES

Upon completion of this module, the trainee will be able to do the following:

1. Identify various types of fixed, sliding, and swinging windows.
2. Identify the parts of a window installation.
3. State the requirements for a proper window installation.
4. Install a pre-hung window.
5. Identify the common types of exterior doors and explain how they are constructed.
6. Identify the parts of a door installation.
7. Identify the types of thresholds used with exterior doors.
8. Install a pre-hung exterior door.
9. Identify the various types of locksets used on exterior doors and explain how they are installed.
10. Install a lockset.

PERFORMANCE TASKS

Under the supervision of the instructor, the trainee should be able to do the following:

1. Install a pre-hung window.
2. Install a pre-hung exterior door.
3. Install a lockset.

MATERIALS AND EQUIPMENT LIST

Transparencies
Markers/chalk
Blank acetate sheets
Transparency pens
Pencils and scratch paper
Overhead projector and screen
Whiteboard/chalkboard
Appropriate personal protective equipment
Manufacturer's catalogs and brochures on windows
Nails:
    4d finish
    6d finish
    8d finish or casing
    16d casing
Pre-hung window unit
Shims
Flashing or drip cap
Pre-hung door unit
Wood shingles for blocking shims
Fiberglass insulation or sill sealer
Lockset with manufacturer's instructions and template (if needed)
Weatherstripping
Screws for attaching weatherstripping
Threshold and manufacturer's installation instructions
Concrete screw anchors and screws
Miter saw
Hand levels
Handsaw
Claw hammer
Framing square
Combination square
Steel tape
30" level
Nail set
Caulking gun and sealer
Boring jig (if available)
Wood chisels
Tin snips
Utility knife

Screwdriver
Drill
Drill bits

Copies of Job Sheets 1 through 6*
Module Examinations**
Performance Profile Sheets**

* Packaged with this Annotated Instructor's Guide.
**Located in the Test Booklet.

SAFETY CONSIDERATIONS

Ensure that the trainees are equipped with appropriate personal protective equipment.

ADDITIONAL RESOURCES

This module is intended to present thorough resources for task training. The following reference works are suggested for both instructors and motivated trainees interested in further study. These are optional materials for continued education rather than for task training.

Window and Door Manufacturers Association. A trade organization representing 145 U.S. window and door manufacturers. **www.wdma.com**

The National Fenestration Rating Council (NFRC). The nation's recognized authority for measuring and evaluating window energy performance. **www.nfrc.org.**

Window & Door magazine. An information source for manufacturers, distributors, and dealers of windows and doors. **www.windowanddoor.net.**

TEACHING TIME FOR THIS MODULE

An outline for use in developing your lesson plan is presented below. Note that each Roman numeral in the outline equates to one session of instruction. Each session has a suggested time period of 2½ hours. This includes 10 minutes at the beginning of each session for administrative tasks and one 10-minute break during the session. Approximately 12½ hours are suggested to cover *Windows and Exterior Doors*. You will need to adjust the time required for hands-on activity and testing based on your class size and resources. Because laboratories often correspond to Performance Tasks, the proficiency of the trainees may be noted during these exercises for Performance Testing purposes.

Topic	Planned Time
Session I. Introduction; Windows	
A. Introduction	_____
B. Windows	_____
1. Window Construction	_____
2. Types of Windows	_____
3. Types of Window Glass	_____
4. Window Installation	_____
5. Glass Blocks	_____
Session II. Laboratory	
A. Laboratory	_____
Hand out Job Sheet 27109-1. Under your supervision, have the trainees perform the tasks on the Job Sheet. This laboratory corresponds to Performance Task 1.	
Session III. Exterior Doors	
A. Exterior Doors	
1. Exterior Door Sizes	_____
2. Thresholds	_____
3. Weatherstripping	_____

   B. Laboratory

      Hand out Job Sheets 27109-2 and 27109-3. Under your supervision, have the trainees perform the tasks on the Job Sheets. (This laboratory is optional based on available time and materials.)

Session IV. Installing an Exterior Pre-Hung Door

   A. Installing an Exterior Pre-Hung Door

      1. Locksets

   B. Laboratory

      Hand out Job Sheets 27109-4 and 27109-5. Under your supervision, have the trainees perform the tasks on the Job Sheets. This laboratory corresponds to Performance Tasks 2 and 3.

Session V. Laboratory; Review; Module Examination and Performance Testing

   A. Laboratory

      Hand out Job Sheet 27109-6. Under your supervision, have the trainees perform the tasks on the Job Sheet. (This laboratory is optional based on available time and materials.)

   B. Review

   C. Module Examination

      1. Trainees must score 70 percent or higher to receive recognition from NCCER.

      2. Record the testing results on Craft Training Report Form 200, and submit the results to the Training Program Sponsor.

   D. Performance Testing

      1. Trainees must perform each task to the satisfaction of the instructor to receive recognition from NCCER. If applicable, proficiency noted during laboratory exercises can be used to satisfy the Performance Testing requirements.

      2. Record the testing results on Craft Training Report Form 200, and submit the results to the Training Program Sponsor.

Windows and Exterior Doors
27109-06

A Carpentry contestant in the SkillsUSA 2005 National Championships prepares to measure and cut a length of lumber for the miniature building framing. Other contestants are shown in the background. Each competitor in the National Championship is a champion of their state contest held earlier in the year, meaning there is a maximum of 50 competitors in the high school division and 50 in the college division.

Instructor's Notes:

Assign reading of Module 27109-06.

27109-06
Windows and Exterior Doors

Topics to be presented in this module include:

1.0.0 Introduction 9.2
2.0.0 Windows 9.2
3.0.0 Exterior Doors 9.14

Overview

Carpenters are regularly called upon to install windows and exterior doors. The doors and windows you encounter will usually be prehung. That is, the window or door will be installed in a frame that is then installed in the rough opening. While this may sound simple, it is not. It requires particular skills that can only be learned through practice.

There are many kinds of windows—double-hung, bay, casement, and jalousie, to name a few. There also several types of exterior doors and many types of door hardware, not to mention thresholds and weatherstripping. The carpenter must learn about all these elements and their unique installation requirements.

Window and door installation is a test of the framing quality. If your wall isn't square, you will have problems when it comes to installing windows and doors.

Instructor's Notes:

Objectives

When you have completed this module, you will be able to do the following:

1. Identify various types of fixed, sliding, and swinging windows.
2. Identify the parts of a window installation.
3. State the requirements for a proper window installation.
4. Install a pre-hung window.
5. Identify the common types of exterior doors and explain how they are constructed.
6. Identify the parts of a door installation.
7. Identify the types of thresholds used with exterior doors.
8. Install a pre-hung exterior door.
9. Identify the various types of locksets used on exterior doors and explain how they are installed.
10. Install a lockset.

Trade Terms

Casing
Curb
Flashing
Glazing
Hipped
Jamb
Lights
Muntin
Rail
Sash
Sill
Stile

Required Trainee Materials

1. Pencil and paper
2. Appropriate personal protective equipment

Prerequisites

Before you begin this module, it is recommended that you successfully complete *Core Curriculum* and *Carpentry Fundamentals Level One*, Modules 27101-06 through 27108-06.

This course map shows all of the modules in *Carpentry Fundamentals Level One*. The suggested training order begins at the bottom and proceeds up. Skill levels increase as you advance on the course map. The local Training Program Sponsor may adjust the training order.

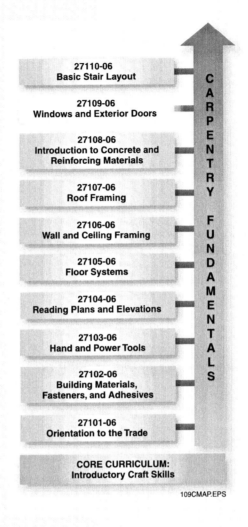

Ensure you have everything required to teach the course. Check the Materials and Equipment List at the front of this module.

Show Transparency 1, Objectives.

Show Transparency 2, Performance Tasks.

Explain that terms shown in bold (blue) are defined in the Glossary at the back of this module.

See the general Teaching Tip at the end of this module.

Discuss some of the basic terms related to windows.

Gather photos of local buildings with various window arrangements. Discuss window aesthetics.

Identify the components of a sash.

1.0.0 ♦ INTRODUCTION

The final step in drying-in a house is closing off the structure with windows and exterior doors. There are so many types and styles of windows and exterior doors, made by so many manufacturers, that all the possibilities cannot be covered in this module. The purpose of this module is to introduce you to the special terms you will need to know and to give you an overview of the various kinds of windows and exterior doors and the important installation practices related to them.

2.0.0 ♦ WINDOWS

In this section, we will introduce the many kinds of windows used in residential construction and will provide an overview of important installation practices that apply to windows. We will start by defining what a window is. When we use the term *window* in this program, it will refer to the entire window assembly, including the glass, **sash,** and frame. These terms are explained in the next paragraph. In this section, we will focus on pre-hung windows; that is, windows that are delivered complete with sash, frame, and hardware.

2.1.0 Window Construction

Window sashes can be made from wood, metal, or vinyl. Aluminum and steel are used in metal window construction. Wood windows, which are subject to decay, must be protected with wood preservatives and paint. Ponderosa pine, carefully selected and kiln-dried to a moisture content of 6% to 12%, is commonly used in making wooden windows. Wood is often preferred over aluminum because it does not conduct heat as readily. Metal window frames are usually filled with insulating material to make them more energy efficient. Although many windows are made of wood, it has become more common to find wooden windows clad on the outside with aluminum or vinyl.

Aluminum windows are much lighter, easier to handle, and more durable than wooden windows.

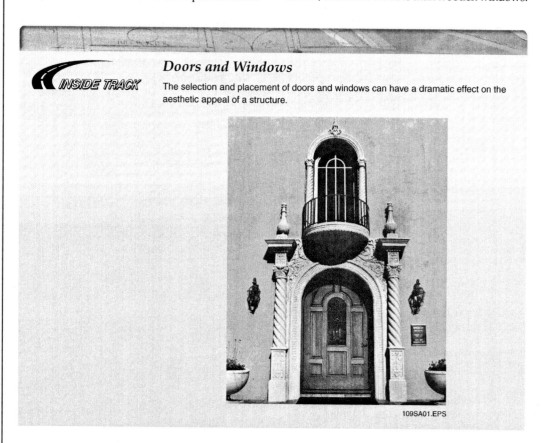

INSIDE TRACK

Doors and Windows
The selection and placement of doors and windows can have a dramatic effect on the aesthetic appeal of a structure.

Aluminum windows are generally coated to prevent corrosion. However, they are not recommended for coastal areas where the corrosive effect of salt air can be extreme.

Steel windows are stronger than aluminum and wood, but are far more expensive. They are more common in commercial buildings than residential construction.

Vinyl window frames are made of impact-resistant polyvinyl chloride (PVC). They are resistant to heat loss and condensation. Vinyl windows cannot be painted, so they must be purchased in a color that is compatible with the building. Vinyl windows may distort when exposed to extreme temperatures. This can cause difficulty in operating the window, as well as increased air infiltration.

The basic component of a window assembly is the sash, which is the framework around the glass. *Figure 1* identifies the component parts of the sash. Note that the sections of glass are called **lights**. The sash may contain several lights or just one light with a false **muntin** known as a *grille*. The grille does not support the glass; it is simply installed on the face of the glass for decorative purposes.

The sash fits into, or is attached to, a window frame consisting of **stiles** and **rails**. The entire window assembly, including the frame and sash (or sashes), is shipped from the factory. The window frame consists of the head **jamb**, side jambs, and **sill**, as shown in *Figure 2*.

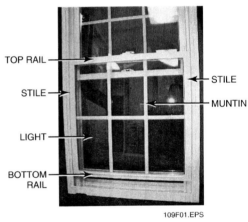

Figure 1 ◆ Parts of a sash.

2.2.0 Types of Windows

There are several types of windows commonly used in residential construction. This section provides an overview of each type.

- *Fixed windows* – In a fixed window, the glass cannot be opened. Fixed windows come in a huge variety of styles and shapes. They are often used to create an attractive, distinctive architectural appearance (*Figure 3*). Decorative **glazing** known as art glass (similar to stained glass) is sometimes used to achieve dramatic effects in these windows.
- *Single- and double-hung windows* – (See *Figure 2*.) These windows contain two sashes. In the double-hung window, the upper sash can be lowered and the lower sash can be raised. In a single-hung window, the upper sash is stationary and the lower sash can be raised. Many double-hung windows have a tilt-and-wash feature that allows the outside of the window to be washed from the inside. In one application of this feature, the jamb can be pushed in enough to allow the top of the sash to pivot inward.
- *Casement windows* – Casement windows are hinged on the left or right of the frame so that they can swing open like a door (*Figure 4*). They are usually operated by a hand crank. A swivel arm prevents the window from swinging wide open. Casement windows can be equipped with a limited ventilation control that limits the opening to a few inches. While this feature provides a certain amount of security from outside entry and accidental falls by occupants, it can also prevent people on the inside from getting out during an emergency. It is therefore not recommended for windows that may be used as emergency exits. Casement windows are often installed side by side and may be used in combination with decorative fixed windows.
- *Awning and hopper windows* – Awning windows are hinged on the top and pushed outward to open (*Figure 4*). They are often operated by a hand crank. Awning windows are commonly used in combination with fixed windows. The hopper window is hinged on the bottom and opens inward from the top. It is commonly used for basement windows. The hopper window is equipped with a locking mechanism. It also has pivot arms on the sides to keep the window from falling.

Identify different types of residential windows, and explain the applications for which they are used.

Show Transparencies 3 through 5 (Figures 2 through 4).

Show the trainees a double-hung window in a local building. Identify its components.

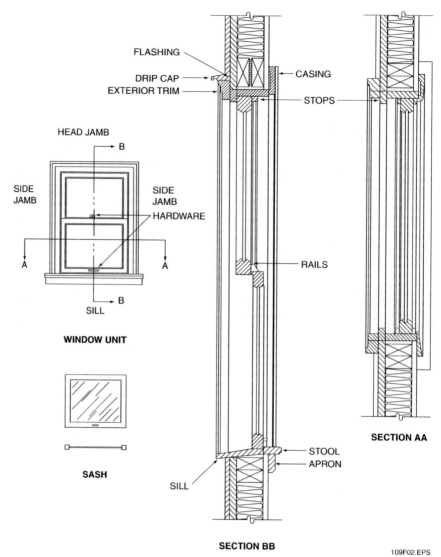

Figure 2 ♦ Parts of a double-hung wood window.

9.4 CARPENTRY FUNDAMENTALS LEVEL ONE ♦ TRAINEE GUIDE

Instructor's Notes:

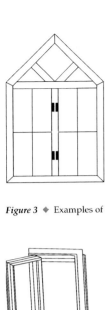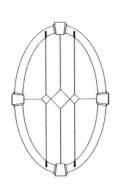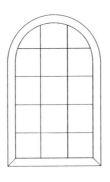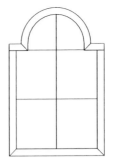

Figure 3 ♦ Examples of fixed window applications.

Discuss various types of bay windows and explain how they are supported by the building structure.

Show Transparencies 6 and 7 (Figures 6 and 7).

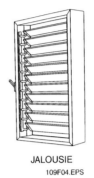

CASEMENT HOPPER AWNING JALOUSIE

Figure 4 ♦ Examples of windows.

- *Jalousie windows* – A jalousie window consists of a series of horizontal glass slats, each in a pivoting metal frame (*Figure 4*). It can provide security while at the same time providing ventilation. The frames of a jalousie window are joined by pivoting arms so that they operate in unison. A hand crank is used to open and close the slats.
- *Bay windows* – A bay window is a three-walled window that projects outward from the structure (*Figure 5*). Bay windows may be constructed from casement, fixed, or double-hung windows (*Figure 6*). Combinations are often used, such as a fixed window in the center and casement or double-hung windows on the sides. The angles of the bay window may be 30°, 45°, or 90°. The latter is known as a box bay window. Under certain load conditions, bay windows must be supported by cables that are secured to the building structure (*Figure 7*). This method is usually recommended for projecting units that do not have a support wall beneath them.

Figure 5 ♦ Bay window.

Explain the difference between a bay window and a bow window.

Show Transparency 8 (Figure 8).

Gather manufacturers' catalogs or photographs of local buildings containing examples of bay windows. Pass them around for the trainees to examine.

DOUBLE-HUNG BAY

CASEMENT BAY

BOX OR GARDEN BAY

109F06.EPS

Figure 6 ♦ Examples of bay windows.

- *Bow windows* – A bow window projects out from the structure in a curved radius (*Figure 8*). A bow window is normally made of several narrow, flat planes set at slight angles to each other. Like the bay window, a bow window is commonly made up of casement windows or a combination of casement and fixed windows.

Transom Windows

Transom windows were common in the high-ceiling buildings erected in the early part of the 20th century. Today, transom windows are becoming popular again because of the growing number of new high-ceiling homes. These decorative windows are usually shaped as a semicircle or rectangle, and may be either fixed windows or awning types. They are often used as decorative "toppers" for double-hung windows, as shown here.

109SA02.EPS

9.6 CARPENTRY FUNDAMENTALS LEVEL ONE ♦ TRAINEE GUIDE

Instructor's Notes:

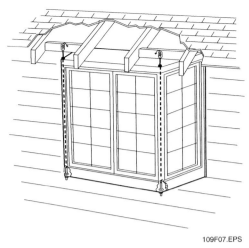

Figure 7 ♦ Supported bay window.

- *Horizontal sliding windows* – A horizontal sliding (gliding) window may be constructed from a number of sashes (*Figure 9*). The most common designs have either two sashes, with one or both sashes movable, or three sashes with the middle sash fixed and the other two movable. Frames are either wood or metal. A locking handle is normally added at installation.
- *Skylights and roof windows* – The difference between a skylight and a roof window is that the roof window can be opened (*Figure 10*). There are many designs and shapes of skylights. Domed and **hipped** skylights are the most common because they more readily shed rain and snow. Skylights and roof windows are often installed using a built-up **curb** (*Figure 11*). Metal **flashing** is installed around the curb to prevent leakage. Another style of skylight is self-flashing. This type has an integral metal flange that is screwed into the roof and sealed on both sides with roofing cement during installation.

INSIDE TRACK

Flashing

Metal flashing is essential to preventing moisture from entering the roof around the seams of roof windows, skylights, and chimneys.

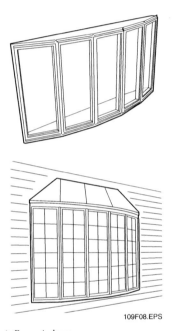

Figure 8 ♦ Bow windows.

Discuss sliding glass windows and skylights.

Show Transparencies 9 through 11 (Figures 9 through 11).

Gather manufacturers' catalogs or photographs of local buildings containing examples of skylights. Pass them around for the trainees to examine.

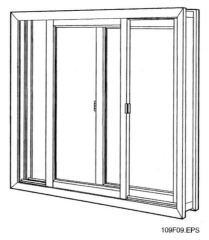

Figure 9 ◆ Sliding (gliding) window.

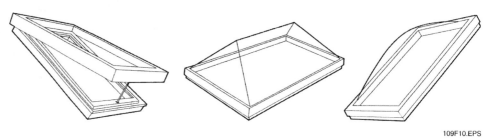

Figure 10 ◆ Roof windows and skylights.

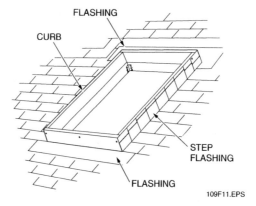

Figure 11 ◆ Skylight curb.

9.8 CARPENTRY FUNDAMENTALS LEVEL ONE ◆ TRAINEE GUIDE

2.3.0 Types of Window Glass

There are several kinds of sheet glass made for use in window glazing. Single-strength (SS) glass is about 1/32" thick. It is used only for small lights of glass. Double-strength (DS) glass, which is about 1/8" thick, can be used for larger lights. Heavy-duty glass ranges in thickness from 3/16" to 7/8".

Glass is a good conductor of heat. In any building, glass accounts for the majority of the heat loss in cold months and heat gain in warm months. When an engineer is sizing the heating/cooling system for a building, window glass normally accounts for 38% of the cooling load and 20% of the heating load (*Figure 12*). The term load refers to the amount of heat that must be added (heating mode) or removed (cooling mode) to keep the building comfortable for occupants.

Money spent on energy-efficient windows is well spent. For example, the use of high-efficiency reflective glass instead of standard single-pane glass could reduce the cooling load by one-third. The load directly affects the size of the heating/cooling system, and therefore directly affects the original cost of the system, as well as the energy cost involved in operating the system.

2.3.1 Energy-Efficient Windows

A single pane of glass provides very little insulation. It has an R-value (insulating value) of less than 1. (The greater the R-value, the greater the insulating value.) Adding another pane with 1/2" of air space more than doubles the R-value. The air space between the panes of glass acts as an insulator. The larger the air space, the more insulation it provides. Windows are commonly designed with two or three layers of glass separated by 3/16" to 1" of air space in order to improve insulation quality (*Figure 13*). To obtain even more insulating value, the space between panes in some windows is filled with argon gas, which conducts heat at a lower rate than air. Where single-pane glass is used, it is common to add storm windows.

A special type of glass known as low-e (low-emissivity) provides even greater insulating

Summarize the physical properties of glass, and identify the impact of glass types on energy use.

Show Transparencies 12 and 13 (Figures 12 and 13).

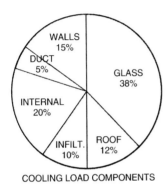
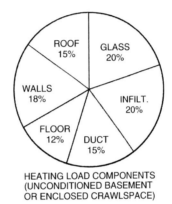

TYPICAL LOAD PERCENTAGES

Figure 12 ◆ Cooling and heating load factors.

Reflective Glass

INSIDE TRACK

Scientists working in the space program developed the first reflective glass. Spacecraft are subjected to high levels of friction when flying through the Earth's atmosphere at great speeds. Reflective glass was developed in an attempt to reflect, transfer, and dissipate high temperatures. Today, reflective glass is used on the space shuttle. This gives astronauts a clear field of vision to navigate the shuttle, and also provides heat protection upon re-entry into the atmosphere.

Point out that some codes require the use of safety glass in certain applications.

Explain how to install a double-hung window.

Energy-Efficient Glass

Low-e glass is coated with a thin metallic substance and is an effective way to control radiant heat transfer. Special heat-absorbing glass is also available. It contains tints that can absorb approximately 45% of incoming solar energy. This energy (heat) is then transferred from the window to the building structure. If these new types of glass aren't available, there are other alternatives that will increase energy efficiency in existing windows. For example, reflective film can be applied to windows to help contain heat loss in the winter and reflect sunlight in the summer. Caulking, weatherstripping, and storm windows, which are discussed in detail later in this module, are also simple approaches to energy conservation.

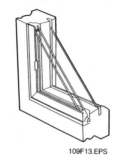

109F13.EPS

Figure 13 ◆ Double- and triple-pane windows.

2.3.2 Safety Glass

Some local and state codes require the use of safety glass in windows with very low sills, and in those located in or near doors. Skylights and roof windows also require safety glass. There are several types of safety glass. Laminated glass contains two or more layers of glass bonded to transparent plastic. Tempered glass is treated with heat or chemicals. When broken, it disintegrates into tiny, harmless pieces. Wired glass has a layer of mesh sandwiched between the panes. The mesh keeps the glass from shattering. Transparent plastic (plexiglass) is also used for safety glazing applications, especially in skylights and doors.

2.4.0 Window Installation

This section contains the basic procedure for installing a pre-hung window. As you know, there are many types of windows and many types of installations. The best approach is to follow the manufacturer's recommended procedure, using the recommended tools, materials, and fasteners.

The basic procedure for installing a pre-hung window is as follows:

Step 1 Make sure the window is shut before you start the installation. Also, ensure that the opening is plumb, level, and square, and is large enough to accommodate the window. You will want to leave ¼" to ½" between the rough header and the window head jamb to account for settling.

properties. Emissivity is the ability of a material to absorb or radiate heat. Low-e glass is coated with a very thin metallic substance on the inside of the inner pane of a double-pane window. In cold weather, radiated heat from walls, floors, furniture, etc., is reflected back into the room by the low-e coating instead of escaping through the windows. This reduces the heat loss, which in turn saves heating costs. In summer months, radiated heat from outdoor sources such as the sun, roads, parking lots, etc., is reflected away from the building by the low-e coating. Although windows with low-e glass are considerably more expensive than standard windows, they usually pay for themselves in reduced heating and cooling costs within three or four years.

Check Opening for Squareness

To check the window opening for squareness, use a framing square. If the opening is not square, make the necessary adjustments and check again. You shouldn't proceed until the opening is square and level.

House Wrap

When house wrap is installed over a rough opening, don't cut the wrap diagonally across the entire opening. Instead, cut diagonally from the bottom corners to the center, then cut straight up from the center to the top of the opening, as shown below. Wrap the material over the sill and sides of the opening and secure it. Cut the wrap flush with the bottom of the header, but do not secure it. This will allow the nailing flange to slide under the wrap for a more secure installation.

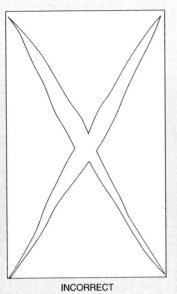

INCORRECT

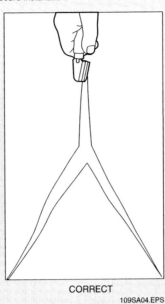

CORRECT

109SA04.EPS

Step 2 Remove protection blocks used to protect the window during shipping. Also remove any horns, which are side jamb extensions that are also used for protection during shipping. If there are diagonal braces on the window, leave them in place temporarily.

Step 3 Install the window in the opening. Make sure you have enough help to avoid injuring yourself or damaging the window.

Step 4 If necessary, insert wedges, wooden shingle tips, or shims under the window to raise it to the correct height. Temporarily nail one corner of the window through the flange or wooden frame.

NOTE

Vinyl and metal windows, as well as clad wooden windows, are equipped with nailing flanges (fins) around the outside (*Figure 14*). Use screws or roofing nails to secure the window.

Step 5 Check the level of the sill. If it is not level, use shims to correct it. On windows with long sills, use shims at intermediate points to ensure that the sill is level with no sag (*Figure 15*). As each side is leveled, secure it with a nail or screw as recommended by the manufacturer.

Show Transparencies 16 and 17 (Figures 16 and 17).

Emphasize the proper methods for securing window units to the building structure.

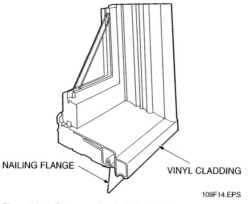

Figure 14 ♦ Cutaway of a vinyl-clad window.

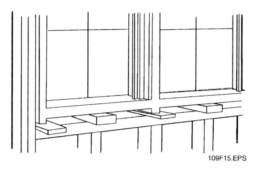

Figure 15 ♦ Leveling a window with shims.

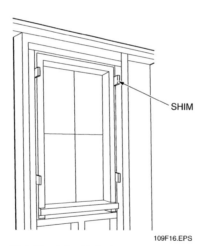

Figure 16 ♦ Plumb the side jambs.

Step 6 Plumb the side jambs using shims (*Figure 16*), but be sure not to shim too tight. When each jamb is plumb, secure that side.

Step 7 Check to see that the unit is plumb, level, and square. Operate the sash to make sure it works smoothly and does not bind. When it is satisfactory, finish securing the unit. If **casing** nails are used, use a nail set to drive the nail in order to avoid denting the frame. Check the manufacturer's instructions to see if there are restrictions on nailing into the header.

NOTE

Casing nails should be used to secure the unit to the building structure. Don't drive the nails all the way in yet. Use a nail set to finish the job and prevent denting of the frame. You can keep your shims permanently in place by nailing through them during this step. Once the basic unit is installed, finish nails should be used on the trim.

Step 8 Pack insulation or expanding foam in the gaps between the trimmer studs and the jambs (*Figure 17*). Check the window manufacturer's recommendations before using expanding foam.

Figure 17 ♦ Insulating between the jambs and trimmer studs.

9.12 CARPENTRY FUNDAMENTALS LEVEL ONE ♦ TRAINEE GUIDE

Instructor's Notes:

2.4.1 Extension Jambs

Modern windows are made for thinner walls than those found in older buildings. When replacing windows in older buildings, it may be necessary to build out the jamb so that it is flush with the wall (*Figure 18*). Jamb extensions are available as optional accessories from the window manufacturer or they can be ripped from a length of 1 × lumber that is wide enough to fill the gap. Also, for 2 × 6 wall construction, most manufacturers supply extension jambs at 4 9/16", 5 1/8", 6 9/16", etc., as options.

2.5.0 Glass Blocks

Glass block panels are sometimes used in place of standard windows to provide light while at the same time providing privacy. Glass blocks are generally 3 7/8" thick and come in three common nominal sizes: 6 × 6, 8 × 8, and 12 × 12. Actual dimensions are 3/4" less than nominal size to account for mortar joints. They are made of two formed pieces of glass fused together in a way that leaves an air space in between. The air space provides excellent insulating qualities. They are available in pre-assembled panels in a range of sizes.

Glass blocks are installed using masonry tools and white mortar or silicone. Prefabricated glass block windows are available in a variety of sizes for direct installation. Some are available with a built-in vent for air circulation. Glass block windows can also be constructed from individual blocks (panels), as shown in *Figure 19*. The illustration shows a glass block window system consisting of spacers, panel anchors, expansion strips, and glass blocks. If the window is larger than 25 square feet, panel reinforcing is also required.

The following considerations are important when framing-in glass block windows. To determine the size of the opening required, multiply the nominal block size by the number of blocks and add 3/8" for expansion strips at the header and jambs. When the size of the opening is less than 25 square feet, the height should be no greater than 7' and the width no greater than 5'. When the opening is greater than 25 square feet, the panels should never be more than 10' wide or 10' high. A built-up sill should be used to protect the bottom row of blocks from damage.

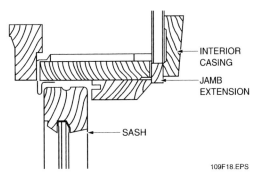

Figure 18 ◆ Jamb extension.

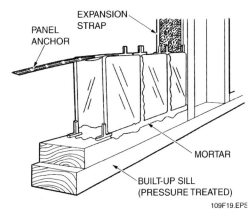

Figure 19 ◆ Glass block installation.

Glass Blocks

Fully assembled glass block windows are available. They are faster to install than building them block-by-block, but are generally more expensive. Regardless of how you choose to install them, glass blocks are approximately four times heavier than other windows of equal size. Check the instructions that come with the window kit for the weight specifications. Extra support for the window may be required.

Identify types of residential entry doors and discuss their applications.

Show Transparencies 20 through 22 (Figures 20 through 22).

3.0.0 ♦ EXTERIOR DOORS

Exterior doors are designed to provide security as well as insulation. Exterior doors, especially main entrance doors, are often designed to give a building an attractive appearance (*Figure 20*). Most residential exterior doors you encounter will be pre-hung in frames with hinges and exterior casings applied. Once the door opening is framed in, all you have to do is install and level the door. The first part of this discussion covers residential doors.

Figure 21 shows the component parts of an exterior door installation. Like windows, doors have headers, side jambs, and sills. There are also several different kinds of exterior doors, as shown in *Figure 22*. You can go to any building supply store and get color brochures containing a wide variety of styles and decorative designs for pre-hung exterior doors.

The panel door is the most common factory-built exterior door. It is made up of vertical members called stiles, cross members called rails, and filler panels. The panels are usually thinner than the stiles and rails. Panel doors are more decorative than flush doors. The doors shown in *Figure 20* are examples of panel doors. Panel doors can be made of wood or metal.

An exterior flush door has a smooth surface made of wood veneer or metal. It usually has a solid core of wood, composition board, or solidly-packed foam. Hollow-core doors consist of a framework of wood strips, metal honeycombs, or other materials that give the door rigidity. Hollow-core doors are not recommended for exterior use because they provide very little insulation and limited security. In addition, local building codes may prohibit their use in exterior entries. A flush door may contain a glass inset or it may come with decorative moldings attached to the surface.

Sash doors may have a fixed or movable window. The window may be divided into lights. Various types of glass are used, including insulated, reinforced, and leaded. Louvered doors are popular as entry doors in warmer climates.

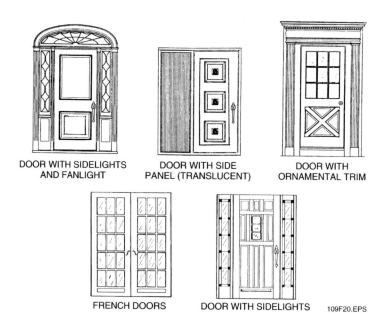

Figure 20 ♦ Entry doors.

9.14 CARPENTRY FUNDAMENTALS LEVEL ONE ♦ TRAINEE GUIDE

Instructor's Notes:

Trimming Door Length

If you need to trim a veneer door to length, lay a straightedge on your marks and score through the layer of veneer with a utility knife. This will prevent the veneer from splintering when you finish the cut with a saw. Before you make the cut, place a piece of tape on top of the veneer so that the saw baseplate doesn't scratch the surface.

Identify the components of an exterior door and frame.

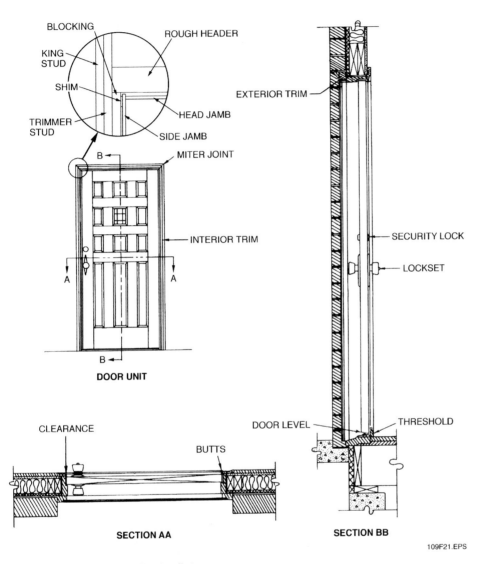

Figure 21 ◆ Parts of a typical exterior door installation.

Sliding Patio Doors

This type of exterior door is normally used for patios and porches. The door shown is solid oak with a painted aluminum exterior and a transom for use in high-ceiling rooms.

Weatherstripping

Weatherstripping systems, such as thresholds, prevent moisture from entering under a door and reduce heat loss by creating a seal between a door and the floor.

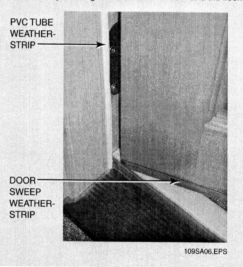

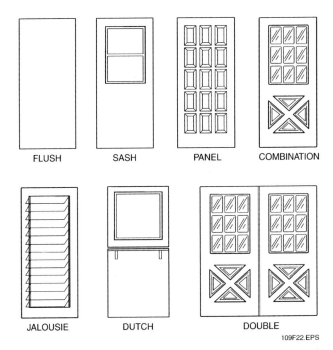

Figure 22 ♦ Types of entry doors.

Discuss the common sizes of exterior doors.

Demonstrate the procedures for installing thresholds and weatherstripping.

Show Transparencies 23 through 25 (Figures 23 through 25).

Hand out Job Sheets 27109-2 and 27109-3. Under your supervision, have the trainees perform the tasks on the Job Sheets. (This laboratory is optional based on available time and materials.)

Assign reading of Section 3.4.0 for the next class session.

Provide an overview of the procedure for installing exterior pre-hung doors.

Show Transparency 26 (Figure 26).

3.1.0 Exterior Door Sizes

Residential exterior doors are generally 1¾" thick and come in a variety of widths and heights. A standard width for a main entry door is 36", although the available range is 18" to 40". Double-width doors are available in 60", 64", and 72" widths. A typical height for an exterior door is 80". Other available heights are 78", 79", 84", 90", and 96". Keep in mind that local building codes sometimes specify the types and sizes of entry doors, as well as the location and minimum number of such doors.

3.2.0 Thresholds

A threshold is a wood or metal piece used to close the gap between the entry floor or sill and the door. It is beveled on both sides. *Figure 23* shows examples of thresholds. Weatherseal thresholds have a rubber or plastic strip in the center that is compressed by the closed door to keep out drafts. An installation detail for a threshold is shown in *Figure 23*.

3.3.0 Weatherstripping

Weatherstripping material is added to the bottom of a door to prevent heat from escaping and moisture from entering. One weatherstripping technique uses a rubber or vinyl sweep attached to the bottom of the door (*Figure 24*). Other methods are shown in *Figure 25*. A wide variety of self-stick and tack-on weatherstripping materials are available to seal off door jambs.

3.4.0 Installing an Exterior Pre-Hung Door

A pre-hung door comes equipped with the frame, threshold, and exterior casing. Often, the lockset is included. Doors can be either left-hand or right-hand swing (*Figure 26*). Different manufacturers use different methods of describing the door. The direction of swing is determined when facing the door from the outside of the building. An example of a manufacturer's product sheet is provided in the *Appendix*.

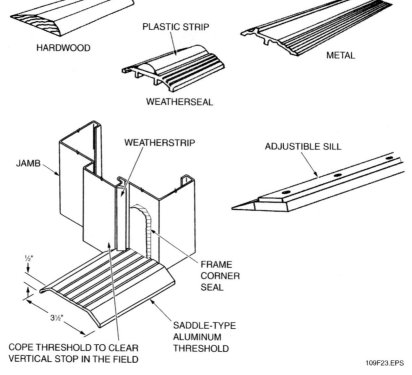

Figure 23 ◆ Thresholds.

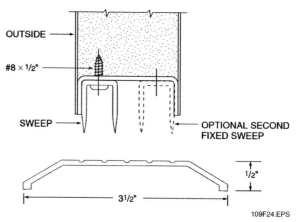

Figure 24 ◆ Fixed bottom sweep.

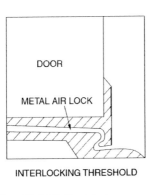 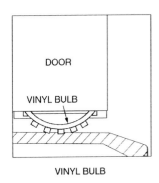 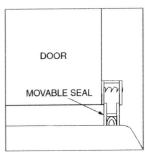

Figure 25 ♦ Weatherstripping.

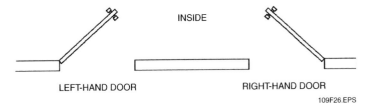

Figure 26 ♦ Door swing.

The following is a typical procedure for installing a pre-hung exterior door:

Step 1 As soon as the door is removed from the packaging, inspect it to make sure there is no damage. Remove the nails if the door was nailed shut for shipping.

Step 2 Make sure the rough opening is the right size.

Step 3 Put the door unit in place and mark the inside of the threshold. Remove the door unit and apply a double bead of caulking to the bottom of the opening inside the line you marked.

Step 4 Center the door unit in the opening. Leave the factory-installed spacer shims between the door and the frame in place until the frame is securely attached to the rough opening.

Step 5 Ensure that the sill is level and the hinge jamb is plumb. Use shims to correct, if necessary.

Step 6 Adjust for correct clearance by inserting shims between the side jambs and rough openings at the bottom, top, and middle. Additional shimming may be required on larger doors.

Step 7 Nail the jamb through the shims using 16d finishing nails. Drive the nails home with a nail set to avoid marring the finish.

Step 8 Adjust the threshold so that it makes smooth contact with the door without binding or leaving a space.

Step 9 Remove the top inner screws on the hinge and install 2½" to 3" screws through the door jamb to the trim.

Step 10 Cut off all shims flush with the edge of the door casing and pack any gaps between the trimmer stud and the door casing with insulation.

3.4.1 Locksets

Figure 27 shows examples of locksets that might be used on residential exterior doors. Mortise locksets (*Figure 28*) are more secure and are therefore considerably more expensive than the other types shown. For that reason, they are more common in commercial buildings than in residential construction. Tubular locksets (*Figure 29*) are less secure than cylindrical locksets (*Figure 30*). Neither is an excellent choice where security is a major concern.

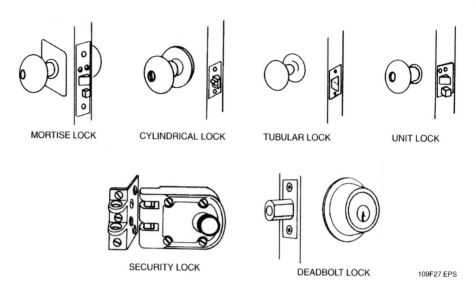

Figure 27 ◆ Examples of exterior door locksets and security locks.

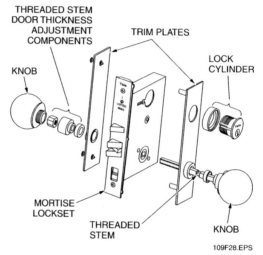

Figure 28 ◆ Exploded view of a mortise lock.

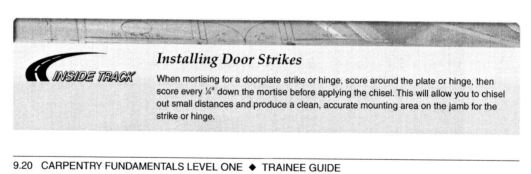

Installing Door Strikes

When mortising for a doorplate strike or hinge, score around the plate or hinge, then score every ¼" down the mortise before applying the chisel. This will allow you to chisel out small distances and produce a clean, accurate mounting area on the jamb for the strike or hinge.

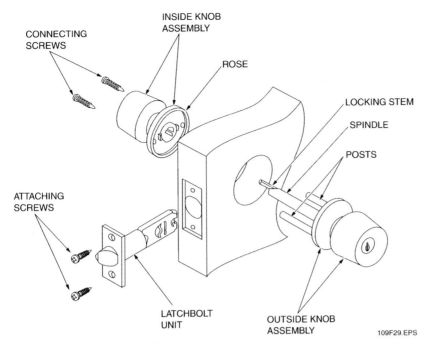

Figure 29 ◆ Disassembled view of a tubular lockset.

Figure 30 ◆ Heavy-duty cylindrical locksets.

In many cases, homeowners will want a security lock or deadbolt lock in addition to the locking mechanism built into the doorknob assembly. Deadbolts are either single- or double-cylinder types. The double-cylinder type requires a key on both sides. It would be used in cases where there is glass close to the lock. If the door is solid, a security lock or single-cylinder deadbolt (requiring a key on the outside only) will suffice.

Lock manufacturers include installation instructions and drilling templates (*Figure 31*) in the packaging for locksets and other locking devices. Professionals generally prefer to use a boring jig and boring bit, along with a mortise marker (*Figure 32*) instead of a template, because this method is faster and more accurate.

Locksets are usually installed at a height of 36" to center from the floor. Correct measuring is extremely important because the bolt on the lockset must fit into a strike plate installed on the door jamb (*Figure 33*). Check the manufacturer's instructions to determine if the backset is 2⅜" or 2¾".

Describe the procedure for installing a lockset.

Show Transparencies 30 through 32 (Figures 31 through 33).

Demonstrate how to install a lockset.

Hand out Job Sheets 27109-4 and 27109-5. Under your supervision, have the trainees perform the tasks on the Job Sheet. This laboratory corresponds to Performance Tasks 2 and 3.

Emphasize the importance of checking the door opening size before attempting to position a door.

Hand out Job Sheet 27109-6. Under your supervision, have the trainees perform the tasks on the Job Sheet. (This laboratory is optional based on available time and materials.)

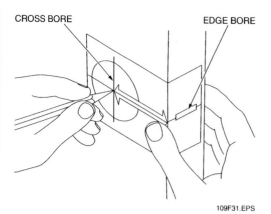

Figure 31 ◆ Using an installation template.

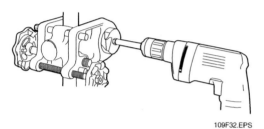

Figure 32 ◆ Boring jig and mortise marker.

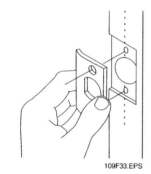

Figure 33 ◆ Installing a door strike.

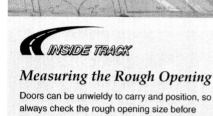

Measuring the Rough Opening

Doors can be unwieldy to carry and position, so always check the rough opening size before attempting to move the door into place.

Shims

Always keep plenty of shims on hand when installing windows and doors. Shims are easy to cut by simply scoring them with a utility knife, then snapping them off along the score.

9.22 CARPENTRY FUNDAMENTALS LEVEL ONE ◆ TRAINEE GUIDE

Instructor's Notes:

Summary

There are many different styles and types of windows and exterior doors, made by many different manufacturers. The overwhelming majority of the windows and exterior doors encountered by carpenters are the pre-hung type. While the basic designs and installation requirements are essentially the same, the specifics will vary from one manufacturer to another. Selection and installation may also depend to some extent on local building codes. The important thing to remember is to follow the instructions provided by the manufacturer for each window and become familiar with local building codes.

Summarize the major concepts presented in the module.

Notes

Have trainees complete the Review Questions, and go over the answers prior to administering the Module Examination.

Review Questions

1. The framework around the glass in a window is known as the _____.
   a. muntin
   b. sash
   c. casing
   d. header

2. A double-hung window contains _____.
   a. a single, fixed sash
   b. two fixed sashes
   c. one fixed sash and one movable sash
   d. two movable sashes

3. A self-flashing skylight is called that because it _____.
   a. has a built-in metal flange
   b. comes with a roll of aluminum tape
   c. comes with a roof curb
   d. reflects sunlight

4. Of the types of glazing listed below, the one that has the greatest insulating value is _____.
   a. DS
   b. low-e
   c. double-pane
   d. argon-filled

5. Extension jambs are most likely to be needed when _____.
   a. the window opening is too high
   b. the new window is not as wide as the old window
   c. a modern replacement window is installed in an older building
   d. casement windows are used

6. When a glass block window opening is greater than 25 square feet, the maximum height permitted is _____.
   a. 7'
   b. 5'
   c. 25'
   d. 10'

7. The most common thickness of a residential exterior door is _____.
   a. 1¼"
   b. 1½"
   c. 1¾"
   d. 2"

8. A flush exterior door usually has _____.
   a. a hollow core
   b. a solid core
   c. decorative work carved into its face
   d. stiles, rails, and filler panels

9. The door shown below is known as a _____ door.
   a. Dutch
   b. combination
   c. panel
   d. sash

10. Which of the following types of locksets is the most secure?
    a. Cylindrical
    b. Mortise
    c. Deadbolt
    d. Tubular

Trade Terms Quiz

1. The installation of a skylight requires the use of _____ to prevent leakage.
2. The glass in a window is known as the _____.
3. A French door uses _____ to secure the panes of glass.
4. The glass inserts in a door are called _____.
5. The lock on a traditional double-hung window would most likely be attached to the top _____ of the window.
6. The lock on a traditional panel door would most likely be attached to the _____ opposite the hinges.
7. A skylight that forms a peak is called a(n) _____ skylight.
8. A(n) _____ is the part of a window that holds the glass.
9. Roller shades would most likely be attached to the window _____.
10. A skylight is mounted on a raised frame known as a(n) _____.
11. When you walk through a door you step over the _____.
12. The trim around a window or door is known as the _____.

Trade Terms

Casing
Curb
Flashing
Glazing
Hipped
Jamb
Lights
Muntin
Rail
Sash
Sill
Stile

Have trainees complete the Trade Terms Quiz, and go over the answers.

Administer the Module Examination. Record the results on Craft Training Report Form 200, and submit the form to the Training Program Sponsor.

Administer the Performance Tests, and fill out Performance Profile Sheets for each trainee. If desired, trainee proficiency noted during laboratory sessions may be used to complete the Performance Test. Record the results on Craft Training Report Form 200, and submit the results to the Training Program Sponsor.

Profile in Success

Gary Humphries
Facilities Services Manager
The Haskell Company
Jacksonville, FL

In college, Gary was a drummer in a band and graduated with a degree in photography. In order to support his music, he answered an ad and took a job as a carpenter's helper in residential framing. He found that he really enjoyed carpentry and has been developing his craft for sixteen years. From this modest beginning, he is now the Facilities Services Manager for one of the foremost design-build organizations in the U.S. with annual sales of over $500 million.

What positions have you held and how did they help you get to where you are now?
My first job in carpentry was helping my boss with some renovation work. After that I worked as a carpenter's helper in residential custom framing. It was really just a gofer position. Over the years I worked with several old-school carpenters who took a lot of pride in their work. I realized that a completed job was a calling card and people noticed the quality of your work.

Out West, I did some exposed post-and-beam and log work. Then I went back to my old crew. After we finished the framing, I would do the finish work and cabinetry. For a few years, I owned my own company and built custom beach houses, but I changed direction when I enlisted in the Marines. I signed up for six years. I don't know why; being a Marine was just something I've always had in the back of my head.

I spent my six years in the Marine Corps as a combat engineer in charge of 82 men. When I retired I started working in my present job.

What are some of the things you do in your job?
I perform superintendent and project manager duties for warranty issues for our company. If there is a problem with one of our buildings during the warranty period, I go out and assess the problem. If there is a leak, I find out where it is leaking and why. For smaller problems we perform corrective action on the spot. Maybe all that is needed is a tube of caulk. In other cases, I go after the subcontractor who maybe did not build according to the plans. I also deal with liability issues.

What do you think it takes to be a success in your trade?
Integrity. I think of it as doing the right thing when no one is looking. It is as simple as doing what you say you are going to do. First, show up on time and give an honest day's work for an honest day's pay. While you are there, don't look for a way out, do the best you can. Finally, when you are done, know that it is done to the best of your ability. In the military we had a motto that is also applicable to the civilian side: Today I have given all I can give, that which I have kept I have lost forever.

If you want to be sought after, know yourself and seek self-improvement. Never be content with where you are, always work toward the next level. If you become complacent you will fall behind.

Instructor's Notes:

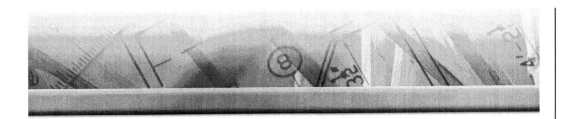

What do you like about the work you do?
When I was doing residential framing I enjoyed having tangible results at the end of the day. When you arrive on site there is nothing. When you leave you can see what you have created. I know that my sweat and brain power is in that home or office building.

Now I enjoy the puzzle, figuring out the cause of a problem and finding a solution. On one project they had been combating leaking windows for years. We conducted systematic tests to find the problem and formulate a solution. The building is now dry. I learned a lot in the process.

What advice would you give someone just starting out?
Be a sponge. Be aware of what is going on and seek out those people who know the tricks of the trade. Ask questions and soak up all the knowledge you can get.

Take pride in being a carpenter. It is not just a way to make a living, it is what you do. It takes many years to become a master carpenter. It is a title that you earn.

Trade Terms Introduced in This Module

Casing: The trim around a window or door.

Curb: A framework on which a skylight is mounted.

Flashing: Sheet metal strips used to seal a roof or wall against leakage.

Glazing: Material such as glass or plastic used in windows, skylights, and doors.

Hipped: The external angle formed by the meeting of two adjacent sloping sides.

Jamb: The top and sides of a door or window frame that are in contact with the door or sash.

Lights: The glass inserts in a door.

Muntin: A thin framework used to secure panes of glass in a door.

Rail: The horizontal member of a window sash or panel door.

Sash: The part of a window that holds the glass.

Sill: The lowest member of a window or exterior door frame.

Stile: The vertical member of a window sash or panel door.

Appendix

Door Manufacturer's Product Data Sheet

HANDING DIAGRAMS

"KS" = KEY SIDE

LH | RH | RHA | LHA
INSIDE
OUTSIDE
RRA | LRA | LHR | RHR

SUFFIX "A" = ACTIVE LEAF OF PAIRS

LOCATION
COL 1: OPENING NUMBER

COL 2: LOCATION

COL 3: FROM or TO
F = FROM
T = TO

COL 4: LOCATION

GENERAL
COL 5: QUANTITY REQUIRED

COL 6: NOMINAL DOOR OPENING DIMENSION

COL 7: ACTUAL HAND & SWING

FRAME DATA
COL 8: SERIES/PROFILE
SU = STANDARD PROFILE UNEQUAL RABBET FRAME
SQ = STANDARD PROFILE EQUAL RABBET FRAME
DU = DRYWALL PROFILE UNEQUAL RABBET FRAME
DO = DRYWALL PROFILE EQUAL RABBET FRAME
BQ = BEFORE DRYWALL PROFILE EQUAL RABBET FRAME (APPLIED ANCHORS)
BU = BEFORE DRYWALL PROFILE UNEQUAL RABBET FRAME (APPLIED ANCHORS)
SE1,2 = DOUBLE EGRESS PROFILE FRAME
SQT = STANDARD PROFILE EQUAL RABBET THERMAL BREAK FRAME
SQW = STANDARD PROFILE EQUAL RABBET WEATHERSTRIPPED FRAME
DQW = DRYWALL PROFILE EQUAL RABBET WEATHERSTRIPPED FRAME
XXX = SPECIAL PROFILE, SPECIAL RABBETS

COL 9: GAGE, MATERIAL AND FINISH
_CRS = _GA. COLD ROLLED STEEL
_A60 = _GA. A60 GALVANNEAL STEEL
_G90 = _GA. G90 GALVANNEAL STEEL
_M40 = _GA. A40 GALVANNEAL STEEL, MADERA
FGFRP = FIBERGLASS, FIBER REINFORCED PLASTIC
_S01 = _STAINLESS STEEL 304-4
_S02 = _STAINLESS STEEL 304-6
_S03 = _STAINLESS STEEL 316-4
_S04 = _STAINLESS STEEL 316-6
_S05 = _STAINLESS STEEL 316-8
_S06 = _STAINLESS STEEL 316-8

"_" represents steel gage (12,14,16,18)
Common codes shown; additional gages, material and finishes available.

COL 10: FRAME DEPTH
300 = 3"
434 = 4-3/4"
578 = 5-7/8"

COL 11: FRAME THROAT
200 = 2"
334 = 3-3/4"
478 = 4-7/8"

COL 12: STRIKE
S = ANSI A115.1 & 2 (4-7/8")
C = CYLINDRICAL A115.2/(3) (2-3/4")
D = DOUBLE DOOR
P = PLAIN (NO PREP)

COL 13: LABEL

FIRE PROTECTION
RATING		AGENCY	
A = 3 HR		U = UNDERWRITERS LABORATORY	
B = 1-1/2 HR		F = FACTORY MUTUAL	
C = 3/4 HR		W = WARNOCK HERSEY	
D = 1-1/2 HR			
E = 3/4 HR			
T = 1/3 HR			

COL 14: TYPE
BL = BORROWED LITE
SL = SIDELITE
TF = TRANSOM
DF = DUTCH
DA = DOUBLE ACTING
CM = COMMUNICATING

COL 15: ANCHORS
WMA = WIRE MASONRY ANCHOR
WS = WOOD STUD
MS = METAL STUD
FFA = FLOOR
EO = EXISTING OPENING
See anchor sheet for more

COL 16: FRAME ELEVATION

DOOR DATA

COL 17: DOOR SERIES AND CORE
RI = REGENT (IMPREGNATED HONEYCOMB CORE, HANDED)
RN = REGENT (NON-IMPREGNATED HONEYCOMB CORE, HANDED)
IU = IMPERIAL (POLYURETHANE FOAM CORE, HANDED)
OI = OMEGA (IMPREGNATED HONEYCOMB CORE, NON-HANDED)
OT = OMEGA DOOR W/250° TEMP. RISE CORE, NON-HANDED)
VU = VERSADOR (POLYURETHANE FOAM CORE, HANDED)
FT = FUEGO (250° TEMP. RISE CORE, HANDED)
MS = MEDALLION (VERTICALLY STIFFENED CORE, HANDED)
MJ = MEDALLION (STILE & RAIL DESIGN, HANDED)
TX = THRULITE (STILE & RAIL DESIGN, HANDED)
AP = ARMORSHIELD (LEVEL III, BULLET RESISTANT POLYSTYRENE CORE, HANDED)
AS = ARMORSHIELD (LEVEL III, BULLET RESISTANT STEEL-STIFFENED CORE, HANDED)
A4 = ARMORSHIELD (LEVEL IV, BULLET RESISTANT CORE, HANDED)
SA = SOUNDTECH (ACOUSTICAL CORE, HANDED)
KU = KHEMPRO (FRP, HANDED)
BI = BETA (IMPREGNATED HONEYCOMB FOAM CORE, NON-HANDED, LIGHT DUTY)
GU = GAMMA (POLYURETHANE FOAM CORE, NON-HANDED, LIGHT DUTY)

COL 18: GAGE, MATERIAL AND FINISH
_CRS = _GA. COLD ROLLED STEEL
_A60 = _GA. GALVANNEAL STEEL
2OT60 = 20 GA. A60 GALVANNEAL TEXTURED STEEL
_G90 = _GA. G90 GALVANNEAL STEEL
_M40 = _GA. GALVANNEAL STEEL, MADERA
FGFRP = FIBERGLASS, FIBER REINFORCED PLASTIC
_S01 = _GA. STAINLESS STEEL 304-4
_S02 = _GA. STAINLESS STEEL 304-6
_S03 = _GA. STAINLESS STEEL 316-4
_S04 = _GA. STAINLESS STEEL 316-6
_S05 = _GA. STAINLESS STEEL 316-8
_S06 = _GA. STAINLESS STEEL 316-8

"_" represents steel gage (12,14,16,18,20)
Common codes shown; additional gages, material and finishes available.

COL 19: DESIGN
F = FLUSH
G = HALF GLASS
G3 = HALF GLASS W/3 LITES
FG = FULL GLASS
N660 = 6 x 60 NARROW LITE
V = 10 x 10 VISION LITE

EMBOSSED PANEL DESIGNS
E601 = 6 PANEL
E801 = 8 PANEL
EC05 = CROSSBUCK W/9 LITES
See door elevation sheet for more

COL 20: LOCK PREP
LC1 = CYLINDRICAL (Gov. 160/161)
LM1 = MORTISE (Gov. 86-4)
LM0 = BLANK (MORTISE) (Gov. 86-5)
LC0 = PLAIN (NO LOCK PREP)

COL 21: LABEL

COL 22: DOOR ELEVATION

COL 23: REMARKS
ASTRAGALS, LOUVER, VISION LITE
See Door Accessories sheet
P_ = COLORSTYLE OR SPECIAL FINISH (NOTE FINISH CODE)
SG_ = MADERA STAIN FINISH (NOTE FINISH CODE)

SYMBOLS
N.B.C. = NOT BY CECO   N.I.C. = NOT IN CONTRACT

DETAIL NUMBER / SHEET NUMBER
QUANTITY OF ANCHORS / TYPE OF ANCHORS

LEGEND COMPUTER INDEX
(FOR USE WITH SCHEDULE SHEET)

Ceco Door Products
An ASSA ABLOY Group company

ASSA ABLOY

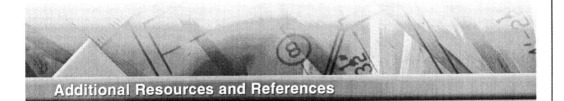

Additional Resources and References

This module is intended to present thorough resources for task training. The following reference works are suggested for further study. These are optional materials for continuing education rather than for task training.

Window and Door Manufacturers Association. A trade organization representing 145 U.S. window and door manufacturers. www.wdma.com

The National Fenestration Rating Council (NFRC). The nation's recognized authority for measuring and evaluating window energy performance. www.nfrc.org.

Window & Door magazine. An information source for manufacturers, distributors, and dealers of windows and doors. www.windowanddoor.net.

MODULE 27109-06 — ANSWERS TO REVIEW QUESTIONS

Answer	Section
1. b	2.1.0
2. d	2.2.0
3. a	2.2.0
4. b	2.3.1
5. c	2.4.1
6. d	2.5.0
7. c	3.1.0
8. b	3.0.0
9. b	3.0.0; Figure 22
10. b	3.4.1

MODULE 27109-06 — ANSWERS TO TRADE TERMS QUIZ

1. Flashing
2. Glazing
3. Muntins
4. Lights
5. Rail
6. Stile
7. Hipped
8. Sash
9. Jambs
10. Curb
11. Sill
12. Casing

MODULE 27109-06 — TEACHING TIPS

The following are suggested activities or instructional methods to help you teach the material in this module.

General

When you call on someone to answer a question, the rest of the class relaxes or even tunes out because they expect that the question and answer will take place only between you and the trainee you called on. Instead, use this technique to involve more trainees in answering questions and to keep them on their toes.

1. Ask trainees to define a term or explain a concept.
2. After one trainee has answered, ask a trainee seated nearby if the answer is right. Then ask whether a trainee in the back of the room agrees.
3. Ask trainees to explain why they think an answer is right or wrong.
4. Use the session to clear up incorrect ideas and encourage trainees to learn from their mistakes.

CONTREN® LEARNING SERIES — USER UPDATE

NCCER makes every effort to keep these textbooks up-to-date and free of technical errors. We appreciate your help in this process. If you have an idea for improving this textbook, or if you find an error, a typographical mistake, or an inaccuracy in NCCER's Contren® textbooks, please write us, using this form or a photocopy. Be sure to include the exact module number, page number, a detailed description, and the correction, if applicable. Your input will be brought to the attention of the Technical Review Committee. Thank you for your assistance.

Instructors – If you found that additional materials were necessary in order to teach this module effectively, please let us know so that we may include them in the Equipment/Materials list in the Annotated Instructor's Guide.

Write: Product Development and Revision
National Center for Construction Education and Research
P.O. Box 141104, Gainesville, FL 32614-1104

Fax: 352-334-0932

E-mail: curriculum@nccer.org

Craft _____ Module Name _____

Copyright Date _____ Module Number _____ Page Number(s) _____

Description

_____

_____

_____

_____

(Optional) Correction

_____

_____

_____

(Optional) Your Name and Address

_____

_____

_____

Basic Stair Layout
27110-06

NCCER STANDARDIZED CRAFT TRAINING PROGRAM

The National Center for Construction Education and Research (NCCER) provides a standardized national program of accredited craft training. Key features of the program include instructor certification, competency-based training, and performance testing. The program provides trainees, instructors, and companies with a standard form of recognition through the National Registry. The program is described in full in the *Guidelines for Accreditation*, published by NCCER. For more information on standardized craft training, contact NCCER by writing to P.O. Box 141104, Gainesville, FL 32614-1104; calling 352-334-0911; or e-mailing **info@nccer.org**. More information is available at **www.nccer.org**.

HOW TO USE THIS ANNOTATED INSTRUCTOR'S GUIDE

Each page presents two sections of information. The larger section displays each page exactly as it appears in the Trainee Module. The narrow column ties suggested trainee and instructor actions to each page and provides icons (detailed below) to call your attention to material, safety, audiovisual, or testing requirements. The bottom of each page includes space for your notes.

The **Audiovisual** icon indicates an appropriate time to show a transparency or other audiovisual aid.

The **Classroom** icon prompts you to define a term, stress a point, ask trainees to explain a concept, or give examples.

The **Demonstration** icon directs you to show trainees how to perform tasks.

The **Examination** icon tells you to administer the written module examination.

The **Homework** icon is placed where you may wish to assign reading for the next class, assign a project, or advise trainees to prepare for an examination.

The **Laboratory** icon is used when trainees are to practice performing tasks.

The **Materials** icon is a reminder for you to gather materials needed for classes, laboratories, and testing.

The **Performance Testing** icon tells you to administer a performance test or a portion thereof.

The **Safety** icon is used to emphasize safety issues. It is often keyed to *Caution* and *Warning!* statements in the Trainee Module.

The **Teaching Tip** icon indicates additional guidance is available, such as how to conduct an exercise, get the most educational value from a field trip, or encourage class participation. Teaching Tips may expand on a feature (*Think About It*, *Did You Know?*) or provide *Quick Quizzes* or similar exercises. You will be referred to the Teaching Tips section at the back of the module if there is additional material.

The **Combination** icon indicates that the laboratory listed corresponds with a performance task. If desired, you can note the proficiency of the trainees during the laboratory, and use it to satisfy performance testing requirements.

PREPARATION

Before teaching this module, you should review the Objectives, Performance Tasks, Materials and Equipment List, and Module Outline. Be sure to allow ample time to prepare your own training or lesson plan and gather all required materials and equipment.

Basic Stair Layout
Annotated Instructor's Guide

Module 27110-06

MODULE OVERVIEW

This module introduces the Carpentry trainee to the materials and methods used to construct interior and exterior wooden stairs.

PREREQUISITES

Prior to training with this module, it is recommended that the trainee shall have successfully completed *Core Curriculum*; and *Carpentry Fundamentals Level One*, Modules 27101-06 through 27109-06.

OBJECTIVES

Upon completion of this module, the trainee will be able to do the following:

1. Identify the various types of stairs.
2. Identify the various parts of stairs.
3. Identify the materials used in the construction of stairs.
4. Interpret construction drawings of stairs.
5. Calculate the total rise, number and size of risers, and number and size of treads required for a stairway.
6. Lay out and cut stringers, risers, and treads.
7. Build a small stair unit with a temporary handrail.

PERFORMANCE TASKS

Under the supervision of the instructor, the trainee should be able to do the following:

1. Lay out and build a small stair unit with a handrail to a given rise.

MATERIALS AND EQUIPMENT LIST

Transparencies
Markers/chalk
Blank acetate sheets
Transparency pens
Pencils and scratch paper
Overhead projector and screen
Whiteboard/chalkboard
Appropriate personal protective equipment
Basic carpenter's toolbox
Framing square
Level
Circular saw and extension cord
Hand saw

Stair gauges
Calculator
2 × 12s for stringers
2 × 12s for treads
1 × 8s for risers
Handrail and brackets
8d box nails
16d box nails
16d casing nails
Stair plans
Copies of Job Sheet 1*
Module Examinations**
Performance Profile Sheets**

* Packaged with this Annotated Instructor's Guide.

**Located in the Test Booklet.

SAFETY CONSIDERATIONS

Ensure that the trainees are equipped with appropriate personal protective equipment.

ADDITIONAL RESOURCES

This module is intended to present thorough resources for task training. The following reference works are suggested for both instructors and motivated trainees interested in further study. These are optional materials for continued education rather than for task training.

Basic Stairbuilding. Newton, CT: Taunton Press, Inc. (Book with companion video or DVD.)
Constructing Staircases, Balustrades & Landings. New York: Sterling Publishing Co., Inc.
For Pros By Pros: Building Stairs. Newton, CT: Taunton Press, Inc.
Framing Floors and Stairs. Berkeley, CA: Publishers Group West. (Book with companion video or DVD.)
A Simplified Guide to Custom Stairbuilding and Tangent Handrailing. Fresno, CA: Linden Publishing.
Stair Builders Handbook. Carlsbad, CA: Craftsman Book Company.
Staircases. New York: Watson-Guptill Publications.
Stair Layout. Homewood, IL: American Technical Publishers.
Stairs: Design and Construction. New York: Birkhauser.
Arcways, Inc. Builders of custom stairways. **www.arcways.com**.
Classic Stairworks, Ltd. Builders of classic custom staircases. **www.classicstairworks.com**.
Coffman Stairs, LLC. Hardwood stair parts manufacturer. **www.coffmanstairs.com**.
L.J. Smith Stair Systems. Manufacturer of stair products. **www.ljsmith.net**.

TEACHING TIME FOR THIS MODULE

An outline for use in developing your lesson plan is presented below. Note that each Roman numeral in the outline equates to one session of instruction. Each session has a suggested time period of 2½ hours. This includes 10 minutes at the beginning of each session for administrative tasks and one 10-minute break during the session. Approximately 12½ hours are suggested to cover *Basic Stair Layout*. You will need to adjust the time required for hands-on activity and testing based on your class size and resources. Because laboratories often correspond to Performance Tasks, the proficiency of the trainees may be noted during these exercises for Performance Testing purposes.

Topic   **Planned Time**

Session I. Introduction; Types of Stairs; Stairway Components and Typical Code Requirements
   A. Introduction
   B. Types of Stairs
   C. Stairway Components and Typical Code Requirements

Session II. Stair Framing
   A. Stair Framing
      1. Headroom
      2. Stringers
      3. Treads and Risers
      4. Width Requirement
      5. Handrails
      6. Stairwells

Session III. Stairway and Stairwell Design and Layout
   A. Stairway and Stairwell Design and Layout
      1. Stairway Design
      2. Stairwells
      3. Laying Out and Cutting a Stringer
      4. Reinforced Cutout Stringers

Session IV. Laboratory

A. Laboratory

Hand out Job Sheet 27110-1. Under your supervision, have the trainees lay out and construct a stairway. This laboratory corresponds to Performance Task 1.

Session V. Forms for Concrete Stairs; Review; Module Examination and Performance Testing

A. Forms for Concrete Stairs

B. Review

C. Module Examination

1. Trainees must score 70 percent or higher to receive recognition from NCCER.
2. Record the testing results on Craft Training Report Form 200, and submit the results to the Training Program Sponsor.

D. Performance Testing

1. Trainees must perform each task to the satisfaction of the instructor to receive recognition from the NCCER. If applicable, proficiency noted during laboratory exercises can be used to satisfy the Performance Testing requirements.
2. Record the testing results on Craft Training Report Form 200, and submit the results to the Training Program Sponsor.

Basic Stair Layout
27110-06

Two Carpentry contestants in the 2005 SkillsUSA National Championships work on miniature building framing. In addition to building a wooden frame and common rafter, contestants were also required to construct a metal frame, stair stringer, and install gypsum wallboard. Contestants were judged based on safety, project quality, and the proper use of tools.

Instructor's Notes:

Assign reading of Module 27110-06.

27110-06
Basic Stair Layout

Topics to be presented in this module include:

1.0.0 Introduction 10.2
2.0.0 Types of Stairs 10.2
3.0.0 Stairway Components and Typical Code Requirements .10.5
4.0.0 Stair Framing 10.8
5.0.0 Stairway and Stairwell Design and Layout 10.10
6.0.0 Forms for Concrete Stairs 10.21

Overview

Although prefabricated stairways are available in a variety of designs, a carpenter will sometimes have to lay out and build stairways, or build a form for concrete stairs. Laying out and cutting stair stringers is an art requiring precise measuring and the ability to perform math calculations. In addition, stairway construction is more code-driven than most other construction tasks because of the potential tripping and falling hazards. When you have mastered stair layout, you will be well on your way to becoming a professional carpenter.

Instructor's Notes:

Objectives

When you have completed this module, you will be able to do the following:

1. Identify the various types of stairs.
2. Identify the various parts of stairs.
3. Identify the materials used in the construction of stairs.
4. Interpret construction drawings of stairs.
5. Calculate the total rise, number and size of risers, and number and size of treads required for a stairway.
6. Lay out and cut stringers, risers, and treads.
7. Build a small stair unit with a temporary handrail.

Trade Terms

Baluster
Balustrade
Closed stairway
Geometrical stair
Guardrail
Handrail
Headroom
Housed stringer
Landing
Newel post
Nosing
Open stairway
Pitch board
Rise
Rise and run
Riser
Run
Skirtboard
Stairwell
Stringer
Tread
Unit rise
Unit run
Winding stairway

Required Trainee Materials

1. Pencil and paper
2. Appropriate personal protective equipment

Prerequisites

Before you begin this module, it is recommended that you successfully complete *Core Curriculum* and *Carpentry Fundamentals Level One*, Modules 27101-06 through 27109-06.

This course map shows all of the modules in *Carpentry Fundamentals Level One*. The suggested training order begins at the bottom and proceeds up. Skill levels increase as you advance on the course map. The local Training Program Sponsor may adjust the training order.

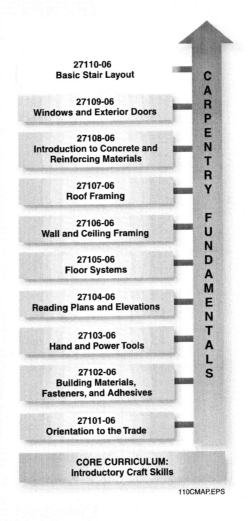

Ensure that you have everything required to teach the course. Check the Materials and Equipment List at the front of this module.

Explain that terms shown in bold (blue) are defined in the Glossary at the back of this module.

Show Transparency 1, Objectives.

Show Transparency 2, Performance Tasks.

See the general Teaching Tip at the end of this module.

Discuss various types of stairs.

Emphasize that safety is the most important concern when designing a stairway. Stress the importance of complying with building codes.

Show Transparencies 3 through 11 (Figures 1 through 9).

1.0.0 ♦ INTRODUCTION

This module deals with the construction of **stairwells**, stair framing, **risers**, and **treads**. The module begins with definitions of stair terminology and a brief introduction to stair construction. Various types of stairs are discussed and procedures for designing, laying out, cutting, and installing the stair framing are detailed.

Stairways result in roughly 4,000 deaths and a million injuries requiring hospital treatment each year; therefore, stairway design and construction is strictly controlled by building codes and regulations.

Many states and localities specify minimum and maximum requirements that may differ from the national codes. Before starting any stair construction, you must be aware of any national code and any superseding state or local codes that govern the construction of the stairs. Furthermore, if construction drawings pertain to the project, they must be followed for the construction of the stairs. If the drawings are incomplete or conflict with national, state, or local codes, a supervisor should be notified to obtain or clarify the needed information.

Normally, as construction of a structure proceeds, the stairs, with temporary stair treads, are put in as soon as possible to make it easier for workers to move themselves and materials from one level to the next with minimum delay and maximum safety. Stairways are not finished until all danger of damage from workers and materials is eliminated.

> **? DID YOU KNOW?**
> **The World's Longest Stairway**
>
> At 5,476 feet, the service stairway for the Niesenbahn funicular railway near Spiez, Switzerland is the world's longest. This stairway, which is used only by employees, has 11,674 steps.
>
> **Source:** *Guinness World Records*

2.0.0 ♦ TYPES OF STAIRS

There are many ways to classify stairs. One way depends on whether or not they are open, closed, or a combination of open and closed. *Figure 1* shows an example of an **open stairway**, which is one with a single wall in the middle. *Figure 2* illustrates an example of a **closed stairway**, and *Figure 3* shows an example of a combined open/

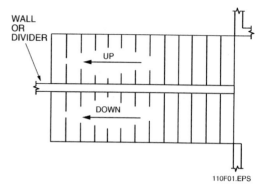

Figure 1 ♦ Open straight-run stairs.

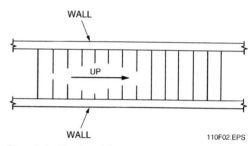

Figure 2 ♦ Closed straight-run stairs.

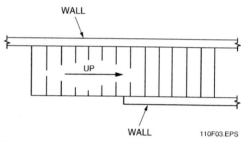

Figure 3 ♦ Combination open/closed straight-run stairs.

closed stairway. Building codes usually have special requirements for closed stairs. Refer to your local codes before construction.

Another classification relates to the shape of the stairway. A few examples are shown in *Figures 4 through 9*.

When a series of steps are straight and continuous without breaks formed by a **landing (platform)** or other construction, they are referred to as a flight of stairs, staircase, or stairway. The various classifications shown in *Figures 1 through 9* can be categorized under one of three headings: straight-run or

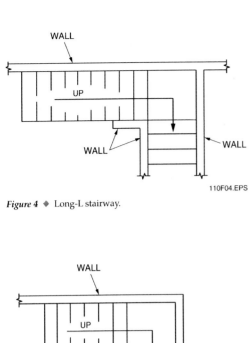

Figure 4 ♦ Long-L stairway.

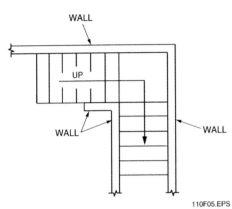

Figure 5 ♦ Wide-L stairway.

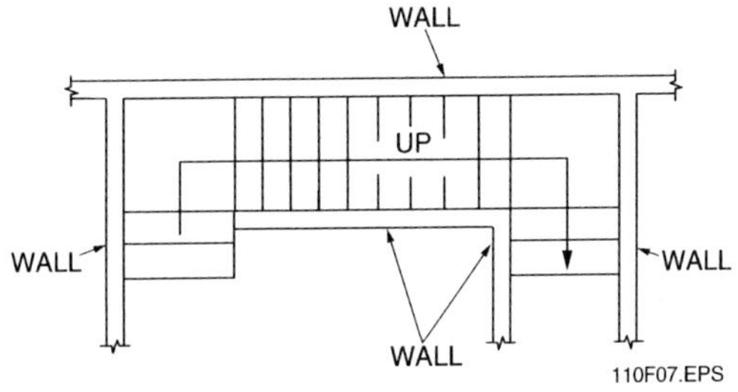

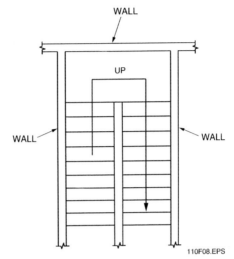

Figure 7 ♦ Double-L stairway.

Figure 8 ♦ Narrow-U stairway.

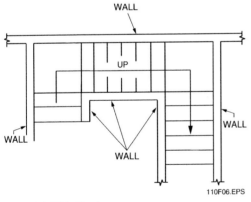

Figure 6 ♦ Wide-U stairway.

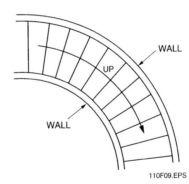

Figure 9 ♦ Winding stairway.

MODULE 27110-06 ♦ BASIC STAIR LAYOUT 10.3

Stairways

Safety is the most important concern in the design of any stairway. Most building codes include detailed requirements for stairway construction. The blueprints for a building provide information on laying out and constructing stairways. This information is normally given on the plan and elevation views, as well as the section and detailed views.

Elegant Stairs

Elaborate custom-built stairs are normally installed in expensive residences. These stairs are usually built to order in a shop from expensive hardwoods and are assembled on site. In rare cases, including some restoration work, they are custom-fabricated on site using specialized tools.

straight-flight stairway, platform stairway, or **winding stairway**. The straight-run stairway is continuous from one level to another without breaks in the progression of steps. The straight-run stairway is the easiest to build. However, it requires a long stairwell, which can cause space utilization problems in smaller structures.

Landing stair design includes a landing where the direction of the stair run usually changes. The landing stair is designed to accommodate a more limited space. **Geometrical stairs** are circular, elliptical, or spiral and gradually change direction in their ascent from one level to another. The geometric change in this type of stairway can be anywhere from 90 to 360 degrees from the starting point. Geometrical stairs can be dangerous. They should be used only where no other method is feasible.

3.0.0 ♦ STAIRWAY COMPONENTS AND TYPICAL CODE REQUIREMENTS

The main components of a flight of stairs are explained in the following paragraphs and shown in *Figures 10* and *11*.

- *Tread* – The horizontal surface of a step.
- *Riser* – The piece forming the vertical face of the step. Commercial stair codes sometimes require a riser piece. Other stair codes may allow the omission of the riser piece (open risers).
- *Cutout stringer* – A cutout **stringer** provides the main support for the stairway. Center cutout stringers may be required by codes when a wooden stairway width exceeds 30", or multiples of 30", or if tread material thinner than 1½" is used on stairs that are 36" wide. Other types of stringers used as side supports for stairs are the cleated, dadoed, or **housed stringers** (see *Figure 11*). A cleated stringer is a stringer with cleats fastened to it to support the treads. A similar design is a dadoed stringer, where the treads fit into slots cut in the stringer. Both of these designs are normally used for open-riser stairs. A housed stringer is similar to a dadoed stringer, except that tapered slots are provided for the treads and risers. The treads and risers are secured with wedges and glue. Housed stringers are usually prefabricated units that are custom-made in a shop.

Identify and discuss the main components of a flight of stairs.

Show Transparencies 12 and 13 (Figures 10 and 11).

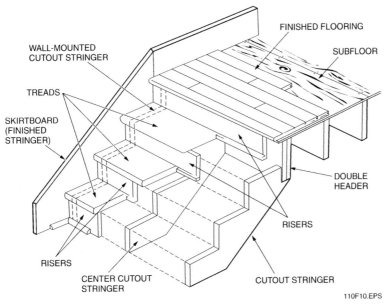

Figure 10 ♦ Main stair components.

Show Transparencies 14 and 15 (Figures 12 and 13).

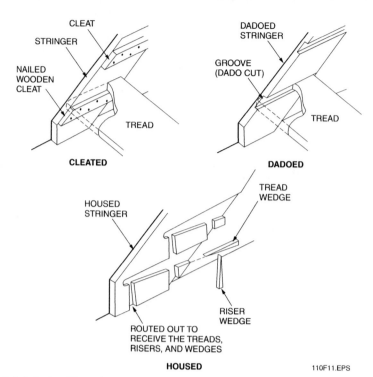

Figure 11 ♦ Cleated, dadoed, and housed stringers.

The terminology used for other components of most interior stairways is described below and shown in *Figure 12*.

- *Ending newel post* – The ending **newel post,** if used, is the uppermost supporting post for the **handrail** or **guardrail.** Sometimes, no newel post or only half of a newel post is used.
- *Landing newel post* – A landing newel post is the main post supporting the handrail (guardrail) at the landing.
- *Gooseneck* – A gooseneck is a bent or curved section in the handrail or guardrail.
- *Spindle* – A spindle or **baluster** is the upright piece that runs between the handrail or guardrail and the treads or a closed stringer. The balusters, handrail or guardrail, and newel posts make up the **balustrade.**
- *Handrail* – A handrail is used for support when ascending or descending a stairway. It may be mounted on the walls or on a guardrail for open stairways. *Figure 13* shows most of the handrail requirements for commercial structures. On the open side of commercial stairs, a guardrail that is 42" in height is typically required or if combined with a handrail, heights up to 42" are usually permitted. For residential use, handrails without extensions at lower ranges of height are usually permitted. In some cases, additional lower handrails are permitted for children. Guardrails used in residential applications are usually permitted at heights of 36".
- *Starting newel post* – A starting newel post is the main post supporting a handrail or guardrail at the bottom of the stairway.

10.6 CARPENTRY FUNDAMENTALS LEVEL ONE ♦ TRAINEE GUIDE

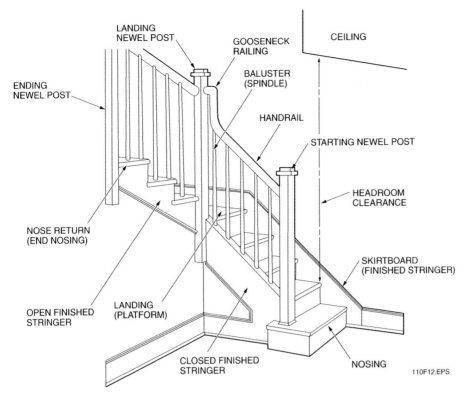

Figure 12 ♦ Components of an interior stairway.

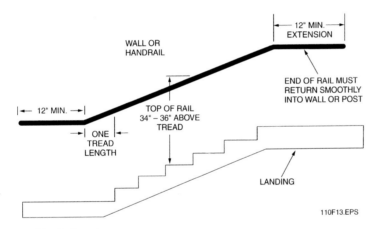

Figure 13 ♦ Commercial handrail requirements.

Assign reading of Sections 4.0.0–4.6.0 for the next class session.

Describe stair framing requirements for headroom, stringers, treads, and risers.

Point out that the minimum width required is specified in building codes.

Show Transparency 16 (Figure 14).

- *Headroom clearance* – **Headroom** clearance is the closest distance between any portion of a step of the stairway and any overhead structure such as a ceiling.
- *Skirtboard* – A **skirtboard** (finished stringer) is a finished board nailed against the wall side of the stairway. The top of the board is parallel to the slope of the stairway, and the ends terminate horizontal with the wall base molding. A housed stringer supporting the treads may be used instead of the skirtboard.
- *Nosing* – **Nosing** is the projection of the tread beyond the face of the riser.
- *Closed finished stringer* – A closed finished stringer is a finished piece fastened to the stair stringer that is also the mounting base for any balusters. A housed stringer supporting the treads may be used instead of the closed stringer.
- *Landing (platform)* – A landing is a flat section that breaks the stairway into two sections between floors. It must be used when the vertical distance (between floors) from the top of a stairway to the bottom exceeds 12' or as specified by code. A platform must be as wide and as deep as the stairway width for each section of the stairs abutting it.
- *Open finished stringer* – An open finished stringer is a finished piece cut to match the stair stringer supporting the stair treads.
- *Nosing return (end nosing)* – On a stairway with an open stringer, the nosing return is the projection over the face of the tread at the end of the tread.

The *International Building Code (IBC)* specifies minimum and/or maximum requirements for stairways and handrails, but makes exceptions for some residential occupancies. The following are examples of the *IBC* requirements:

- *Stairway width* – 44 inches minimum (Exception: 36 inches minimum for occupancies serving fewer than 50 people)
- *Stair tread depth* – 11 inches minimum (Exception: 10 inches minimum in specified residential occupancies)
- *Stair riser height* – 7 inches maximum, 4 inches minimum (Exception: 7.75 inches maximum in specified residential occupancies)
- *Headroom* – 80 inches minimum (Exception: 78 inches minimum in specified residential occupancies)
- *Nosing* – Minimum of 0.75 inch, maximum 1.25 inches (applies only to specified residential occupancies)
- *Vertical rise* – 12 feet maximum between floor levels or landings
- *Handrail height* – 34 inches minimum, 38 inches maximum

4.0.0 ♦ STAIR FRAMING

The procedure for building stairs will vary from one locality to another due to differences in building codes. Rough framing of stairs is normally done during the framing of the building. Regardless of procedure, the rough opening in the floor, in combination with the height from the floor level, will decide the length and width of a stairway. Check local building codes for special requirements in your area. Keep in mind that code requirements for residential construction may vary from commercial requirements.

4.1.0 Headroom

Headroom is defined as the closest vertical distance between any stair tread nosing and any structure or ceiling above the stairs. The standard headroom clearance is 6'-6" for residential stairs and 6'-8" for commercial stairs. However, 7' is desirable, if at all possible.

4.2.0 Stringers

All stringers should be constructed in accordance with local building codes. The size of the stringer material will be found in the construction drawings. All material used for stringers should be selected at one time and the crowns should be matched. Normally, cutout center stringers are required on wooden stairways wider than 30", in multiples of 30", or if tread material thinner than 1½" is used on stairs that are 36" wide.

4.3.0 Treads and Risers

For optimum safety and comfort, it is important that the dimensions of stair treads and risers be uniform within any stairway. As a person walks down a flight of stairs, the stride is uniform. Therefore, the system to support the person (the risers and treads) should also be uniform. If a tread and/or riser is constructed of a different size than the others, this will create an unsafe condition and will most likely cause the person to lose his or her balance and fall.

4.4.0 Width Requirement

The minimum width required is also specified in building codes. The standard is a minimum stair width of 36" to 44", depending on the type and occupancy of the building. See *Figure 14*.

10.8 CARPENTRY FUNDAMENTALS LEVEL ONE ♦ TRAINEE GUIDE

Instructor's Notes:

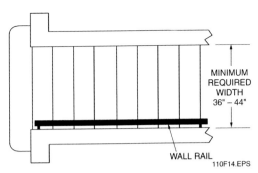

Figure 14 ♦ Minimum stairway width.

4.5.0 Handrails

A handrail is used on stairways to assist people when ascending or descending a stairway by furnishing a continuous rail for support along the side. This differs from a guardrail in that a guardrail is erected on the exposed sides and ends of stairs and platforms and incorporates a handrail. Open stairways have a low partition or banister. The handrail in a closed stairway is called a wall rail because it is attached to the wall with special brackets. As a rule, standard stairs will have a rail only on one side; however, stairs wider than 44" usually require handrails on both sides. Always refer to the local codes for specific requirements.

4.6.0 Stairwells

Stairwells must be constructed so that they are wide enough and long enough to provide the code-required width and headroom clearance for the stairs below them. If the architectural drawings do not specifically detail the stairs, carpenters can be confronted with several situations at a job site.

In one situation, the stairwells may not have been constructed and the carpenter must lay out and frame any stairwell openings as well as the stairs. In another circumstance, someone else may have already framed the openings and the task will be to frame the stairs to provide at least the minimum headroom clearance with an optimum **rise and run** for the stairs.

In a one-story structure, the stairway would be a set of service stairs to the basement. The stairwell opening should be constructed to allow code-required headroom clearance. When framing any stairwell, the construction must include double trimmers and headers to support weight placed on or around the stairwell opening. See *Figure 15*.

The stairwell framing opening will vary with the type of stairway being installed. As stated earlier, building codes have restrictions on headroom clearance, riser height, and tread width. Always check local codes.

Discuss the requirements for handrails and stairwells.

Show Transparency 17 (Figure 15).

Assign reading of Sections 5.0.0–5.4.0 for the next class session.

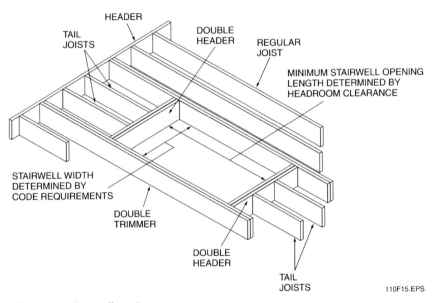

Figure 15 ♦ Typical framed stairwell opening.

Explain the requirements for stairwell construction.

Point out that OSHA requires placement of guardrails around any stairwell opening during construction.

Explain the mathematical relationship between the riser and tread dimensions.

Discuss riser height.

Building Code Variations

Local building codes for stairs vary widely from one location to another. Always check on the rules for the type of stairways permitted, stair headroom clearance, tread depth/riser height relationship, and handrail quantity and height before construction of any stairway.

5.0.0 ♦ STAIRWAY AND STAIRWELL DESIGN AND LAYOUT

If architectural building plans are being used for a project, the **rise** and **run** of the stairways in the building have usually been designed by the architect, and the **unit rise** (riser height), **unit run** (tread width), and number of risers and treads for each stairway have already been determined, along with the stairwell opening. In these cases, the stairwell openings are framed first, in accordance with the plans. Then, the carpenter lays out and constructs the rough stairways using the architect's plan. If there is no architect's plan, the carpenter may be required to design and construct any stairways and stairwells.

5.1.0 Stairway Design

An important element in stair design is the mathematical relationship between the riser (unit rise) and tread (unit run) dimensions. The ratio of these two elements (or similarly, the ratio of the total rise and total run of the stairway) determines the slope of the stairway. One of three simple, generally-accepted rules is normally used for determining the riser-to-tread (unit rise-to-run) ratio:

- *Rule 1* – Unit run (tread) + (2 × unit rise) = 24" to 25"
- *Rule 2* – Unit run + unit rise = 17" to 18"
- *Rule 3* – Unit run × unit rise = 70" to 75"

You must check the code to determine which of the above is the governing rule. The most common is Rule 1, which allows a maximum rise of 8" for the usual code-minimum tread depth of 9". Rules 2 and 3 allow a maximum rise of 8¼" for the usual code-minimum tread depth of 9".

Given the unit rise of the stairway, the applicable rule can be used to determine the longest and shortest recommended unit run (tread) length. Using any recommended tread length equal to or greater than the code minimum (usually 9"), the total run length of the stairway can then be determined. The stairway riser height (unit rise) is normally determined first.

5.1.1 Determining Riser Height

Normally, the preferred slope of a stairway is between 30 and 35 degrees for maximum ease in climbing and safe descent. Most codes limit the tread depth to a minimum of 9" and the riser height to a maximum of between 7¾" and 8¼", which, if used, can result in maximum slopes of 40 to 42 degrees.

Optimizing Design

Be aware that because of space limitations in residential and light commercial structures, most architectural drawings will detail or specify the steepest stairs (maximum unit rise) with the minimum amount of headroom clearance (minimum unit run) permitted by the applicable codes for the area.

OSHA Regulations

OSHA requires that guardrails be placed around any stairwell opening during its construction. In addition, the normal access point to the stairs must be roped off or barricaded until the temporary or permanent stairs are installed.

Instructor's Notes:

Commercial Stairways

Commercial stairways typically use a 7" rise and 11" tread depth. These dimensions are generally regarded as the most comfortable for customer use; however, always check local codes for specific requirements.

Whenever possible, a unit rise between 7" and 7½" and a unit run between 10½" and 12" is recommended; this will result in slopes of 30 to 35 degrees. The maximum unit rise and minimum unit run permitted by code are often used in building plans or by contractors to minimize stairway space.

The riser height (unit rise) can be determined as outlined in the following procedure:

Step 1 From the building drawings, determine the final thickness, in inches, of the floor assembly above the stairs. Include the anticipated or actual finished floor, subfloor, floor joists, any furring, and the anticipated or actual finished ceiling. As shown in *Figure 16*, the floor assembly thickness for our example stairway is 8½".

Step 2 Add the floor thickness (Step 1) to the vertical distance, in inches, from the anticipated or actual finished ceiling to the anticipated or actual finished floor used at the base of the stairs. This sum is the total rise of the stairway. For our example:

$$\begin{aligned}\text{Total rise} &= 8.5" + 7'\text{-}6" \\ &= 8.5" + (7' \times 12") + 6 \\ &= 98.5"\end{aligned}$$

NOTE

The total rise of a straight stairway normally must never exceed 12' (144") or as specified by applicable codes. If the limit is exceeded, an intermediate landing must be used.

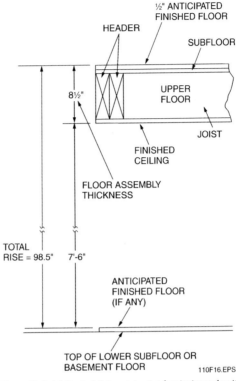

Figure 16 ◆ Method of determining total stair rise and unit rise.

Point out that a 7" rise and 11" tread are generally regarded as the most comfortable for commercial use.

Explain the procedure for determining riser height.

Show Transparency 18 (Figure 16).

Determining Stairway Rise

Instead of using building plans to determine the rise of a stairway, an alternate method is to simply measure the vertical distance from the anticipated or actual finished floor at the top of the stairs to the anticipated or actual finished floor at the base of the stairs.

MODULE 27110-06 ◆ BASIC STAIR LAYOUT   10.11

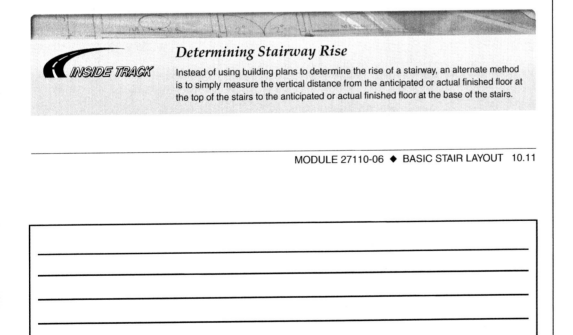

Define tread run and total stair run.

Show Transparency 19 (Figure 17).

> ### Determining Riser Height
> A simple summary of the procedure for determining riser height is to take the total unit rise and divide it by 7". This gives the number of risers for the stairs. By dividing the number of risers into the total unit rise, the exact height of each riser can be determined.

Step 3 Divide the total rise of the stairway determined in Step 2 by 7" (the optimum unit riser height). The result is usually a whole number with a decimal remainder. The whole number normally represents the optimum number of risers desired.

$$98.5" \div 7" = 14.07 \text{ or } 14 \text{ risers}$$

Step 4 Divide the total rise by the number of risers determined in Step 3. This results in the unit rise (riser height) in inches and, usually, a decimal remainder. Examine the result to determine if it falls within the desired range of 7" to 7.5" or is equal to or below the maximum riser height limitation for the applicable code. If not, or if a different height is required, use the next higher or lower number of risers and repeat Step 4. Increasing the number of risers decreases the riser height. Conversely, decreasing the number of risers increases the riser height.

Unit rise = 98.5" ÷ 14 = 7.036"
= 7" + decimal remainder of 0.036"

In this case, the riser height falls within the desired range of 7" to 7.5". However, if the number of risers is decreased to 13, the riser height would be 7.58", which is only a small fraction of an inch more than the desired range of 7" to 7.5". Although 13 risers is within code, 14 risers would yield a rise closer to 7", which would be more comfortable for use by the occupants.

Step 5 Convert the decimal remainder to fractions of an inch (32nds) by multiplying the decimal remainder by 32 and rounding up or down to the nearest whole number. This whole number represents the number of 32nds of an inch equal to the decimal remainder. Reduce the fraction to 16ths or 8ths if possible and combine with the whole number of inches of the unit rise determined in Step 4.

Unit rise = 7" + 0.036"
= 7" + (0.036" × 32 = 1.15 = $\frac{1}{32}$")
= 7$\frac{1}{32}$"

5.1.2 Determining Tread Run and Total Stair Run

The tread run (unit run) or depth is measured from the face of one riser to the face of the next and does not include any nosing (*Figure 17*).

To find the unit run, use the unit rise determined by either Method 1 or 2 and apply one of the appropriate rules (Rule 1, 2, or 3) as defined by the applicable code. Normally, the minimum unit run allowed is used to conserve the space required for the total run of the stairway.

To find the total run if the stringer uses the stairwell header as the last riser, subtract one from the number of risers (one less tread required) and multiply the result by the unit run.

If the stringer is mounted flush with the top of the floor as shown in the alternate configuration on the figure, multiply the number of risers by the unit run (same number of treads as risers). For our example, we will use *Figure 17*.

Step 1 Using the unit rise of 7$\frac{1}{32}$" determined from the previous example and the relationship defined in Rule 1, solve for the unit run as follows:

Maximum unit run =
25" − (2 × unit rise or 7$\frac{1}{32}$")

Minimum unit run =
24" − (2 × unit rise or 7$\frac{1}{32}$")

The maximum and minimum allowable unit runs are:

Maximum unit run = 25" − 14$\frac{1}{16}$" = 10$\frac{15}{16}$"
Minimum unit run = 24" − 14$\frac{1}{16}$" = 9$\frac{15}{16}$"

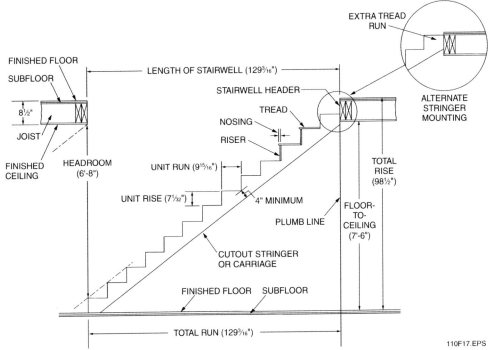

Figure 17 ♦ Example stairway and stairwell with terminology and dimensions.

Step 2 For our example, a minimum unit run of 9 15/16" will be used. If the stairwell header is used as the last riser, we will require one less tread than the number of risers and the total run will be:

Total run = (number of risers − 1) × unit run

Total run = (14 − 1) × 9 15/16" = 129 3/16"

If the top end of the stringer will be flush with the top of the floor as depicted in the alternate mounting shown in *Figure 17*, then the same number of treads as risers are required, and the total run would be:

Total run = number of risers × unit run

Total run = 14 × 9 15/16" = 139 1/8"

5.2.0 Stairwells

If a stairwell has been framed but the unit rise and run of the stairway are not specified, they must be determined. If the stairwell hasn't been framed, the stairwell opening should be determined as follows for the desired headroom clearance.

5.2.1 Determining Stairwell Openings

The stairwell opening width for a straight stairway is determined by the desired width of the stairway plus the thickness of any skirtboard(s). The stairwell width for an open U-shaped stairway could be as wide as both stairways, plus the skirtboards, plus the amount of space required to turn the U-shaped handrails.

The length of the stairwell depends on the slope and total rise of the stairway. Stairs with a low angle and/or low total rise require a longer stairwell to provide adequate headroom clearance. To find the required length of a stairwell, use the previously determined unit rise and run for the stairway along with the desired headroom clearance and the following procedure.

Explain how to determine stairwell openings.

Discuss pre-framed stairwells.

Explain how to determine the unit rise and run used for a pre-framed stairwell.

Riser/Tread Combinations

To construct a staircase having the proper angle for comfortable climbing and descending, a riser height of between 7" and 7½" and a tread width of between 10" and 11" is recommended. In addition, a rule of thumb states that the sum of one riser and one tread should equal between 17" and 18". Some examples include:

Riser Height	Tread Width	Total
7"	11"	18"
7¼"	10"	17¼"
7⅜"	10½"	17⅞"
7½"	10½"	18"

Step 1 Determine the final thickness of the floor assembly above the stairs in inches. Include the thickness of the anticipated finished floor, subfloor, floor joists, any furring, and the anticipated finished ceiling (from *Figure 17* and the previous examples).

Floor thickness = 8½" = 8.5"
Headroom required = 6'-8"
= (6' × 12") + 8" = 80"
Unit rise = 7½₂" = 7.03"
Unit run = 9¹⁵⁄₁₆"

Step 2 Add the floor thickness (Step 1) to the code-required or desired headroom clearance in inches and divide the sum by the riser height (unit rise) in inches. Round the answer up to the next whole number. This number represents the number of risers from the top of the stairs down to, or slightly below, the headroom clearance point.

Number of risers required =
(8.5" + 80") ÷ 7.03"
= 12.59 (round up to 13)

Step 3 Multiply the tread depth (unit run) by the whole number obtained in Step 2 to obtain the stairwell length in inches. This length will be correct if the header at the top of the stairway will be the top riser. If the stringers will be framed flush with the top of the header, add one additional tread width (a unit run) to the overall length of the stairwell. In this example, assume that the stairwell header is used as the last riser and the addition of one unit run to the length will not be required.

Stairwell length = unit run × number of risers
Stairwell length = 9¹⁵⁄₁₆" × 13 = 129³⁄₁₆"

In the example shown in *Figure 17*, the stairwell length works out to be the same length as the total stairway run because when the unit rise of 7½₂" (bottom step) is subtracted from the floor-to-ceiling height of 7'-6", the result is essentially the minimum headroom clearance required. In stairways involving higher floor-to-ceiling heights, the stairwell will usually be shorter than the total stairway run.

5.2.2 Determining Stairway Unit Rise and Run Used for a Pre-Framed Stairwell

In some instances, you may arrive at a job site and find that someone has framed the stairwell openings but the stairway unit rise, run, and headroom clearance used are not available. In these rare cases, you must determine the stairway unit rise and run, along with the total run, based on the length of the stairwell opening, the total stairway rise, and, initially, the minimum headroom clearance allowed by the applicable code. This can be accomplished using the following process:

Step 1 Determine the optimum unit rise (riser height) in inches, as previously described. Using the values shown in *Figure 17* as an example, the total rise of the stairway is determined to be 98½" and the optimum unit rise is 7½".

Step 2 Determine the final thickness of the floor assembly above the stairs in inches. Include the thickness of the anticipated finished floor, subfloor, floor joists, any furring, and the anticipated finished ceiling.

10.14 CARPENTRY FUNDAMENTALS LEVEL ONE ♦ TRAINEE GUIDE

Instructor's Notes:

Apparent Run Length Errors

If the unit run appears to be too large and the resultant total run appears to be too long for the space allowed, the stairwell may have been calculated using more than the minimum required headroom or it may have been lengthened by an extra tread width to accommodate for flush-mounting the stringers with the floor above. In that case, repeat the required calculations using one extra riser or greater headroom clearance (7' or more, if possible) to determine if either method provides the correct total run.

Referring to *Figure 17*, the floor thickness is 8½" and the minimum headroom is 6'-8".

Step 3 Add the floor thickness (Step 2) to the code-required headroom clearance in inches and divide the sum by the unit rise determined in Step 1. Round the answer up to the next whole number. This number represents the number of risers of optimum height from the top of the stairs down to, or slightly below, the headroom clearance point.

Number of risers required =
(8.5" + 6'-8") ÷ 7½₂" = 88.5 ÷ 7.03"
= 12.59 (round up to 13)

Step 4 Divide the stairwell length by the number of risers calculated in Step 3 to obtain the unit run (tread width) that may have been used to determine the length of the stairwell.

Unit run = 129³⁄₁₆" ÷ 13 = 9.93" or 9¹⁵⁄₁₆"

Step 5 Check the unit run-to-rise ratio using the applicable rule to determine if the calculated tread width meets the minimum requirement. If it does not, the maximum allowable unit rise may have been used in the original calculations for the stairs. In that case, repeat all the steps using the maximum allowable unit rise. In this example, the unit run (and consequently, the unit rise) appears to be valid because it will satisfy Rule 1 for the minimum tread width.

5.3.0 Laying Out and Cutting a Stringer

Marking a cutout stringer is a simple task. It can be done with either a framing square or a **pitch board**. When using a framing square, the blade will represent the unit run and the tongue will represent the unit rise. See *Figure 18*. To make the stair layout go faster, a set of stair gauges (*Figure 19*) can be used to set the unit rise and run measurements on the framing square.

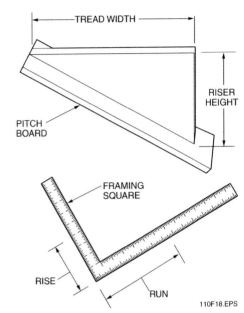

Figure 18 ♦ Pitch board and framing square.

Figure 19 ♦ Stair gauges.

Explain how to use two framing squares to determine stringer length.

Show Transparency 21 (Figure 20).

To obtain the approximate length of the stringer, the principles of the right triangle can be used. The rise and run are known; therefore, they form two sides of a right triangle. The stringer is the hypotenuse or third side of the right triangle. Using the outside back of the framing square, locate the total run on the blade and the rise on the tongue. Each inch increment on the outside back of the framing square represents 1' and each small increment represents 1". Mark the points on paper. Measure the hypotenuse (diagonal) between the two points along the same side of the blade. This will be the approximate length of the stringer. See *Figure 20*.

For example:

Rise = 9'-3"
Run = 12'-6"
Stringer = 15'-7" (use 16')

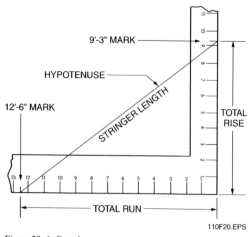

Figure 20 ♦ Framing square.

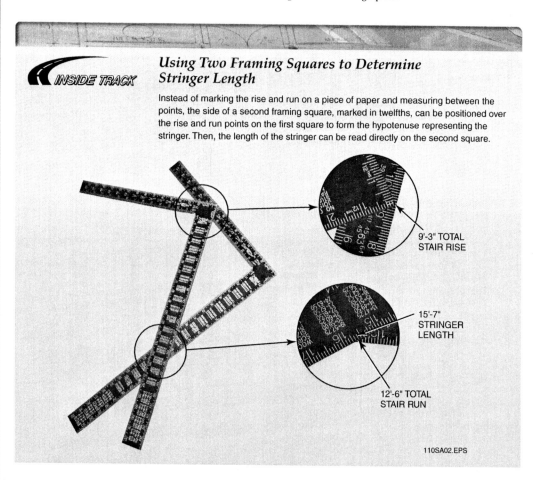

Using Two Framing Squares to Determine Stringer Length

Instead of marking the rise and run on a piece of paper and measuring between the points, the side of a second framing square, marked in twelfths, can be positioned over the rise and run points on the first square to form the hypotenuse representing the stringer. Then, the length of the stringer can be read directly on the second square.

This measurement can also be calculated mathematically by applying the Pythagorean theorem: $(a^2 + b^2 = c^2)$. In this case: $(9'\text{-}3")^2 + (12'\text{-}6")^2 = (15'\text{-}7")^2$ or $(16')^2$.

The cutting and installation of stringers must be done with caution so that the carpenter finishing the stairs will not encounter difficulty in fitting the risers and treads. When laying out a stringer, all lines drawn must be thin and accurate. Cutting the risers and treads from the stringer can be done with a hand saw or power saber saw, or with a circular saw, provided you do not allow the circular saw to cut farther than the riser line. Then, a hand or saber saw can be used to finish the cuts. A stringer must be cut accurately. If the stringer is too long, the treads will slant backwards; if the stringer is too short, the treads will slant forward. Laying out a stringer can be accomplished using one of the following two methods.

Method One:

Step 1 The stock should be 2 × 10, 2 × 12, or some other structurally sound size and of a species and grade of wood approved by the local code. The stock must be as straight as possible.

Step 2 Lay the stock on sawhorses or a work table. Place the top edge (good edge) away from you. All layout work is done from the top edge. See *Figure 21*.

Step 3 Place the framing square at the end of the stringer, as shown in *Figure 22(A)*. Mark the unit rise and run on the stringer using the inside edge of the square. Remove the square and check that there is a minimum of 4" between the junction of the rise/run and the bottom of the stringer, as shown in *Figure 22(B)*. If not, a larger stringer must be used. Place the square back on the stringer, as shown in *Figure 22(C)*, and extend the unit run mark to the bottom of the stringer. This will be the portion of the stringer that rests on the floor. Identify the extended unit run mark as a cut line.

WARNING!
Overcutting at the junctions of the rise/run can significantly weaken the stringer.

Step 4 Move the framing square to the next tread and mark the rise and run. Proceed to mark the rise and run for all the other steps. If the riser is to be installed as shown in *Figure 23*, then the stringer will be cut as it was laid out because the header will be the top riser (one more riser than the number of treads). If the stringer is to be placed so that the top tread is at the level of the finished floor, the stringer will be cut with one more tread. See *Figure 24*.

Step 5 Starting at the left side, cut the first line that goes from the top to the bottom. All cuts must be done cautiously. Be sure not to overcut. See *Figure 25*.

Step 6 Cut the rest of the treads and risers.

Step 7 If the stringer is to sit on top of a finished floor, the stringer must be dropped (cut) the thickness of a tread so that the top and bottom step of the stairs will be at the same height. See *Figure 26*. The stringer may need to be scribed. If the floor that the stringers will rest on will have some type of finished floor added later, the bottom of the stringer must be cut to compensate. The amount to be cut would be the difference between the finished flooring thickness and the thickness of the tread. See *Figure 27*.

Step 8 Use the completed stringer as a pattern to lay out the remaining stringer(s). Lay out the opposite wall stringer so that any bow faces in toward the stairway.

Step 9 If a stairway is 30" or less in width, only two stringers are required. For added strength and stability of the stairway, a third stringer may be centered between the two stringers. A stairway that is 30" or wider usually requires three stringers. Check local codes for requirements.

Step 10 Install the skirtboards, if required. Then install the stringers, one on each side of the stairway. Nail the stringers to the stud wall with 16d common nails. Complete any other related construction activities as shown on the drawings.

Step 11 Install rough treads of 2" nominal stock. These treads will stay in place until the finished risers and treads are ready to be installed. The rough treads can be reused, if applicable.

Describe the procedure for stringer layout using a framing square.

Show Transparencies 22 through 28 (Figures 21 through 27).

Point out that overcutting at the junctions of the rise/run can weaken the stringer.

Show the trainees how to lay out and cut a stringer.

Figure 21 ♦ Positioning of a stringer for layout.

Figure 22 ♦ Marking a stringer for the floor level cut line.

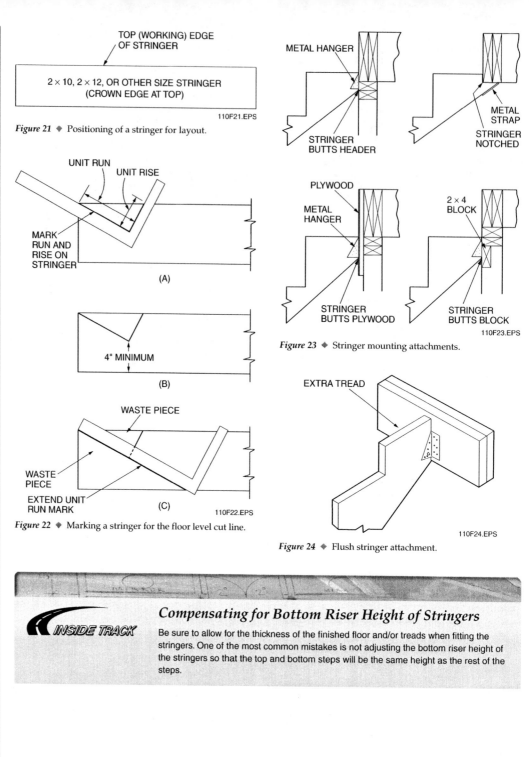

Figure 23 ♦ Stringer mounting attachments.

Figure 24 ♦ Flush stringer attachment.

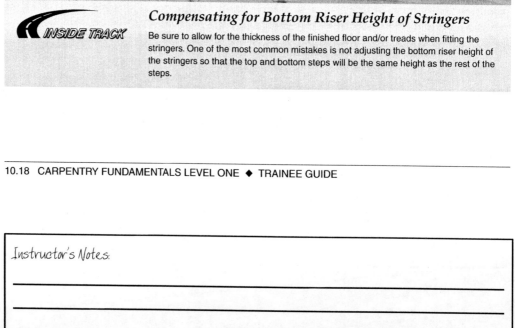

Compensating for Bottom Riser Height of Stringers

Be sure to allow for the thickness of the finished floor and/or treads when fitting the stringers. One of the most common mistakes is not adjusting the bottom riser height of the stringers so that the top and bottom steps will be the same height as the rest of the steps.

Instructor's Notes:

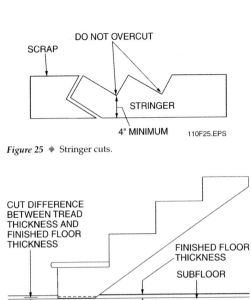

Figure 25 ♦ Stringer cuts.

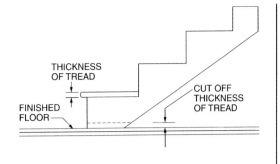

Figure 26 ♦ Dropping the stringer for tread thickness.

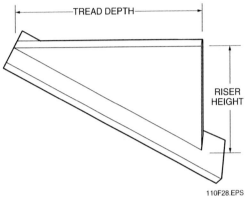

Figure 27 ♦ Dropping the stringer for tread thickness minus the finished floor thickness.

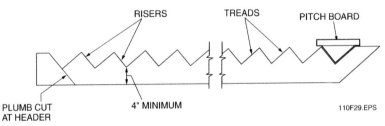

Figure 28 ♦ Pitch board.

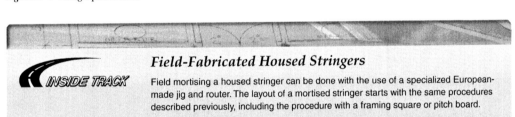

Figure 29 ♦ Using a pitch board.

Field-Fabricated Housed Stringers

INSIDE TRACK

Field mortising a housed stringer can be done with the use of a specialized European-made jig and router. The layout of a mortised stringer starts with the same procedures described previously, including the procedure with a framing square or pitch board.

Describe the procedure for stringer layout using a pitch board.

Show Transparencies 29 through 32 (Figures 28 through 31).

Show the trainees how to lay out and construct a stairway according to a specific plan.

Hand out Job Sheet 27110-1. Under your supervision, have the trainees lay out and construct a stairway. This laboratory corresponds to Performance Task 1.

Describe how to use a strongback to reinforce cutout stringers.

Show Transparencies 33 and 34 (Figures 32 and 33).

Method Two:

Laying out stringers with a pitch board is similar to laying out stringers using a framing square.

The pitch board is made by marking a piece of 1" stock with the correct rise and run for the stairway being constructed. After cutting the stock, a piece of 1 × 2 stock is nailed on the cut piece. This enables the pitch board to slide on the stringer (*Figure 28*). *Figure 29* illustrates the use of a pitch board.

Stringers that are cut and installed correctly will cause no problem to the carpenter cutting and assembling the finished risers and treads. Service and main stair stringers can be prepared in several ways:

- Stringers can be cut to receive finished risers and treads or finished treads only.
- Stringers can be dadoed to receive treads only (*Figure 30*).
- Stringers can be cleated to receive treads only.

Figure 31 illustrates a stringer with cleats to receive treads. Cleated stringers have limited applications and, if used, are typically in hidden areas, such as utility stairs for basements. Installing cleats on a stringer is done by first laying out the stringer as previously described, cutting 1" stock to the proper length or using a metal angle cleat, and fastening the cleat even with the tread mark drawn on the stringer (*Figure 31*). The method or option of cutting stringers for the construction is usually the responsibility of the contractor and/or the architect.

5.4.0 Reinforced Cutout Stringers

Some carpenters will add a 2 × 4 or 2 × 6 reinforcement known as a strongback to all longer cutout stair stringers that are supported only at the header and floor, not fastened to an adjacent wall. This strongback is secured to one side of the cut stringer below the cut notches to add strength and rigidity. See *Figure 32*.

It is also a common practice to secure a 2 × 4 ledger to the floor and/or header to add strength to the stair assembly. The stair stringers are notched to fit over the 2 × 4 ledger. See *Figure 33*.

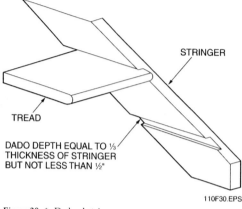

Figure 30 ♦ Dadoed stringer.

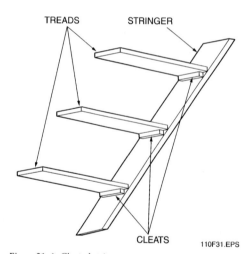

Figure 31 ♦ Cleated stringer.

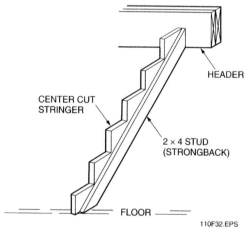

Figure 32 ♦ Reinforced cutout stringer.

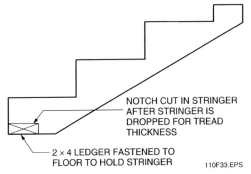

Figure 33 ♦ Reinforced stringer mounting.

NOTE

If the ledger is fastened to a concrete floor, treated lumber must be used.

6.0.0 ♦ FORMS FOR CONCRETE STAIRS

Figure 34 shows a basic concrete stair form. Note that the size and position of the riser boards establish the height, depth, and spacing of the stairs. When the stairs are wide, a center brace is needed for additional support.

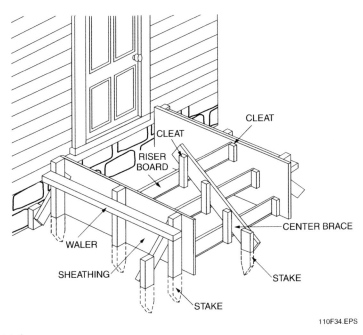

Figure 34 ♦ Basic stair form.

Assign reading of Section 6.0.0 for the next class session.

Discuss the use of concrete stair forms.

Show Transparency 35 (Figure 34).

Summarize the major concepts presented in the module.

Summary

A variety of stairs are used in residential and commercial construction. This module covered stair terminology and general building code requirements. It also described the methods used when designing, laying out, cutting, and installing the framing for basic wooden stairs. As a carpenter, you will be involved in the framing and finishing of various types of stairs, so you must be thoroughly familiar with the stair construction information and techniques covered in this module.

Notes

Instructor's Notes:

Review Questions

1. The easiest type of stairway to build is a _____ stairway.
   a. straight-run
   b. platform
   c. wide-U
   d. winding

2. The stairway in *Figure 1* would be classified as a(n) _____.
   a. open stairway
   b. closed stairway
   c. combination open/closed stairway
   d. landing

Figure 1

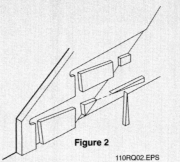

Figure 2

3. Which of the following is a geometrical stairway?
   a. Long-L
   b. Wide-U
   c. Elliptical
   d. Double-L

4. Center cutout stringers may be required on stairs that are wider than _____.
   a. 30"
   b. 36"
   c. 40"
   d. 45"

5. The stringer in *Figure 2* is known as a _____ stringer.
   a. dadoed
   b. housed
   c. common
   d. cleated

6. The main support for a handrail at the bottom of a stairway is called a(n) _____.
   a. baluster
   b. main newel post
   c. ending newel post
   d. starting newel post

7. A typical minimum headroom specified by the *IBC* for residential use is _____.
   a. 6'-4"
   b. 6'-6"
   c. 6'-8"
   d. 6'-10"

8. Stairways wider than _____ usually require handrails on both sides.
   a. 32"
   b. 36"
   c. 44"
   d. 88"

9. Which of the following is used to establish the rise and run of a staircase?
   a. Unit run + (2 × unit rise) = 17" to 18"
   b. Unit run × unit rise = 24" to 25"
   c. Unit run + (2 × unit rise) = 24" to 25"
   d. Unit run + (2 × unit rise) = 70" to 75"

10. The total rise of a straight stairway should *not* exceed _____.
    a. 10'
    b. 12'
    c. 14'
    d. 16'

Have trainees complete the Trade Terms Quiz, and go over the answers.

Administer the Module Examination. Record the results on Craft Training Report Form 200, and submit the form to the Training Program Sponsor.

Administer the Performance Tests, and fill out Performance Profile Sheets for each trainee. If desired, trainee proficiency noted during laboratory sessions may be used to complete the Performance Test. Record the results on Craft Training Report Form 200, and submit the results to the Training Program Sponsor.

Trade Terms Quiz

1. Circular, elliptical, and spiral stairs are all types of _____.
2. A(n) _____ has solid walls on both sides.
3. The _____ is the main post supporting the handrail of a staircase.
4. The projection of the tread beyond the face of the riser is called the _____.
5. The height of each riser is the _____.
6. The depth of each tread is the _____.
7. When you walk up several flights of stairs, you are likely to pause on the _____ between flights.
8. An open fire escape stairway is likely to be enclosed by a(n) _____.
9. An enclosed stairway will have a(n) _____ along the wall to serve as a handhold.
10. The spindles of a staircase are also called _____.
11. The distance between any portion of a step and the ceiling is the _____.
12. A(n) _____ provides the main support for a stairway.
13. The horizontal part of a step is known as the _____.
14. The vertical part of a step is known as the _____.
15. A(n) _____ is a vertical compartment in a building into which stairs are placed.
16. The _____ is the horizontal distance from the face of the first riser to the face of the last riser.
17. The _____ is the vertical dimension of a set of stairs.
18. A(n) _____ is a curved stairway with a newel post at the bottom only.
19. A(n) _____ is sometimes used against the wall in a closed staircase.
20. The complete handrail, newel post, and spindle assembly is known as the _____.
21. A stringer with grooves cut into it to hold the risers and treads is a(n) _____.
22. A stairway that can be viewed from within a room is called a(n) _____.
23. A(n) _____ can be used as a pattern for marking cuts for stairs.
24. The degree of incline of a set of stairs is the _____.

Trade Terms

Baluster	Headroom	Pitch board	Stairwell
Balustrade	Housed stringer	Rise	Stringer
Closed stairway	Landing	Rise and run	Tread
Geometrical stair	Newel post	Riser	Unit rise
Guardrail	Nosing	Run	Unit run
Handrail	Open stairway	Skirtboard	Winding stairway

10.24 CARPENTRY FUNDAMENTALS LEVEL ONE ♦ TRAINEE GUIDE

Instructor's Notes:

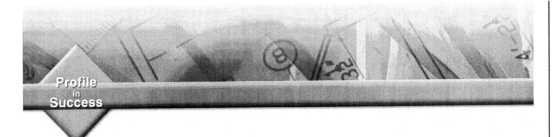

Profile in Success

Curtis McLawhorn
2004 SkillsUSA Carpentry Championship Gold Medal Winner

Curt has participated in many carpentry competitions. He competed in three regional SkillsUSA competitions and won several medals. Winning at the regional level allowed him to go to the national competition where he won the gold medal in carpentry. He is now studying civil engineering in college and plans to continue working in the construction industry.

How did you become interested in carpentry?
My father is a carpenter and general contractor and I have been working with him for as long as I can remember. I helped him on many smaller projects over the years. When I was 12 or 13, I started working for his company during the summer. I liked the work and took some classes to improve my skills.

What kind of formal training have you had?
I took some carpentry classes in high school and that is how I got involved with the SkillsUSA competitions. When I was working for my father, I also enrolled in an apprenticeship program through the North Carolina Department of Labor. They also sponsor competitions. I won a second place in the state competition. Now I am studying civil engineering at North Carolina State University.

What do you think it takes to be a success?
I would say that it takes a good background in math to be a successful contractor. You have to be able to read plans and prepare estimates. If your estimates are off, you won't be able to make a profit.

What do you do in your job?
As a laborer working for my father, I do it all. We build residential houses, so I have done framing, roofing, plumbing, electrical work, and everything else involved in building houses.

What do you like about carpentry?
I really enjoy framing houses. I like the satisfaction of building something. When you start there is nothing there, just an empty lot. When you finish, there is the basic framework of a house. Now, when I drive by, it feels pretty good, knowing that I helped build those houses.

I also enjoyed the carpentry competitions. I was never really good at sports but I do have some trade skills. It was pretty cool to be able to show off my skills in the competitions. I won some nice prizes. I won some tools and a trip. On the trip I had the chance to meet Matt Kenseth, a NASCAR driver. It is great that SkillsUSA has competitions where you can win medals that are not sports related.

What advice would you give someone just starting out?
Try hard and don't be afraid to ask other people with more experience for help. If you get into it, get all the knowledge and experience you can.

Trade Terms Introduced in This Module

Baluster: A supporting column or member; a support for a railing, particularly one of the upright columns of a balustrade.

Balustrade: A stair rail assembly consisting of a handrail, balusters, and posts.

Closed stairway: A stairway that has solid walls on each side.

Geometrical stair: A winding stairway built around a well. Examples include circular, elliptical, and spiral stairs.

Guardrail: A rail secured to uprights and erected along the exposed sides and ends of platform stairs, etc.

Handrail: A member supported on brackets from a wall or partition to furnish a handhold.

Headroom: The vertical and clear space in height between a stair tread and the ceiling or stairs above.

Housed stringer: A stair stringer with horizontal and vertical grooves cut (mortised) on the inside to receive the ends of the risers and treads. Wedges covered with glue are often used to hold the risers and treads in place in the grooves.

Landing: A horizontal area at the end of a flight of stairs or between two flights of stairs.

Newel post: An upright post supporting the handrail at the top and bottom of a stairway or at the turn of a landing; also, the main post about which a circular staircase winds or a stone column carrying the inner ends of the treads of a spiral stone staircase.

Nosing: The portion of the stair tread that extends beyond the face of the riser.

Open stairway: A stairway that is open on at least one side.

Pitch board: A board that serves as a pattern for marking cuts for stairs. The shortest side is the height of the riser cut, and the next longer side is the width of the tread. This is used mainly when there is great repetition such as in production housing.

Rise: The vertical dimension of a set of stairs. Also called the total rise.

Rise and run: A term used to indicate the degree of incline.

Riser: A vertical board under the tread of a stair step; in other words, a board set on edge for connecting the treads of a stairway.

Run: The horizontal distance from the face of the first or upper riser to the face of the last or lower riser. Also called the total run.

Skirtboard: A baseboard or finishing board at the junction of the interior wall and floor. Also called a finished stringer.

Stairwell: A compartment extending vertically through a building into which stairs are placed.

Stringer: The inclined member that supports the treads and risers of a stairway.

Tread: The horizontal member of a step.

Unit rise: The vertical distance from the top of one stair tread to the top of the next one above it; also called the stair rise.

Unit run: The horizontal distance from the face of one riser to the face of the next riser.

Winding stairway: A type of geometrical staircase that changes direction by means of winders or a landing and winders. The stair opening is relatively wide, and the balustrade follows the curve with only a newel post at the bottom.

Instructor's Notes:

Additional Resources and References

This module is intended to present thorough resources for task training. The following reference works are suggested for further study. These are optional materials for continued education rather than for task training.

Basic Stairbuilding. Newton, CT: Taunton Press, Inc. (Book with companion video or DVD.)

Constructing Staircases, Balustrades & Landings. New York, NY: Sterling Publishing Co., Inc.

For Pros By Pros: Building Stairs. Newton, CT: Taunton Press, Inc.

Framing Floors and Stairs. Berkeley, CA: Publishers Group West. (Book with companion video or DVD.)

A Simplified Guide to Custom Stairbuilding and Tangent Handrailing. Fresno, CA: Linden Publishing.

Stair Builders Handbook. Carlsbad, CA: Craftsman Book Company.

Staircases. New York, NY: Watson-Guptill Publications.

Stair Layout. Homewood, IL: American Technical Publishers.

Stairs: Design and Construction. New York, NY: Birkhauser.

Arcways, Inc. Builders of custom stairways. www.arcways.com.

Classic Stairworks, Ltd. Builders of classic custom staircases. www.classicstairworks.com.

Coffman Stairs, LLC. Hardwood stair parts manufacturer. www.coffmanstairs.com.

L.J. Smith Stair Systems. Manufacturer of stair products. www.ljsmith.net.

MODULE 27110-06 — ANSWERS TO REVIEW QUESTIONS

Answer	Section
1. a	2.0.0
2. a	2.0.0; Figure 1
3. c	2.0.0
4. a	3.0.0
5. b	3.0.0; Figure 11
6. d	3.0.0
7. b	3.0.0
8. c	4.5.0
9. c	5.1.0
10. b	5.1.1

MODULE 27110-06 — ANSWERS TO TRADE TERMS QUIZ

1. Geometrical stairs
2. Closed stairway
3. Newel post
4. Nosing
5. Unit rise
6. Unit run
7. Landing
8. Guardrail
9. Handrail
10. Balusters
11. Headroom
12. Stringer
13. Tread
14. Riser
15. Stairwell
16. Run
17. Rise
18. Winding stairway
19. Skirtboard
20. Balustrade
21. Housed stringer
22. Open stairway
23. Pitch board
24. Rise and run

MODULE 27110-06 — TEACHING TIPS

The following are suggested activities or instructional methods to help you teach the material in this module.

General

When you call on someone to answer a question, the rest of the class relaxes or even tunes out because they expect that the question and answer will take place only between you and the trainee you called on. Instead, use this technique to involve more trainees in answering questions and to keep them on their toes.

1. Ask trainees to define a term or explain a concept.
2. After one trainee has answered, ask a trainee seated nearby if the answer is right. Then ask whether a trainee in the back of the room agrees.
3. Ask trainees to explain why they think an answer is right or wrong.
4. Use the session to clear up incorrect ideas and encourage trainees to learn from their mistakes.

CONTREN® LEARNING SERIES — USER UPDATE

NCCER makes every effort to keep these textbooks up-to-date and free of technical errors. We appreciate your help in this process. If you have an idea for improving this textbook, or if you find an error, a typographical mistake, or an inaccuracy in NCCER's Contren® textbooks, please write us, using this form or a photocopy. Be sure to include the exact module number, page number, a detailed description, and the correction, if applicable. Your input will be brought to the attention of the Technical Review Committee. Thank you for your assistance.

Instructors – If you found that additional materials were necessary in order to teach this module effectively, please let us know so that we may include them in the Equipment/Materials list in the Annotated Instructor's Guide.

Write: Product Development and Revision
National Center for Construction Education and Research
P.O. Box 141104, Gainesville, FL 32614-1104

Fax: 352-334-0932

E-mail: curriculum@nccer.org

Craft _____ Module Name _____

Copyright Date _____ Module Number _____ Page Number(s) _____

Description _____

_____

_____

_____

(Optional) Correction _____

_____

_____

(Optional) Your Name and Address _____

_____

_____

Glossary

ACI: American Concrete Institute.

Admixtures: Materials that are added to a concrete mix to change certain properties of the concrete such as retarding setting time, reducing water requirements, or making the concrete easier to work with.

Aggregates: Materials used as filler in concrete; may include mixtures of sand, gravel, crushed stone, crushed gravel, or blast-furnace slag.

Axle-steel: Deformed reinforcing bars that are rolled from carbon-steel axles used on railroad cars.

Baluster: A supporting column or member; a support for a railing, particularly one of the upright columns of a balustrade.

Balustrade: A stair rail assembly consisting of a handrail, balusters, and posts.

Barge rafter: A gable end roof member that extends beyond the gable to support a decorative end piece. Also known as a fly rafter.

Bench mark: A point established by the surveyor on or close to the building site. It is used as a reference for determining elevations during the construction of a building.

Bevel cut: A cut made across the sloping edge or side of a workpiece at an angle of less than 90 degrees.

Blocking: A wood block used as a filler piece and a support between framing members.

Brace: A diagonal supporting member used to reinforce a form against the weight of the concrete.

Butt joint: The joint formed when one square-cut edge of a piece of material is placed against another material.

Cantilever: A beam, truss, or slab (floor) that extends past the last point of support.

Casing: The trim around a window or door.

Catalyst: A substance that causes a reaction in another substance.

Closed stairway: A stairway that has solid walls on each side.

Compound cut: A simultaneous bevel and miter cut.

Contour lines: Imaginary lines on a site/plot plan that connect points of the same elevation. Contour lines never cross each other.

Cripple stud: In wall framing, a short framing stud that fills the space between a header and a top plate or between the sill and the soleplate.

Crosscut: A cut made across the grain in lumber.

Crown: The high point of the crooked edge of a framing member.

Curb: A framework on which a skylight is mounted.

Cured concrete: Concrete that has hardened and gained its structural strength.

Dado: A rectangular groove or rabbet cut that is made partway through and across the grain in lumber.

Dead load: The weight of permanent, stationary construction and equipment included in a building.

Door jack: A holder or stand used to hold a door on edge while planing, routing, or installing hinges.

Double top plate: A plate made of two members to provide better stiffening of a wall. It is also used for connecting splices, corners, and partitions that are at right angles (perpendicular) to the wall.

Dovetail joint: An interlocking joint with a triangular shape like that of a dove's tail.

Drying-in: Applying sheathing, windows, and exterior doors to a framed building.

Easement: A legal right-of-way provision on another person's property (for example, the right of a neighbor to build a road or a public utility to install water and gas lines on the property). A property owner cannot build on an area where an easement has been identified.

Elevation view: A drawing giving a view from the front or side of a structure.

False fascia: The board that is attached to the tails of the rafters to straighten and space the rafters and provide a nailer for the fascia. Also called sub fascia and rough fascia.

Finish carpentry: The portion of the carpentry trade associated with interior and exterior trim, cabinetry, siding, wall finishes, and decorative work.

Firestop: An approved material used to fill air passages in a frame to retard the spread of fire.

Flashing: Sheet metal strips used to seal a roof or wall against leakage.

Flatwork: Work connected with concrete slabs used for walks, driveways, patios, and floors.

Footing: The base of a foundation system for a wall, column, and chimney. It bears directly on the undisturbed soil and is made wider than the object it supports to distribute the weight over a greater area.

Forms: Wood or metal structures built to contain plastic concrete until it hardens.

Foundation: The supporting portion of a structure, including the footings.

Front setback: The distance from the property line to the front of the building.

Furring strip: Narrow wood strips nailed to a wall or ceiling as a nailing base for finish material.

Gable roof: A roof with two slopes that meet at a center ridge.

Gable: The triangular wall enclosed by the sloping ends of a ridged roof.

Galvanized: Protected from rusting by a zinc coating.

Geometrical stair: A winding stairway built around a well. Examples include circular, elliptical, and spiral stairs.

Glazing: Material such as glass or plastic used in windows, skylights, and doors.

Green concrete: Concrete that has hardened but has not yet gained its structural strength.

Guardrail: A rail secured to uprights and erected along the exposed sides and ends of platform stairs, etc.

Gypsum: A chalky material that is a main ingredient in plaster and drywall.

Handrail: A member supported on brackets from a wall or partition to furnish a handhold.

Header joist: A framing member used in platform framing into which the common joists are fitted, forming the box sill. Header joists are also used to support the free ends of joists when framing openings in a floor.

Header: A horizontal structural member that supports the load over a window or door.

Headroom: The vertical and clear space in height between a stair tread and the ceiling or stairs above.

Hip roof: A roof with four sides or slopes running toward the center.

Hipped: The external angle formed by the meeting of two adjacent sloping sides.

Housed stringer: A stair stringer with horizontal and vertical grooves cut (mortised) on the inside to receive the ends of the risers and treads. Wedges covered with glue are often used to hold the risers and treads in place in the grooves.

Hydration: The catalytic action water has in transforming the chemicals in portland cement into a hard solid. The water interacts with the chemicals to form calcium silicate hydrate gel.

Jamb: The top (head jamb) and side members of a door or window frame that come into contact with the door or window.

Jamb: The top and sides of a door or window frame that are in contact with the door or sash.

Japanned: Coated with a glossy finish.

Joist hanger: A metal stirrup secured to the face of a structural member, such as a girder, to support and align the ends of joists flush with the member.

Joist: Generally, equally spaced framing members that support floors and ceilings.

Kerf: The width of the cut made by a saw blade. It is also the amount of material removed by the blade in a through (complete) cut or the slot produced by the blade in a partial cut.

Kickback: A sharp, uncontrolled grabbing and throwing of the workpiece by a tool as it rejects material being forced into it.

Kip: An informal unit of force that equals one thousand (kilo) pounds.

Landing: A horizontal area at the end of a flight of stairs or between two flights of stairs.

Let-in: Any type of notch in a stud, joist, etc., which holds another piece. The item that is supported by the notch is said to be let in.

Lights: The glass inserts in a door.

Live load: The total of all moving and variable loads that may be placed upon a building.

Lookout: A structural member used to frame an overhang.

Material safety data sheets (MSDS): Information that details any toxic, chemical, or potentially harmful substances that are contained in a product.

Millwork: Manufactured wood products such as doors, windows, and moldings.

Miter box: A device that is used to cut lumber at precise angles.

Miter cut: A cut made at the end of a piece of lumber at any angle other than 90 degrees.

Monolithic slab (monolithic pour): Concrete placed in forms in a continuous pour without construction joints.

Monuments: Physical structures that mark the location of a survey point.

Mortise: A rectangular cutout or depression made in a piece of wood to receive something such as a hinge, lock, or tenon.

Muntin: A thin framework used to secure panes of glass in a door.

Nail set: A punch-like metal tool used to recess finishing nails.

Newel post: An upright post supporting the handrail at the top and bottom of a stairway or at the turn of a landing; also, the main post about which a circular staircase winds or a stone column carrying the inner ends of the treads of a spiral stone staircase.

Nominal size: Approximate or rough size (commercial size) by which lumber, block, etc., is commonly known and sold, normally slightly larger than the actual size (for example, 2 × 4s).

Nosing: The portion of the stair tread that extends beyond the face of the riser.

Open stairway: A stairway that is open on at least one side.

Pascal: A metric measurement of pressure.

Pier: A column of masonry used to support other structural members, typically girders or beams.

Piles: Column-like structural members that penetrate through unstable, nonbearing soil to lower levels of loadbearing soil. They provide support for grade beams or columns that carry the structural load of a building.

Pitch board: A board that serves as a pattern for marking cuts for stairs. The shortest side is the height of the riser cut, and the next longer side is the width of the tread. This is used mainly when there is great repetition such as in production housing.

Plan view: A drawing that represents a view looking down on an object.

Plastic concrete: Concrete when it is first mixed and is in a semiliquid and moldable state.

Plyform: American Plywood Association's tradename for a reusable material for constructing concrete forms.

Pocket (plunge) cut: A cut made to remove an interior section of a workpiece or stock (such as in a countertop for a sink) or to make square or rectangular openings in floors and walls.

Pozzolan: The name given by the ancient Romans to describe the volcanic ash they used as a type of cement. Today, the term is used for natural or calcined materials (including fly ash and silica fume) or air-cooled blast furnace slag.

Property lines: The recorded legal boundaries of a piece of property.

Purlin: A horizontal roof support member parallel to the plate and installed between the plate and the ridgeboard.

Quantity takeoff: A procedure that involves surveying, measuring, and counting all materials indicated on a set of drawings.

Rabbet cut: A rectangular cut made along the edge or end of a board to receive another board that has been similarly cut.

Rafter plate: The top or bottom horizontal member at the top of a wall.

Rafter: A sloping structural member of a roof frame to which sheathing is secured.

Rail: The horizontal member of a window sash or panel door.

Rail-steel: Deformed reinforcing bars that are rolled from selected used railroad rails.

Rebars: Abbreviation for reinforcing bars. Also called rerod.

Resins: Protective natural or synthetic coatings.

Ribband: A 1 × 4 nailed to the ceiling joists at the center of the span to prevent twisting and bowing of the joists.

Rip cut: A cut made in the direction of the grain in lumber.

Rise and run: A term used to indicate the degree of incline.

Rise: The vertical dimension of a set of stairs. Also called the total rise.

Riser diagram: A schematic drawing that depicts the layout, components, and connections of a piping system.

Riser: A vertical board under the tread of a stair step; in other words, a board set on edge for connecting the treads of a stairway.

Rough carpentry: The portion of the carpentry trade associated with framing and other work that will be covered with finish materials.

Run: The horizontal distance from the face of the first or upper riser to the face of the last or lower riser. Also called the total run.

Sash: The part of a window that holds the glass.

Scab: A length of lumber applied over a joint to strengthen it.

Scarf: To join the ends of stock together with a sloping lap joint so there appears to be a single piece.

Screeding: Leveling newly placed concrete to an established grade. Also called striking off.

Sheathing: Sheet material or boards to which roofing material or siding is secured.

Shiplap: A method of cutting siding in which each board is tapered and grooved so that the upper piece fits tightly over the lower piece.

Shoring: Temporary bracing used to support above-grade concrete slabs while they harden.

Sill plate: A horizontal timber that supports the framework of a building on the bottom of a wall or box joist. It is also called a sole plate.

Sill: The lower framing member attached to the top of the lower cripple studs to form the base of a rough opening for a window.

Sill: The lowest member of a window or exterior door frame.

Skirtboard: A baseboard or finishing board at the junction of the interior wall and floor. Also called a finished stringer.

Slab-on-grade (slab-at-grade): A ground-supported concrete slab 3½" or thicker that is used as a foundation system. It combines concrete foundation walls with a concrete floor slab that rests directly on an approved base that has been placed over the ground.

Slag: The ash produced during the reduction of iron ore to iron in a blast furnace.

Slump: The distance a standard-sized cone made of freshly mixed concrete will sag. This is known as a slump test.

Soleplate: The bottom horizontal member of a wall frame.

Span: The distance between structural supports such as walls, columns, piers, beams, or girders.

Stairwell: A compartment extending vertically through a building into which stairs are placed.

Stile: The vertical member of a window sash or panel door.

Stringer: The inclined member that supports the treads and risers of a stairway.

Strongback: An L-shaped arrangement of lumber used to support ceiling joists and keep them in alignment.

Studs: Vertical members of a form panel used to support sheathing.

Subgrade: Soil prepared and compacted to support a structure or pavement system.

Tail joist: Short joists that run from an opening to a bearing.

Takeoff: A list of building materials obtained by analyzing the project drawings (also known as a material takeoff).

Tenon: A tongue that is cut on the end of a piece of wood and shaped to fit into a mortise.

Tongue-and-groove: A joint made by fitting a tongue on the edge of a board into a matching groove on the adjoining board.

Top plate: The upper horizontal framing member of a wall used to carry the roof trusses or rafters.

Topographical survey: An accurate and detailed drawing of a place or region that depicts all the natural and man-made physical features, showing their relative positions and elevations.

Tread: The horizontal member of a step.

Trimmer joist: A full-length joist that reinforces a rough opening in the floor.

Trimmer stud: A vertical framing member that forms the sides of rough openings for doors and windows. It provides stiffening for the frame and supports the weight of the header.

True: Accurately shaped or fitted.

Truss: An engineered assembly made of wood, or wood and metal members, that is used to support floors and roofs.

Underlayment: A material, such as particleboard or plywood, laid on top of the subfloor to provide a smoother surface for the finished flooring.

Unit rise: The vertical distance from the top of one stair tread to the top of the next one above it; also called the stair rise.

Unit run: The horizontal distance from the face of one riser to the face of the next riser.

Vaulted ceiling: A high, open ceiling that generally follows the roof pitch.

Walers: Horizontal pieces placed on the outsides of the form walls to strengthen and stiffen the walls. The form ties are also fastened to the walers.

Water-cement ratio: The ratio of water to cement, usually by weight (water weight divided by cement weight), in a concrete mix. The water-cement ratio includes all cementitious components of the concrete, including fly ash and pozzolans, as well as portland cement.

Winding stairway: A type of geometrical staircase that changes direction by means of winders or a landing and winders. The stair opening is relatively wide, and the balustrade follows the curve with only a newel post at the bottom.

Figure Credits

Module 27101-06

Arcways, Inc., 101F05
Benson Industries LLC, 101F03
Digital Imagery ©2000 Photodisc, Inc., 101SA02
SkillsUSA, module divider
Topaz Publications, Inc., 101F04
Gary Wilson, module overview photo, 101SA01

Module 27102-06

Michael Anderson, 102F03, 102SA04
APA—The Engineered Wood Association, Table 4
Benson Industries LLC, 102F13, 102F14
Used with permission of the Brick Industry Association, Reston, Virginia, **www.gobrick.com**, 102F12
Easi-Set Industries, 102SA06
National Gypsum Company, 102SA07
Portland Cement Association, 102F15B
SkillsUSA, module divider
Tilt-Up Concrete Association, 102F16
Topaz Publications, Inc., 102SA01, 102SA02, 102F10, 102SA05, 102F22, 102F25 (photo), 102SA08, 102F33, 102F43B
Trus Joist, A Weyerhaeuser Business, module overview photo, 102F07, 102F08
Western Wood Products Association, Appendix B

Module 27103-06

American Clamping Corp./Bessey, 103F04 (hand-screw clamp)
Associated Builders and Contractors, Inc., module divider
Cooper Hand Tools, 103F04 (locking C-clamp), 103F05 (backsaw, compass saw, coping saw)
DeWalt Industrial Tool Co., 103F08, 103F13
Donald Dixon, 103F06
John Hoerlein, 103F01 (automatic builder's level)
IRWIN Industrial Tool Company, IRWIN is a registered trademark of IRWIN Industrial Tool Company, 103F04 (quick clamp)
Milwaukee Electric Tool Corporation, 103F11A
Ridge Tool Company (RIDGID®), 103F05 (hacksaw), 103F11B
Daniele Stacey, 103SA06, 103SA09
The Stanley Works, 103F01 (line level), 103SA01, 103F04 (web clamp, spring clamp), 103F05 (dovetail saw)
Topaz Publications, Inc., module overview photo, 103SA02, 103SA03, 103SA04, 103SA05, 103F07, 103SA07, 103SA08, 103F14, 103F15, 103F17
David White, 103F01 (laser detector, rotary laser)
Zircon Corporation, 103F01 (water level)

Module 27104-06

Construction Specifications Institute (CSI) and Construction Specifications Canada (CSC)*, 104F28
John Hoerlein, 104F04, 104F06, 104F07, 104F08, 104F09, 104F13
Ryan Homes, 104SA02
Scalex Corporation, 104SA05
SkillsUSA, module divider
Staedtler USA, 104SA04
Topaz Publications, Inc., 104SA01

Module 27105-06

APA—The Engineered Wood Association, Table 1
Associated Builders and Contractors, Inc., module divider
Topaz Publications, Inc., module overview photo, 105SA02, 105SA04, 105SA05, 105SA06, 105SA07
Trus Joist, A Weyerhaeuser Business, 105SA03, 105F20

---

*The Numbers and Titles used in this textbook are from MasterFormat™ 2004, published by the Construction Specifications Institute (CSI) and Construction Specifications Canada (CSC), and are used with permission from CSI. For those interested in a more in-depth explanation of MasterFormat™ 2004 and its use in the construction industry visit **www.csinet.org/masterformat** or contact:

The Construction Specifications Institute (CSI)
99 Canal Center Plaza, Suite 300
Alexandria, VA 22314
800-689-2900; 703-684-0300
http://www.csinet.org

Module 27106-06

Associated Builders and Contractors, Inc., module divider

Topaz Publications, Inc., module overview photo, 106SA01, 106SA02, 106SA03, 106F31, 106F32

Module 27107-06

SkillsUSA, module divider

Southern Forest Products Association, 107F30

The Stanley Works, 107F19

Topaz Publications, Inc., module overview photo, 107SA01, 107SA02, 107SA03, 107F27, 107F28, 107SA05, 107F31, 107F32

Reprinted with permission from the Wood Truss Council of America (WTCA). For more information, visit **www.woodtruss.com**, 107F24, 107SA04

Module 27108-06

Associated Builders and Contractors, Inc., module divider

Benner-Nawman, Inc., 108SA05

Portland Cement Association, 108SA01, 108SA03, 108SA04, 108F03

Rosamond Gifford Zoo, 108SA02, 108SA08, 108SA09

Topaz Publications, Inc., module overview photo, 108SA06, 108SA07

Module 27109-06

Ceco Door Products, Appendix

Ingersoll Rand Safety & Security, 109F30

SkillsUSA, module divider

Topaz Publications, Inc., module overview photo, 109SA01, 109F01, 109F05, 109SA02, 109SA03, 109SA05, 109SA06, 109SA07, 109SA08

Module 27110-06

Classic Stair Works, Ltd., module overview photo, 110SA01

SkillsUSA, module divider

Topaz Publications, Inc., 110F19, 110SA02

Index

Note: The designation "*f*" following a page number refers to a figure.

A

Abbreviations
   drawing sets and, 4.35, 4.56-4.59
   on elevations, 4.56-4.57
   in lumber industry, 2.13, 2.61-2.62
   on plan views, 4.57-4.59
Abrasive saws, 3.14-3.15, 3.15*f*
Absenteeism, 1.14-1.15
ACI (American Concrete Institute), 8.5, 8.6
   defined, 8.38
Addenda and change orders, 4.41
Adhesive applicators, 2.48, 2.48*f*
Adhesives, 2.46-2.49
Adjustable T-squares, 3.4*f*
Admixtures, 8.2, 8.4-8.5
   defined, 8.38
Aesthetics, of doors and windows, 9.2
Aggregates, 8.2, 8.4
   defined, 8.38
Alcohol, 1.19
Aligning, of walls, 6.16-6.19, 6.17*f*
Amber, 2.6
American Concrete Institute (ACI), 8.5, 8.6, 8.38
American National Standards Institute (ANSI), 8.2
American Plywood Association (APA), 2.13
American Society for Testing and Materials (ASTM), 8.2
Anchor bolts, 2.39, 2.40*f*, 5.12, 5.13*f*
Anchor straps, 5.12, 5.14*f*, 5.15
Animal glue, 2.47
Annular rings, 2.3, 2.3*f*
ANSI (American National Standards Institute), 8.2
APA (American Plywood Association), 2.13
Apprenticeship Committee, 1.11, 1.15-1.16
Apprenticeship programs
   Apprenticeship Committee, 1.11, 1.15-1.16
   Bureau of Apprenticeship and Training, 1.10-1.11
   employee responsibilities, 1.12-1.15
   employer responsibilities, 1.15
   sample apprentice training recognitions, 1.26-1.28
   standards, 1.11-1.12
   training programs and, 1.9, 1.15
   youth apprenticeships, 1.11
Architect/engineer (A/E), and project organization, 4.45
Architects, 1.8, 1.10
Architect's scales, 4.34
Architectural drawings. *See also* Drawing sets
   floor systems and, 5.7-5.8, 5.10-5.12
   information in, 4.2, 4.2*f*, 5.8*f*
Architectural plans, 4.2, 4.3
Architectural terms, drawings and specifications and, 4.35, 4.60-4.61

Area, formulas for finding, 8.40
Asbestos, 2.18
As-built drawings, 4.20
ASTM (American Society for Testing and Materials), 8.2
ATELS (Office of Apprenticeship, Training, Employer and Labor Services), 1.10
Attitudes, and human relations, 1.17
Awning windows, 9.3, 9.5*f*
Axle-steel, 8.13
   defined, 8.38

B

Backsaws, 3.8, 3.8*f*
Balloon-frame construction, 1.3, 1.4*f*, 5.2, 5.4, 5.5*f*, 6.4
Balusters, 10.6, 10.7*f*
   defined, 10.26
Balustrades, 10.6
   defined, 10.26
Barge rafters, 7.8, 7.14, 7.14*f*
   defined, 7.35
Bar joists, 7.28, 7.28*f*
Bark, 2.3, 2.3*f*
Bar supports, 8.16
BAT (Bureau of Apprenticeship and Training), 1.10-1.11
Bay windows, 9.5, 9.5*f*, 9.6*f*
Beams. *See also* Girders; Wood I-beams
   grade beams, 8.21, 8.22*f*
   lumber grades, 2.10
Beams and stringers (B&S), 2.10
Bench marks, 4.5
   defined, 4.55
Bevel cuts, 3.3, 3.10
   defined, 3.27
Bird's mouths, 7.7, 7.7*f*, 7.10, 7.11*f*, 7.39, 7.39*f*
Blocking
   defined, 6.35
   wall framing and, 6.2
Block planes, 3.6, 3.6*f*
Blueprints
   drawing sets and, 4.2
   floor systems and, 5.11-5.12
Board feet, 2.20, 2.21*f*, 2.64
Boards (BD)
   appearance grades, 2.9
   size classifications, 2.10
Board subfloors, 5.26-5.27
Bolt anchors, 2.41, 2.41*f*
Bolts, 2.39, 2.39*f*
Boring jigs, and locksets, 9.21, 9.22*f*
Bow windows, 9.6, 9.7*f*
Box beam headers, 6.6, 6.7*f*
Boxnails, 2.30, 2.32*f*
Box sills, 5.33, 5.34*f*
Braced frames, 5.2, 5.4

Braces
  defined, 8.38
  for wall forms, 8.25, 8.26f
Bracing
  diagonal bracing, 6.25-6.26
  metal wall bracing, 6.18, 6.18f
  permanent bracing, 6.18, 7.24, 7.24f
  temporary bracing, 6.18, 7.24, 7.24f, 7.25
Break lines, 4.22
Bridging
  estimating material for, 5.41
  floor systems and, 5.24-5.25, 5.25f
  installing, 5.35-5.36, 5.36f
Builder's levels, 3.2, 3.2f
Building boards, 2.16-2.17
Building codes, 4.43, 10.10
Building materials
  for commercial construction, 1.6, 1.6f, 2.2, 2.23-2.30
  concrete blocks, 2.21-2.22
  fire-retardant, 2.20
  introduction, 2.2
  lumber, 2.3-2.21
  for residential construction, 1.6, 1.7f, 2.2, 2.2f
  stacking and unloading of, 2.7-2.8
Building owners, and project organization, 4.44-4.45
Built-in-place forms, 8.28
Built-up headers, 6.6, 6.6f
Built-up lumber girders, 5.15-5.16, 5.15f
Bureau of Apprenticeship and Training (BAT), 1.10-1.11
Butt joints, 2.13
  defined, 2.55

C

Calculating
  jack rafter lengths, 7.42
  lumber quantities, 2.20-2.21, 2.21f
  rafter length, 7.9-7.10, 7.9f
  rise and pitch, 7.8
  stud length, 6.12-6.14, 6.13f, 6.14f, 6.15
  volume calculations for concrete, 8.9-8.10, 8.11f, 8.12, 8.12f
Cambium layer, 2.3, 2.3f
Cantilever applications
  defined, 2.55
  glue-laminated lumber and, 2.19
Cantilevered floors, installing joists for, 5.38, 5.39f, 5.40
Carbide-tipped saw blades, 3.16f, 3.17
Carpenters
  career opportunities of, 1.8, 1.10
  requirements of trade, 1.8
  responsibilities of, 1.12-1.15
  working with other trades, 4.47
Carpentry
  history of, 1.2-1.3, 1.4f, 1.5f, 1.6
  modern carpentry, 1.6-1.8, 1.6f, 1.7f
Carriage bolts, 2.39, 2.39f
Casein glue, 2.47
Casement windows, 9.3, 9.5f
Casing, 9.12
  defined, 9.28
Casing nails, 2.31, 2.32f, 9.12
Catalyst, 2.47
  defined, 2.55
Ceiling joists
  cutting and installing, 6.20, 6.21, 6.22f
  estimating material for, 6.26
  layout, 6.19-6.20, 6.20f
  reinforcing, 6.21, 6.22f
  splicing, 6.20-6.21, 6.21f

Ceilings
  framing, 6.19-6.21, 6.20f, 6.21f, 6.22f
  framing spacing and, 2.29
  laying out, 6.19-6.21, 6.20f
Cellular floors, 2.24-2.25, 2.25f
Center lines, 4.22
Chalklines, 6.9
Change orders, 4.41
Changing saw blades, 3.11
Chisel point staples, 2.34, 2.34f
Chisel-tooth combination saw blades, 3.16, 3.16f
Chop saws, 3.14-3.15, 3.15f
Cinder blocks, 2.21-2.22
Circular saws, 3.10-3.17, 3.10f
Clamps, 3.6-3.7, 3.7f
Classification
  defects in lumber, 2.12-2.13
  lumber, 2.4, 2.5
  size classification of lumber, 2.11
Cleated stringers, 10.5, 10.6f, 10.20, 10.20f
Closed finish stringers, 10.7f, 10.8
Closed stairways, 10.2, 10.2f
  defined, 10.26
Code of Federal Regulations (CFR), 1.10
Colored concrete, 8.8
Columns, 5.17, 5.18
Column spacing, 5.18, 5.18f
Combination rip saw blades, 3.16
Combination squares, 3.3
Commercial construction
  building materials for, 1.6, 1.6f, 2.2, 2.23-2.30
  concrete blocks in, 2.22
  concrete construction, 2.23-2.24, 2.23f
  construction methods, 2.23-2.30
  exterior walls and, 2.25, 2.26f
  floors, 2.23, 2.24-2.25, 2.25f
  metal framing materials and, 2.27-2.28, 2.27f, 2.29f
  partition walls and, 2.25-2.26, 2.26f, 2.27f
Commercial interior walls, 2.25-2.26
Commercial stairways, 10.11
Common nails, 2.30, 2.32f
Common rafters
  laying out, 7.18, 7.19f, 7.20
  roof layout and, 7.3, 7.4f
Common roofing nails, 2.32f, 2.33
Common site/plan topographical survey symbols, 4.27, 4.30f
Common studs, measuring and cutting, 6.12-6.13, 6.13-6.14f
Compass saws, 3.8, 3.8f
Competency-based training, 1.11
Complete specifications, 4.41
Compound cuts, 3.13
  defined, 3.27
Compound miter saws, 3.13-3.14, 3.13f
Concrete. *See also* Concrete forms
  admixtures, 8.2, 8.4-8.5, 8.38
  aggregates, 8.2, 8.4, 8.38
  curing methods and materials, 8.8
  estimating volume, 8.9-8.10, 8.11, 8.12
  mix proportions and measurements, 8.5-8.6
  portland cement, 8.2, 8.3
  reinforcement materials, 8.12-8.17
  slump testing, 8.8, 8.9f
  special types, 8.6
  water and, 8.4, 8.8
Concrete blocks, 2.21-2.22
Concrete calculators, 8.9, 8.9f

Concrete construction
  commercial construction methods and, 2.23-2.24, 2.23f
  concrete block construction, 2.21-2.22, 2.22f
  tilt-up concrete construction, 2.24, 2.24f, 2.25
Concrete floors, 2.24-2.25
Concrete forms
  edge forms, 8.27-8.28, 8.30f, 8.31
  footings, 8.2, 8.18-8.22, 8.20f, 8.21f, 8.23f, 8.24
  removing, 8.31
  safety and, 8.17-8.18
  slabs, 8.27-8.28, 8.29f, 8.30f, 8.31
  for stairs, 10.21, 10.21f
  wall forms, 8.24-8.27, 8.25f
Concrete/masonry screws, 2.37, 2.37f
Concrete masonry units (CMUs), 2.21
Concrete tables, 8.9, 8.10f
Concrete walls, 2.25, 2.26f
Coniferous trees, 2.4
Construction adhesives, 2.48
Construction industry
  opportunities in, 1.8-1.16, 1.9f
  types of carpentry in, 1.7-1.8, 1.7f
Construction managers (CM)
  project organization and, 4.45
  responsibilities of, 1.8, 1.10
Construction training. *See also* Apprenticeship programs
  contractors and, 1.11-1.12
  formal training, 1.10-1.11
Contact cement, 2.47, 2.48
Continuous footings, 8.19, 8.20, 8.20f, 8.24
Contour lines, 4.22
  defined, 4.55
Contractors/owners
  on-the-job training and, 1.11-1.12
  project organization and, 4.44-4.45
  responsibilities of, 1.8, 1.10
Conversion tables
  decimal conversion tables, 8.39
  lumber, 2.63-2.64
  metric conversion tables, 8.39, 8.41
Cooling loads, and window glass, 9.9, 9.9f
Cooperation, 1.14
Coping saws, 3.8, 3.8f
Cordless nailers and staplers, 3.20-3.22, 3.21f, 3.22f
Cordless power tools, 3.17, 3.20-3.22, 3.21f, 3.22f
Cores, in plywood, 2.14
Corners, wall framing, 6.3, 6.3f, 6.4f
Corrugated steel forms, 2.24, 2.25f
Crane deliveries, 7.16
Cripple studs, 6.2, 6.2f, 6.6, 6.7, 6.14
  defined, 6.35
Crosscut chisel staples, 2.34, 2.34f
Crosscuts, 3.10
  defined, 3.27
Crowns, 5.15
  defined, 5.50
Curbs
  defined, 9.28
  and skylights, 9.7, 9.8f
Cured concrete, 8.2
  defined, 8.38
Curing blankets, 8.8
Curing compounds, 8.8
Curing mats, 8.8
Curing methods and materials for concrete, 8.8
Curing paper, 8.8
Curtain walls, 2.23, 2.23f, 2.24, 2.24f
Customer relations, 1.14

Cut nails, 2.32, 2.32f
Cutout stringers, 10.5, 10.5f
Cutting
  king studs, 6.12-6.13, 6.13-6.14f
  lumber, 2.3-2.4, 2.3f, 2.4f
  overcutting, 10.17, 10.19f
  planes, 4.22
  stringers, 10.15-10.17, 10.15f, 10.16f, 10.18f
  studs, 6.12-6.13, 6.13-6.14f
Cylindrical locksets, 9.19, 9.21f

D

Dado cuts, 3.12
  defined, 3.27
Dadoed stringers, 10.5, 10.6f, 10.20, 10.20f
Deadbolt locks, 9.21
Dead loads, 5.20
  defined, 5.50
Deciduous trees, 2.4
Decimal conversion tables, 8.39
Deck screws, 2.37, 2.38f
Defects, in lumber, 2.5-2.8, 2.6f, 2.12-2.13, 6.15, 8.26
Demolition saws, 3.14, 3.15f
Detail drawings
  drawing sets and, 4.2f, 4.11, 4.16f
  floor systems and, 5.8
Diagonal bracing, estimating material for, 6.25-6.26
Dimension and extension lines, 4.22
Dimension lumber
  appearance grades, 2.9, 2.10
  size classification, 2.11
Dimensions, 4.32, 4.33f, 4.34, 5.11
Divergent staples, 2.34, 2.34f
Door jacks, 3.20
  defined, 3.27
Doors. *See also* Exterior doors
  in masonry walls, 6.28
  symbols, 4.23, 4.25f
Door sizes, 9.17
Door strikes, 9.20, 9.21, 9.22f
Door swing, 9.17, 9.19f
Dormers, 7.26, 7.27f
Double-end-trimmed lumber (DET), 2.12
Double-hung windows, 9.3, 9.4f
Double joists, 5.33, 5.33f
Double-L stairways, 10.3f
Double-strength (DS) glass, 9.9
Double top plates, 6.2, 6.2f, 6.17, 6.18f
  defined, 6.35
Dovetail joints, 3.8
  defined, 3.27
Dovetail saws, 3.8, 3.8f
Drawing sets, 4.2-4.20. *See also* Reading and interpreting drawings
  as-built drawings, 4.20
  blueprints, 4.2
  detail drawings and, 4.2f, 4.11, 4.16f
  elevation drawings, 4.2f, 4.11, 4.12f
  floor systems and, 5.7-5.12, 5.8f
  lines used in drawings, 4.20, 4.21f, 4.22
  plan view drawings, 4.2f, 4.3-4.11, 4.6f, 4.7f, 4.8f, 4.10f, 4.55
  plumbing mechanical and electrical plans, 4.2f, 4.18-4.19, 4.19f
  schedules, 4.2f, 4.16, 4.17f, 4.18f
  section drawings and, 4.2f, 4.11, 4.13-4.15f
  shop drawings, 4.20

Drawing sets, *continued*
 structural drawings, 4.18, 5.10
 title sheets, title blocks and revision blocks, 4.2-4.3, 4.2*f*, 4.4*f*
 title sheets and, 4.2-4.3
Dressed lumber, 2.11
Dressed sizes, 2.11
Drilling wooden joists, 5.22, 5.23*f*
Drill presses, 3.17-3.18, 3.17*f*
Drive screws, 2.38, 2.38*f*
Driving wood screws, 2.37
Drop-in anchors, 2.41, 2.41*f*
Drugs, 1.19
Drying, of lumber, 2.7
Drying-in, 6.19
 defined, 6.35
Drywall adhesive, 2.48
Drywall (ratchet) nails, 2.31, 2.32*f*
Drywall screws, 2.38, 2.38*f*
Drywall squares, 3.4
Dull saw blades, 3.16
Duplex or scaffold (doublehead) nails, 2.31, 2.32*f*

E
Earthquakes, 5.7
Easements, 4.5
 defined, 4.55
Edge forms, 8.27-8.28, 8.30*f*, 8.31
Edge-grained lumber, 2.4, 2.4*f*
Electrical plans, 4.2*f*, 4.18-4.19, 4.19*f*
 and floor systems, 5.11
Electrical symbols, 4.25, 4.26*f*, 4.27, 4.27*f*
Elevation drawings, 4.2*f*, 4.11, 4.12*f*
Elevation views, 4.23
 defined, 4.55
Employees
 human relations and, 1.16-1.17
 responsibilities of, 1.12-1.15
 safety obligations of, 1.17-1.19, 1.18*f*
Employers
 responsibilities of, 1.15
 safety obligations of, 1.17-1.19, 1.18*f*
Ending newel posts, 10.6, 10.7*f*
End nosing, 10.7*f*, 10.8
Energy efficient windows, 9.9-9.10
Engineered lumber girders, 5.16
Engineered wood products, 2.18-2.19, 2.18*f*, 2.20, 6.6
Entry doors. *See* Exterior doors
Epoxies, 2.48
Epoxy anchoring systems, 2.46, 2.46*f*
Escutcheon pins, 2.34
Estimates, 6.25
Estimating material
 for bridging, 5.41
 for ceiling joists, 6.26
 for diagonal bracing, 6.25-6.26
 for floor joists, 5.40
 floor plans and, 6.23, 6.23*f*
 for floor systems, 5.40-5.41, 5.40*f*
 for girders, 5.40
 for headers, 6.24-6.25
 for rafters, 7.26
 for ridges (ridgeboards), 7.26
 for roof framing, 7.25-7.26
 for sheathing, 7.26
 for sills, 5.40
 for soleplates, 6.24
 for studs, 6.24
 for subflooring, 5.41
 for top plates, 6.24
 for wall framing, 6.23-6.26
Estimating volume of concrete, 8.9-8.10, 8.11, 8.12
Estimators, 1.8, 1.9
Ethical principles, 1.12
Extension jambs, 9.13, 9.13*f*
Exterior doors
 component parts of installation, 9.14, 9.15*f*
 door strikes, 9.20, 9.21, 9.22*f*
 installation of pre-hung exterior doors, 9.17, 9.19, 9.19*f*
 locksets, 9.19, 9.20*f*, 9.21, 9.21*f*
 sizes of, 9.17
 types of, 9.14, 9.14*f*, 9.16, 9.16*f*, 9.17, 9.17*f*
 weatherstripping, 9.16, 9.16*f*, 9.17, 9.19*f*
Exterior walls, 2.25, 2.26*f*
Eye injuries, 2.44

F
Fabrication, steel stud walls, 6.28
Face veneers, in plywood, 2.14-2.15
False fascia, 7.7
 defined, 7.35
Fasteners
 bolts, 2.39
 epoxy anchoring systems, 2.46, 2.46*f*
 hammer-driven pins and studs, 2.38
 mechanical anchors, 2.39-2.46
 nails, 2.30-2.34, 2.32*f*
 screws, 2.34-2.38, 2.35*f*
 staples, 2.34, 2.34*f*
Featherboards, 3.13
Felt installation, 7.17
Fiberboard core plywood, 2.14, 2.14*f*
Field-fabricated housed stringers, 10.19
Fill-in specifications, 4.41
Finish carpentry, 1.7-1.8, 1.7*f*
 defined, 1.25
Finished floors, and riser height, 10.18
Finished stringers, 10.7*f*, 10.8
Finish lumber, 2.8, 2.11
Finish nails, 2.30, 2.32*f*
Fire-retardant building materials, 2.20
Firestops, 5.4, 5.5*f*, 6.15, 6.16*f*
 defined, 5.50
Fitzgerald Act, 1.10
Fixed windows, 9.3, 9.5*f*
Flashing, 9.7, 9.8*f*
 defined, 9.28
Flat-grained lumber, 2.3, 2.4*f*
Flatwork, 8.27
 defined, 8.38
Floor joists
 estimating material for, 5.40
 floor systems and, 5.20-5.24, 5.20*f*
 installing, 5.33, 5.34*f*
 laying out sills and girders for, 5.31-5.32, 5.32*f*
 sizing of, 5.43, 5.43*f*
Floor joist transitions, 5.37
Floor loads, 5.41, 5.42*f*
Floor plans
 drawing sets and, 4.2*f*, 4.9, 4.10*f*
 estimating material and, 6.23, 6.23*f*
 floor systems and, 5.7, 5.10
 laying out wall openings and, 6.9, 6.9*f*

Floors
- board subfloors, 5.26-5.27
- cantilevered floors, 5.38, 5.39f, 5.40
- cellular floors, 2.24-2.25, 2.25f
- finished floors, 10.18
- framing openings in, 5.35, 5.35f
- plywood subfloors, 5.26

Floor systems
- bridging, 5.24-5.25, 5.25f
- cantilevered floors and, 5.38, 5.39f, 5.40
- drawings and specifications and, 5.7-5.12, 5.8f, 5.9f, 5.10f
- estimating material quantities, 5.40-5.41, 5.40f
- floor joists, 5.20-5.24, 5.20f
- girders and supports, 5.15-5.16, 5.15f, 5.17-5.18, 5.17f, 5.30-5.31, 5.31-5.32, 5.31f
- house framing methods and, 5.2, 5.3f, 5.4, 5.5f, 5.6f, 5.7
- layout and construction, 5.27-5.38
- materials and components, 5.12-5.27
- sills, 5.12, 5.13f, 5.14, 5.15
- sizing girders and joists, 5.41, 5.42f, 5.43
- subflooring, 5.25-5.27, 5.26f

Flush stringer attachment, 10.17, 10.18f
Flush top surfaces, 5.21
Footing forms, 8.21, 8.23f, 8.24f
Footings
- concrete forms, 8.2, 8.18-8.22, 8.20f, 8.21f, 8.23f, 8.24
- defined, 8.38

Foreman/lead carpenters, 1.8
Formats, of specifications, 4.41, 4.42f, 4.43
Form pressures, 8.18
Forms, 8.2
- defined, 8.38

Forms for concrete
- edge forms, 8.27-8.28, 8.30f, 8.31
- footings, 8.18-8.22, 8.20f, 8.21f, 8.23f, 8.24
- removing, 8.31
- safety and, 8.17-8.18
- slabs, 8.27-8.28, 8.29f, 8.30f, 8.31
- wall forms, 8.24-8.27, 8.25f

Forms for concrete stairs, 10.21, 10.21f
Form spreaders, 8.26, 8.26f
Form ties, 8.26-8.27, 8.26f, 8.27f
Formulas for finding areas, 8.40
Formulas for finding volumes, 8.40
Foundation plans
- drawing sets and, 4.2f, 4.7, 4.7f, 4.8f
- wall systems and, 5.7

Foundations
- checking for squareness, 5.14, 5.27-5.28, 5.28f
- defined, 5.50

Foundation walls, as girder supports, 5.19f
Frame and trim saws, 3.14, 3.14f
Framing. *See also* Roof framing; Wall framing
- balloon-frame construction, 1.3, 1.4f, 5.2, 5.4, 5.5f, 6.4
- ceilings, 6.19-6.21, 6.20f, 6.21f, 6.22f
- gable ends, 7.12, 7.13f
- house framing methods, 5.2, 5.3f, 5.4, 5.5f, 5.6f, 5.7
- metal framing, 2.2, 2.27-2.28, 2.27f, 2.29f, 7.28, 7.28f
- plank-and-beam framing, 7.27-7.28, 7.28f
- post-and-beam framing, 5.2, 5.4, 5.6f, 5.7, 7.27-7.28, 7.28f, 7.36f
- stairs, 10.8-10.9
- western platform framing, 1.3, 1.5f, 5.2, 5.3f, 6.4

Framing lumber, 2.11
Framing openings
- in floors, 5.35, 5.35f
- in masonry construction, 6.28
- in roofs, 7.13, 7.14, 7.15f
- for stairwells, 10.13-10.14
- in steel stud walls, 6.28, 6.29f

Framing squares
- cutting stringers and, 10.15-10.17, 10.15f, 10.16f, 10.18f
- description of, 3.4, 3.4f
- hip rafter length and, 7.38-7.39, 7.38f
- rafter framing squares, 7.4-7.5, 7.5f, 7.6f

Front setback, 4.5
- defined, 4.55

Frost, and footings, 8.19
Fuel cells, 3.22
Furring strips, 6.26, 6.27f, 6.28f
- defined, 6.35

G

Gable and valley roofs, 7.2, 7.2f
Gable dormers, 7.26, 7.27f
Gable ends, framing, 7.12, 7.13f
Gable overhangs, 7.13, 7.14f
Gable roofs
- defined, 6.35
- erecting, 7.11-7.13
- as roof type, 7.2, 7.2f
- wall framing and, 6.19

Gables, 7.2, 7.2f
- defined, 7.35

Galvanized coatings, 2.30
- defined, 2.55

Gambrel roofs, 7.2, 7.2f
General contractors
- project organization and, 4.45
- responsibilities of, 1.8, 1.10

Geometrical stairs, 10.4
- defined, 10.26

Ghost wood, 2.4
Girder hangers, 5.18, 5.19f
Girder pockets, 5.18, 5.19f
Girders
- estimating material for, 5.40
- example girder and support column data, 5.31f
- floor systems and, 5.15-5.16, 5.15f
- installing, 5.30-5.31
- laying out, 5.31-5.32
- sizing of, 5.41, 5.42f, 5.43

Girder supports, 5.17-5.18, 5.17f, 5.19f
Glass blocks, 9.13, 9.13f
Glazing, 9.3
- defined, 9.28

Glue-laminated lumber (glulam), 2.18, 2.19, 2.19f, 5.15f, 5.16
Glue-laminated lumber (glulam) headers, 6.6
Glues
- in plywood, 2.14
- types of, 2.47

Goosenecks, 10.6, 10.7f
Grade beams, 8.21, 8.22f
Grade stamps, 2.8, 2.9f
Grading
- of lumber, 2.5, 2.8-2.13, 2.9f
- of plywood, 2.13-2.15
- terms, 2.10-2.12

Green concrete, 8.2
- defined, 8.38

Grilles, 9.3
Growth rings, 2.3-2.45, 2.3f
Guardrails, 10.6, 10.7f
- defined, 10.26

Gypsum, 2.17
  defined, 2.55
Gypsum walls, framing spacing, 2.28, 2.29f

H

Hacksaws, 3.8, 3.8f
Hammer-driven pins and studs, 2.38, 2.38f
Hammer-set anchors, 2.40f, 2.41
Handrails, 10.6, 10.7f, 10.9
  defined, 10.26
Hand-screw clamps, 3.6, 3.7f
Hand tools, 3.2-3.9
  clamps, 3.6-3.7, 3.7f
  levels, 3.2-3.3, 3.2f
  planes, 3.6, 3.6f
  saws, 3.8, 3.8f
  squares, 3.3-3.4, 3.4f, 3.5
Hardboard, 2.16
Hardboard siding nails, 2.33
Hardwoods, 2.4-2.5, 2.5f, 2.58-2.59
H-clips, 7.15, 7.15f
Header joists, 5.20, 5.33, 5.34f
  defined, 5.50
Headers
  defined, 6.35
  estimating material for, 6.24-6.25
  types of, 6.4, 6.6, 6.6f, 6.7f
  as wall components, 6.2, 6.2f
Headroom, 10.8
  defined, 10.26
Headroom clearances, 10.7f, 10.8
Heartwood, 2.3, 2.3f
Heating loads, and window glass, 9.9, 9.9f
Heavyweight (high-density) concrete, 8.6
Hidden lines, 4.22
Hide glue, 2.47
High-density overlay (HDO) plywood, 2.17
High fire/noise resistance partitions, 2.25-2.26, 2.26f, 2.27f
High-performance concrete, 8.6
High-strength concrete, 8.6
Hip and valley roofs, 6.19, 7.2, 7.2f
Hipped, 9.7
  defined, 9.28
Hip rafter position, 7.38, 7.38f
Hip rafters
  laying out with speed square, 7.42-7.43, 7.43f
  layout, 7.3, 7.4f, 7.36, 7.37f, 7.38-7.39, 7.41f
Hip roofs, 6.19, 7.2, 7.2f
  defined, 6.35
Hollow-ground plane (miter) saw blades, 3.17
Hollow-wall anchors, 2.43-2.46
Honesty, 1.12-1.13
Hopper windows, 9.3, 9.5f
Horizontal sliding windows, 9.7, 9.8f
Housed stringers, 10.5, 10.6f
  defined, 10.26
House framing methods, 5.2, 5.3f, 5.4, 5.5f, 5.6f, 5.7
House wrap, and window installation, 9.11
Human relations, 1.16-1.17
HVAC symbols, 4.27, 4.29f
Hydration, 8.2
  defined, 8.38

I

Impulse cordless nailers, 3.22, 3.22f
Incorrect anchoring, 2.46
Inside chisel staples, 2.34, 2.34f

Installation
  of bridging, 5.35-5.36, 5.36f
  of ceiling joists, 6.20, 6.21, 6.22f
  of felt, 7.17
  of floor joists, 5.33, 5.34f
  of joists for cantilevered floors, 5.38, 5.39f, 5.40
  of pre-hung exterior doors, 9.17, 9.19, 9.19f
  of rafters, 7.11, 7.11f
  of sheathing, 7.15-7.16, 7.17
  of sills, 5.28-5.30, 5.29f, 5.30f
  of subflooring, 5.36-5.38, 5.37f, 5.38f
  of trusses, 7.21, 7.23
  of windows, 9.10-9.13, 9.12f
Instant-bond glue, 2.48
Insulated siding nails, 2.33
International Building Code (IBC), 4.43
International Building Code (IBC) requirements for stairs, 10.8
International Code Council, 4.43

J

Jack planes, 3.6, 3.6f
Jack rafters, layout, 7.3, 7.4f, 7.39-7.40, 7.40f, 7.41f, 7.42
Jalousie windows, 9.5, 9.5f
Jambs, 6.10, 9.3, 9.4f
  defined, 6.35, 9.28
Japanned coatings, 2.30
  defined, 2.55
Jointer planes, 3.6, 3.6f
Joist framing, 5.20, 5.21f
Joist hangers, 5.20, 5.21f
  defined, 5.50
Joist headers, 5.20, 5.33, 5.34f
Joist layout, 5.32, 5.33f
Joist nails, 5.22
Joists, 2.9. *See also* Ceiling joists; Floor joists; Header joists
  defined, 2.55
Joists and planks (J&P), 2.10
Journeyman carpenters, 1.8

K

Kerfs, 3.12
  defined, 3.27
Keyhole saws, 3.8, 3.8f
Keyways, 8.22, 8.23f
Kickback, 3.12, 3.13
  defined, 3.27
King studs, 6.2, 6.2f
  measuring and cutting, 6.12-6.13, 6.13-6.14f
Kips, 8.13
  defined, 8.38
Knock-down (KD) door frames, stacking, 2.8

L

Lag screws, 2.36-2.37, 2.37f
Lally columns, 5.17, 5.17f
Laminated strand lumber (LSL), 2.18, 2.18f, 2.19, 2.20
Laminated veneer lumber (LVL), 2.18, 2.18f, 2.20, 5.15f, 5.16
Laminated veneer lumber (LVL) headers, 6.6
Laminate trimmers, 3.18-3.19, 3.19f
Landing newel posts, 10.6, 10.7f
Landings, 10.2, 10.7f, 10.8, 10.11
  defined, 10.26
Laser detectors, 3.2f, 3.3
Laser levels, 3.2f, 3.3, 3.5
Laying out
  ceilings, 6.19-6.21, 6.20f
  continuous footing forms, 8.24
  floor systems, 5.27-5.38

hips and valleys, 7.36, 7.38-7.40, 7.42-7.43
joist locations, 5.32
rafters, 6.19, 6.20f, 7.5, 7.7-7.10, 7.7f, 7.11f, 7.18, 7.18f, 7.19f, 7.20
roof framing, 7.3-7.14, 7.4f, 7.5f, 7.6f, 7.7f, 7.9f, 7.11f, 7.12f, 7.13f, 7.14f
sills and girders for floor joists, 5.31-5.32, 5.32f
stairways, 10.10-10.13
stringers, 10.15-10.20
studs, 6.8, 6.8f, 6.11, 6.12f
wall framing, 6.8-6.11
wall openings, 6.9-6.11, 6.11f, 6.12f
Leader lines, 4.22
Ledgers, and concrete floors, 10.21
Let-in, defined, 5.50
Light framing (L.F.), 2.10
Lights, 9.3, 9.3f
  defined, 9.28
Lightweight concrete, 8.6, 8.7f
Lignin, 2.3
Line levels, 3.2, 3.2f
Lines used in drawings, 4.20, 4.21f, 4.22
Live loads, 5.20
  defined, 5.50
Loadbearing and nonbearing partitions, and estimating, 6.24
Locking C-clamps, 3.6, 3.7f
Lock installation templates, 9.21, 9.22f
Locksets, 9.19, 9.20f, 9.21, 9.21f
Long-L stairways, 10.3f
Lookouts, 7.13, 7.14
  defined, 7.35
Low-e glass, 9.9-9.10
Loyalty, 1.13
Lumber
  building boards, 2.16-2.17
  calculating quantities, 2.20-2.21, 2.21f
  classification of defects, 2.12-2.13
  classifications of, 2.4, 2.5
  conversion tables, 2.63-2.64
  cutting, 2.3-2.4, 2.3f, 2.4f
  defects in, 2.5-2.8, 2.6f, 2.12-2.13, 6.15, 8.26
  drying of, 2.7
  engineered wood products, 2.18-2.19, 2.18f, 2.20, 6.6
  grading, 2.5, 2.8-2.13, 2.9f
  plywood, 2.13-2.15
  pressure-treated, 2.20
  selection of, 2.2
  storage of, 2.2, 2.6, 2.7-2.8, 2.15, 2.16f
Lumber core plywood, 2.14, 2.14f

M
Machine bolts, 2.39, 2.39f
Machine screws, 2.36, 2.36f
Mahogany, 2.5
Mansard roofs, 7.2, 7.2f, 7.3
Manufactured board panel subfloors, 5.26
Masonry nails, 2.32, 2.32f
Masonry wall framing, 6.26-6.28, 6.27f, 6.28f
Mass concrete, 8.6
Master carpenters, 1.8
Master combination saw blades, 3.16, 3.16f
MasterFormat, 4.41, 4.42f, 4.43
Mastics, 2.48-2.49
Matched lumber, 2.11, 2.12, 2.12f
Material safety data sheets (MSDS)
  defined, 2.55
  glues and, 2.47
  plywood and, 2.16, 2.17

Material symbols, 4.23, 4.24f
Mathematics, 1.8, 1.12
Measurements
  concrete mix proportions and, 8.5-8.6
  floor systems and, 5.32
  measure twice cut once, 2.21, 6.13
  measuring and cutting studs, 6.12-6.14, 6.13f
  metric measurements, 4.33, 8.39, 8.41
Mechanical anchors, 2.39-2.46
  anchor bolts, 2.39, 2.40f, 5.12, 5.13f
  bolt anchors, 2.41, 2.41f
  drilling anchor holes in concrete or masonry, 2.42-2.43, 2.44f
  hollow-wall anchors, 2.43-2.46
  one-step anchors, 2.39-2.41, 2.40f
  screw anchors, 2.42, 2.42f
  self-drilling anchors, 2.42, 2.43f
Mechanical drawings, 4.19
Mechanical plans
  drawing sets and, 4.2f, 4.18-4.19, 4.19f
  floor systems and, 5.11
Medium-density overlay (MDO) plywood, 2.17
Medullary rays, 2.3, 2.3f
Metal framing
  commercial construction and, 2.27-2.28, 2.27f, 2.29f
  popularity of, 2.2
  and roofs, 7.28, 7.28f
Metal roof trusses, 2.27, 2.27f, 7.28, 7.28f
Metal studs, 2.27, 2.27f, 2.28f
Metal tiedowns, 7.23, 7.23f
Metal wall bracing, 6.18, 6.18f
Metric conversion factors, 4.33
Metric conversion tables, 8.39, 8.41
Metric measurements, 4.33, 8.39, 8.41
Millwork, 2.11
  defined, 2.55
Mineral fiberboards, 2.17
Minimum coverage of steel reinforcement, 8.15
Minimum stairway widths, 10.8, 10.9f
Miter boxes, 3.8
  defined, 3.27
Miter cuts, 3.3, 3.10
  defined, 3.27
Miter saws, 3.8, 3.8f
Miter squares, 3.3, 3.4f
Modern carpentry, 1.6-1.8, 1.6f, 1.7f
Moisture, warped lumber and, 2.6
Monolithic slabs (monolithic pours), 8.28
  defined, 8.38
Monuments, 4.4
  defined, 4.55
Mortise, defined, 3.27
Mortise and tenon joints, 3.8, 3.9f
Mortise locksets, 9.19, 9.20f
Mortise markers, and locksets, 9.21, 9.22f
Multiple pane windows, 9.9, 9.10f
Muntins, 9.3, 9.3f
  defined, 9.28

N
Nailing flanges, 9.11, 9.12f
Nails
  casing nails, 2.31, 2.32f, 9.12
  kinds of, 2.30-2.34, 2.32f
  penny designations of, 2.30, 2.31f, 2.32
  selection of, 2.33
Nail set, 2.30
  defined, 2.55

Narrow-U stairways, 10.3f
National Apprenticeship Act of 1937, 1.10
National Fire Protection Association (NFPA), 4.43
National Registry, 1.11
NCCER (National Center for Construction Education and Research), and apprenticeship, 1.10, 1.11, 1.26-1.28
NCCER Standardized Craft Training, 1.11
Neoprene adhesive, 2.48
Newel posts, 10.6, 10.7f
   defined, 10.26
NFPA (National Fire Protection Association), 4.43
N-grade veneers, 2.14-2.15
Nominal sizes
   defined, 4.55
   dimensions and, 4.32
   grading lumber and, 2.11
Nosing, 10.7f, 10.8
   defined, 10.26
Nosing returns, 10.7f, 10.8
Notching wooden joists, 5.22, 5.23f
Notes, on drawings, 4.36

O
Object lines, 4.22
Occupational Safety and Health Act of 1970, 1.18-1.19
Occupational Safety and Health Administration (OSHA)
   concrete equipment and tool requirements, 8.18
   employee responsibilities and, 1.18
   employer responsibilities and, 1.18-1.19
   exposed rebar requirements, 8.16
   guardrails requirements, 10.10
Office of Apprenticeship, Training, Employer and Labor Services (ATELS), 1.10
One-hour rated walls, 2.26, 2.27f
One-piece anchors, 2.40, 2.40f
One-step anchors, 2.39-2.41, 2.40f
On-the-job safety training, 1.18
On-the-job training (OJT), 1.10, 1.11-1.12
Open/closed stairways, 10.2, 10.2f
Open finished stringers, 10.7f, 10.8
Open stairways, 10.2, 10.2f
   defined, 10.26
Oriented strand board (OSB), 2.17, 2.17f, 5.16, 5.26, 5.27
OSHA. *See* Occupational Safety and Health Administration (OSHA)
OSHA Standards 1926, 1.19
Outline specifications, 4.41
Outside chisel divergent staples, 2.34, 2.34f
Outside chisel staples, 2.34, 2.34f
Overcutting, 10.17, 10.19f

P
Panelized walls, 6.19
Panel nails, 2.33
Panel pinch, 5.38
Parallel strand lumber (PSL), 2.18-2.19, 2.18f, 2.20
Parallel strand lumber (PSL) headers, 6.6
Particleboard, 2.16-2.17
Particleboard core plywood, 2.14, 2.14f
Partitions
   commercial construction and, 2.25-2.26, 2.26f, 2.27f
   high fire/noise resistance partitions, 2.25-2.26, 2.26f, 2.27f
   laying out joist locations for, 5.32
   loadbearing and nonbearing partitions, 6.24
   masonry construction and, 6.26, 6.27f, 6.28
   nailing surfaces for, 6.5f
   partition backing, 6.26, 6.27f
   partition intersections, 6.4, 6.5f, 6.6f
   as wall components, 6.2, 6.2f
Party walls, 2.28
Pascals, 8.6
   defined, 8.38
Patterned concrete, 8.8
Patterned lumber, 2.11, 2.12
Pegboard, 2.16
Penny designations, of nails, 2.30, 2.31f, 2.32
Permanent bracing
   roof trusses and, 7.24, 7.24f
   wall framing and, 6.18
Personal protective equipment, 1.8
Phantom lines, 4.22
Pier forms, 8.22, 8.24f
Piers, 5.4, 5.17, 5.17f, 5.18, 8.20, 8.21f
   defined, 5.50
Pigtails, 8.16, 8.16f
Piles, 8.21
   defined, 8.38
Pilot holes, 2.36, 2.37
Pipe clamps, 3.7, 3.7f
Pitch, 7.3, 7.5
Pitch boards
   defined, 10.26
   laying out stringers and, 10.15, 10.15f, 10.19f, 10.20
Plain-sawed lumber, 2.3, 2.4f
Planes, 3.6, 3.6f, 3.20, 3.20f
Plank-and-beam framing, 7.27-7.28, 7.28f. *See also* Post-and-beam framing
Plan view drawings
   defined, 4.55
   drawing sets and, 4.3-4.11
   floor plans, 4.2f, 4.9, 4.10f
   foundation plans, 4.2f, 4.7, 4.7f, 4.8f
   roof plans, 4.9, 4.11
   site plans, 4.2f, 4.4-4.5, 4.6f, 4.7
Plan views, 4.23
Plastic concrete, 8.2
   defined, 8.38
Plastic resin glue, 2.47
Plastic sheeting, and curing concrete, 8.8
Plates, 7.3, 7.4f
Platform floor systems, 5.12, 5.13f
Platform framing, 1.3, 1.5f, 5.2, 5.3f, 6.4
Platforms. *See* Landings
Platform stairways, 10.4
Plumbing plans
   drawing sets and, 4.2f, 4.18-4.19, 4.19f
   floor systems and, 5.11
Plumbing symbols, 4.27, 4.28f
Plunge cuts, 3.10
   defined, 3.27
Plyform, 8.23, 8.25
   defined, 8.38
Plywood, 2.8, 2.13-2.15, 2.14f, 2.16f
Plywood forms, 8.23
Plywood Specification and Grade Guide, 2.15
Plywood subfloors, 5.26
Pneumatic/cordless nailers and staplers, 3.20-3.22, 3.21f, 3.22f
Pocket cuts, 3.10
   defined, 3.27
Polyvinyl or white glue, 2.47
Portable power planes, 3.20, 3.20f
Portable table saws, 3.12-3.13, 3.12f
Portland cement, 8.2, 8.3

Positioning stringers for layout, 10.17, 10.18f
Positive attitudes, 1.17
Post anchors, 5.17-5.18, 5.17f
Post-and-beam framing, 5.2, 5.4, 5.6f, 5.7, 7.27-7.28, 7.28f, 7.36f
Post caps, 5.17-5.18, 5.17f
Posts, 5.17, 5.17f
Posts and timbers (P&T), 2.10
Powder-actuated fasteners, 2.28, 2.30
Power metal shears, 3.20, 3.20f
Power miter saws, 3.13-3.14, 3.13f
Power saw blades, 3.16-3.17, 3.16f
Power saws, 3.10-3.17
Power tools, 3.9-3.22
    care of, 3.10
    drill presses, 3.17-3.18, 3.17f
    guidelines for use, 3.9-3.10
    pneumatic/cordless nailers and staplers, 3.20-3.22, 3.21f, 3.22f
    portable power planes, 3.20, 3.20f
    power metal shears, 3.20, 3.20f
    power saws, 3.10-3.17
    routers/laminate trimmers, 3.18-3.19, 3.18f
    safety and, 3.9
Pozzolan, 8.5
    defined, 8.38
Precision-end-trimmed lumber (PET), 2.12
Precut studs, 6.12
Pre-framed stairwells, 10.14-10.15
Preplaced aggregate (prepacked) concrete, 8.6
Presentation drawings, 4.3
Pressure-treated lumber, 2.20
Product data sheets, for doors, 9.30
Productivity, and human relations, 1.16-1.17
Professionalism, 1.12, 1.14
Profiles in success
    Caldwell, Berry, 6.34
    Deeds, Adam, 5.49
    Giardina, Vincent, 3.26
    Gomez, Luis Alexander, 4.54
    Humphries, Gary, 9.26-9.27
    Hunter-Parker, Erin M., 2.54
    Kubicek, Clay, 1.23-1.24
    McLawhorn, Curtis, 10.25
    Payne, John, 8.37
    Van Holten, Alan, 7.34
Projections, installing joists for, 5.38-5.40
Project managers/administrators, 1.8, 1.9
Project organization, 4.44-4.45, 4.46f
Project schedules, 4.47, 4.48f
Property lines, 4.4
    defined, 4.55
Proportions, for concrete, 8.5-8.6
Purlins, 7.11
    defined, 7.35
Pythagorean theorem, 10.17

Q

Quantity takeoffs, 4.43-4.44
    defined, 4.55
Quarter-sawed lumber, 2.4, 2.4f
Quick clamps, 3.6, 3.7f

R

Rabbet cuts, 3.12
    defined, 3.27
Rafter framing squares, 7.4-7.5, 7.5f, 7.6f
Rafter marks, 7.12
Rafter plates, 5.4, 5.5f
    defined, 5.50
Rafters
    defined, 2.55
    estimating material for, 7.26
    installing, 7.11, 7.11f
    laying out with speed square, 7.18, 7.19f, 7.20
    layout, 6.19, 6.20f, 7.5, 7.7-7.10, 7.36, 7.37f, 7.38-7.39
    lumber grades and, 2.10
    wood I-beams and, 2.19
Rafter tables, 7.4-7.5, 7.5f
Rafter tails, 7.16
Rails, 9.3, 9.3f
    defined, 9.28
Rail-steel, 8.13
    defined, 8.38
Raising walls, 6.17
Reading and interpreting drawings
    abbreviations, 4.35, 4.56-4.59
    blueprints, 5.11-5.12
    dimensioning, 4.32, 4.33f, 4.34, 5.11
    guidelines, 4.35-4.36
    lines used in drawings, 4.20, 4.21f, 4.22
    symbols used in drawings, 4.23-4.32
Rebar cutters and benders, 8.15
Rebars
    bar supports for, 8.16
    as concrete reinforcement material, 8.12-8.14, 8.14f, 8.15
    defined, 8.38
    identifying rebar, 8.13, 8.14f, 8.15
    splicing, 8.16
Rectangular volume calculations, 8.9-8.10, 8.11f
Reflective glass, 9.9
Reinforced cutout stringers, 10.20, 10.20f, 10.21f
Reinforced footings, 8.19
Reinforced stringer mounting, 10.20, 10.21f
Reinforcement materials for concrete, 8.12-8.17
Reinforcing bars. *See* Rebars
Reinforcing ceiling joists, 6.21, 6.22f
Residential construction
    building materials for, 1.6, 1.7f, 2.2, 2.2f
    concrete blocks for, 2.22
Residential material specifications, 4.36, 4.37-4.40f
Resins, 2.16
    defined, 2.55
Resorcinal resin glue, 2.47
Responsibility, of employees, 1.14
Revision blocks, 4.2, 4.3, 4.4f
Ribbands, 6.21, 6.22f
    defined, 6.35
Ridges (ridgeboards), 7.3, 7.4f
    estimating material for, 7.26
Rigging trusses, 7.21, 7.21f, 7.23
Rip cuts, 3.10
    defined, 3.27
Rise, 7.3, 7.4f, 10.10, 10.11
    defined, 10.26
Rise and run, 10.9
    defined, 10.26
Riser diagrams, 4.18
    defined, 4.55
Riser heights, 10.10-10.12
Risers, 10.2, 10.5, 10.8
    defined, 10.26
Riser/tread combinations, 10.14
Roller-compacted concrete, 8.6, 8.7

Roof components, 7.3, 7.4f
Roof cut. *See* Slope
Roof framing
    dormers, 7.26, 7.27f
    estimating materials for, 7.25-7.26
    installing sheathing, 7.15-7.16, 7.17
    laying out hips and valleys, 7.36-7.43
    layout of roofs, 7.3-7.14, 7.4f, 7.5f, 7.6f, 7.7f, 7.9f, 7.11f, 7.12f, 7.13f, 7.14f
    metal roof framing, 7.28, 7.28f
    plank-and-beam framing, 7.27-7.28, 7.28f
    rafter layout using speed square, 7.18, 7.19f, 7.20
    truss construction, 7.20-7.21, 7.20f, 7.21f, 7.23, 7.24
    types of, 7.2, 7.2f
Roofing nails, 2.32-2.33, 2.32f
Roofing panel nails, 2.32f, 2.33
Roof openings, 7.14, 7.16
Roofs. *See also* Roof framing
    types of, 7.2, 7.2f
Roof windows, 9.7, 9.8f
Rough carpentry, 1.7
    defined, 1.25
Rough lumber, 2.11
Rough openings, 6.2, 6.2f, 6.9f, 6.11, 9.10, 9.22
Rough sills, 6.2, 6.2f
Router bits, 3.19
Routers, 3.18-3.19, 3.18f
Routing, and straightedges, 3.19
Rules and regulations, employee responsibilities for, 1.14
Run, 7.3, 7.4f, 10.10
    defined, 10.26

S

Safety
    adhesives and, 2.47, 2.48
    asbestos and, 2.18
    bracing and, 7.23
    carpentry and, 1.8, 1.12
    cement and, 8.2
    chop saws and, 3.15
    circular saws and, 3.10, 3.11
    concrete forms and, 8.17-8.18
    demolition saws and, 3.14, 3.15
    drilling concrete and, 2.43
    employer and employee obligations, 1.17-1.19, 1.18f
    erecting roofs and, 7.11
    frame and trim saws and, 3.14
    joist installation and, 6.20
    laser levels and, 3.3
    nails and, 2.32
    on-the-job training for, 1.18
    plywood and, 2.15
    pneumatic nailers and staplers and, 3.21-3.22
    portable table saws and, 3.12-3.13
    powder-actuated fasteners and, 2.30
    power metal shears and, 3.20
    power planes and, 3.20
    power tools and, 3.9
    pressure-treated lumber and, 2.20
    rebar and, 8.16
    roof openings and, 7.16
    routers and laminate trimmers and, 3.19
    sawing and, 3.15
    sheathing and, 7.17
    stairways and, 10.4
    trenches and, 8.19
    trusses and, 7.21
    weather conditions and, 1.8
Safety committees, 1.17-1.18
Safety glass, 9.10
Safety managers, 1.8, 1.9
Sammy anchors, 2.40f, 2.41
Sand, concrete and, 8.3, 8.4
Sapwood, 2.3, 2.3f
Sashes, 9.2, 9.3, 9.3f
    defined, 9.28
Saw blades
    changing, 3.11
    dull saw blades, 3.16
Sawbucks. *See* Frame and trim saws
Scabs, 5.33
    defined, 5.50
Scale, drawings and, 4.33, 5.11
Scarf, 5.33
    defined, 5.50
Schedules
    drawing sets and, 4.2f, 4.16, 4.17f, 4.18f
    laying out wall openings and, 6.9-6.10, 6.10f
Screeding, 8.31
    defined, 8.38
Screw anchors, 2.42, 2.42f
Screw heads, 2.35, 2.35f
Screws, 2.34-2.38, 2.35f, 2.37f, 2.38f
Section drawings
    drawing sets and, 4.2f, 4.11, 4.13-4.15f
    floor systems and, 5.8, 5.9f
Section lines, 4.22, 4.22f
Select lumber, 2.11
Self-drilling anchors, 2.42, 2.43f
Service-grade hardboard, 2.16
Sheathing
    defined, 2.55
    estimating material for, 7.26
    installing, 7.15-7.16, 7.17
    safety and, 7.17
    storage of, 2.8
    wall framing and, 6.18
Shed dormers, 7.26, 7.27f
Shed roofs, 7.2, 7.2f
Sheet metal screws, 2.35, 2.36f
Shelf life, of adhesives, 2.49
Shields, for lag screws, 2.36, 2.37f
Shims, 9.22
Shiplap, 2.11, 2.12, 2.12f
    defined, 2.55
Shop drawings, 4.20
Shoring, 8.18
    defined, 8.38
Sill plates, 2.20
    defined, 2.55
Sills
    defined, 6.35, 9.28
    estimating material for, 5.40
    floor systems and, 5.12, 5.13f, 5.14, 5.15
    installing, 5.28-5.30, 5.29f, 5.30f
    laying out for joists, 5.31-5.32, 5.32f
    and sashes, 9.3, 9.4f
Sill sealers, 5.13f, 5.15, 5.40
Simplex fasteners, 2.32, 2.32f
Single- and double-expansion anchors, 2.41, 2.41f
Single-hung windows, 9.3, 9.4f
Single-strength (SS) glass, 9.9
Site plans, 4.2f, 4.4-4.5, 4.5f, 4.6f, 4.7
Skirtboards, 10.7f, 10.8
    defined, 10.26
Skylights, 9.7, 9.8f

Slab-on-grade (slab-at-grade), 8.27
    defined, 8.38
Slabs, 8.27-8.28, 8.29f, 8.30f, 8.31
Slabs with foundations, 8.28, 8.29f
Slabs with thickened edges, 8.28, 8.29f
Slag, 8.2
    defined, 8.38
Sleeve anchors, 2.40, 2.40f
Sliding patio doors, 9.16
Sliding T-bevel squares, 3.3, 3.4f
Slope, 7.3, 7.4f, 7.5
Slump, 8.2
    defined, 8.38
Slump testing, 8.8, 8.9f
Smoothing planes, 3.6, 3.6f
Snow, George W., 1.3
Softwoods, 2.4-2.5, 2.5f, 2.9, 2.11, 2.58-2.59, 2.60
Soil reports, 4.20
Soleplates
    defined, 5.50
    estimating material for, 6.24
    floor systems and, 5.2
    layout, 6.8, 6.8f, 6.11, 6.12f
    as wall components, 6.2, 6.2f
Solid lumber girders, 5.15, 5.15f
Solid-web trusses, 5.22
Span, 5.4, 7.3, 7.4f
    defined, 5.50
Spear point staples, 2.34, 2.34f
Specifications
    floor systems and, 5.12
    format of, 4.41, 4.42f, 4.43
    organization and types of, 4.36, 4.41, 4.43
    typical materials specification, 4.37-4.40f
Speed squares, 3.4, 3.4f
    and hip rafter layout, 7.42-7.43, 7.43f
    and rafter layout, 7.18, 7.18f, 7.19f, 7.20
Spindles, 9.6, 9.7f
Spindle sanders, 3.18
Splicing rebar, 8.16
Split lumber, preventing, 2.7-2.8
Spread footings, 8.20, 8.20f
Spring clamps, 3.6, 3.7f
Square-end-trimmed lumber, 2.12
Squareness, and foundations, 5.27-5.28, 5.28f
Squares, 3.3-3.4, 3.4f, 3.5
Stacking building materials, 2.7-2.8
Stacking interior finish material, 2.8
Stair gauges, 10.15, 10.15f
Stair indicator lines, 4.22
Stairs
    components of, 10.5-10.6, 10.5f, 10.7f, 10.8
    forms for concrete stairs, 10.21, 10.21f
    framing, 10.8-10.9
    International Building Code requirements for, 10.8
    laying out and cutting stringers, 10.15-10.20
    reinforced cutout stringers, 10.20, 10.20f, 10.21f
    stairway design and layout, 10.10-10.13
    stairwells, 10.2, 10.9, 10.13-10.15, 10.26
    terms on plans, 4.60-4.61
    types of, 10.2, 10.3f, 10.4
Stairway design and layout, 10.10-10.13
Stairwells, 10.2, 10.9, 10.13-10.15
    defined, 10.26
Stakes, and forms, 8.31
Standard A lumber, 2.12
Standard B lumber, 2.12
Standard C lumber, 2.12

Standard D lumber, 2.12
Standard E lumber, 2.13
Standard F lumber, 2.13
Standard G lumber, 2.13
Standards, for apprenticeship programs, 1.11-1.12
Staples, 2.34, 2.34f
Starting newel posts, 10.6, 10.7f
Steel I-beam girders, 5.15f, 5.16
Steel I-beam headers, 6.6
Steel stud channels, 6.28, 6.29f
Steel studs, 6.28, 6.29f
Stepped continuous footings, 8.20, 8.21f
Stiles, 9.3, 9.3f
    defined, 9.28
Storing plywood, 2.15, 2.16
Stove bolts, 2.39, 2.39f
St. Paul's Cathedral, 1.6, 1.6f
Straightedges, and routing, 3.19
Straight-flight stairways, 10.4
Straight-run stairways, 10.2, 10.4
Strap clamps, 3.7, 3.7f
Strength, of concrete, 8.4, 8.5-8.6, 8.5f
Stress-grade lumber, 2.11
Stretcher blocks, 2.22f
Stringer mounting attachments, 10.17, 10.18f
Stringers, 2.9, 10.5, 10.8
    defined, 10.26
    laying out and cutting, 10.15-10.20
Strongbacks, 6.21, 6.22f
    defined, 6.35
    and stringers, 10.20, 10.20f
Structural drawings, 4.18, 5.10
Structural member symbols, 4.27, 4.31f
Structural terms, 4.61
Stub nails, 5.22
Stud bolt anchors, 2.40, 2.40f
Studs
    defined, 8.38
    estimating material for, 6.24
    identifying structural studs, 2.30
    layout, 6.8, 6.8f, 6.11, 6.12f
    measuring and cutting, 6.12-6.14, 6.13f
    as wall components, 6.2, 6.2f
    for wall forms, 8.25
Subflooring
    estimating material for, 5.41
    floor systems and, 5.25-5.27, 5.26
    installing, 5.36-5.38, 5.37f, 5.38f
Subgrades, 8.18
    defined, 8.38
Supervisors, 1.8
Surfaced lumber, 2.11
Sweeps, 9.17, 9.18f
Symbols
    common site/plan topographical survey symbols, 4.27, 4.30f
    doors and windows, 4.23, 4.25f
    in drawings, 4.23-4.32
    electrical symbols, 4.25, 4.26f, 4.27, 4.27f
    HVAC symbols, 4.27, 4.29f
    material symbols, 4.23, 4.24f
    plumbing symbols, 4.27, 4.28f
    welding symbols, 4.27, 4.31f, 4.32, 4.32f

T

Tail joists, 5.20
    defined, 5.50
Takeoffs, 1.8
    defined, 1.25

Taper form ties, 8.27
Tardiness, 1.14, 1.15
Teamwork, 1.16
Technical aspects, of specifications, 4.41
Temporary bracing
  roof trusses and, 7.24, 7.24f, 7.25
  wall framing and, 6.18
Tenons, 3.8, 3.9f
  defined, 3.27
Termite shields, 5.13f, 5.14, 5.40
Thresholds, 9.17, 9.18f
Tie wire, 8.16, 8.16f
Tilt-Up Concrete Association, 2.25
Tilt-up concrete construction, 2.24, 2.24f, 2.25
Timbers, appearance grades, 2.9, 2.10
Title sheets, title blocks and revision blocks, 4.2-4.3, 4.2f, 4.4f
T-nails, 2.31, 2.32f
Toggle bolts, 2.43, 2.44-2.45, 2.45f
Tongue-and-groove, 2.12, 2.12f
  defined, 2.55
Tools
  hand tools, 3.2-3.9
  history of, 1.2-1.3
  operation of, 1.8
  power tools, 3.9-3.22
Topographical surveys, 4.22
  defined, 4.55
Top plates
  defined, 6.35
  estimating material for, 6.24
  layout, 6.8, 6.8f, 6.11, 6.12f
  as wall components, 6.2, 6.2f
Total stair run, 10.12-10.13, 10.13f, 10.15
Training programs, 1.10-1.12. *See also* Apprenticeship programs
Transit levels, 3.2-3.3, 3.5
Transom windows, 9.6
Tread run, 10.12-10.13, 10.13f
Treads, 10.2, 10.5, 10.8
  defined, 10.26
Tread thickness, and stringers, 10.17, 10.19f
Tread widths, 10.10
Trench ducts, 2.25, 2.25f
Trim, 2.12
Trimmer joists, 5.20
  defined, 5.50
Trimmer studs, 6.2, 6.2f, 6.14
  defined, 6.35
Trimming doors to length, 9.15
True, 3.6
  defined, 3.27
Trusses
  bracing for, 7.24, 7.24f, 7.25
  components of, 7.20, 7.20f
  construction of, 7.20-7.21, 7.20f, 7.21f, 7.23, 7.24
  defined, 5.50
  erecting trusses, 7.21, 7.21f, 7.23
  floor systems and, 5.24, 5.24f, 5.25f
  metal roof trusses, 2.27, 2.27f, 7.28, 7.28f
  truss placement diagrams, 7.21, 7.22f, 7.23
  types of, 7.20f
Truss headers, 6.6, 6.7f
Truss rigging, 7.21, 7.21f, 7.23
Truss storage, 7.21, 7.23
Try squares, 3.3, 3.4f
Tubular locksets, 9.19, 9.21f
Two-hour rated walls, 2.26, 2.27f

U

Underlayment, 5.25, 7.15-7.16, 7.16f
  defined, 5.50
Uniform Building Code, nailing requirements, 5.30
Uniform Construction Index, 4.41
Unit rise, 10.10, 10.14-10.15
  defined, 10.26
Unit run, 10.10, 10.14-10.15
  defined, 10.26
Unloading building materials, 2.7-2.8
Urea formaldehyde glue, 2.47
U.S. Department of Labor
  Bureau of Apprenticeship and Training, 1.10
  OSHA and, 1.18

V

Valley jacks, 7.39, 7.39f
Valley rafters, layout, 7.3, 7.4f, 7.39, 7.39f, 7.40f
Vaulted ceilings, 2.19
  defined, 2.55
Veneer core plywood, 2.14, 2.14f
Veneers, 2.13
Vertical-grained lumber, 2.4
Vinyl-clad windows, 9.11, 9.12f
Volume calculations for concrete
  circular, 8.12, 8.12f
  rectangular, 8.9-8.10, 8.11f
Volumes, formulas for finding, 8.40

W

Walers, 8.22, 8.27
  defined, 8.38
Wall anchors, 2.45-2.46, 2.45f
Wall forms, 8.24-8.27, 8.25f
Wall framing
  assembling walls, 6.14-6.15
  components of, 6.2-6.7, 6.2f
  erecting walls, 6.16-6.19
  estimating material for, 6.23-6.26
  laying out, 6.8-6.11
  masonry wall framing, 6.26-6.28, 6.27f, 6.28f
  measuring and cutting studs, 6.12-6.14
  steel studs, 6.28, 6.29f
Wall lifting jacks, 6.16, 6.16f
Wall openings, layout, 6.9-6.11, 6.11f, 6.12f
Walls. *See also* Wall framing
  aligning of, 6.16-6.19, 6.17f
  assembling of, 6.14-6.15
  commercial interior walls, 2.25-2.26
  components of, 6.2-6.3, 6.2f
  curtain walls, 2.23, 2.23f, 2.24, 2.24f
  erecting of, 6.16-6.19
  plumbing and aligning, 6.16-6.19, 6.17f
Wall ties, 8.27, 8.27f
Warped lumber
  moisture and, 2.6
  preventing, 2.7-2.8
Waste, and calculating concrete volume, 8.11
Water, and concrete, 8.4, 8.8
Water-cement ratio, 8.5
  defined, 8.38
Water levels, 3.2, 3.2f, 3.3
Waterproof glue, 2.47
Weather conditions
  lumber storage and, 2.7-2.8
  safety and, 1.8
Weatherstripping, 9.16, 9.16f, 9.17, 9.19f

Web clamps, 3.7, 3.7f
Wedge anchors, 2.40, 2.40f
Welded-wire fabric, 8.16, 8.17, 8.17f
Welding symbols, 4.27, 4.31f, 4.32, 4.32f
Western platform framing, 1.3, 1.5f, 5.2, 5.3f, 6.4
White glue, 2.47
Wide-L stairways, 10.3f
Wide-U stairways, 10.3f
Willingness to learn, and employee responsibilities, 1.13
Winding stairways, 10.3f, 10.4
   defined, 10.26
Window glass, 9.9-9.10
Windows
   construction of, 9.2-9.3, 9.3f, 9.4f
   glass blocks, 9.13, 9.13f
   installing, 9.10-9.13, 9.12f
   in masonry walls, 6.28, 6.29f
   symbols, 4.23, 4.25f
   terms on plans, 4.60
   types of, 9.3, 9.5-9.7, 9.5f, 9.6f, 9.7f
   types of window glass, 9.9-9.10, 9.10f
Wire brads, 2.34
Wire tie, 8.16, 8.16f
Wood building materials, 2.2. *See also* Lumber
Wood I-beams
   as engineered wood products, 2.18, 2.19, 2.19f, 2.20
   floor systems and, 5.16, 5.22, 5.23, 5.23f
   headers and, 6.6
Wood screws, 2.34-2.35, 2.35f, 2.37
Working drawings, 5.7-5.12, 5.8f

Y

Youth Apprenticeship Program, 1.11